Optics

Johannes Kepler

Optics

Paralipomena to Witelo
&
Optical Part of Astronomy

Translated by

William H. Donahue

Green Lion Press
Santa Fe, New Mexico

Manufactured in the United States of America

Published by Green Lion Press, Santa Fe, New Mexico, USA

www.greenlion.com

Green Lion Press books are printed on fine quality acid-free paper of high opacity. Both softbound and clothbound editions feature bindings sewn in signatures. Sewn signature binding allows our books to open securely and lie flat. Pages do not loosen or fall out and bindings do not split under heavy use by students and researchers. Clothbound editions meet the guidelines for permanence and durability of the Committee on Production Guidelines for Book Longevity of the Council on Library Resources. The paper used in all Green Lion Press books meets the minimum requirements of American National Standard for Information Sciences—Permanence of Paper for Printed Library Materials, ANSI Z39.48-1984.

Printed and bound by Sheridan Books, Chelsea, Michigan.

Cover design by William H. Donahue. The front cover text is a somewhat free adaptation of Kepler's very long title page text (see pp. 2–3).

Cataloguing-in-publication data:
Johannes Kepler
Optics: Paralipomena to Witelo & Optical Part of Astronomy
Translated with notes by William H. Donahue

Includes indexes, bibliography

ISBN-13: 978-1-888009-12-5 (printed hardcover)

1. Kepler, Johannes. 2. History of Optics
3. History of science. I. William H. Donahue (1943–). II. Title

QC353.K46 A3 2000

Library of Congress Control Number: 00-109402

Contents

Kepler's Optics

The Green Lion's Preface

The book you have before you, reader, is not merely the first English translation of the *Optics*: it is the first complete translation into any language. This is astonishing, in view of the esteem in which Kepler's optical work is held. This book is the culmination of the perspectivist tradition in optics, which began with the ancient Greeks, was augmented and redirected by Arabic writers such as Alhazen, and taken up by Europeans in the thirteenth century. At the same time, Kepler's *Optics* inaugurates the modern approach to optics, with a clear understanding of how the eye works, a physical account of refraction using something very much like a wave front, and a new level of understanding of how lenses form images. Descartes based his optical work largely on Kepler, and every optical theorist since then has been indebted to this book. Green Lion Press is glad to be able to present this epochal treatise in a form accessible to modern readers.

In this edition, as in other Green Lion books, care has been taken to make the book reader-friendly and durable.

Kepler's marginal notes have been kept in the margins, with the translator's notes at the page bottoms, as footnotes, not the annoying end notes that are often found in scholarly works. Further, Kepler's end notes have been included as footnotes, or, when too long to fit at the bottom of the page, as appendices directly following the passages to which they apply.

Page numbers of the 1604 edition have been placed in the inner margins, to allow easy reference either to that edition or to the text in *JKGW* II, which also includes the 1604 page numbers. The marginal page numbers also facilitate the use of Kepler's index, in which the original page numbers have been retained.

The 100-odd diagrams have been redrawn, with the exception of two instances where a reproduction of the original woodcut was preferable (these have been reproduced from the 1604 edition). In almost every instance, diagrams, repeated as necessary, are placed adjacent to all the text that refers to them.

The book is constructed in the traditional way that fine books have been made, with sewn signatures. The paper is acid-free and the cover boards and binding were chosen with strength and durability in mind.

Green Lion Press prefers to use typefaces and sizes that are elegant and easy on the eye. However, the length of this book led us to use ten point type size and a font, Times, which, despite being somewhat inelegant, is very efficient in the use of space. It also occurred to us that this book, arising as it did from practical concerns, and intended by its author to be a compendium of practical observational techniques, might most appropriately be presented in a font that smacked more of the workshop than the study.

Dana Densmore and William H. Donahue
for the Green Lion Press

Translator's Introduction

In the summer of 1600, Kepler assembled a large wooden instrument (described at the beginning of chapter 11 of the present work), and set it up in the Market Square in Graz to observe the solar eclipse of June 30/July 10. His experience with this instrument drew his attention to certain optical matters that had been troubling astronomers. From a few notes written down at the time of the eclipse, his optical project grew to encompass metaphysical and theological speculations about the nature of light (Chapter 1), a thoroughgoing attempt to account for refraction, involving a novel treatment of conic sections (Chapter 4), a study of the anatomy and physiology of the eye (chapter 5), a historical study of solar eclipses (Chapter 8), and many other matters relating to the light of the heavenly bodies and how it behaves. The result is one of the most important optical works ever written, which, even when it is wrong, is wrong in an interesting and fruitful way.

The problem that Kepler had initially encountered, that initiated his optical studies, involved the way light flows through pinholes. Astronomers had been puzzled about why the moon seemed smaller in a solar eclipse than it did at other times. Kepler figured out why. In a characteristic series of numbered sentences jotted down on a piece of paper at the time,[1] he set out point by point the way a finite luminous object and a small but finite opening interact to form a slightly enlarged image. (An expanded version of these notes constitutes chapter 2 of the present work.)

At first, Kepler had it in mind to publish this important discovery in a separate small work, which he had actually written during the following month.[2] But his exile from Styria, move to Prague, and involvement with the work of Tycho Brahe intervened. Much of his time in the next year was taken up with the orbits of Mars and of the earth.

With the death of Brahe in October, 1601, everything changed. As his successor in the post of Imperial Mathematician, Kepler was expected to produce works that would reflect well on his patron, Rudolph II. In particular, he was to complete the work unfinished by Brahe and to produce the astronomical tables that would bear

---

[1] Mss. Pulkovo XV, fol. 248v, transcribed in Latin in Straker, *Kepler's Optics*, pp. 526–8. A German translation of this document is included in *JKGW* II pp. 400–1.

[2] Letters to Michael Maestlin, 9 September and 6/16 December 1600, in *JKGW* XIV, nos. 175 and 180, pp. 150–1 and 159.

the name of Rudolph. Kepler filled sheet after sheet with computations and ruminations about the orbit of Mars, which, he believed, held the key to the deeper astronomy. But in the spring of 1602, his investigation took a surprising turn: he discovered that the orbit could not be perfectly circular, but had to be squeezed in slightly at the sides. The unexpected complications that this introduced, together with difficulties with the Brahe family regarding use of the Tychonic observations, led him to realize that the Mars book would not be finished as soon as he had hoped, and that he had better find something else to fill in. His thoughts returned to the book on pinholes.

However, the project turned out not to be as simple as it had seemed at first. As Kepler outlined it in his dedication to the Emperor, he thought he should also discuss the other main optical problem in astronomy, the refraction of light in the atmosphere. That, in turn, required an understanding of refraction itself, which demanded a study of the nature of light. And since there was some uncertainty about how the eye itself interacted with astronomical sights, it seemed that he needed to understand how the eye functioned. To fill these needs, he worked his way through Witelo's *Perspectiva*,[3] itself no insignificant feat, and compared Witelo's treatment of vision with the anatomical and physiological approaches of Jessenius and Platter. Further, his study of refraction suggested to him that curves other than circles would be required to explain it, and that led to a study of Apollonius's *Conics*, which he recast in Keplerian style. And, having gone this far, he thought he would round it out with a few chapters on "the light, the place, and the motion of the heavenly bodies," and the theory of parallax, so important for the moon's motion. As a result, although he confidently promised the book for Christmas of 1602, he was still hard at work on it in the summer of 1603. In May he wrote, "measuring refractions: here I get stuck. Good God, what a hidden ratio! All the *Conics* of Apollonius had to be devoured first, a job which I have now nearly finished."[4] He at last sent the finished manuscript to the Emperor in January, 1604.

---

[3] Witelo (latinized "Vitelo" or "Vitellio") was a Polish scholar of the thirteenth century who studied at Padua and Paris. The little that is known of his life is presented in detail by Lindberg in his introduction to the *Thesaurus*. His principal work, titled *Perspectiva* or *Opticae*, was the foremost European optical treatise in the Middle Ages and Renaissance, its only rival being Alhazen's treatise. It was published in three editions in the sixteenth century. The edition Kepler used was Risner's *Opticae thesaurus*, which included Alhazen and Witelo in a single huge volume (with two separate paginations).

[4] Letter to Herwart von Hohenburg, no. 256 in *JKGW* XIV p. 396.

The book everywhere shows signs of the haste with which it was written. Syntax is often jumbled, words misspelled, numbers wrong, citations incorrect, and demonstrations vague. Once the book was in press, Kepler had second thoughts about many passages, for which he wrote notes, often extensive, which he advised the reader to consult first. Production was difficult owing to Kepler's having entrusted the book to a printer in Frankfurt, perhaps hoping that, in this great center of publishing and marketing, production would be quick and efficient and sales brisk. As it turned out, the pace was far from quick, and it was not until the autumn of 1604 that the book finally appeared, with a lengthy (and incomplete) errata sheet appended.

The resulting work is historically important in many ways, some of which, at least, have not yet been explored. Attention has been disproportionately directed to those parts of the *Optics* that are seen as points of departure for later scientific theories, covering, for example, the inverted retinal image and the way lenses work. Keplerian studies are now entering a phase in which scholars are trying to understand Kepler's thought as a unified whole, in the context in which he lived and wrote. In this regard, there is much in the *Optics* that is worth studying. In particular, the way in which Kepler's astronomical works are related to his work in optics has received little attention. Chapter 10, in which Kepler argues, on the basis of their optical works, that the ancient optical writers were heliocentrists, is a notable example of this, as is his stated intention of building his unfinished *Hipparchus* on the foundation of his *Optics*.[5]

As the book's translator, I can easily sympathize with Kepler's frustration at the difficulties of the task and the amount of time it took. The range of subjects and materials to be covered was enormous. I have spent considerable time in rare book collections researching Kepler's sources, with only partial success. Kepler's labor of gathering and weighing the reports and opinions of various authors, and of trying out various mathematical models of his own, evidently left little time for the actual writing.

My approach to the text has therefore been somewhat different from my treatment of the *Astronomia nova* in my translation of that work (see *JKNA*). This is not really a finished work, and to smooth off the rough places would misrepresent it. I have accordingly followed the Latin more closely than I did in the *Astronomia nova*, at the expense of fluidity in the prose. The only point on which I have chosen to depart from a strict rendering of Kepler's text has been

---

[5] Cf. p. 433, below.

to break up some of his long sentences, which could not be clearly reproduced in English because of our language's lack of syntactic markers. The entire translation has been reviewed by classicist Bruce M. Perry of St. John's College, whose help has made it much better than it otherwise would have been.

In writing the footnotes, I have aimed primarily at clarifying obscurities and providing references to sources which, though perhaps familiar in Kepler's day, are obscure to us. The editions of Frisch (*JKOO*) and Hammer (*JKGW*) have more explanatory material than is present here; readers who are interested may consult those works. Some passages, notably Kepler's physical explanation of refraction and his resulting refraction law, cry out for study and interpretation, but this is too much to burden a translation with. Such matters deserve separate studies, such as those that have already been done by Stephen Straker, David Lindberg, and others. What readers need is a few notes to get them past the difficulties that the text presents at the level of simple comprehension, and that is what I have endeavored to provide.

The title, like many of Kepler's titles, presents peculiar problems. It appears to be the title of two books, one of which is about Witelo's optics, and the other about astronomical optics, an impression that is confirmed in the text (for example, pp. 91, 171, and in the running heads themselves, which are different in the two halves of the book). Nevertheless, the book does have a coherent theme: it is about light, and therefore its proper subject is Optics. Kepler tacitly acknowledged this in referring to it later as "mea *Optica*." The title he gave it is too modest, suggesting only that it has a few things to say about Witelo, and then will go on to talk about some astronomical matters. So I believe that, in calling it *Optics*, I am only restoring to it its rightful name.

The other problem with the title is how to translate "paralipomena." This latinization of the Greek παραλιπόμενα literally means "things that are left on the side," which can either denote things omitted or neglected, or things that are ancillary or supplementary to some central matter.[6] One might then think that Kepler means to discuss things that Witelo neglected. However, his use of the preposition "*ad*," "to," shows that he had the latter meaning in mind. So the opening lines of the title become "Supplement to Witelo." I was still unhappy with this because it lost the odd flavor of the original. Then it was suggested to me that this word is an early example of the German fondness for using Greek words in titles, exemplified,

---

6 I am indebted to Eva Brann for her thoughtful discussion of this term with me.

for example, by Kant's *Prolegomena to Any Future Metaphysics*. If so, then perhaps the best approach is simply to carry it over from the Latin, as Kepler had from the Greek.

Acknowledgements

This translation was supported by a translation grant from the National Endowment for the Humanities, whose support is gratefully acknowledged. There were also many individuals who contributed to the project in one way or another.

Special thanks go to Peter Barker and Katherine Webb, who provided food, lodging, and friendship during two extended periods of research in the remarkable History of Science Collection at the University of Oklahoma. Thanks also to Tom and Haydie Callaghan, gracious hosts during a stay in the Boston area for work at the Houghton Library. The staff at the respective libraries were most helpful, especially Marilyn Ogilvie and Steven Wagner of the University of Oklahoma.

The staff of the Meem Library at St. John's College, Santa Fe, helped out in more ways than I can acknowledge here, especially Inga Waite, Director, and Laura Cooley and Heather MacLean. Lisa Richmond, Director of the Greenfield Library at the Annapolis campus of St. John's, was energetic and helpful in arranging for a loan of the facsimile of the 1604 edition of the *Optics*.

Bruce M. Perry of St. John's College, Santa Fe, reviewed the entire translation, catching many misinterpretations and infelicities of expression. Special thanks to David C. Lindberg for lending me a copy of his reprint of the *Thesaurus* and reading and commenting on parts of the translation. Peter Barker, Eva T. H. Brann, Bernard R. Goldstein, Owen Gingerich, William Kerr, Peter Machamer, Thomas North, Bruce Stephenson, Stephen M. Straker, and James R. Voelkel all helped in various ways with comments and answers to queries, for which I am grateful.

The typesetting and layout were done in TEX, using the superb ZzTEX book layout software written by Paul Anagnostopoulos.

William H. Donahue
Santa Fe, New Mexico
September 19, 2000

Addendum, August 2021
I am grateful to Alan Shapiro for recommending four additions to the Bibliography.

Johannes Kepler

Paralipomena to Witelo

whereby

The Optical Part of Astronomy is Treated

Paralipomena to Witelo

whereby

The Optical Part of Astronomy is Treated;

Above all

on the Technically Sound Observation
and Evaluation of the Diameters
and Eclipses of the Sun and Moon.
With Examples of Important Eclipses.

In this Book, Reader, you have, among many other new things,
A Lucid Treatise on the Means of Vision, and the Use
of the Humors of the Eye, against the Opticians and Anatomists,
by the author

Johannes Kepler, H. I. M.'s

Mathematician.

AD VITELLIONEM PARALIPOMENA,

Quibus

ASTRONOMIÆ PARS OPTICA TRADITVR;

Potißimùm

DE ARTIFICIOSA OBSERVATIONE ET ÆSTIMATIONE DIAMETRORVM deliquiorumq; Solis & Lunæ.

CVM EXEMPLIS INSIGNIVM ECLIPSIVM.

Habes hoc libro, Lector, inter alia multa noua,

Tractatum luculentum de modo visionis, & humorum oculi vsu, contra Opticos & Anatomicos,

AVTHORE

IOANNE KEPLERO, S. C. M[tis]

Mathematico.

FRANCOFVRTI,

Apud Claudium Marnium & Hæredes Ioannis Aubrii

Anno M. DCIV.

Cum Priuilegio S. C. Maiestatis.

Dedication

)(2^r To Rudolph II

Ever August Emperor of the Romans

P. F. P. P.

King of Hungary, Bohemia, etc.; Archduke of Austria, etc.

Since this our time, most August Caesar, has brought forth that extraordinary astronomer Tycho Brahe, who, in his most meticulous observations and incredible precision rivalled the very nature of things, and since Germany has, as well as those ancient lights of the mathematical faculty, Peurbach, Regiomontanus, Apianus, and those whom the Viennese Academy can enumerate in a lengthy list, also now this new exemplar, sublime and admirable, of carefulness and perfection: I therefore think it fitting, as the Professors of this whole art are energetically
)(2^v following in the footsteps of this preeminent artist, and compare themselves with his example, and especially since the Germans labor in it, each in his own capacity, that they may yet continue to retain the prize for this glory, hitherto set before them, within their nation. For the treasury of the secrets of Nature is unexhausted, its richness beyond telling, and anyone who brings forth some new thing about it into the light does no more than open the path to others for subsequent investigations. Moreover, for me, who have devoted considerable time to optical studies, the Tychonic astronomy, in which this Your Majesty's most brilliant court made me a participant while Tycho was still alive, provided the occasion, with the result that I considered certain theorems of optics, trifling in appearance, but bearing the seeds of great things, worth pursuing and unfolding at greater depth. For I thought it unworthy in optical science, since senses and instruments are required in astronomy, while in optics, geometrical certainty is not lacking, that optics be surpassed by astronomy, and that one cannot establish in the former by demonstrations what in the latter the eyes have comprehended; it is much more unworthy that when summoned by astronomers to assist them, optics does not appear, and is unable of itself to soften the fault of hindering the precision of astronomy, with which it is charged. And so I have considered it no small honor if it might fall to me to undo the knots that have been in the way,
)(3^r by a good method and the clarity of demonstrations, and to lead the optical science through to that degree of subtlety that it might satisfy the astronomer. Which plan of mine, after Your Imperial Majesty approved it, I began to follow through with the greatest industry, by the collection and publication in book form of those things which either I had discovered once in Styria, while drawing a stipend from the Provincial Governors, or had to be added to the discoveries later, which were in fact the most difficult of all, and required the most work.

And there were two things that Tycho Brahe's accurate carefulness first brought out into the light, that pertain to optical science: one, on the refraction of the light of the heavenly bodies; the other, on the diminution of the moon's diameter in solar eclipses. Although several other passages also appear throughout that man's writings in which he sent the astronomers back and exhorted them to probe the mysteries of optics, these two main points are nonetheless of the

greatest weight. And in fact Brahe did carefully investigate the measure of refractions at all degrees of altitude, but since the cause of the measure had not yet been made evident by optics, there arose involved discussions about the universe and about the elements, and it was not sufficiently clear whether the refractions were the same or different at all times and in all places. From this, it finally follows that sometimes an observer, attributing too much to his own diligence and carefulness that he had applied among the things to be observed, and supposing *)(3v*
the place of the celestial body to be known with complete precision, throws the demonstrations into difficulty: the nature of light, beset by the inconstancy of optical causes, does not always allow such precision of instruments. And so I have devoted Chapter 4 to this matter, before which, because of the method of the investigation, Chapter 3 and part of Chapter 1 had to be presented. Moreover, the earth's shadow is drawn into the discussion, once the refractions have been found, and once it is known that the sun's rays do not spread out in the same straight lines from the source all the way to the earth, and much less so beyond it. And further, since this matter is in doubt, we make a test of the universal proportion of the measurement of the celestial bodies (since they depend on the earth's shadow), that is, of the thing, foremost in astronomy, that the public admires and the philosophers too praise and revere. And so, to preserve astronomy's dignity, and to storm this hostile fortification of doubt, although in Chapters 4 and 5 several steps in the big undertaking were made, and were, so to speak, cut into the extremely hard rocks by the sharpness of intellect, there nevertheless remained so much work as to require Chapter 7 especially for it. Now, in the diameters of the luminaries, which was the other subject, if the sense of sight commits any error, this affects not only the whole theory of eclipses, but also most of all in what I have just spoken of, the measurement of the celestial bodies. And thus, so that the tables of the motions, which, broken off by the author's death, Your *)(4r*
Majesty proposed for completion at this time, and is providing for with a not insignificant expenditure—so that these tables, I say, may achieve due soundness in their most conspicuous part, the labors of the sun and moon, I have resolved that every stone must be turned and nothing must be overlooked that might leave any hesitation. Therefore, in Chapter 2 I have explained the extent to which instruments can deceive us in investigating the dimensions of the luminaries, and in Chapter 5 I have begun to speak of the extent to which the sense of sight is itself beset with errors. And since the complete account of vision had to be set out as a whole, which is brought to completion with refractions, and with the images of things seen, and with color, it should not appear strange to anyone if I digress at some length into the sections of the cone in Chapter 4, which give form to the refractions of the eye, the optical entertainments of Joannes Baptista Porta in Chapter 2, the nature of light and colors in Chapter 1, and other things elsewhere. For although these matters might be not at all concerned with astronomy, they are in themselves worth knowing. Again, to the discoveries in Chapter 11, a matter that is commonly said to be easy,[1] I have added discoveries, and have

---

[1] *Rursum capite XI inventis, quod vulgo fertur facile, inventa addidi,...* The syntax

shown how to measure the diameters of these luminaries without chance of error, while in Chapter 8, on the occasion of eclipses of the sun, I have shown how to compare them with each other. And this much Tycho Brahe had provided me with the occasion of writing about.

)(4v Further, since the things just enumerated includes the greatest part of what can be said about astronomy from optics, and the book would now almost deserve the title, *Optics of Astronomy*, I thought I would make a compendium of the work if, whatever little were remaining, I should add it too. And these were about the light, the place, and the motion of the heavenly bodies, which I have gone through in Chapters 6, 9, and 10. For both the sixth and the ninth are ancillary to the eleventh, the former for estimating the moon's diameter, the latter for demonstrating optically the remarkable appearances of eclipses, and the paradoxical[2] changes of directions. Therefore, those things that I have found out in other chapters about these three remaining attributes of the heavenly bodies, I have, as much as they allowed, amplified and set forth in a much more polished manner. Especially in Chapter 9, the theory of parallax, which alone of all astronomy is the most difficult and vexing: this, I say, I have aided with the most easy short cuts, and have thus improved it so, that it might be taken as almost new, and I have constructed a new parallactic table, which does indeed take its name from this theory of parallaxes, but which has a very general and absolutely compendious application in the whole remaining theory of the secondary mobile bodies, which I shall set forth, God willing, in its own time.

And so, at this time, I have given this out to the public, however much work it involves, so that, because this most beautiful science lies neglected, I may arouse some people from this lengthy sleep to embrace it, and to navigate
(:)1r this practically new channel usefully, which I was the first to open with risk and perhaps with some expense; that is, to untie more dexterously the knots over which I have sweated, if they shall perhaps find anything lacking, while distinguishing those things which I have proposed as certain, provided only that they not fly into these most pure rites of Apollo with, as they say, unwashed hands, rashly taking on something greater than their powers and grasp.

But that I should bring this book of mine forward into Your Imperial Majesty's view, and give it in trust under the protection of this august Name, all the highest Offices persuaded me; for in these times Your Majesty, though distracted by the war against the Turks, the greatest of all and extremely costly, has not passed astronomy over unfunded; and has hitherto supported me, devoted to the Tychonic astronomy, with a liberal salary, by whose help I have carried these things out; and I in turn, by whatever means is in my power, wish to show my gratitude—indeed, my truest deference—and pay homage to this regal affection

---

is opaque here; Chevalley writes, "De plus en chapitre XI, j'ai ajouté de nouvelles découvertes à ce qui est couramment connu, . . . ," which makes some sense but doesn't fit the Latin.

2 παραδόξους.

for the liberal arts, and, according to what is due, make commendation to posterity; while those present should also understand through me that it is possible for no one to hide how much Your Majesty has done for these arts of peace, or which hopes of Your Majesty's eventual triumphs in this lengthy war, and upon the recovery of peace, they ought to cultivate.

If there is anything further in this work which should give me confidence in *(:)1v*
speaking in Your Majesty's presence, it is undoubtedly this, that I have expended a huge and disagreeable amount of work in explicating matters that have lain neglected for so many ages, and these are all different things, about which others have put together separate books for each, with as many patrons; nor have I filled up my soul with speculations of abstract geometry, that is, with pictures "both of those things that are and of those things that are not",[3] which are almost the only things on which the most celebrated geometers of today spend their lives, but I have tracked geometry through the cosmic bodies portrayed through her, following the Creator's imprints with sweat and panting. Finally, that both these works and others of this kind which I have in hand even now, I have taken up to do honor to my profession, which I hitherto follow by Your Majesty's order and salary, at the expense, not so much of health, but of other additional studies, by which I was able, and, in the judgment of my friends, was obliged, to create security for my old age, if such be destined, and for my family. If any inconvenience redounds to me because of this situation, this one thought, worthy of a German man, makes it light and pleasant for me: that it is beautiful even to die in honest service of such a Prince, and in this not to be overcome in valor by soldiers, but to defend vigorously an assigned task, as if it were some castle. Should Your Majesty most clemently approve of this most deferent affection of *(:)2r*
mine, I shall consider myself sufficiently happy; more happy, where he shall also have found in this work that which is adequate to His divine judgment on all the arts, and that which his mind, previously sated with the most outstanding discoveries, may yet desire from this work; but most happy by far if the same Clemency that I have thus far experienced should continue to favor me. For it will thus be by no means a cause of fear in me that that most pernicious enemy of arts, Indigence, might cast me down, conquered by starvation, from this task of mine, as from a citadel entrusted to my loyalty, nor will there be any doubt but that Your Majesty will have sent me that assistance and those provisions in time, upon the receipt of which I might be able to bear up under siege, and finally, with all difficulties overcome, happily to complete the rest of my studies for the glory of God, for the celebration of Your Majesty, and for the use of the human race.

May God Most Good and Most Great preserve Your Majesty unharmed for as long as possible, and, with your victorious arms, may you drive forth

---

[3] καὶ τῶν ὄντων καὶ μὴ ὄντων. This is an allusion to Protagoras's famous saying, "Man is the measure of all things: of the things that are, that they are, and of the things that are not, that they are not." (πάντων χρημάτων μέτρον ἄνθρωπον εἶναι, τῶν μέν ὄντων, ὡς ἔστι, τῶν δέ μή ὄντων, ὡς οὐκ ἔστιν.) These words were quoted by Plato in the *Theaetetus*, 152 A. However, Plato discusses being and nonbeing in the context of pictures in the *Sophist*, beginning about 235 D, so this may be what Kepler was thinking of when he wrote these Greek words.

from Christian necks, and do so to as great a distance as possible, the tyrant of
the house of Ottoman, advanced in years, odious in barbaric pride, insufferable
(:)2v in blasphemies. At Prague, 28 August, which day has brought Your Imperial
Majesty's fifty-third birthday, famous for the most honor-bearing embassy of the
King of the Persians. In the year of human salvation 1604.

Your Imperial Majesty's
Most Humble
Mathematician
Johannes Kepler.

Epigrams

(:)3r On the Optical Books of
Johannes Kepler, His Holy Roman Majesty's Mathematician

Epigram

of Ioannes Seuss the Elder, Secretary to Christian II, Elector of Saxony.

He who desires to observe the fire-belching eye of the world,
Or who wishes to perceive the eyes with his own eyes:
Let him, flashing with your mind's eyes, O Kepler,
Hide himself right in the bowels of your book;
Nor should he fear the rays of the radiant, and flying atoms,
And the wondrous phantasms that multifarious color bears;
[Nor] the mirror, be it a simple line or compound;
Since those things never seen by the eyes, you will give for seeing.

The Author's Epigram

upon his eyes, and his treatise on the eyes.

Eyes:
O dear mind, we have lost our sharpness, while the lights of the true
We have sent to your threshold, through our glazing.
Without this marriage, you would remain blind: of the work,
Give some return to your partners; give, sweet sister!
Mind:
What should I do for those in distress? When did an implacable hour
separate me far from your hospitality?
Eyes:
Snatch us away from darkness, lead us to whatever light you will go to;
And from whatever fear of death you lack, deliver us also.
Mind:
As far as possible, I shall do it; only let fame favor the speaker;
I shall make you mortals eternal with my writings.
Here also I bring back the losses that you have borne for me;
And here I shall cast rays upon its brilliance, even with blemishes.

Another with the same sense

The eye speaks.

I devalue life in exchange for fame; for a name, perception:
Teach, O soul, how to die more fruitfully, so as not to die.

Preface

1 Astronomy, which deals with the motions of the heavenly bodies, principally has two parts. One consists of the investigation and comprehension of the forms of the motions, and is mainly subservient to philosophical contemplation. The other, arising from it, investigates the positions of the heavenly bodies at any given moment, and has a practical orientation, laying the foundations for prognosis. These two parts fly up into the heavens, supported, as Plato used to say, by (as it were) a pair of wings, *geometry and arithmetic*, each of which serves both of the parts referred to, though geometry serves the contemplative part more, and arithmetic the practical part. And thus the highest apex of astronomy culminates in the *arithmetical* part, which embraces the *tables of motions* and the ephemerides derived from them.

Division of astronomy into parts

The geometrical part has many others beneath it, according to the variety of subject matter which an honest evaluation applies to geometrical astronomy. For, to trace it from the beginning, in astronomical demonstrations there are two kinds of principles: one, the *observations*, and the other, the *physical or metaphysical axioms*. Now around the observations, there arise the three parts of astronomy. The first is the *mechanical* part, dealing with instruments, suitable for observing the celestial motions, and the way of using them, which that phoenix of astronomers, the late *Tycho Brahe*, published five years ago.[1] The second is the *historical* part, comprising the observations themselves. The reader should be informed that 24 books of the most meticulous observations of this sort, embracing the past 40 years or so, were left by Tycho Brahe, which I hope will come forth into the light at some opportune time.[2] But because all celestial observation takes place through the mediation of light or shadow, and because the media between the stars and the eye have a variety of modifications, and because those things that we observe in the heavens are either motions (whose kinds include retrogra-
2 dation, station, and so on), or arcs (that is, angles at the observer), or luminous bodies; and because all these are considered in optical science: hence arises the third, *optical*, part of astronomy, which I am treating now, through a brief recounting, as if included among the principles, of the old things that Witelo treated methodically, or the new things that Tycho Brahe treated here and there, on this subject.

The other kind of principles in astronomical demonstrations, the physical or metaphysical principles, together with the subject matter itself of astronomy, which is principally the motions of the heavens, a physical matter, makes up

---

[1] Tycho Brahe, *Astronomiae instauratae mechanica,* Wandesburg 1598. In *TBOO* V pp. 3–162.

[2] Although Kepler had hoped to publish these observations, (see his comment in Vol. 5 of the Kepler Manuscripts in St. Petersburg, quoted by Frisch in *JKOO* I p. 192) he never succeeded in doing so. The were first published by the Jesuit Albertus Curtius in a book dated 1672 on its title page (though the colophon has the date 1666), under the title *Historia coelestis*. They appear in *TBOO* vols. 10–13.

the fourth part of astronomy, namely, *the physical part*, which deals with the efficient causes of the motions, or the movers; the formal causes, or figures that the movers strive for;[3] the material causes or orbs; and the physical intension or remission of motions; which part, if God should grant me life, I shall encompass by means of *Commentaries* on the motions of *Mars* (built out of the Tychonic observations, and upon the adamantine foundation of astronomy of the motion of the sun, reestablished by Tycho Brahe, and of the fixed stars, set in order by him); which I think I can call *the key to a deeper astronomy*.[4]

Status of the Tychonic astronomy

You see, dear reader, the position that is occupied by the Tychonic (that is, the truest and most accurate) astronomy. He collected most ample material for future building in the observation books; he showed the soundness of that material in the *Mechanica*; he laid two very solid foundations for the house, as I have just said: through the Catalog of Fixed Stars, described very accurately and truly, which has played the role of the best cement, and will serve to glue together the material of the observations; and through the theory of the sun that he established, which has the firmness of a foundation, and is the central post, embracing all the arches, right to their tops, the door too, with groined roof belonging to all the chambers.[5] But he brought the front of the house to competion, I mean the theory of the moon, the primary palace and portico, so that the house might be suited to habitation. To this, in the present book, for this purpose, windows and stairs are in part added and in part replaced where they were broken out. The armory, or the theory of Mars, was constructed; doors, or publication, will be added in the near future. There remains the office, the oratory, the dining hall with bedroom, the study, above all of which is built a platform, in the place of a watchtower, for catching a view of the ages, the theory of the eighth sphere, and the apogees of the planets.[6] These are indeed mixed with some

---

[3] Kepler originally conceived of the oval orbit as resulting from the combination of two movers each "striving" to accomplish a circular motion. See *JKGW* XX.2 pp. 576–9, and W. H. Donahue, "Kepler's First Thoughts on Oval Orbits: Text, Translation, and Commentary", *Journal for the History of Astronomy* **75** (1993) pp. 71–100, and "Kepler's Approach to the Oval of 1602, from the Mars Notebook," *Journal for the History of Astronomy* **89** (1996) pp. 281–295.

[4] This was published in 1609 under the title *Astronomia nova*; it is in Vol. 3 of *JKGW*, and has been translated into English by the present translator (*New Astronomy*, Cambridge 1992; new edition: *Astronomia Nova,* Green Lion Press, 2015).

[5] Kepler is alluding to the presence of the sun's (that is, the earth's) motion in the apparent motions of all the planets. There is an irony here, however: largely because of his erroneous estimation of the solar parallax, Brahe's solar eccentricity, the central parameter of his solar theory, was incorrect. See Curtis Wilson, "The Error in Kepler's Acronychal Data for Mars," *Centaurus* **13** (1969), 263–8, reprinted in Wilson's *Astronomy from Kepler to Newton* (London: Variorum, 1989).)

[6] The "eighth sphere" is the sphere of the fixed stars. Kepler is using conventional terminology here, not so much in placing the stars on a spherical surface (which he also did), but in ascribing the motion of precession to this sphere rather than to a change in the orientation of the earth's axis. The very slow motions of precession and of the planetary apsides required observation over very long periods of time.

timbers of ancient observations, rougher and consumed with rot; nevertheless, these too, as far as possible, will be smoothed and strengthened, so that in the end it will be possible to place on top of it the roof or highest pinnacle, the *Tables of Rudolph*. That this may also come to pass at the first possible time, pray with me
3 to God All-great and All-good, if in fact you admire the Tychonic astronomy, that by means of the arms of Emperor Rudolph he may push back that sworn enemy of the name "Christian," and that he may grant his people victory. For if that should happen, there is no doubt that the Most Wise Emperor will then provide suitable outlays in the greatest abundance, since even in these upheavals of war he does not interrupt them.

Division of astronomical topics.

Those things that come to be considered optically in astronomy are either the objects themselves set before the sense of sight, where images of things, or rather, light and shadow, are considered; or the medium through which vehicle the light of the images passes, because of which some light comes to us refracted; or finally, the visual instrument, or the eye.

Again, the objects themselves are either the celestial bodies (the sun, the moon, and the stars), or the motions, or the places of the bodies. We properly begin with the bodies. Now in bodies, in astronomical terms, we consider nothing but their image, which they send down to us by the aid of their light, with which they are endowed, and of that image, the shape and quantity, chiefly of the sun and the moon, but, regarding the earth, that of its shadow.

The natural purpose of eclipses of the sun and moon.

For the most noble and ancient part of astronomy is the eclipse of the sun and the moon, a subject that, as Pliny says,[7] is in the entire study of nature the most wondrous, and most like a portent. Anyone who ponders this carefully will find (if he will refuse to have recourse to faith in holy scripture) both that there is a God, founder of all nature, and that in the very mechanics of it he had care for the humans that were to come. For this theater of the world is so ordered that there exist in it suitable signs by which human minds, likenesses of God, are not only invited to study the divine works, from which they may evaluate the founder's goodness, but also are assisted in inquiring more deeply.

For, I implore you: what is the cause, if it is not this, for nature's playing such games in the sun's and moon's bodies, by which not only humans, as history bears witness, are turned to wonder and stupefaction, so long as they are ignorant of the causes, but even the quadrupeds, by Pliny's testimony, commonly take fright?

Furthermore, the extent to which humans are assisted by eclipses of the luminaries in all of astronomy, all the books of the astronomers teach. For, as regards the motions of the sun and moon, and the lengths of years and months, this entire theory first arose solely from the observation of eclipses, nor could it be constructed otherwise. Moreover, it cannot be smoothed and polished further except by considering eclipses of the luminaries more accurately and finely, which is the aim of this book.

Now anyone who will consider how closely all the rest of astronomy is linked to the sun's motion, and how much the moon, participant in day and

[7] Pliny, *Historia naturalis* II vi 46, trs. Rackham, vol. I pp. 196–7.

night, assists us, when all other means fail us, will rightly believe that all of astronomy is supported by these obscurations of the luminaries, so much so that these *darknesses* are the astronomers' *eyes*, these *defects* are a cornucopia of *theory*, these *blemishes* illuminate the minds of mortals with the most precious *pictures*. O excellent theme, praiseworth among all people, of the *praises of shadow*! 4

And thus the quantity of the image which the moon or the sun, whether whole or eclipsed, shows us, and of the shadow which the earth stretches out to the moon, must be carefully investigated by the astronomer. The diameters of the other stars are sought out to the extent that, if neglected, they will render the observations uncertain, and to the extent that we are busying ourselves to know the same things about them as about the bulk of the sun's and moon's bodies.

Next, those images come down to us thanks to the light in the sun, which is direct and intrinsic, and in the moon, which is reflected and extrinisic.

Even though these images are obvious to everyone's eyes, all practicing astronomers complain that it is with difficulty that they are measured. This is partly because the bodies have a very narrow apparent size, and partly because they constrict the eyes with their exceptional light, so as to prevent their fulfilling their function in seeing. But even in this place, nature has not forsaken those desirous of learning, for she has shown a procedure by which we may accomplish in darkness, without detriment to the eyes, what is completely impossible in clear light, with the eyesight directed towards the sun. Since this method is worthy of admiration, and also was grasped most ingeniously by the practitioners of the art, it is fitting that it not be held as contemptible and neglected among astronomers, but adorned with geometrical demonstrations, and illustrated with instances, which was done by me three years ago, on the occasion of the eclipse of the sun of 1600, in that manner which follows shortly, although I shall preface a few things first.[8] For because many things, not only about the direct ray, but also about the reflected and refracted ray, were overlooked by Witelo, and many things that should have been explained *a priori* were only brought in extraneously from experience and set up in place of axioms, I thought it a good idea to look a little more deeply into the whole nature of light, and to relate to their principles those things that appear, insofar as possible at present, in case readers should come along whose mental powers for seeking out the arcana of light might be able either to be more aroused or even more assisted by these discussions. Although these things are not a little remote from the proposed subject, there are also not a few things that are to be cleared up for astronomers by all the kinds or images of rays.

---

[8] An example of Keplerian wit: the solar eclipses of 1600 and 1601 are not considered until the final pages of this lengthy treatise.

Chapter 1
On the Nature of Light

5 Albeit[1] that since, for the time being, we here verge away from Geometry to a physical consideration,[2] our discussion will accordingly be somewhat freer, and not everywhere assisted by diagrams and letters or bound by the chains of proofs, but, looser in its conjectures, will pursue a certain freedom in philosophizing—despite this, I shall exert myself, if it can be done, to see that even this part be divided into propositions.

In the place of Common Notions[3] I am giving a preliminary admonition on vocabulary: what the Greeks denote with the single term "bending back" (ἀνα-´κλασις)[4] of rays, the Latins for the sake of distinction divide into two classes, establishing one class of *reflections* and the other of *refractions*. Now if words are to be diversified because of the different natures of things, it is preferable to choose those in which the nature of any particular thing is expressed as nearly as possible to the usage of the common people. Hitherto, this does not seem to have been brought happily to pass by the Latin optical writers.

For this much is clear to me, that the proper usage of the Latin word "flectere" (to bend), κάμπτειν[5] to the Greeks, concerns things that are bowed, and which, if the force were to cease, would return to themselves. Hence, the stretching of the ligaments and the bowed appearance of the member is also called "bending the knee".[6] Other uses are analogous to these.

The Greek word ἀνακλάσθαι,[7] however, most properly expresses that

---

[1] *Caeterum*, in Latin. This is a very odd word with which to begin. It is the adverbial use of the neuter of *caeterus* (the rest, the remainder), and usually means "moreover" or "apart from this" or something similar, indicating a transition from what went before. It therefore seems appropriate to use an archaic English word and keep Kepler's sentence unbroken, to suggest the odd flavor of the original.

[2] According to the prevailing Aristotelian division of the sciences, mathematics and physics were sharply distinguished in that the former treats abstract forms while the latter treats the concrete. "The geometer deals with physical lines, but not *qua* physical, whereas optics deals with mathematical lines, but *qua* physical, not *qua* mathematical." (Aristotle, *Physics* II 2, 194 a 10–12.) Kepler carefully preserved this distinction, using geometry as an intermediate between observation and physical theory. See especially *New Astronomy* ch. 6, trans. W. H. Donahue (Cambridge 1992) p. 162.

[3] Kepler here alludes to Euclid's introduction of what he calls "common notions" (κοιναὶ ἔννοιαι) immediately following the postulates. They are concepts that are common to all of mathematics, as distinguished from the postulates, which are proper to geometry.

[4] This is the noun formed from the verb ἀνακλάω, from ἀνα- (up or back) and κλάω (bend).

[5] Kepler gives the infinitive form. The present indicative is κάμπτω. Kepler's remarks are correct, except that the word does not imply the presence of stress in the thing bent.

[6] *Genu flecti*.

[7] Kepler confuses the matter by giving the infinitive of the middle voice rather than the active. This is ἀνακλάω, for which see note 5.

which is "frangere" (to break) in Latin, which presupposes a dissolution of continuity (most especially of wood) and a loud noise. Thus κλάδος[8] in Greek is a small branch torn from a tree. And κλάζειν[9] is used to describe the noise which in Latin is called "fragor" (noise as of breaking), from "frangendo" (breaking). If this be compared with the characteristics of rays, the word "flectere", if you look to its proper sense, clearly does not square with them; for nothing of the sort pertains to rays, whether acted upon by mirrors or by water. On the contrary, both of these are most correctly described by the same word: "fractio" (breaking). For both are truly broken, the one by the surface of the mirror, the other by the surface of the water, and the broken parts constitute a rectilinear angle.[10] Thus although in a broader sense something is also said to "bend aside"[11] which, while it should have been single, has gone off into two contiguous straight lines, 6
exactly like the rays; nevertheless, for the sake of avoiding the undesirable implication, *the word* flectere *is clearly to be eschewed*. For in this signification, *reflexus* (bending back) would pertain to both classes of rays, exactly like the Greek word ἀνακλάσθαι.[12] But because rays are affected in one way by mirrors and in another by water (for the one rebounds from the mirror towards those parts whence it came, while the other bends down from the surface of the water into the depths, and towards those parts opposite those whence it came), let us therefore follow here the convention of the optical writers, and use diverse names: let the one be denoted by the Virgilian and special word "Repercussion,"[13] and the other by the class name "Infraction,"[14] so that the propositions themselves may allude to the nature of the thing.

Principles and suppositions for arguments concerning the nature of light.

For demonstrations of characteristics of this kind, however, all philosophers and optical writers establish a certain comparison between physical bodies (and their motions), and light, which comparison we shall explicate somewhat more broadly.[15].

---

[8] There is, *pace* Kepler, no sense of tearing in this word, which simply means "branch" or "twig". Nor must the branch be separated from the tree.

[9] This word (κλάζω), though it somewhat resembles κλάω, for which see note 3, is actually derived from κλαγγή, and means "make a sharp, piercing sound". The two roots are unrelated, nor does κλάζω have anything to do with κλάδος, as Kepler implies (although the latter is related to κλάω). See H. Frisk, *Griechisches Etymologisches Wörterbuch* (Heidelberg, 1960), vol. 1 pp. 863–867.

[10] Kepler believed light to be a propagated *species* (form, image) and not a moving corporeal substance. Had he held the latter opinion, he, like Newton (in Query 20 of the *Opticks*), would probably have argued that refraction does not occur at a single point.

[11] *Deflectere*.

[12] See footnote 7, above.

[13] *Repercussus*. Cf. *Aeneid* VIII 23, where the participle form *repercussum* is found.

[14] *Infractus*, an action noun apparently coined by Kepler from *infringo* (to break, refract, or disjoint).

[15] The following cosmogony is a recasting of the theory sketched out in Kepler's letters to his teacher Michael Maestlin in 1595 (letters 21, 22, and 23 in *JKGW* vol. 13 pp. 28–38) and first published in chapter 2 of the *Mysterium Cosmographicum* (Tübingen,

First, it was fitting that the nature of all things imitate God the founder, to the extent possible in accord with the foundation of each thing's own essence. For when the most wise founder strove to make everything as good, as well adorned and as excellent as possible, he found nothing better and more well adorned, nothing more excellent, than himself.[16] For that reason, when he took the corporeal world under consideration, he settled upon a form for it as like as possible to himself. Hence arose the entire category of quantities, and within it, the distinctions between the curved and the straight, and the most excellent figure of all, the spherical surface. For in forming it, the most wise founder played out[17] the image of his reverend trinity. Hence the point of the center is in a way the origin of the spherical solid, the surface the image of the inmost point, and the road to discovering it. The surface is understood as coming to be through an infinite outward movement of the point out of its own self, until it arrives at a certain equality of all outward movements. The point communicates itself into this extension, in such a way that the point and the surface, in a commuted proportion of density with extension, are equals.[18] Hence, between the point and the surface there is everywhere an utterly absolute equality, a most compact union, a most beautiful conspiring, connection, relation, proportion, and commensurateness. And since these are clearly three—the center, the surface, and the interval—they are nonetheless one, inasmuch as none of them, even in thought, can be absent without destroying the whole.

The origin of light.

The spherical is the image of the Holy Trinity.

7 This, then, is the authentic, this is the most fitting, image of the corporeal world, which anything that aspires to the highest perfection among corporeal created things takes on, either simply or in some respect. The bodies themselves were confined separately within the limits of their surfaces and could not by themselves have multiplied themselves into an orb. For this reason, they were endowed with various powers, which, though they do have their nests in the bodies, nevertheless, being somewhat freer than the bodies themselves and lacking corporeal matter (though they do consist of their own kind of matter which is subject to geometrical dimensions), may proceed forth and might try to achieve an orb, as appears chiefly in the magnet, but appears plainly in many other instances. What wonder, then, if that principle of all adornment in the world, which the divine Moses introduced immediately on the first day into barely created matter, as a sort of instrument of the Creator, for giving form and growth to everything—if, I say, this principle, the most excellent thing in the whole corporeal world, the matrix of the animate faculties, and the chain linking the corporeal and spiritual world,

The spherical is the archetype of light (and likewise of the world).

In praise of light.

1596). The present account differs not only in being more succinct, but also in the dynamic notion of the point unfolding itself into a sphere.

16 Cf. Plato's *Timaeus* 30 c–d.

17 *Lusit*. In the *Tertius Interveniens* (Frankfurt, 1610), Section 126, Kepler writes, "Now, as God the creator has played [*gespielt*], he has also taught nature, his image, to play; and the game is just the same as the one He had played [*vorgespielt*] for her." (*JKGW* 4 p. 246.)

18 As the density increases, the extension decreases, so that the quantity compounded of the two remains constant.

has passed over into the same laws by which the world was to be furnished. The sun is accordingly a particular body, in it is this faculty of communicating itself to all things, which we call light; to which, on this account at least, is due the middle place in the whole world, and the center, so that it might perpetually pour itself forth equably into the whole orb. All other things that have a share in light imitate the sun. From this consideration there arise, in a way, certain propositions, which are among the principles in Euclid, Witelo, and others.

The sun's place in the world.

Proposition 1

To light there belongs an outflowing or projection from its origin towards a distant place. For it was said that light had to be communicated to all bodies. This communication had to come to pass through a conjunction of dimensions, for we said that light falls under geometrical laws, and that it is considered in place as a geometrical body. Therefore, it will be communicated either by the approach of its source to the object, which is absurd, and is no communication: therefore, it remains that it be communicated through a local egress, and an outflowing from its body.

Proposition 2

Any point whatever flows out in lines infinite in number. That is, in order to
illuminate all of the surrounding orb, which is what we said must happen. But the 8
spherical has infinite lines.

Proposition 3

Light, in itself, is fitted for moving forward into infinity. For since it partakes of size and density, by the above, it will be able to vanish into nothing with no size, because size, and thus density also, can vanish through division to infinity. So much concerning the essence. But also, the force projecting it is infinite, because, by the above, light has no matter, weight, or resistance. Therefore the ratio of power to weight is infinite.

Proposition 4

The lines of these projections are straight, and are called "rays." For we have said that light strives to attain the configuration of the spherical. However, its true geometrical genesis lies in the equality of the intervening spaces, through which the middle point is spread out into a surface. But these are straight lines. If light were to make use of curves, there would be no equality in the spreading, and therefore nothing similar to the spherical.

The same thing is also proved, or rather declared, in this manner. The ends of motion are various. For nature strives after either the unity of the parts, or their separation, and both take place most easily by a straight motion. And things that are closer together are understood as being more united, and straight lines are the shortest of all lines between the same points. Therefore, the motion that unites a thing, such as the motion of heavy things towards the earth, or of iron to a loadstone, must necessarily occur along a straight line; otherwise, not all parts of the motion would tend to the same end, but, somewhere in mid path, that which was to unite with the other would turn aside from this striving for union. The same is to be understood of the contrary motion of separation, which is called "violent" in the realm of nature. For to it also there belongs a motion that is the contrary of the motion of uniting, and therefore it is straight, since to the straight only the straight is contrary.

To light belongs, not uniting, but something similar to separating, and a certain most violent projection or flowing out. Therefore it is also a straight motion. Or, if you prefer, call it the union of that light in the lucid body with the object to be illuminated. For the same thing will follow.

Nor does a curved line follow from the nature of light. For by Prop. 3 it, in
9 itself, is fitted for being called forth to infinity, while curved lines, inasmuch as they are curved, return to themselves, and are bounded.

Proposition 5

The motion of light is not in time, but in a moment. For as is demonstrated by Aristotle in the books on motion,[19] there is a certain proportionality of time to that ratio that exists between the moving power and the weight or movable bulk, or to the ratio of the weight to the medium. But this moving force has an infinite ratio to the light to be moved, because light has no matter, and therefore no weight. So the medium does not resist light, because light lacks matter by which resistance could occur. Therefore, the swiftness of light is infinite.

Proposition 6

With departure from the center, light receives some rarefaction along the breadth. For by 2 and 4, light goes forth in infinite straight lines. But these are closer together at the center, because there are the same number in a narrow place as there are in a greater space. But this is the definition of rarity and density. Therefore, it is rarefied along the breadth.

Proposition 7

With departure from the center, a ray of light receives no rarefaction along the length; that is, it is not the case that the longer the ray, the more rare or diffuse, at least not because of that length itself. For in the geometrical genesis of the spherical, nothing of the sort can be conceived. In contrast, light strives for that genesis by its communication of itself. Besides, the ratio of the projecting power to that which flows out is infinite, since it lacks matter, as was said above; and, by Prop. 5, the motion occurs in a moment; and, by Prop. 3, it goes to infinity. Therefore, the projecting strength is equal out to infinity, namely, that very strength which is also at the origin. So also, the strength of a ray is equal along the length.

---

[19] Aristotle, *Physics* V 8, 215 a 25: "For we see that the same weight and body is carried faster through two causes, either by a difference in the through-which, as through water rather than earth or through air rather than water, or by a difference in what is carried, if the other things are equal, because of an excess of heaviness or lightness." (Translation by Joe Sachs.) Note Kepler's change of Aristotle's speed into time. Kepler's ratio should, of course, be inverted.

Proposition 8

A ray of light is no part of the light itself flowing forth. For, by Prop. 4, a ray is nothing but the motion itself of light. Exactly as in physical motion, where its motion is a straight line, while the physical movable thing is a body; likewise, in light, the motion itself is a straight line, while the movable thing is a kind of surface. And just as in the former instance the straight line of the motion does not belong to the body, so in the latter the straight line of the motion does not belong to the surface.

Proposition 9

The ratio that holds between spherical surfaces, a larger to a smaller, in **10**
which the source of light is as a center, is the same as the ratio of strength or density of the rays of light in the smaller to that in the more spacious spherical surface: that is, inversely.[20] For, by 6 and 7, there is the same amount of light in a smaller spherical surface as there is in a more spread out one, and therefore it is that much more compact and denser in the former than in the latter. If, however, the density of a linear ray were different depending upon position in relation to the center (which is denied in Prop. 7), the situation would be different.

Proposition 10

Light is not impeded by the solidity of bodies as such, so as to prevent its passing through them. For whatever is impeded is impeded or repelled by something of the same kind, as a body by a body. Solids, as such, have three dimensions. To light, by 6 and 7, there belong only two dimensions. Therefore, light, or its rays, are unaffected by solids as such, nor do they interact as regards solidity.

Proposition 11

The pellucid[21] *is one body whose geometrical constitution, or the positions occupied by the internal parts, is established by means of a certain flow.* For the geometrical definition of the humid is found in Aristotle: that is called "humid" which is not bounded by itself.[22] Therefore, all the least parts are mutually bounded by themselves, and the whole is entirely united by the internal parts, not actually divided by surfaces. But now, by Prop. 10, the solidity, as such, does not impede light in passing through. Therefore, whatever is one, is pellucid.[23] And if

---

[20] This is a correction of the linear inverse relation stated by Kepler in Chapter 20 of the *Mysterium cosmographicum*. See *JKGW* I p. 71.

[21] In his appendix to this chapter (p. 46), Kepler says, "For something is transparent (*perspicuum*) when it is seen through, and pellucid (*pellucidum*) when shone through by some lights."

[22] Aristotle, *On Generation and Corruption*, II 2, 329 b 31.

[23] Here Kepler refers to the following endnote.

To p. 10. The logical connection is in the origin of the word "pellucid." However, things that are black in the highest degree are excluded, for, although they are one, they are nonetheless not pellucid, because of their property of the highest blackness, as

there are things that are partly liquefied, and partly not, such mixed things do not constitute one pellucid body. Further, this definition is to some degree extended to all natural bodies, an extension which Aristotle clearly approved in the book *On Sensibles*.[24] Moreover, hard as well as ringing bodies are pellucid. Nor is it an impediment for them to be colored, for pellucid bodies should be so, as is shown below.

Proposition 12

Light is acted on by the surfaces of whatever bodies it encounters. For those
11 which fall under the same kind are fitted for acting on each other mutually. Light, by Prop. 7, is of the same kind as bodies because of the surfaces by which they are bounded. Therefore, it is suited to being acted on. But it is also actually acted on. For to be bounded is to be acted on, and a line is bounded by its point, and hence is acted on by it. The motion is a straight line, whose boundary is to some extent a point on the encountered surface. And the boundaries of the infinite rays of light are the infinite points, that is, the surface, which is, as it were, composed of them.

Proposition 13

The surfaces of dense bodies, or bodies of which many parts of matter fill a small solidity, are also in a sense dense; namely, in that respect in which light and bodies mutually act on each other. For although surfaces, as quantities, are assigned no thickness, which belongs to bodies, nevertheless, inasmuch as they are surfaces of dense bodies on account of this material geometrical characteristic, they too are considered dense, since they are material surfaces. More clearly, density is an affect of matter, which assumes three dimensions: of these, two belong to the surface. Therefore, they participate in the density of bodies according to their measure. But above, a sort of density of the surface was ascribed to light in the same respect.

Proposition 14

Light passes through the surfaces of dense bodies with greater hindrance, to the extent that they are dense. For since, by Prop. 1, motion is a property of light, the properties of rectilinear motion also are properties of light. Therefore, being

---

follows in Prop. 17. And you should take the word, "colored," with the same limitation, in line 30 following. Here, too, the supremely white color is excluded, not insofar as it is color, but through a property, insofar as it can only be represented by rough bodies (which are not pellucid). When I attribute this opinion to Aristotle (on line 28 and on p. 13 line 4), it is because in the passage cited, after he had spoken of water, air, and other bodies thus described (that is, as pellucid), he added, "it is in them, and participates in the other bodies, to a greater or lesser extent." [*On Sense and Sensibles* 3, 439 a 24.] He seems to have been saying that aside from the things that are usually called "pellucid," this nature is also in the rest of the bodies to greater or lesser extent. Should a truer parsing instruct you otherwise, strip this opinion of the authority of the Philosopher, and send it back to me naked.

24 Aristotle, *On Sense and Sensibles* 3, 439a 22. See the preceding note.

hindered by a denser medium is also a property of light. However, by Prop. 10, this is not insofar as the medium is a solid, and therefore, it is hindered insofar as it is bounded by a dense surface. More clearly, the motion of light occurs naturally with spreading out, by Prop. 6, since it always goes from a single source to all regions. Therefore, just as the surface, because of its infinite points, resists motion, which is in lines, so a dense surface resists a spreading motion, since density and spreading come under the same genus.

Proposition 15

Notes. White and black appear to be opaque in its highest degree; the rest can be in the pellucid in its highest degree as well. And while, by this account, white and black are in a certain way material or corporeal colors, the difference is nonetheless this, that white is more like a shining body, such as the sun, and black is more like a dark body. It is possible to see the emerald and the ruby and the sapphire and the amber, pellucid bodies, by the colors green, blue, red, and yellow, that no such body is in the highest degree white and black that is at the same time pellucid.

Color is light in potentiality, light entombed in a pellucid material,[25] *if it now be considered apart from vision. Different degrees in the arrangement of matter, by reason of rarity and density, or of pellucidity and darkness, and likewise, different degrees of the spark of light,*[26] *which is condensed into matter, bring about the distinctions of colors*. For since the colors seen in the rainbow are
of the same kind, whence the colors in things are as well, the origin of the two will 12
be the same. But the former originate from just these causes that have now been mentioned. For when the eye is removed from its position, the color changes. And indeed, at the common boundary of light and shadow, all the colors seen in the rainbow rebound, so that it is certain that they exist from the rarefaction of light and from the scattering of a watery material above it. Therefore, colors in things, too, will arise from the same source. And there will be this much difference, that in the rainbow the light is extrinsic, while in colors it is inherent, in the same way that in parts of many animals there are certain lights that are at work. And to the degree that there are differences in the potential heat in ginger from the actual heat in fire, so it appears that light in colored matter differs from light in the sun. For something exists potentially, which does not communicate itself, but is contained within the bounds of its subject, as is light, which lies hidden in colors, as long as they are not illuminated by the sun. However, you do not know whether colors too might not scatter their sparks of light from the depth of night. But this subject has variously exercised the wits of the most shrewd philosophers, and is

---

[25] Here Kepler refers to the following endnote.

To p. 11, to Prop. 15. Since this definition belongs to colors according to greater and less, it does not at all fit the perfectly black. For this is the limit of all colors, and is related to colors as a point is related to a line, pertinent to quantities, although it is not a quantity. So also the perfectly black color is lacking all potential light, and consists in pure materialized darkness. And radiating into a dark chamber, it does not paint the wall black, but gray, nor would it be noticed unless the images of other colors were set about it. It therefore comes pretty much only under the account of something lacking almost all radiation; it is noticed in a picture of the wall. It remains for you to wonder, however, that matter has so great a power, that through it light produces its enemy, the color black—that is, nothing but rays of darkness, painting themselves to some extent on the surface presented.

[26] *Lucula*. This word appears to have been coined by Kepler. It is the word *lux* with a diminutive ending, and evidently means "a small bit of light." Since the English word "lucule" is already taken (it refers to a bright speck on the sun's surface), "spark of light" has been chosen instead.

of such obscurity that it cannot be cleared up in the present account, particularly since it is not particularly relevant. Should you object that darkness is a privation, and can therefore not become something positive, an active principle (namely, one that radiates and colors the walls), I would likewise raise cold in objection, which is a mere privation, and nonetheless also becomes an active quality in matter.

Proposition 16

Light passing through colored bodies is affected everywhere, both at the surface and in the solid, to the extent that it is colored. For those things that fall beneath the same genus are suited to act upon each other mutually. But light and color fall under the same genus, by Prop. 15. However, by the same, color inheres in the matter of the pellucid body. And matter has three dimensions, so therefore, for light also, the action is upon that color which is in the depth of the medium.

Proposition 17

The opaque is something that is broken up into many surfaces, and something that possesses great density, and something that possesses much color, whether in quantity or in quality. For the opaque is that which does not transmit the rays of light. But, by Prop. 12, many surfaces greatly impede the rays, just as, by Prop. 14, the surface of something dense greatly impedes them. So also, by
13 Prop. 16, the surface of a colored body greatly impedes them, whether the depth of the medium adds quantity to the color, or the color itself recedes greatly from brightness, and participates greatly in darkness.

Further, that nothing is absolutely opaque, Aristotle also allows in the book on perceptibles [27]

Corollary on Method

Therefore, since light has two aspects—that of the essence, by virtue of which it is light, and that of the quantities which it acquires—it also possesses two operations,[28] the prior that of local motion, and the posterior in the order of nature (which is prior in its goal or form) which is illumination (with which heating is conjoined). Of these, local motion belongs to light because of of quantities, while illumination belongs to it because of its essence, by virtue of which it is light. In conformity with these two operations, it also has two objects—quantities and color— with which matter is conjoined. And because local motion also has two kinds of objects—the medium through which it passes and the thing towards which it is carried—there also exist two aspects of this operation in light: in the

---

[27] Aristotle, *On Sense and Sensibles* 3, 439a 22.

[28] *Energias*. This is a Latinization of the Greek ἐνέργεια, used by Aristotle and Galen to mean something like "operation" or "function." Cf. Galen, *Opera*, ed. C. G. Kühn, Leipzig 1821–33, VI 21. In Aristotle's philosophy, it is also the technical term usually (but misleadingly) translated "actuality," as opposed to δύναμις, "potentiality." Joe Sachs, who discusses the term in the glossary to his translation of the *Metaphysics* (*Aristotle's Metaphysics*, Green Lion Press 1999, p. li) translates it "being at work." Cf. Aristotle's use of it in *Metaphysics* 1048a 26.

case of the medium, light penetrates the medium and is refracted by the medium if it is denser; in the case of the thing, light strikes the surface which it encounters, and is made to rebound by it. These two aspects are distinguished from each other in different ways. For by motion, light is both spread through the medium, and dashed against a surface, and in turn is either picked up by the surface of a pellucid medium, or is made to rebound by the surface of the boundary. Finally, as regards illumination, which is the primary task of light: colors are diluted by light, and light is tinted and dyed by colors. All of these will be treated in order in what follows.

The reader should note here the origin of the fourth species of light, which the other optical theorists treat ineptly.[29]

Proposition 18

Light that has fallen upon a surface is made to rebound in the direction opposite to that whence it approached. For motion is an attribute of light, by 1, and therefore so is a particular species of motion, namely impulse [pulsus]. For what in physical motion is hardness of the colliding bodies, which consists of the permanence of the surfaces, in light is the bare surface, or the bounding and shaping of bodies. For by the very fact that physical bodies are even themselves bounded, they are understood to be hard. Soft and moist things, on the other hand, are defined by way of their not having a proper boundary. But the cause **14**
of rebounding's occurring in both physical motion and light is in the violence[30] of the motion. So since the motive force cannot all be destroyed by the collision, the motion will accordingly continue beyond the end of its line, which is the surface. But it cannot do this straight forward; for in the former case, body is in the way of body; and in the latter, surface is in the way of surface; in the former the motion would go into the solid, in the latter partly so, as we shall hear. What is left, therefore, is that it move in the opposite direction. These things may perhaps be put more concisely thus. Impulse is an action, and is between contraries, but with action there is a corresponding effect [passio]: therefore, this is also the case in impulse. But the contrary to the impact by which light strikes a surface in one direction is a rebound in the other direction. For both the power that drives a movable towards something, and the power that drives it back from something in the way, are the same, because the collision is considered to occur at a point.

---

[29] Earlier optical theorists distinguished three kinds of light: direct, reflected, and refracted. See, for example, Alhazen IV 1, *Thesaurus* I p. 102. The three branches of optics corresponding to these species are Optics proper, catoptrics, and dioptrics, respectively. To these Kepler adds communicated light: see Prop. 22 below.

[30] Aristotle distinguished between natural motion and forced (βίαιος) motion (see, for example, *Physics* IV 8, 215 a 1). The standard Latin translation of βίαιος was *violens*, which adds a sense (quite foreign to Aristotle's thought) that the force involved might be unnecessarily great or destructive. Although Kepler uses the standard term, his understanding of it seems closer to Aristotle's meaning than to the sense of the Latin word.

Proposition 19

Rebounding occurs at equal angles, and the rebound of that which strikes obliquely is to the other side. This is V 10 in Witelo.[31] For an equal power also presupposes an equal motion. But the power at the point of collision is one and the same: therefore, the motion is also equal. But that which is at equal angles is equal. Therefore, the rebound of movables will be at equal angles. I say also that it will be to the other side from that whence the movable was brought into collision, if this applies. For if it were always to rebound in the same line, it would always have to have approached directly, by 18 preceding. But most often it approaches obliquely. Accordingly, then too the direction from which it approaches is set obliquely in the way of the surface, not directly: therefore, the rebound is also in the opposite direction obliquely, not directly. Or, more clearly, when something moves obliquely towards a surface, that motion is compounded of motion perpendicular and motion parallel to the surface. But the surface is only turned to face that part which is perpendicular to it, not that part which is parallel to it. Accordingly, it does not impede that part which is parallel to it, but allows the movable, in rebounding, to head in the other direction, just as it had approached. Let CDF be the surface, BD the motion of the light. Let BD be extended to E, intersecting CDF at D, and let CDE be equal to CDA. Therefore, since the surface CD is supposed not to impede the motion B in the direction from B towards AE, it will not change the angle CDE. For if B were not to rebound at D, it would make the angle CDE equal to BDF, because the line BDE would be straight. But, while not affecting the angle CDE, B nonetheless rebounds to the side of BA. Therefore, the amount of motion that would have been allowed to B in the direction from B towards AE had it not rebounded, is the same as the amount allowed to it even though it does rebound: not, however, towards E, because it rebounds, but therefore towards A, so that CDE and CDA are equal. But BDF and CDE would also have come out equal without the rebound. Therefore, the angles of incidence BDF and of reflection ADC are equal.

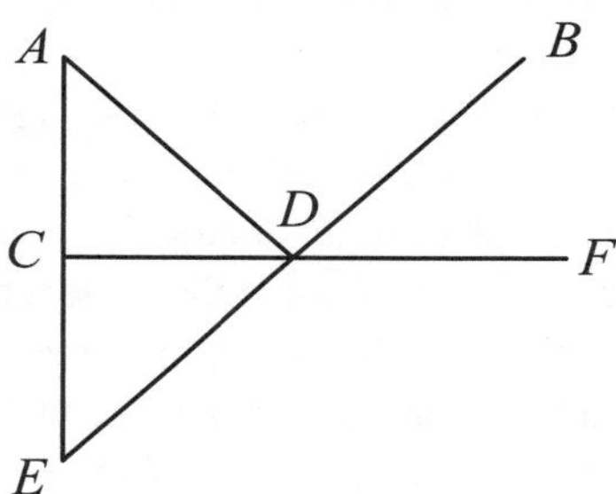

Proposition 20

Light that has approached the surface of a denser medium obliquely, is refracted towards the perpendicular to the surface. For by Prop. 14 and 16, the effects upon light here are opposite to its motion, and similar to the medium. But motion spreads light, and the medium is posited as denser. Therefore, the medium impedes light so as to prevent its spreading. But the evidence [argumentum] of spreading is oblique incidence, because when light

---

[31] *Thesaurus* II pp. 195–6.

falls obliquely upon a surface, it also falls directly upon the same surface extended, and therefore between the oblique and direct rays an angle is interposed, and by the angle the rays are spread. Let the oblique ray be AB falling upon the surface at the point B, let a line touching the surface be drawn (let it be BC), and from A let the perpendicular AC be dropped. Thus this light from A is spread to one side through the angle BAC, and the amount of light that was within the angle BAC in parts closer to A is the same amount as is also on the line BC. Let AB be extended to D, AC to E, so that DE is parallel to BC. If a denser surface does not stand in the way at BC, the light will be spread further, and there will now again be the same amount of light at DE as there was at BC. But if BC is the surface of a denser medium, it will check this spreading out, by Prop. 14, and this will occur in greater or less degree. But the light BC, if it moves further without spreading out, will have to occupy as much space on the surface DE as on BC, and therefore will have to cut off from DE some space equal to BC: let this be FE.

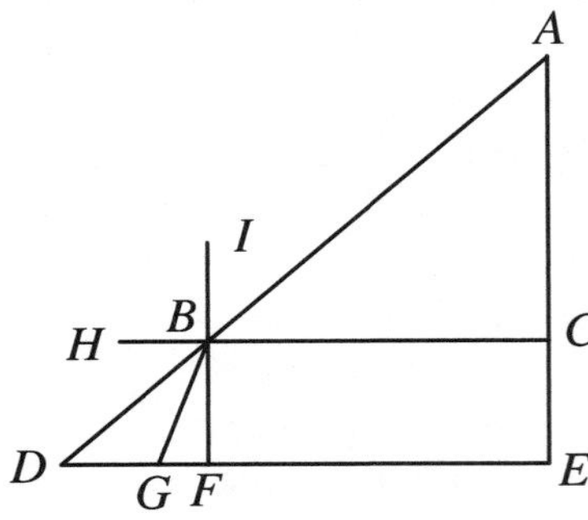

Now on ACE is the limit of all the spreading out of light, because the ray AC is at a right angle to CB, and consequently E will be the common limit of the 16
lines EF and ED. So BF will be parallel to CE, and therefore will also itself be perpendicular to BC. So light, coming as far as ED without any spreading out, would occupy the space EF, and the same light descending as far as the same place without any disturbance, would occupy the space ED, spreading and thinning itself out in the same ratio. Therefore, when a denser medium BC comes in between, hindering the spreading out, it makes the light occupy a space that is intermediate between EF and ED: let this be EG. Therefore, the ray AB will be refracted at B and below the surface of the denser medium will become BG, approaching the perpendicular BF, which was to be demonstrated. The question arises, however, by what faculty could it happen that the influence of a pellucid surface be imposed upon light? I answer: motion belongs to light, by Prop. 1, and therefore also the species of motion, and the rest of the accidents, namely impact upon a denser surface, and the overcoming of it, and a certain amount of resistance by the medium overcome. Moreover, that this also happens necessarily in physical movables, whenever a globe is spun into the water, provided that it goes beneath the water, is shown thus. For it may be permissible here for me to use the words of the optical writers in a sense contrary to their own opinion, and to carry them over into a better one.

Let BC be the water, AB the motion of a small sphere, let CB be extended to H and FB to I. Now since the motion AB of the small sphere is, from a certain perspective, composed of IB and BH, it also will happen that the depth BF resists it as well as the breadthwise thickness BH. The prior impediment makes its descent slower, and blunts it, as long as it descends; the latter, on the other hand, pushes it back from its line; so that the motion that was going to move along BD, is pushed back from BH and becomes BG. This happens of necessity

in projectiles, which are impeded by a denser body. In light, that portion of the motion which is in the direction BH, is that motion of thinning out, while that which is in the direction of IB or BF would not vary the light's own matter (so to speak), but only transports it. To this transporting, by Prop. 10, the body or thickness BF does not present any hindrance, but only the surface BC does so, and this, indeed, not otherwise than to the extent that it participates in density, by Prop. 14. It does not, however, participate in density in depth, for then it would be a body, not a surface. But it is dense in length and breadth, as in the parts BH
17 (for with respect to this line alone it is most greatly oblique to AB). And the light is carried or thinned out not towards the part BC but towards BH. Therefore, the surface from the part BH resists this motion, and hence there in a way exists a kind of bending back of AB into BG, entirely similar to those which occur in projected natural bodies.

I shall make an attempt at the same comparison of physical motion with the motion of light using other, denser proofs. For what in physical motion is a denser medium that is nonetheless yielding, or a weight that is in itself at rest, but is in motion through the impulse of another body, is in the present consideration a denser surface, but one which nonetheless may transmit light and may to some extent yield to the transmission.

Now as in physical motion a weapon sometimes collides with the object at which we were aiming it, and sticking to each other, they proceed along the same path with a single motion, so the same thing happens here with light and the denser surface which light penetrates, but without matter or the dimension of solidity.

On the principle of the balance, by way of digression.

I see, however, that with regard to physical motion, even up to the present day it has not yet been adequately explained why here too there should occur a refraction, or rather (since this word is more appropriate to physical motion), a deflection, from the direct line of motion towards the line that is perpendicular to the surface that is in the way, whenever something that makes an impulse strikes upon it obliquely. And because this whole matter depends upon the principle [ratio] of the balance, it should be derived from its source. For I don't think there is going to be any other place given me where I can more conveniently go through with this demonstration.[32]

In the *Mechanics*, Aristotle asks what the cause is, why the beam of a balance, which was previously inclined, returns to equilibrium when the support [trutina] is above the balance and the pans are empty; and why, on the other hand, when the support is below, to the extent that the beam is inclined by the weights, it stays there, not returning when the weights are removed.[33] Jordanus

---

[32] For general information on the ancient and medieval traditions regarding the balance, on which Kepler draws in what follows, see David Lindberg, editor, *Science in the Middle Ages* (Chicago: University of Chicago Press, 1978), Chapter 6, "The Science of Weights," by Joseph E. Brown, pp. 179–205, as well as the sources cited below.

[33] Aristotle, *Mechanica*, Chapter 2, 850 a 2–29. Present-day scholars agree that this work is not by Aristotle, though it is likely to have been a product of his school. The Pseudo-Aristotle solves the problem by noting the proportions of the beam lying on

did not repeat Aristotle's demonstration,[34] nor did Cardano give his approval in the book *De subtilitate*,[35] for he would not have substituted another. This is evidently because Aristotle's demonstration is obscure.

Skirmish with Guidubaldo, from the Marches of Monte Ferrato.

But Cardano, who was not much clearer, so displeased Guidubaldo[36] that he inveighed against him in an entire book which he wrote on the balance. The reason for the controversy seems to me to be this, that one element of the Aristotelian proposition seems absurd and contrary to sense perception. Jordanus accordingly
ignored it; Cardano also rather candidly set out even to demonstrate what was **18**
falsely admitted. Guidubaldo, noting this, and undermining it with sense perceptions, appeared (which was remarkable) to deny even that part that is beyond controversy true, laying the blame for Aristotle and others having erred in the attempt upon the slipperiness of the material.

Let the upright support be AB, the beam CD, and let CA, AD weigh equally. I say that anyone who denies that CD is going to return to horizontal equilibrium,[37] or at right angles with BA, declares war, not just on antiquity, but on the nature of things, but on the interest of the human race.

On the other hand, let the support be AE, inverted and underneath; let the beam be CD, having its center at A, the end of the support, from the depiction [imaginatio] of Cardano and Guidubaldo; and let CA, AD, weigh equally. I say again that the beam is going to return to equilibrium. If Aristotle affirmed of this arrangement that the beam CD, once inclined, is going to remain so, it will by all means have to be said that this is a lapse in experience, although Guidubaldo

---

the two sides of the perpendicular through the fulcrum. See Aristotle, *Minor Works*, tr. W. S. Hett (Cambridge and London: Loeb Classical Library, 1955), pp. 327–355.

34 Jordanus Nemorarius, *De ponderibus propositiones XIII et earundem demonstrationes* (Ingolstadt 1533 and other editions), Prop. 2, in Ernest A. Moody and Marshall Clagett, *The Medieval Science of Weights* (Madison: The University of Wisconsin Press, 1960), pp. 130–131 (Latin text and English translation); English translation of selections in Edward Grant, *A Source Book in Medieval Science* (Cambridge: Harvard University Press, 1974), pp. 214–15. Jordanus was actually trying to prove a somewhat different proposition: he states, "When a balance has an equal position, with equal weights hung upon it, it does not depart from that position, and if it be separated from equidistance above the earth, it returns to the position of equality." His proof depends on seven suppositions (Moody and Clagett, pp. 128–9; Grant, pp. 212–13.) which allow one to compare downward tendencies of bodies.

35 Girolamo Cardano, *De subtilitate rerum* (Nürnberg 1550 etc.), ed. Basel 1554, Book I p. 24. Cardano notes that Jordanus neither proved nor understood what the Pseudo-Aristotle was trying to demonstrate. However, Cardano also seems to argue that there is some power in the angle itself between the support and the vertical that gives power to one side or the other, and incorrectly ascribes this argument to the author of the *Mechanica*.

36 Guidubaldo del Monte (1545–1607) was from Monte Baroccio (Mombarocio) in Umbria. I have not found the "Mons Ferratus," mentioned by Kepler in the margin. Guidubaldo wrote *Mechanicorum liber* (Pesaro 1577 etc.). The argument cited by Kepler is in Book I Proposition 4.

37 Literally, "equilibrium of the horizon" (*Horizontis aequilibrium).*

took it upon himself to demonstrate the falsity of this in both arrangements. But the experts should see whether the Philosopher's words might not admit of some interpretation; for I suspect that Aristotle's contemporaries had sought some easy method in the balance, using a curved beam with the support clasped in the middle of the arc. In this way, whether the balance hang from the support or rest upon it, it can be considered now above and now below, according as the arc curves upward or downward. And then if in this way the center of the beam be above the support, I grant that what Aristotle seeks happens. For the cause is nearly the same as that which makes a cone standing upon its angle not remain
19 there, but turn over, though it meanwhile remains hanging from its angle.

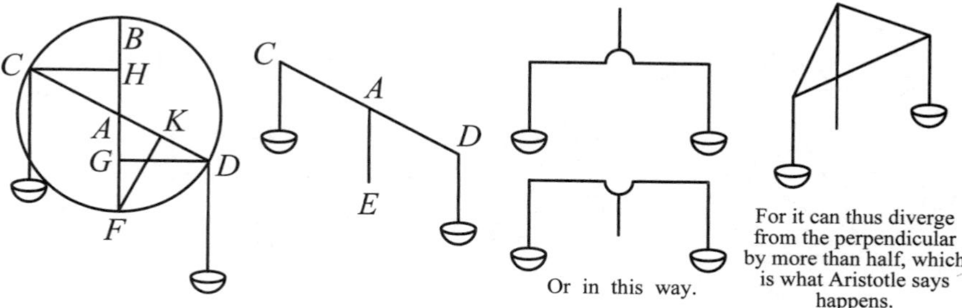

Now the error originated for Guidubaldo from a false principle: he was defining equally weighing things [aequiponderantia], which would stay in place however they might be located, by means of equal lines from the center.

But let us come nearer to the subject. Cardano, following Aristotle, says that CA is heavier since it is higher (namely, by the smaller angle CAB), while AD is lighter since it is lower (namely, by the greater angle BAD), and in demonstrating this he stumbles.[38] The whole matter comes back to the assigning of angles to gravity, a support derived from mechanics. For CAB is acute, and its sides are pulled apart more easily by the weight at C than the sides of the obtuse angle BAD are by the weight at D: therefore, the former weight wins out. If this were indeed the true cause, the effect would follow the proportion of the cause, and the ratio of the weights on the inclined balance would be the same as that of the angles. But this is false. And so Cardano shows us another cause, from a distance, but does not work out the demonstration. Although C is heavy, he says, and, being so fitted by nature, is able to be carried towards the center of the

Skirmish with Cardano.

[38] This is not quite what Cardano argues, and not at all what Aristotle said. Cardano writes, "Aristotle says that this happens when the support is above the beam because the angle FAC of the cone is greater than the angle FAD. . . . but a greater angle makes the weight greater." (*De subtilitate* (Basel 1554), Book I p. 24; the letters have been changed to match Kepler's diagram). Aristotle, however, wrote, "Is it because, when the cord is attached above, there is more of the beam on one side of the perpendicular than on the other, the cord being the perpendicular?" (I have used E. S. Forster's translation, from *The Works of Aristotle*, Vol. 6 (Oxford: The Clarendon Press, 1913 etc.) The Greek is rather terse, and evidently both Cardano and Kepler misunderstood Aristotle's sense.

earth very nearly through the extent of AC, nevertheless, it does not descend beyond equilibrium, because the equal weight AD would be forced to ascend, contrary to its nature. This begs the question. Indeed, Cardano, I seek to know, not just this, but something further: if in all cases that which is heavier strives to approach closest to the center, why then, when there are unequal weights in the pans, does the heavier one not simply seek the lowest place, while the lighter is lifted to the very top? The cause, then, is this, namely, the same as in the unequal armed balance [statera].[39] About center A, with radius AC, let the circle AD be described, and in it the perpendicular BAF. It is evident that neither of the weights at C and D can either descend lower than F or be raised higher than B.

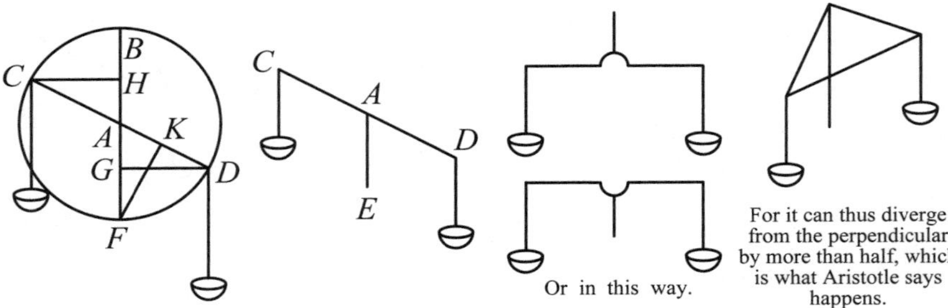

And since both are of this nature, that they tend to the bottom, and they mutually compete with each other, they divide up the descent BF between themselves in that ratio in which they themselves are. From D and C let the perpendiculars DG, CH be drawn. Now from what has been said, BH, the descent of the weight C, will be to BG, the descent of the weight D, as the weight C is to D. I say that this is the proportion of the unequal armed balance. For also, because HAC, GAD are equal, and CA, AD are equal, and H, G are right, AH, AG will also be equal, and therefore also the remainders of the equals HB, GF. Therefore, as C is to D, so is FG to GB. From F let a perpendicular be drawn to CD,
and let this be FK. Therefore, since CAH, FAK are equal, and CA, AF are **20**
equal, and H, K are right, CH, FK will also be equal. Likewise also AH, AK. Consequently, the remainders of the equals AB, AD, AF, that is, HB, GF, and KD, are also equal. And again, the remainders GB, KC of the equals FB, DC, are equal. Therefore, as C is to D, so is DK to KC. And if the beam CD, thus loaded, be suspended from the support at K, it will be the ratio of the unequal armed balance, and C, D will weigh equally, as is demonstrated in mechanics.

---

[39] It is not certain exactly what sort of balance is meant, though Kepler's argument makes it clear that he has an unequal armed balance in mind. Historically, it may be the same as Thabit ibn Qurra's *Karaston*, or "Roman balance." See Ernest A. Moody and Marshall Clagett, *The Medieval Science of Weights* (Madison: The University of Wisconsin Press, 1960), pp. 77–117. The derivation of the word *statera* from the Greek στατήρ, which denotes a coin of a standard weight, and its use by Cicero (*De Oratione* 2.159), suggest that it is a fine scale and not a "steelyard" or coarse balance, as is suggested in the *Oxford Latin Dictionary* (p. 1814). Cf. H. G. Liddell and R. Scott, *A Greek-English Lexicon* (Oxford: The Clarendon Press, 1843 etc.), p. 1634.

The proposition is therefore evident. Therefore, by subsumption, it is evident why the arms of the balance rotate to equilibrium. For since they weigh equally, it is fitting that they also make equal descents on the circle. This is a digression and stands as a sort of protheorem.[40] Now to physical motion, which is common to light.

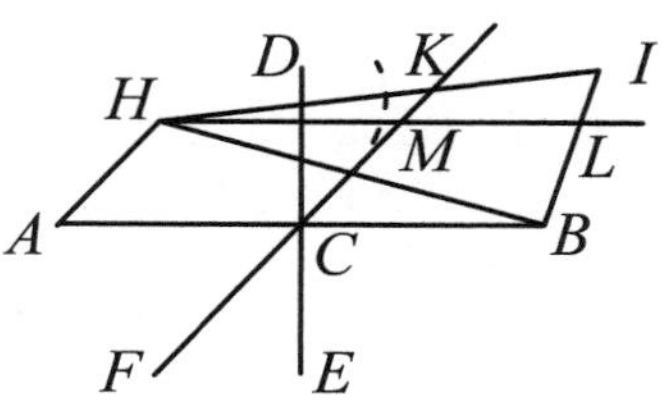

Let *AB* be a panel,[41] *C* the center, *ED* perpendicular through it; as, if a globe or missile were carried from *E* into the panel *AB*, it would drive it forward towards *D*; or as if *AB* were oars, of equal length on both sides, and *ED* were a river.[42] For since *ECA*, *ECB* are right, the arms *AC*, *CB* are placed in equal balance, and meet the impact of the mobile body with an equal power. Now let the oblique *FC* strike at *C* and let it be extended to *K*. And let the missile or the stream rush in from *F* to *AB*. Since the angle *ACF* is less than the angle *FCB*, the parts *AC*, *CB* will not be impelled with equal force, but the one that resists more will feel the blow more. And the one that faces at an obtuse angle resists more than that facing at an acute angle.[43] Therefore, the exterior part *CB* will resist more. For also, if *KC* should become a support, *CB* has a stronger turning power[44] towards *F* than has *AC*. And there is the same ratio of resistance of equal powers here in violent[45] motion, as there was of unequal weights above in motion that is in accord with nature: that is, the lines, not the angles, measure it. Therefore, there is

---

40 *Protheorema*, in Latin. The word was used by Martianus Capella (5th century AD), *The Marriage of Philology and Mercury* II.138, most likely as a direct transliteration from the Greek. If so, it would have meant something like "preliminary consideration," It is clear from the context, however, that Kepler meant it to denote a preliminary or subordinate proposition (the term is also used in a similar way in the *Astronomia nova* ch.59). I have chosen to call them "protheorems," as I did in the *Astronomia nova*, rendering Kepler's Latin neologism with an English one.

41 *Tabula*. Kepler evidently wished to use a term not usual in geometry, to remind the reader that this is a physical surface and not a geometrical plane.

42 Kepler used the analogy of oars paddling in a river later in the *Astronomia nova* (1609), to explain how a planet might move itself towards and away from the sun through the aethereal air. See chapters 38 and 57, in *JKGW* 3 pp. 255 and 349; *New Astronomy*, trans. W. H. Donahue, pp. 405 and 549; Green Lion Press ed. pp. 299–300 and 270–71.

43 Kepler may have been thinking here of Jordanus's Theorem 7 of Part 4, which states that "the shape of a heavy body changes the power of its weight." (Moody and Clagett, pp. 216–17).

44 Here Kepler uses the Greek word ῥοπή, defined in H. G. Liddell and R. Scott, *A Greek-English Lexicon* (Oxford: The Clarendon Press, 1843 etc.), p. 1575, as "turn of the scale, fall of the scale-pan, weight." It is thus not simply "weight" or "force," but something analogous to our notions of torque or moment. In order to avoid the anachronistic use of such terms, "turning power" is suggested as a translation.

45 Motions that do not arise from the nature of a body are called "enforced" or "violent"

a greater impression of violent motion on CB. So when AB is moved in position, B advances more than A. I say that it will happen in such a manner that the center C is not carried on the line FK, but turns aside towards the perpendicular CD. And that the oar AB will at length be pushed forth to the shore, if it be artificially held back in this position AB. And that the arrow CK, because of the oblique transversal AB, will go wide of the target to the left. For from A let a small part of the path of A be described, and let it be made parallel to CK (for it will not turn
aside farther to the right, since the violent motion itself is not more oblique than 21
FK), and let it be AH. And, HB being connected, let a triangle be constructed above it, with one side HI being equal to AB and the other BI greater than AH. And let HI intersect FC at K. Finally, let HL be drawn, equidistant from AB, intersecting FC at M. Now since AH, FK are equidistant, they cut off from the equidistant lines AB, HL the equal parts AC, HM, and AC is half AB. Therefore, HM is also half AB. But HK is greater than HM. Therefore, HK is greater than half AB, that is, than half HI. Therefore, the center is between H and K, and is not at K. It has therefore turned aside to the parts D, which is what was to be demonstrated. If AH were so drawn as to come together with CF towards F, so that A would thus also have turned aside from the path towards the left, HM will be made greater than AC, and consequently HK will be made much greater than half AB or HI. This happens at all inclinations, so that even in the extreme case, where the movable body moves in a line that is close to equidistant to the surface itself, the surface itself, lightly struck, rebounds almost on its perpendicular, unless an enormous force of motion move it forward considerably. And so this behavior of violent physical motion also flows back to what is analogous in light.

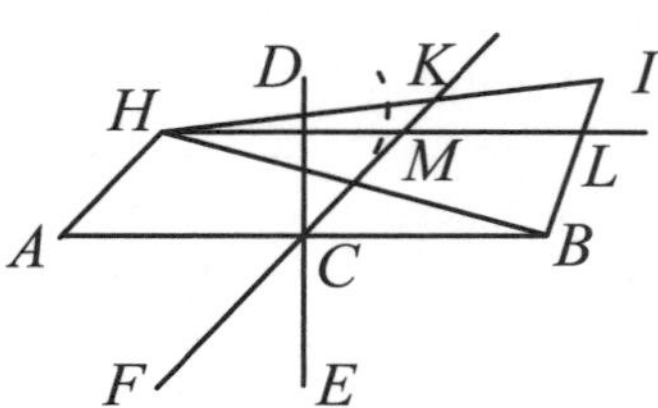

Proposition 21

Both the reflected [repercussus] and refracted rays are straight after the place where they are affected. For the nature of light is to move in straight lines, as long as it is not at all affected by the interposition of surfaces, by Prop. 4. Nor are the rays any longer affected by the solidity of the medium after passing through the surface, by Prop. 10. But physical motions are curved into arcs: reflected motion, because its power is finite; bent motion, because the medium contributes something by its solidity as well. These do not apply to light.

Proposition 22

Light illuminating colors is reflected in all directions, and the illuminated colors radiate in an orb, as light itself; more strongly, however, on the perpendicular. For all things, even those that are colored, are to a certain extent transparent,

---

by Aristotle (cf. *Physics* viii 4, 255a 3). The Greek word is βία. See the note on p. 26 above.

by Prop. 11, and light illuminates colors through the solidity, by Prop. 16. Therefore, it illuminates from all sides. But now color is a correlate of light for mutual action.[46] But light strikes upon color, and does so from all directions through the solidity. It is therefore reflected as if from surfaces in all directions. But at the same time it is also tinted, by Prop. 16.[47] Therefore, the relected ray has the color
22 of the medium or the object: so color radiates, and so on. Hence, those colored bodies that are also in the highest degree smoothed off, nevertheless radiate in all directions, which would not happen if all the light were reflected from that one surface in one direction, and nothing were to penetrate the genuine seat of color, or it would not be tinted. For light is in the matter and internal arrangement of a body, not in its mere surface. One may say, if it is more acceptable to him, that the potential lights of colors are brought over and aroused to act, exactly as that heat which is in ginger is stimulated by the approach of moisture, and itself takes fire, and begins to communicate itself, which same thing all seeds[48] do. On this fourth kind [*species*] of light, the optical writers say little. We call it "communicated light". For both reflected and refracted light is nonetheless the light of that thing from which it has approached, undergoinging that action. But this communicated light now in a certain way becomes the light of that surface which it had illuminated. Moreover, it is in the highest degree necessary for the astronomer to take it into account.

Proposition 23

Light descending through the substance of pellucid colored bodies is refracted at whatever point, and the colors of the illuminated body radiate into

---

[46] Cf. Propositions 15 and 16, above.

[47] Actually, the corollary to 17 is where this word (*tingo*) is used.

[48] Here Kepler refers to the following endnote.

To Prop. 22 pp. 21–2. How this fourth kind of light arises—I mean communicated light, where the light of the sun falls from a single direction, and is so taken over by the colored surfaces, whether smooth or rough, that it is not only spread out in an orb (while it ought to have been spread only on the side turned away from the sun, if it had remained simple and reflected), but also takes on the color of its surface—this, I have said, originates in two ways; by which I do not know how I can fail to satisfy myself fully. One makes use of reflection, refraction, pellucidity going to some extent through all bodies, penetration of the solar light, or of the day, to some depth, all of which are not yet seen to be fully sufficient. The other does indeed state something, but does not point out the way. Thus an occult account has been adopted, of how light and darkness may be bound by the chains of matter. So it seems that this too still has to be found out: how they may again be summoned forth from matter by the extraneous light, and ignited, like one torch from another; and whether this can be done through principles either previously assumed or previously demonstrated, or through others in addition. Let the problem be proposed for optical theorists and philosophers. They shall also consider whether the radiances that are constituted in this way can be held to be among the number of those things possessing the highest degree of density. For if things of the highest density should have individual luminous points, each of these radiates in an orb in its own right.

the region from the light in an orb, but more strongly on the perpendicular. For since light illuminates colors through the solidity of the substance, by Prop. 16, it therefore illuminates from all sides. And at the same time it passes through, because it is pellucid. But now color is a correlate of light for mutual action. And so, in passing through, in spreading itself out, it gathers up colors from the substance of the colors, which is what "being refracted" is. For here too, as before, color assumes the nature of the surface. But at the same time it is also tinted, by Prop. 16. Thus the refracted ray has the color of the medium. So color radiates past the medium. And because that which is refracted less is stronger, while that is less refracted which is nearer to the perpendicular, therefore those rays which are nearer to the perpendicular are stronger. So in this way light comes to be the property of the colored medium, and is communicated to it.

Corollary

The cause was to have been stated why, when the sun illuminates the air everywhere equally, nevertheless anyone perceives the greatest brightness of that air through which the sun radiates most nearly, so that one and the same region of air is at its brightest to one person, and to another, to whom the sun makes its way differently, it is less bright, for air has its own albedo[49] or
Halos, Parhelia, Twilights.
color. If the cause has not been clearly enough stated, the reader should think **23**
up another. I believe that a basis for *halos and twilights* is established at the same time.

Proposition 24

Light reflected from the surface of a body, insofar as it is body, is not colored.[50] For by 15 color is in bodies through their matter, which has three dimen-

---

49 This Latin word did not acquire its technical astronomical sense until the nineteenth century; nonetheless, Kepler's meaning seems consistent with it.

50 Here Kepler refers to the following endnote.

To Prop. 24–5, p. 23. That light is not colored in reflection is evident from a simple example, if you set out in the sun, in some order, a number of dishes of variously colored liquids (equally pure, however), in such a way that the reflection falls in a shadowy place, and upon a white surface. It thus becomes evident by these examples that the colors are not on the surface, but in the depth, and accordingly shine forth, and therefore inhere in matter that is to some extent pellucid. In all cases, colors require body, no less than light (below, Ch. 6). But that reflecting bodies nonetheless mingle their own communicated light, spread out in an orb, with the reflected light, is evident in weaker radiations or illuminations. For example, certain metallic mirrors represent the face as ruddy, which is because they illuminate and tint the retina with the simple ray of their color not much more weakly than the ray of the face, which is doubled or reflected by the mirror. Hence, by Prop. 28, there arises a mixture of the two with the new color, in the eye.

Accordingly, in Prop. 25, color in a surface is established no differently than is density in Prop. 13 above. And the sense is that hitherto light, since it is a kind of surface, is affected by the density of the surface alone, because the density of the corporeal bulk [*corpulentia*; cf. the footnote on p. 97] had nothing in common with it.

sions; a surface has only two. Therefore, there is no color in it. But that something of color is mixed in, this the body to which the surface belongs adds on its own—a small amount indeed which it spreads about, not just towards that place, but also round about.

Proposition 25

Light passing through a colored medium is more and more colored, and that which penetrates the medium more deeply leaves with more ruddiness.[51] This is evident from Prop. 16, because color, in standing in the way of light, is not only in the surface, but also in the body.

Proposition 26

The rays of light neither mutually color each other, nor mutually illuminate each other, nor mutually impede each other in any way. For the rays, by Prop. 4, are nothing other than the motion itself of light and color, nor does light exist in them, by Prop. 8, but now has passed through. This is just like one physical motion's not impeding another.

Proposition 27

Light in the same medium is partly reflected, partly refracted, and also partly adheres to the color of the medium, or is propelled back by the color, and thus is divided into more tenuous lights. For to the extent that it strikes upon a surface, it is reflected, by Prop. 18; but to the extent that the surface belongs to a pellucid body, it passes through, by Prop. 10, and it is refracted, by Prop. 20. Therefore, while it is illuminating two objects, one through reflection, the other through refraction, the two together are equal to the one through which it would have illuminated directly: it is accordingly attenuated, since this is the definition of tenuity. The dual nature of the medium itself brings this about. It is assisted by the essence of light, participant in density, which can consequently be divided into more tenuous parts. In physical motion this is not so evident, because physical movables are all materiate and hard, while here there is nothing of the sort. Witelo proves this by experiments.[52]

Proposition 28

Different lights falling upon the same object are both accumulated and intermingled, among themselves as well as with the color of the object, each in proportion to its strength or density, whence there arises a new color, or rather,

---

Now, however, all color, through the whole depth of the body, is partly akin to light, because of its potential light, partly contrary, because of the intermingled darkness. Therefore, light is affected by color through the whole corporeal bulk of the pellucid body, not just in passing across the surface, when it is refracted.

51 Here Kepler again refers to the endnote already referred to in Prop. 24 above. See the preceding footnote.

52 Witelo, *Optics* X 10, in *Thesaurus* II p. 414.

a generated light different from all the others. For because all are participants in quantity, just as they have hitherto been capable of attenuation, they are now also capable of accumulation. However, this does not occur in a pellucid medium, because the lights are not in it, as such, but have now passed through it. Instead, they are in the object, because it is in this that they are first established and fixed after passing out from their body. They are, moreover, intermingled, because, by Prop. 16, either of them is mutually affected by the color of the object. And there arises a new color, radiating, because, by Prop. 13, colors differ only in degrees of light and darkness, while accumulated lights vary this degree, and therefore vary the color itself. **24**

Proposition 29

When the mutual ratio between lights flowing to the same object is excessive, the sense does not discern the weaker light. For to discern is to prepare or distinguish the function of the visual sense. But as the ratio is, so also is the preparation. Therefore, if all ratios of colors or brightnesses could be discerned, since they are infinite in respect to the increment of magnitude, the faculty of discrimination would have to be infinite, and this the natural philosophers deny. For all sensory faculties have certain prescribed powers.

Proposition 30

Colored lights on those surfaces which have colors that are more nearly akin to light, as on white surfaces, appear brighter than on black ones. For the task of light is illumination, in which (like all agents) it makes what it acts on more like itself, and light has its contrary color or darkness. Therefore, where colors more approach darkness, as black ones, illumination becomes more difficult. And by Props. 28 and 29, blackness wins out in color, which radiates from the illuminated surface. Thus it radiates less from a black than from a white surface, and so is also less seen.

Proposition 31

It being posited that we are able to perceive one particular color out of many equally clearly on black surfaces as on white ones,[53] *which can be brought about by the accommodation of a stronger light to a black surface: now the differences of those colors that are radiating mutually side by side will be more rightly noted on the black surface than on the white.* For because the white surface has great brightness, the radiating colors closer to the light become violently bright, by Prop. 28. And they will thus efface the colors closer to black, by Prop. 29. This
is not so with black surfaces, because they are more a privation of light. There **25**
follows hence a kind of corollary to Props. 30 and 31: that the rays that have flowed to black surfaces are perceived most distinctly, and to white ones most evidently; and if a surface be a mean between black and white, such as blue,

---

[53] Here Kepler refers to the following endnote.

To p. 24. Proposition 31 with its corollary is properly carried to Ch. 5 and should be alloted to the cause of color on the retina.

white washed with red, and the like, they will stand about equally in rendering both the individual colors and their differences.

Proposition 32

Heat is a property of light. It could be proved that light is hot from the principles we have assumed. For if the life of things is dependent upon heat, and light has been ordained to nurture that life, it therefore ought to give heat. But every agent strives to make what it acts on similar to itself. Therefore light, striving to make matter hot, will itself be hot. But I will establish what I have said by experience. For light alone is always and everywhere accompanied by some heat, according to the measure of its brightness. Concerning the solar light this is most brilliantly obvious, because this is also the most brilliant.

All heavenly bodies give heat.

Of the light of the heavenly bodies, the proportion of brightness bears witness that the heat of all of them is in a lesser ratio to the heat of the one sun, than to allow it at the same time to be both perceived and evaluated by a human, who possesses the sure measure of heat. However, it appears from the effects of the heavenly bodies that there is in the innermost parts of the earth a perception of their heat. For no outpouring of vapors, whether cold or hot (in comparison with us) can be aroused without some heat to thin them out and raise them up. But all the planets are suited to arousing vapors. Therefore each one has its own effect of heat. Of things containing fire, the matter is again evident.

Fireflies.

Concerning the extremely diluted light of fireflies, you cannot deny that it is accompanied by heat. For they live and move, and this is not carried out without heating.

Rotten wood.

But the light of rotting wood is also not without its heat, for the rotting itself is a kind of gentle fire. Although Aristotle draws a distinction, allowing that it shines [λάμπειν] while denying that it produces light [φῶς ἐμποιεῖν], if you understand this most soundly, it is to say nothing other than this, that that light exists in its least intense degree.

Carbuncle.

If in fact it is true that the carbuncle is self-luminous, they do indeed also attribute powers, both to it and to all gems, of which powers it will be fitting that those that are in the carbuncle come forth in some manner through the mediation of heat.

Now it must be proved, on the other hand, that in all other things heat is extrinsic, and depends upon the heat of light, thus being rightly called "passive".
26 It is evident because there is nothing that has heat from itself, since it is material. This is manifest in those things that are made hot by the sun or by fire. For when they are freed from the presence of heat or of fire they again become cold.

Heat of animate things.

As for the heat that is in animate things, it is certain that it comes into the body from the arteries. For the bodies immediately become cold when the arteries are blocked. It comes into the arteries from the heart, and I have no apprehension about asserting that there exists in the heart the very thing itself of which Fernel said there is a likeness, namely, an enduring flame.[54]

Flame in the heart.

bags.

smoke chambers.

For to what end are the bags of the lungs, breathing the air, lest life be blocked off for want of air, as happens in fire; the smoke chambers and pulse of the arteries, or the emission of smoke,

---

[54] Jean Fernel (1497–1558), physician to the French court, wrote a very widely reprinted textbook of medicine (*Medicina* (Paris 1554), later issued in expanded form under the title *Universa medicina* (Paris 1567 etc.).

lamp.

oil.

Ventilation and Exhalation.

Is there a spark of light in the heart?

lest this little fire be smothered by its own waste, the hidden lamp of the heart, the blood drawn across into the heart through a special channel from the very trunk of the vena cava, like oil, whence this flame might live? I do not maintain this, that the air of the lungs and the blood of the arterial vein[55] is nothing more than just food for this flame; nor do I maintain this, that through the arteries there passes nothing more than just the wastes from this little flame, so that all would thus be subservient to the one heart, as to a chief. For I grant that the heart itself has been assigned to the service of the entire animal. I grant that these returns, so to speak, are brought into the heart as if into a workshop, in order that, being worked out here, they be apportioned in a different way to the body as a whole through the artery, for the purpose of indispensable usefulness. Meanwhile, the form itself of drawing in and driving out, and the mechanism of nature in the valves, cries out with a raised voice that there is in the heart a creative flame, which would support life from the revenues of the things that enter, and would expel the wastes through the same path with these its works, to the carrying out of which it is ordained—that is, with the vital spirit.[56] It cries out, indeed, that the nature of the body as a whole, and of the heart in it, is so established that the body is most rightly maintained by the waste products of the heart. For if the necessity of conserving this little flame were not to have impelled nature, it might have carried the blood into the heart, and have driven it thence, more gently and calmly, as in the liver, nor was there any need for the motion of systole and diastole. If, to the contrary, you raise the question, by what means some kind of fire or flame might most directly be preserved in a closed vessel, such as the heart had to be, I answer that this takes place in no other way that that by which the heart both is shaped and is opened and closed. I consequently have no doubt that we might also catch this spark of light with the eyes, if it were to happen to us to look into the hiding places of the heart by means of this unimpaired flame, with **27**
the light of day excluded. Thus the animal warmth depends upon light. Not to mention that it befits the soul, itself invisible, to have an essence that is akin to light: in which guise it will come, along with light, into the same connection with warmth, to the extent that light is the offspring of the soul.

Heat of plants.

There is indeed a heat of their own, both in plants and in shoots, but that too comes from a small fire, which, while not so manifest as that in animals, nonetheless exists in plants. This is evident because their body is at length devoured, lying hidden in seeds, and it generates little worms that glow at night, and so ignites the wood, when rot has finally set in, that at night one can even see the spark of light with the eyes.

Heat of the earth. The animate faculty in the globe of the earth.

That there is heat in the earth, everyone knows. That there is an animate faculty in it, akin to light, the guardian of heat, not everyone will admit. I accordingly appeal to the fires of Etna, and to the innumerable hot springs, fetching

---

55 *Vena arterialis*, pulmonary artery.

56 The three "spirits," natural, vital, and animal, had become fundamental to the Galenic system of physiology, though Galen himself did not unequivocally espouse them. The vital spirit was related to the heart and to the faculty of locomotion. For the development of the "Galenic" system, see Owsei Temkin, *Galenism*, esp. pp. 95–108.

up the fiery heat. The same source that produces the heat therefore produces the fire. In oils, sulphur, coal, there is heat as well as fire, but potentially. For when they are made hot in such a way as not to catch fire, they undergo a remission of heat when the fire is removed, but when they are ignited, they show manifestly the source of their heat, when they burn up. Some things create heat through the presence of a humor, and at the same time ignite themselves, such as hay. To sum up: the soul precedes heat; fire or light accompanies it.

Proposition 33

The heat of light is immaterial. For we have made it the companion of light, and light has no matter, as we now consider it.

Proposition 34

The action of light for producing heat is directed toward matter. For all matter is cold, by Prop. 32, light is hot, so they are under the same genus. Thus opposite is acted upon by opposite. This also occurs because of the essential contrariety of the material and the immaterial. In Prop. 10 above, we were concerned with geometrical action and subjection to action[57] by reason of position, where light and the surface were equivalent. Here, now, the action is physical, not reciprocal, for the matter is only acted upon. But the action that appears in turn to carry over to light works through geometrical action; as, for example, a crystal globe exposed to the sun for a long time at last becomes hot, while a ray passing
28 for a long time through the globe is no colder than it was at first.

Proposition 35

Heat in matter is aroused in time. For even if, with its accompanying heat, light is present in a moment, nevertheless, matter, by the fact that it is matter, is subject to time. Therefore, heat, now materialized and passive, can only be generated in time. This is not true of colors, lying hid in the depth of matter. For they were being considered as surfaces, from the nature of light and illumination, and hence their illumination by light also continues to be instantaneous.

Proposition 36

Light destroys and burns things. For by Prop. 34 it acts upon matter. Further, it strives to make the things acted upon similar to itself, in the manner of all agents. Therefore, attacking the matter, in which the essence of all things consists, it destroys it. This happens by dissipation and inflammation, so that everything might become light.

Proposition 37

Light bleaches the colors of things in time. For by Prop. 36 it destroys matter. But by Prop. 15 the essence of colors resides in matter, and if this passes away the color passes away for them. Further, it bleaches them because (by the same proposition) white things are more akin to light, black things are those having

---

[57] *passio.*

more of a share in the darkness and density of matter: therefore, there is more for light to take away in black things. But this occurs in time, by Props. 34 and 35.

Proposition 38

Light ignites black things more easily than white things. For by 30 less of the rays is reflected by black things, and thus more is consumed in them. So light places more of its action in black things, and by 36 this consists of destroying and igniting. This is the origin of the opinion that rays are concentrated by black things, dissipated by white.

In the place of a conclusion, the sixth proposition of the third book of Witelo should be noted here.[58] Experience testifies that the image of a vivid visual perception remains in the vision to some extent even when the bright body, from which the image had come down to the eye, is removed. This happens to such an extent that that surviving image is blended with the colors of other things to the viewing of which the vision, so permeated, is transferred. This is introduced here from experience: we are unable to prove it *a priori*. For more principles have to be brought in than just those of optics. This alone should be noted against the common manner of speaking, that those images do not come to inhere in the humors of the eye, nor are they images of light or colors. For this is inconsistent ***29***
with the nature of transparent things and of light, and with the principles of optics. That is, an image is always placed with its body, of which it is the image, and when a body is separated off by something opaque, the image is destroyed by the opposing shadow. Thus, by the very fact that humors are transparent, they never receive images, but transmit them. Further, whatever image there is does not inhere in the opaque coverings of the eye. For again, no color, no opaque surface, is excited and rendered radiant except by a present, uncovered luminous body. The remaining possibility, therefore, is that what inheres in the eye is an image of the action and effect, not of light, but of illumination. In like manner, the sensation of pain from a blow persists, and this is understood as a kind of image of that effect. In like manner also, in violent motion, a kind of image of it carries over into the projectile, and carries it forth to some place even after the one who had given the motion removed his hand. And since all sensation is carried out through the nerves, and through the spirits that are borne by them, this image of vision will therefore reside in the spirits, not in the humors. It will become necessary below to refer to this relationship.

[58] Witelo, *Optics* III 6, in *Thesaurus* II pp. 87–8. The proposition states, "Vision takes place from the action of a visible form on the sense of vision, and from the sense of vision's being affected by this form."

Appendix to Chapter One, and Airing of Aristotle's Arguments on Vision in *De Anima* Book 2 Chapter 7.

The foolish studies of humans have come to such a pitch of vanity that no one's work becomes famous unless he either builds up or burns down the temple of Diana—unless, I say, he either fortifies himself with the authority of Aristotle, or takes a stand in battle against him, seeking to show off. This is indeed the reason why the most true axioms of the optical theorists (amplified upon in this chapter) have hitherto been held in neglect, and, through this paucity of opticians, have undeservedly been regarded as inferior to the Aristotelian darkness, since Aristotle reigns everywhere, while the optical writers turn a blind eye and privately remain content with their liberty. Therefore, in order to make opposites illuminate opposites by placing them together, and to lure the Aristotelians at last into the school of the opticians for the aim of either learning or refuting, it seemed right here to discuss explicitly Aristotle's comments on vision. The subject matter is in fact appropriate to the fifth chapter, if you wish to observe the literal and topical niceties, since the subject matter of the first chapter is light and colors, by nature prior to vision and the eye, which I have postponed to the fifth chapter. But Aristotle's arguments are so constructed that very little must be borrowed from the fifth chapter, and the rest appear to be most appropriately treated in this place.

First, I shall consider the individual opinions, and those conclusions that
30 *follow from them. Next, I shall construct the whole sequence of arguments, and shall respond to them. These, then, are the main opinions.*[1]

1. *Color, properly and in itself, constitutes the subject of vision, and contains in itself the cause for the existence of the visible.*[2]
2. *Light is the activity of the transparent so far as it is transparent.*[3]
3. *It is in a way a kind of color proper to the transparent itself, when that is truly transparent.*[4]
4. *It is not fire, nor a body, nor anything that flows down from a body; but it is the presence of fire, or of something luminous, and so on, in the transparent.*[5]
5. *It is the presence in the body of that disposition, on account of which the body is called "transparent".*[6]

---

[1] The following list of "opinions" consists largely of direct quotations from Aristotle in Latin translation. Those that are not (numbers 9 and 11) appear to be conclusions drawn by Kepler from these quotations. The source of each quotation is given below.

[2] *De Anima* 418a29.

[3] *De Anima* 418b9.

[4] *De Anima* 418b11.

[5] *De Anima* 418b14.

[6] *De Anima* 418b20.

6. *And these things are to be comprehended thus, that we may know that Aristotle is speaking against Empedocles, who had said that light is carried and spread out in straight lines between us and what surrounds and encloses us (the heaven), even though we are not aware that this occurs.*[7]

7. *The nature of the body that is now light and now darkness, is the same.*[8]

8. *And when that body is [only] potentially transparent, then darkness persists there.*[9]

9. *And so it is not when it is actually transparent, but when it is potentially so, that it is then also as a consequence both dark and able to receive colors; especially because only then is it without color.*[10]

10. *The same thing is also affirmed of things that are not seen at all, and of things that are seen with difficulty: that they are able to receive color.*[11]

11. *Further, that which is actually transparent is to be held to be among the visibles, not, however, in itself, but through foreign and extrinsic color.*[12]

12. *Furthermore, vision (or the movement that is prior to vision in nature, which I would call "the illumination of the eye") occurs in this way. Color moves that which is actually transparent, such as air, while by this being thus moved, because it is a continuous body, the instrument of vision, or the eye, is in turn also moved.*[13]

13. *And this is one form of vision, namely, when color is seen, i. e., in light, never by itself; because light is the being-at-work*[14] *of the transparent.*[15]

14. *And so vision (the movement of the instrument which vision follows) occurs when the sensory instrument receives something from what lies in between.*[16]

15. *For the eye is not affected in any way (I mean, the wall*[17] *of the eye is not moved or altered) by the color itself that is seen.*[18]

---

7 *De Anima* 418b21.

8 *De Anima* 418b31.

9 *De Anima* 418b10.

10 This is not a quotation from Aristotle, but appears to be an expansion upon no. 8.

11 *De Anima* 418b27.

12 Not from Aristotle directly, but evidently a consequence of no. 3 above.

13 *De Anima* 419a13.

14 *Energia*. This is not a Latin word, but a transliteration of Aristotle's ἐνέργεια (though the word Aristotle used in the passage being paraphrased was ἐντελέχεια). I have accordingly chosen Joe Sachs's translation of the Greek term as used in the *Metaphysics* (Santa Fe: Green Lion Press, 1999). Sachs discusses the meaning of this term on p. li.

15 *De Anima* 419a9 and 419a12.

16 *De Anima* 419a18.

17 Kepler is drawing implicitly on the analogy between the eye and the camera obscura, in which the image is formed on the back wall of the chamber.

18 *De Anima* 419a19.

16. And accordingly if it were to happen that the space lying in between were empty of body, nothing might be seen.[19]

17. For there is an analogy between vision, hearing, and smell, by reason of the [space] in between.[20]

18. There is, furthermore, another form of vision, by which we perceive, not color, but other things. Under this category, there is one and the same something in fire and in the sun.[21]

19. Nor are all things seen in light: some bring about sense perception (or this motion of the instrument [of perception] that precedes it) even in darkness.[22]

20. For it is also through fire that the transparent (in potency) becomes transparent (in act).[23]

21. And of those things that are perceived at night or in darkness, some, while they do indeed shine, do not give light. λάμπουσι μὲν αλλ' οὐ φῶς ἐμποιεῖ.[24]

From these aphorisms there appears what I say is the cause of the major premise's being evident, that the motion of the instrument [of vision] that is
31 *necessary for vision may occur. For the seeing of color, Aristotle requires two motions of the air or of a body of this sort: one by light, to bring the transparent into action (which motion is sufficient for seeing a luminous body); and another by the color of the object seen.*

Now, as regards the first opinion, it is indeed true, provided that a proper definition of color also be supplied. For the reason why color has the power of moving the instrument of vision, is because it is of the nature of light. And so it is a primary and inherent property of light to alter the walls (and therefore the eye).

But for this it does not suffice color to be akin to light: it also must have been actually illuminated by light, and must have imbibed a certain light, which in this chapter is called "communicated light".

In the second aphorism, Aristotle defines light, not, I think, in its nature, but to the extent that it is is characteristic of the process of vision.[25] *However, even if it appears impossible to scrutinize deeply the very nature of light itself,*

---

[19] *De Anima* 419a21.

[20] *De Anima* 419a26.

[21] *De Anima* 418a27; cf. also 419a3.

[22] *De Anima* 419a2.

[23] *De Anima* 419a24. Kepler apparently put "in potency" and "in act" in parentheses because they are not explicitly present in Aristotle's Greek.

[24] This does not appear in *De Anima*, but is very close to a phrase from *On Sense and Sensible Objects*, Ch. 2, 437a33: ἐν τῷ σκότει λάμπειν, οὐ μέντοι φῶς γε ποιεῖ. Perhaps Kepler was quoting from memory.

[25] *Ad videndi negotium concurrit.* I am reading *concurrit* as a literal translation of the Greek συμβαίνω. In a philosophical context, this can mean "*to have an attribute necessarily resulting from* the notion of a thing" (H. G. Liddell and R. Scott, *A Greek-English Lexicon* (Oxford: The Clarendon Press, 1843 etc.), p.1674); emphasis in the original.

it is nonetheless best to scrutinize some things that more nearly concern the true nature of light, before passing on to how light functions. For it is certain that we know most rightly what anything can do to something else, when we have understood what it is in itself.

The same is also to be said of the transparent: that Aristotle defines this no differently than if he were giving an explanation of the word. For something is transparent when it is seen through, and pellucid when shone through by some lights. But neither happens except in light. And he approaches much nearer to the nature of things who points out the characteristics by whose presence bodies are suited to being shone through, whether light be present or absent. For it is certain that by the approach of light the nature of bodies is not changed; and nonetheless some bodies remain dark in both cases, while others [remain dark] only when light is absent.

Thus in the third opinion it happens that the two have power, both light and the pellucid. For the denotation of the words is as if light were in nature primarily for the sake of the pellucid, which is wrong. For light exists for the sake of the colors to be illuminated, while the pellucid exists for the sake of both [light and color]; that is, in order that the colors might be illuminated by a single sun that is absent.[26] *And so light is not the activity of the transparent, but rather the activity of colors, insofar as they are seen, or radiate.*

In the fourth, I wish to learn, if light (or, if one would draw distinctions with Scaliger, illumination)[27] *is not a downward flow from the luminous body, how it is present in the pellucid. If the sun is present in the air, and is nonetheless attached to the heavens, it will therefore be present in the air through an outflowing. Unless perchance something of occult philosophy lies hidden in these wondrous words; at which, indeed, one for whom the unknown will be pleasing may marvel, [though] those desirous of learning are not satisfied by empty words. Further, according to what has been demonstrated in this chapter, the ray is not in the transparent (insofar as it is transparent), but it was there, or was as if there. It is, on the other hand, only in the colors and surfaces of things.*

Thus I deny the fifth in all respects. For if we hunt out the fine points of the term [transparent], it is indeed not actually transparent unless light in passing through it strikes upon the wall. Light is thus then not the presence of that disposition, but something greater than that presence. For the term does not originate from light itself, but from the motion of light through a body, or of vision through a body. Or if the nature of the transparent is more acceptable [as a criterion], light and the form of the transparent are distinct to such a degree that they cannot be under the embrace of the same physical category. For color—that is, light **32**
of any kind—would be denied of the transparent; surface would be denied, which

[26] That is, the sun is not in the same place as the colors; hence, the pellucid medium is necessary.

[27] *Lumen*. This is the term used by J. C. Scaliger, *Exercitationum exotericarum adversus Cardanum liber XV*, Paris 1557, Ex. 71; ______, Frankfurt 1592, p. 259.

is not denied of light; and finally density would be denied, which is itself also left to light.

Nor do I think anything different of the sixth, than this: that under the person of Empedocles the truth itself is spoken, provided only that you leave out this, that the light that is thus extended from the sun all the way to the earth is present in the intermediate [space] according to the proportion of the density and opacity of matter; that is, that it is not present in the transparent to the extent that it is transparent; and that this is the reason why it is not perceived by us.

Let the seventh and eighth be judged by the preceding.

Now in the ninth, it does not therefore follow that the transparent is capable of receiving color, if it lacks color. For the transparent, to the extent that it keeps this form of corporeal nature, does not contain a potency for receiving color; and without such a potency a simple negation of the thing, as is known, does not posit a disposition. Rather, to the extent that the transparent happens to be more colored (since nothing is absolutely transparent), it will be that much less transparent.

The tenth flatly denies that things lying hidden in darkness have colors. Again, the complaint is the same as it was with the transparent: [he argues] as if there are no colors unless they are seen. Common usage, as Aristotle himself admits, is the master of speech; and common usage does not take colors so narrowly. If you say that Aristotle did not wish here to extend the definition of color beyond what pertains to vision, I say in turn that the most correct line of enquiry is that first of all the nature of the thing be considered in itself, and then what it could do in something else. For by this procedure [of Aristotle's], things are indeed confused and made obscure. Color, then, is really in the things themselves, even if they are not illuminated, and if they likewise do not radiate and are not seen. Nor does the illumination of the pellucid make the colors move the pellucid, but the illumination of colors makes the medium be traversed, and makes it be truly said to be pellucid.

The eleventh also includes the transparent [perspicuum] among the visibles, which are the things seen [conspicua]. But things that are seen [conspicua] and things that are seen through [perspicua] are opposites, as the force of the terms imply. As for the matter in question, color is seen in itself, as Aristotle testifies, but color, to the extent that it is actually seen, is a characteristic of the surface, not of the body, to which Aristotle also agrees. Therefore, only surfaces become visible through color. But the pellucid is a body, not a surface; therefore, by Aristotelian principles, it cannot become visible through color. Unless you should take the word "color" here with another concept, which would give rise to extreme ambiguity.

Now in the twelfth you may indeed find many things lacking. First, if for the motion or alteration of the sensory instrument, no more is required than that whatever is between the color and the instrument be actually transparent, that is, be actually carried over into the possession of light, then no illuminated colors whatever will be seen. But it cannot be proven by any experience, that vision of color occurs in air that is everywhere passable by light but not illuminated by

color. An instance is given.[28] *Let the sun alone give light, and let it touch at the forward edge a surface in which there is color; and let it be in the rest of the body after that surface, if it be supposed to be extended. It will therefore not be able to illuminate the surface which it grazes with the outermost ray, but it will illuminate all the air placed in front of it. Nevertheless, vision will not follow, even by an eye set in the space of the illuminated air. Experience, on the other hand, testifies the contrary: to the extent that some color is illuminated more strongly, it is seen (that is, it shines back) more evidently, and that happens continuously. Therefore, where there is no illumination of color, there is also no vision of it, whatever may happen because of the transparent. So also, experience testifies concerning the transparent, that to the extent that the light in it is more notice-* **33**
able, the vision of colors that are beyond it is that much more hindered, because the presence of light in the transparent destroys the definition of the transparent.

If you should ask Aristotle what kind of motion it is, when color moves the actually transparent, and when this moves the eye, he will say, I think, "alteration" (ch. 5)[29] *If you should ask, in what quality [it is a change], he would answer, in color. Therefore, the transparent is carried over from non-color to color, and again from green to red and black. Therefore, the same transparent, in the same part of itself, will be permeated with absolutely all colors, and nonetheless will move different senses of vision, not with that confusion of colors that would make all see the same confusion of colors, but, as experience testifies, it will make this one see green, another black, a third red. I ask, moreover, by what procedure will the pellucid be able to accomplish this? And how, in a moment, will such a depth of air be altered; and indeed how will it support so many colors in the same part of itself? And so, since this fabricated cause is not commensurate with the effect, and since it also may not itself admit various modifications at one and the same moment, according to the variety of viewings, is to be held as nothing. Therefore, let us embrace the true opinion described in this chapter, and established by irrefutable examples [experimenta]: that from the sun, and from the colors illuminated by the sun, there flow out forms [*species*] similar to each other; and that in this flow itself they are diluted, until they strike upon a medium that is in some proportion opaque, and there they represent their source; and that vision occurs (as will be said in ch. 5 below) when the opaque wall of the eye is colored in this manner, the vision being confused when the images of different colors are intermingled, and distinct when they are not intermingled.*

---

[28] Here Kepler refers to the the following endnote.

To p. 32. I am saying that the case is given, not ever in actuality, but on the fiction that the sun alone gives light. For it never really happens that the sun alone gives light, for all the air and all the earth shines along with it.

[29] Kepler appears to mean *On the Soul* Book II ch. 5. Near the beginning of that chapter (416b34), Aristotle says, "[Sensation] appears to be some sort of alteration (ἀλλοίωσις)".

For unless such an outflowing, and the dilution of the outflowing form, be adopted, the eye will never be made by Aristotelian principles to be affected in one way by the vision of a distant thing, in another by that of a near thing, with the quantity seen and the color seen being set at an equal level by way of compensation. On this, see also ch. 3.

Opinion 13 has this right, that colors are only seen in light, but it assigns a false cause, when it brings in light for the sake of the transparent, which, as I have said, does injury to the transparent instead. On the contrary, the reason why light is required for seeing colors is because colors do not radiate or emit a form in a hemisphere, unless they are illuminated by the light of the sun or of torches.

Thus opinion 14 must be completely reversed. To the extent that the eye is more affected (in respect to vision) by what lies in between, things will be seen through the medium less rightly and transparently. Accordingly, the vision is most perfect when the eye is completely unaffected by what lies in between.

And 15, plainly the eye is altered by the form of color passing through the pellucid body, without the assistance of the pellucid.

And 16, if the region between heaven and earth were entirely empty, vision of those things that are now seen in heaven would occur most accurately of all (not, however, the vision of the ant in the heaven, of which that Philosopher spoke, since the weakness of the [visual] instrument hinders this).[30]

And so others should see what is to be thought of the analogy of the senses introduced in opinion 17. For what if the analogy were formed this way instead: that once a determination about vision is made by us, we now use that as a norm to reason about hearing and about smell? And in fact, concerning smell no one will deny that something of the odor flows out from the actual substance of the object and is received in the nostrils; and that this is perceived all the more strongly to the extent that the medium and the distance is diminished. Therefore,
34 *the medium here contributes nothing to informing the sense. In fact, this flow persists in time, when the substance has been removed and the source is drying up. And as for the proportion of this effluvium to the medium (that is, to air), the odor has the proportion of fire, and is expelled upwards by the thickness of the air. This has been noticed most evidently in the extremely high and completely bare mountains of Carinthia, as well as other places. For indeed, those who are walking through them are met with a most fragrant and plainly ambrosial odor from the flower-laden valleys lying below.*

About hearing, the way in which the form of a beat goes forth is certainly more complex than that of odor, but easier than that of color. For the motion of light is in a moment. The form of a beat, however, runs out in time, and therefore accomodates itself more to material explanations. And as for the medium, I again ask what I asked before: if there is a medium for giving a form to hearing, why is sound spread farther in pure air, if not because matter hinders sound; while the more any medium deserves to be called a medium, the more material it is?

---

30 Aristotle, *On the Soul* 419a17, ascribed to Democritus the opinion that if the intervening space were empty, an ant in the heaven would be visible.

Why again, if the medium gives form to hearing, do we draw in sounds more correctly from nearby than from a distance? For I think it fitting that when the cause is augmented the effect should be augmented. Thus in general, all sensations[31] *happen when the instrument is affected. But in order that the senses may be affected by things that are absent, the explanation of effluvia has been brought in, so that what could not be done by the things themselves could be accomplished by the forms. Media, on the other hand, exist only by the necessity of nature to excluding a vacuum. Since these were destined to become a hindrance to effluvia, they were purified and spread out; that is, they were made pellucid (i. e. diluted, stretched out, and decolored) so that light, and the things standing in air that give off odors and sounds, might pass through them, so that these forms too might be spread around.*

The next four opinions contain a contradiction, which I leave to the Aristotelians to discuss. In 18, I find nothing at all wanting: fire and the celestial light do have something in common. And opinion 19 was taken from experience. The heavens, fire, rotten wood, and things of this sort are perceived at night. And what Aristotle attributes to fire in his way in opinion 20, I also attribute in my way. For him, too, fire gives [its] form to the medium (for which see above); for me, too, fire illuminates colors. But, for Aristotle, it does so because φῶς ἐμποιεῖ *[light acts within], and* φῶς *[light] is the actualization of the transparent. In fact (in the final opinion) rotten wood and salted kids do not do this: they only shine (*λάμπουσι*). Therefore, these things are perceived when the air is not actually transparent, for light is the actuality of the transparent, and this is absent. Here, therefore, the eye is not affected at all by the medium. Neither is it affected by the object itself, which is at a distance. But it is also not affected by an outflowing, which Aristotle denies. Therefore, vision occurs when the eye receives nothing. The same can be objected to opinion 14. If in the complete absence of an external medium, vision is not carried out, how does it happen that at night one see a flashing of his own eyes, which is inside the eye? If you say that the humor of the eye serves as a medium, I retort that it can also serve as a medium for seeing something external, when that is in contact with the eye. But the cause proposed by Aristotle that prevents this from happening is the lack of an external medium.*

From this it is clear enough that the optical writers differ from Aristotle on nearly all the opinions. There must accordingly be a flaw in his arguments, which I shall now weigh.

The sequence is this:

I. In every sense, the instrument of perception [αἰσθητήριον] *is moved by some perceived thing* [αἰσθητός].[32] *For sense is an affect, not one that leads to a ceasing to be* [φθορά], *but one that leads from potency to actuality, similar to*

---

31 In the first edition and in *JKGW* (Vol. 2 p. 43 l. 3), this word is given as *sensiores*. This does not appear to be a Latin word, and Frisch (*JKOO* 2 p. 150) changed it to *sensiones*, which is also not in any dictionary or word list, but is a plausible formation from the verb *sentio*. I have adopted Frisch's emendation.

32 Here, and in the rest of the summary of Aristotle's argument, Kepler includes certain Greek words without translation. I have therefore included the Greek in square brackets after the English translation of that word.

35 *alteration* [ἀλλοίωσις] *but nevertheless not alteration. Ch. 5.*[33]

Vision is a sense. Therefore, the instrument of perception [το αἰσθητήριον] *is moved into vision too by some perception* [αἰσθητός].

II. Color, if it be moved to the eye, is not seen. Therefore, a medium is required to come in between in vision. This is corroborated by the analogy of the other senses.

III. Therefore, it is either color or the medium that moves the vision. But it is not color. (I understand that this is because, since the medium lies in between, color does not touch the instrument of perception [αἰσθητήριον], *while there is no motion without contact.) It therefore remains that it is the medium that moves the instrument of perception* [αἰσθητήριον].

IV. That which does not move [something] by itself, moves it because it is previously moved by another. The medium does not move the vision by itself, because it is itself without color. It therefore moves after being previously moved by another; i. e., colored by color; or, what is the same thing, color (admittedly visible) moves the medium.

V. Color moves the medium towards vision, which [medium] we call "pellucid," but color does not move the medium at all if it exists in darkness (I understand that this is because in darkness no vision follows). Therefore, a dark medium is not pellucid; or, what is the same thing, air and things of that sort are not yet pellucid in darkness.

VI. If darkness is a privation of something, light is an active state [habitus] of that thing, because these appear to be opposites [δοκεῖ ἐναντία]. *But darkness is the privation of the pellucid, as has just been concluded: therefore light is an active state or actuality of the pellucid.*

Reply to these arguments:

The first I endorse. I only add, by the way, that Aristotle need not have so carefully avoided having vision pretend to be some sort of passing away [φθορά], *as if this were unworthy to think about something that has been given to living things for their well being. For in all instances, as has been shown in this chapter, light practices deadly enmity with all matter, chiefly with the family of the blacks, which are as it were materialized darkness. The eye, however, both consists of matter and is black. Therefore, it is gradually spoiled by light. Hence, the pain from looking at the sun; hence, in part, that of the eyes of the elderly.*

To II: On the vision of colors, an induction[34] *was made. However, the instance is given of seeing flashes of the eyes; of the vision of an image remaining*

---

[33] *On the Soul* II 5: "Sensation consists, as has been said, in being moved and being acted upon; for it is held to be a sort of alteration." (416b33–4); "Even the term 'being acted upon' is not used in a single sense, but it is either a kind of destruction by something contrary, or, instead, a preservation of that which exists potentially by something that exists in actuality and is similar, according as potentiality is related to actuality." (417b2–5).

[34] This is Aristotelian induction (ἐπαγωγή) rather than a statistical sampling. For Aristotle, this word meant the presentation of the universal common to particulars through the use of an example. See *Posterior Analytics* 71 a 7–9, *Physics* 247 b 5–7, and Sachs's discussion of this term in *Metaphysics* (Santa Fe: Green Lion Press, 1999) p. liii–liv.

in the spirits after the removal of the visible object; of the vision of impurity of the humors of the eye after headaches. Therefore the conclusion cannot be drawn for all vision; and so the conclusion is not universal. But there is also more in the conclusion than in the premises. For even if it be granted that there is always some medium between the color and the vision, it nevertheless still does not follow that a medium is absolutely necessary for giving form to seeing or for moving the sense of vision.

Now other causes are given why colors that are moved right up to the eye are less perceived.

1. If color, that is, the surface of the visible object, were to touch the eye, the eye would feel pain. 2. Only one eye could be used. 3. No more color could be perceived than can fit within the circle of the opening of the pupil. 4. The chief point is, that by the contact of the eye and the surface to be seen, all illumination of color, without which vision is impossible, is excluded, except for the amount of the spark of light that the eyes of certain animals maintain. 5. That there is another additional underlying cause, alleged either by what has been said hitherto or by Aristotle, that explains why an object touching the eye is not seen, is clear from this: that even when the object is excessively close to the eye, howsoever much of **36**
the external medium remains, it is nevertheless perceived confusedly, and is almost imperceptible, but is surrounded by an indeterminate and nearly shadowy edge, the cause of which is explained in chapter five below. And thus mere closeness also does harm, even without consideration of the medium. The eyes of old people also prove this, which see things more correctly from a distance than from nearby.

Once these causes of the given experience present themselves, one cannot restrict the conclusion to one particular cause, much less to one other than these. This is all the more so, in that here the foundations for explaining vision are being laid.

To III: There is an insufficient enumeration in the major premise. For it is neither the body of the sun or of color, nor the medium, that moves the eye, but the forms or lights, or the rays of the sun and of the colors descending through the medium, and vibrated spherically onto the hemisphere. For the way of all vision is one and the same, whether of the sun, or of color, or of fire, or of things that shine [τῶν λαμπόντων], *as far as the alteration of the wall of the eye is concerned. The only difference is that colors must be illuminated first, while the others are bright by themselves. Also, the sun and fire make other things bright.*

Indeed, Aristotle seems to take it as generally admitted that a form, like a mere property, cannot stir up any motion except insofar as it adheres to the underlying pellucid air. This, however, is false, as has been shown in this chapter. For light or rays descending from illuminated things is affected in certain ways, insofar as it is light, not insofar as it adheres to the pellucid air. Among these affects are emission and spreading out, and their contraries, reflection and refraction or condensation. Therefore, nothing prevents there also being certain activities of the same light or rays, namely (as shown by experience), illumination and alteration of the walls, by which colors and light are not only poured forth, but are also impressed and their contraries destroyed.

But that light, in flowing down, undergoes these affects without benefit of a host or of the pellucid, I prove as follows. Reflection occurs with respect to place, and without time; but local motion without time does not belong to any material body.

But what if one should say that Aristotle proposed this motion or affect of the pellucid analogically, and not entirely physically? I answer that in that case the Aristotelian meaning coincides with the meaning of the optical writers, though the words differ. But why did he not rather imitate the nature of things and set up an analogical motion in light? Now what I hear from others is that light or form, a mere accident, cannot support any motion. I reply, from the same philosophy, that a body cannot be moved in a moment. And so, the motion of light (without time) is such as to square with the quality of its body (without matter), the two being analogical.

In sum, Aristotle seems to distinguish between the form of colors and the light [τὸ φῶς], *while nonetheless, from every color, when it is seen, light is spread.*

So, since the major premise has been dissolved, the conclusion also does not follow. Thus it is easy to reply to IV, whose minor premise was the conclusion in III, and was found to be false. For the medium does not move the sense of vision either in itself or moved by anything else. Nor is the conclusion in any sense true once the premise is destroyed, unless one should say that being passed through is being acted upon. However, to the extent that the medium is passed through more freely, it is that much less acted upon; nor is it acted upon in any part but at their mutual interface, in which respect it cannot be a medium. For in order
37 *to carry the form of colors all the way across to contact with the eye, Aristotle interposed a solid body, and wanted all of it to be moved by color. However, the pair of bounding surfaces is not an interposed body.*

Hence, V is also refuted, because it depends on the fourth. But it will also not be good if it be cast according to my principles, in this way: "Light passes through a pellucid body, but does not pass through a dark one; therefore, those things that are in darkness are not pellucid." For it begs the question, since the very thing that makes darkness is the absence of light from the pellucid. Nor does the pellucid take its name from the actuality itself, as if that which is not in actuality pellucid were not of that nature. For it would be pellucid if light were present.

Now VI is moot. A reply is made to the minor premise by distinction. The essence of the pellucid, insofar as it is pellucid, is not light, whether present or absent, but the internal disposition of the body alone. Darkness does not deprive the pellucid of this essence, and consequently light does not bestow it. But if "pellucid" means the same as "acting pellucid", and likewise "darkness" means "not acting pellucid," illumination accordingly obtains for itself the name "pellucid". But upon this interpretation, illumination is nothing else but a form emanating from colors, nor does it serve Aristotle's purpose, who makes light [τὸ φῶς], *the actuality of the pellucid, prior to the form of color by nature and in the understanding; for, in his view, first the pellucid is induced into actuality by the form-giving light* [φωτὶ], *and afterwards color moves this actually pellucid, and impresses upon it its own form.*

I expect that the Academics are going to bring up something against this, and are going to focus on how to place the honor of their master (who himself never ever sought it) before the truth. For the rest, you, whoever you are, whom it pleases to contend with me, let it be known that you are going to be held unworthy in this ring unless you enter into my chamber [camera][35] *described in Chapter 2 following, which was the only thing that Aristotle lacked. If you ignore this after being warned, the same excuse that saved Aristotle will not save you.*

In Ch. 5 Sect. 4 below, when the opinion of J. B. Porta is examined, you will find a summary and abridgement of this disputation.[36]

[35] There is a pun here: Chapter 2 presents the theory of the pinhole camera, or *camera obscura*.

[36] See below, p. 224

Chapter 2
On the Shaping of Light

That a ray of the sun cast in through any kind of crack, falls in the form of a circle upon a plane object, is a fact obvious to everyone.[1] This is seen beneath roofs that are split open, in shrines, perforated windowpanes, and beneath any tree. Led on by the wonder of this thing, the ancients had devoted their energies to finding out the causes. Aside from that, no one has yet appeared to me to have
38 carried out a proper demonstration of the problem. Witelo, three hundred years ago, claimed that this happened because of some sort or other of equidistance of the rays. Thus he demonstrates Proposition 39 of Book II by Prop. 35 of the same book. But he himself does not cover up the flaw of this his own demonstration, saying in Prop. 35, "it may be that a property of the rays contributes much to this."[2] In dealing in these ambiguities, he shows that he had not understood the true cause, which is gathered obscurely from another part of his demonstration. Following him, Ioannes Pisanus, Bishop of Cambrai,[3] rejected the opinions of others, and among them (you may be surprised to note) the actual true cause, and withdrew with Witelo into the hidden recesses of the arcane nature of light; in Book I Prop. 5, Hartmann, who published Pisanus, and set about removing the flaws in the proofs, left this hesitation. Pisanus, though, introduces the words of the others, through which the true cause is set out: "Others," he says, "assume as a remote cause the roundness of the sun, and the intersection of the rays as the proximate cause." Nevertheless, the demonstration itself, which he immediately appends, without doubt following the opinion of those same authors, again hides the truth, and attributes to these very words a sense that they cannot have. It is credible that nearly the same thing happened to Aristotle. For when, in Sect. 15 Ch. 10 of the *Problems*,[4] he has raised the question, "Why, if the rays of the sun during an eclipse should pass through the leaves of a plane tree, or through the

---

[1] Here Kepler refers to the following endnote.

To p. 37. From now on, the word "ray" is almost always taken to mean that portion or shape of light that has fallen upon a surface after passing through, and being configured by, some window or hole.

[2] *Thesaurus* II pp. 73–75. Proposition 35 states, "Rays that go forth from a single point of a luminous body approach closer to apparent equidistance as the lines get longer." Proposition 39 states, "Every light that comes through angular openings is rounded off."

[3] Ioannes Pisanus, or John Peckham, author of the *Perspectiva communis*, was Archbishop of Canterbury. However, Georg Hartmann's edition of the *Perspectiva* (Nuremberg 1542 and other editions), which Kepler evidently used (see below), erroneously described Peckham as *Episcopus Cameracensis*. The passage cited is in the demonstration of Book I Prop. 5.

[4] The *Problems*, or *Problemata*, is not a work by Aristotle himself, though it may incorporate some of his thoughts. The passage cited is in Book 15 Ch. 11 (912 b 11) of modern editions. Similarly, the passage cited below is in Ch. 6, not Ch. 5. See W. D. Ross, ed., *Works of Aristotle* (Oxford: Clarendon Press, 1927 etc.), Vol. 7.

fingers of both hands placed crosswise to each other, crescent shapes would be made upon the pavement?" he in fact adduces the true cause when he says "That two cones come together at their vertices in the narrow space of the opening, of which the base of one is in the sun, and that of the other on the pavement; and that it is consequently necessary that when the sun has its upper part removed in the shape of the crescent moon, the same thing happens to the ray below on the opposite side." But in Ch. 5, when he explicitly asked what he had assumed as a principle in Ch. 10, [namely,] "Why a ray that has entered through four cornered figures does not make four cornered shapes, but circular ones?", he ascribes the cause in part to the weakness of vision "that one who is unable to catch those rays that withdraw into the corners of the opening, in view of the brightness of those that pass through the middle of the opening," and in part he appeals to the actual form of vision, "because vision may arise through a cone of rays going forth from the eye, whose base is a circle."[5] By this introduction of an irrelevant cause, and one which is not accepted in optics, he again brings darkness upon those things that he deduces from these in ch. 10, to such an extent that neither Witelo, nor **39**
Pisanus, nor any subsequent person that I know of, understood Aristotle.

Several years ago, some light shone forth upon me out of the darkness of Pisanus. Since I was unable to understand the very obscure sense of the words from a diagram drawn in a plane, I had recourse to seeing with my own eyes[6] in space. I set a book in a high place, which was to stand for a luminous body. Between this and the pavement a tablet with a polygonal hole was set up. Next, a thread was sent down from one corner of the book through the hole to the pavement, falling upon the pavement in such a way as to graze the edges of the hole, the image of which I traced with chalk. In this way a figure was created upon the pavement similar to the hole. The same thing occurred when an additional thread was added from the second, third, and fourth corner of the book, as well as from the infinite points of the edges. In this way, a narrow row of infinite figures of the hole outlined the large quadrangular figure of the book on the pavement. It was thus obvious that this was in agreement with the demonstration of the problem, that the round shape is not that of the visual ray but of the sun itself, not because this is the most perfect shape, but because this is generally the shape of a luminous body. This is the first success in this work.[7]

In addition, when both Aristotle and the man I have called "Pisanus", to the end of untangling the subject, introduced the most beautiful example[8] of the ray of the eclipsed sun being similarly eclipsed when it is taken through a small hole, they afforded Reinhold, Gemma, and my teacher Maestlin,[9] the opportunity to

---

[5] Aristotle, *Problems* XV 6, 911b 3–25. For the most part, Kepler paraphrases Aristotle rather than quoting him.

[6] Kepler uses the Greek word αὐτοψία here.

[7] David Lindberg has suggested that Kepler drew his inspiration for this book-and-thread procedure from Albrecht Dürer's *Underweysung der Messung* (Nürnberg, 1525). See *Theories of Vision*, p. 186.

[8] *Experimentum*.

[9] Cf. Erasmus Reinhold's scholia to his edition of Georg Peurbach's *Theoricae novae*

apply the theorem to a use that is no less noble. For these authors I have named had taught astronomers how to use a compass to measure the magnitudes of solar eclipses, the ratios of the diameters of the sun and moon, and the inclinations to the vertical of the circle drawn through the centers of the luminaries, avoiding the inadequacy of the eyes, and avoiding the error which generally occurs in a bare estimation. And so, from that time, however many solar eclipses were documented by eminent mathematicians, it is likely that they were observed in the way just now described. For aside from this, no other sure procedure can be established for measuring something that happens in the sky.

It is indeed well worth while here to see how much detriment would result from the ignorance of the proof of this theorem. For since it escaped a number of authors, the result was that in believing in the theorem without restrictions
40 they fell into a large error. For however many eclipses were observed in this way, they all had come out much greater in the sky than it appeared in the ray: all showed a much greater lunar diameter in the sky than in the ray. Hence it is that that Phoenix of astronomers, Tycho Brahe, in his wonder, was driven to such straits as to pronounce that the lunar diameter is always a fifth part smaller in conjunctions than it appears to be in oppositions, even though it is the same distance from us in both instances.[10] I would nonetheless not deny that there are other underlying causes for the lunar diameter's in fact appearing somewhat smaller at conjunctions, for which see below.

It is my hope in these pages to remove these considerable difficulties, which wall off our entry to the prediction of eclipses and to an exact reconstruction of the moon's motion, by a straightforward demonstration of the theorem, and by laying bare the sources of the errors which displayed themselves for me to examine through a careful consideration of the solar eclipse that occurred in 1600.[11]

Definition

All points from which straight lines can be drawn to some single distant point, making a pyramid, are said to belong to one luminous surface, whether they in fact belong to the same or to different surfaces, because that pyramid is understood as being cut by a single perpendicular plane.

Postulate

A pair of lines of illumination arising from the same luminous point, even though they actually come together at this their origin, are to be regarded as equivalent to equidistant lines within the limits of sense perception provided that they have a very large ratio to the base by which the two are connected.

---

planetarum (Wittenberg 1553, etc.), fol. 198 (this is the edition Kepler had); also Rainer Gemma Frisius, *De radio astronomica et geometrico* (Antwerp and Louvain, 1545), ch. 18. For Maestlin's contribution to the observation of solar eclipses with a *camera obscura*, see ch. 11 below.

[10] *Progymnsamata*, p. 135, in *TBOO* Vol. 2 p. 147.

[11] See below, Ch. 11 Problem 22, p. 423

Proposition 1

If there were a single point within the limits of sense perception from which light is radiated, the ray upon a wall interposed perpendicularly would have a shape similar to the window through which it had entered in a perpendicular incidence, and the ratio of the dimensions of the illuminated wall and the window would be the same as that of the departures of each from the luminous point. This is Witelo I 99 and 100.[12]

Let there be the wall or the plane $ABCD$, and the luminous point E on high; and let there be a window of any shape whatever, which shall now be
FGH, interposed between the two. Now let a perpendicular descend through the 41
transparent body, such as the air, from E to both the center O of the window FGH and the plane $ABCD$, and let this be EI. I say that the part of the surface $ABCD$ illuminated by the luminous point E will have a shape similar to the window FGH, that is, also a triangle, in the present case.

Demonstration

Now since the whole region of the window FGH and all its points lie exposed to the luminous point E without anything opaque lying in between, the rays from E will fall upon all these points, by 1 2. Therefore, the radiating straight lines will likewise be led through the infinite points of the edges of the window. Let the window FGH have any number of bounding straight lines. Let one of them be FG, whose ends are thus F and G, to which let two straight lines be joined from E: thus, by Euclid XI 2, FGE are in the same surface. And by Euclid XI 1, all the points of the straight lines FG, EM, EK, and however many descend from E to FG, are in one surface. Now the ray EOI is by supposition perpendicular to both of the surfaces FGH and MKL. Therefore, by Euclid XI 14, they will be parallel planes. Consequently, by Prop. 16 of the same book, the two lines of intersection FG and MK of the same plane MEK with the two parallel planes FGH and AC will be parallel. The same will be demonstrated in the same manner for GH and KL, for HF and LM, as well as for an infinity of sides, if there were any. Therefore, by Euclid

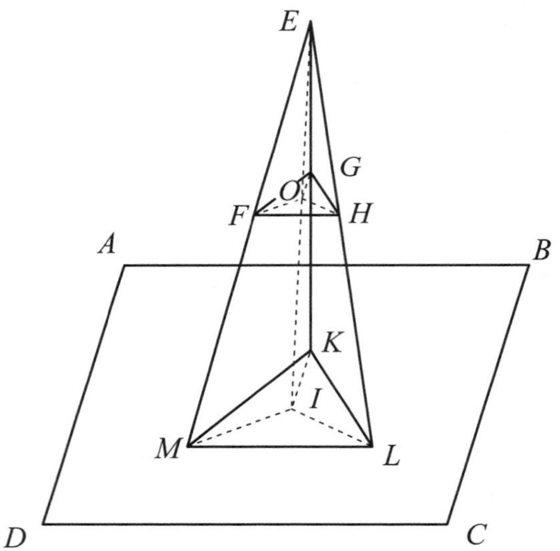

12 *Thesaurus* pp. 37–8. Witelo's propositions are purely mathematical, and concern only cones and cylinders. Kepler extends them considerably here, and applies them to a concrete situation.

XI 10, the angle FHG of the intersecting lines FG and HG will be equal to the angle MLK of lines parallel to the former and intersecting. In the same way all the angles of the one will be equal to all of the other. But the sides are also proportional to the sides. For by Euclid XI 17, the planes AC and FHG cut the intersecting straight lines EL, EM, EK, and however many others, in the same ratios. Thus as ME is to EF, so is KE to EG, and alternately, as ME is to EK, so is FE to EG, while the angle at E is common. Therefore, by Euclid VI 6, FEG and MEK are equiangular. Consequently, by Euclid VI 4, FG and MK
42 are homologous, and so are all the sides of the one to all the sides of the other. Therefore, by Euclid VI Def. 1, the figures are similar. But beyond the limits MKL of the lines descending through the edges of the window, there falls no ray, but a shadow, by 4 of the first chapter, because the parts surrounding the window FGH are by supposition opaque. Therefore, the portion of the illuminated wall MKL has a figure similar to the window FGH, which is what was to be demonstrated. Now let the centers O, I be joined with any angles whatever, such as with F, G, H, and M, K, L here. I say that EO is to OF as EI to IM. But since OF and IM, or any such others, have an equal ratio to their diameters, because the figures are similar, and since EO is the distance of the window from the luminous point, while EI is the distance of the wall from the same point, the other part of the proposition is clear as well.

Corollary

It follows hence that from any point of a luminous surface a pyramidal ray is cast upon an interposed wall, the base of which ray is similar to the shape of the window. Thus the ray descending from the whole luminous surface to the illuminated wall consists of shapes that are potentially infinite, similar to the window, mutually overlapping, and falling upon approximately the same region of the wall. These shapes individually would nonetheless have their own proper boundaries, if they were separated.

Proposition 2

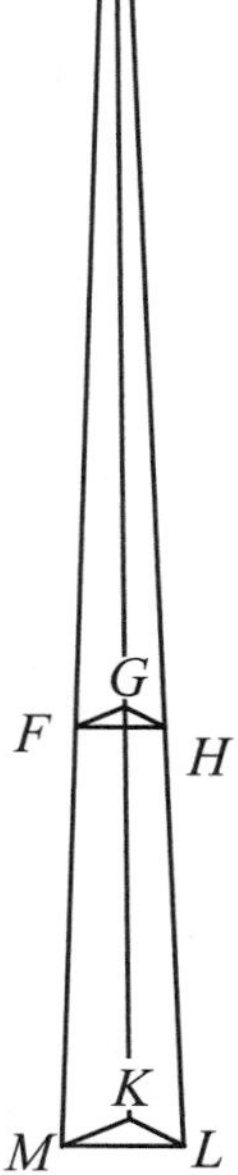

If there shines a single point, remote from a wall and a window (which are near each other) by an incalculably great distance, the light upon the perpendicularly interposed wall will resemble not just in shape but also the quantity of the window through which it passed in a perpendicular path. In the former diagram, let the luminous point be E, the window FGH, the illuminated wall MKL, and let the ratio EO to OF be incalculably great, such as it would be if EO were to measure the enormous distance between the window and the sun or moon, while OF measures the tiny diameter of the window. I say that FGH and MKL are equal to each other within the limits of sense perception. For since FO or any line whatever from the center of the window FGH has a perceptible ratio to OI (because the surfaces are supposed near each other), but OF has an imperceptible ratio to OE, OI will also have an imperceptible ratio to OE.

Consequently, by the postulate, the rays EF and EO, joined by the base FO, having a magnitude imperceptible in comparison to them, are equidistant within the limits of sense perception. But the angles EOF, EIM are equal, and are in the same plane of the triangle EIM, so by Euclid I 27, OF and IM are **43**
equidistant, and by 33 of the same book, are equal within the limits of sense perception. Further, the same can be demonstrated concerning any lines similarly drawn from the centers O and I. Therefore, the whole FGH is equal within the limits of sense perception to the whole MKL, but it is also similar by the first proposition of this chapter. Therefore, "If there shines a single point...," which is what was to be demonstrated.[13]

Proposition 3

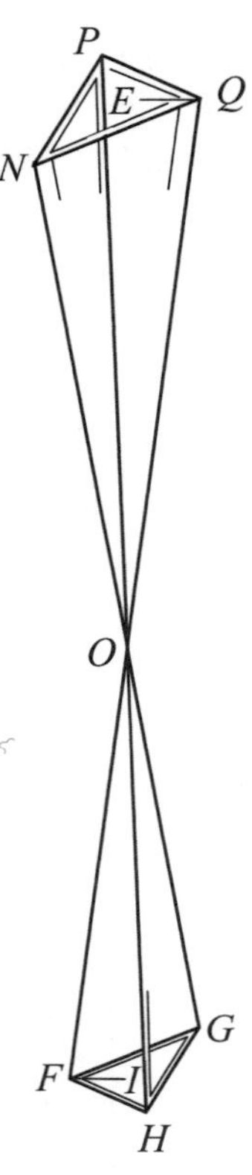

If an window could be a mathematical point, the illumination of the squarely interposed wall would precisely assume the shape of the illuminating surface, but inverted; and the ratio of the diameters of the luminous surface and the illuminated wall would come out the same as that of the distances of each from the point of the window. Let FGH be the surface to be illuminated, NQP the luminous surface equidistant from it, and let the window be at point O. In accord with what was previously demonstrated, let straight lines be drawn from N, Q, P, and any other points, to O and beyond it to the surface FGH, which lines represent the rays of the luminous surface, and let them be QF, PH, NG. Now, since all meet at O, they will thus cut each other when produced, and the ones on the right will become those on the left, and vice versa. Further, because two straight lines QF and PH intersect each other, they are consequently in one plane, by Euclid XI 2, and now by Prop. 16 of the same book, because two equidistant planes NPQ and GHF are cut by the plane $PQOFH$, the common sections PQ and HF will be equidistant. In the same way it can be proved that NP and GH are equidistant, and likewise NQ and GF. Thus by Prop. 10 of the same book, the angle FHG of the meeting lines FH, HG, is equal to the angle QPN of the meeting lines QP, PN, equidistant from the former ones, and GFH is equal to angle NQP, and FGH to angle QNP, to their opposites, respectively. Further, because the planes NPQ and GHF are parallel, they will cut POH and QOF in the same ratios. Accordingly, as PO is to OH, so is QO to
OF, while the angles POQ and HOF are equal, since they are vertical angles. **44**
Consequently, by Euclid VI 6, these triangles are equiangular [with each other],

[13] It was Euclid's procedure to restate at the end of the proposition (in what was called the *conclusio*) what was to have been proved, in words identical to the enunciation. In Kepler's day it was becoming common (though not universal) practice to abbreviate the restatement, which is what is happening here.

and PQ and FH are homologous.[14] And thus also are all sides of the one to all of the other. Therefore, the whole figure FGH is similar to the whole QNP, by Euclid VI Def. 1. Further, let the centers I and E be joined to F and Q or to any points whatever of opposite extremities. Accordingly, IF and EQ will also be equidistant by Euclid XI 16. And because IE and QF cut each other at O, the angles IOF, EOQ will be equal by Euclid I 15. But FIO, QEO are also equal, being by supposition right angles, and consequently the remaining angles IFO, QEO are also equal, by Euclid I 32. Therefore, the sides are proportional, and as the distance of the wall OI is to the line on the illuminated surface IF, or any other line whatever, so is the distance of the luminous surface OE to the corresponding line EQ. Which was to be demonstrated.

Corollary

It follows hence that through the individual points of any window, which are infinite, individual (and thus infinite) inverted images of the luminous surface are transmitted to the illuminated surface, following each other in the same order which the points of the window themselves have.

Proposition 4

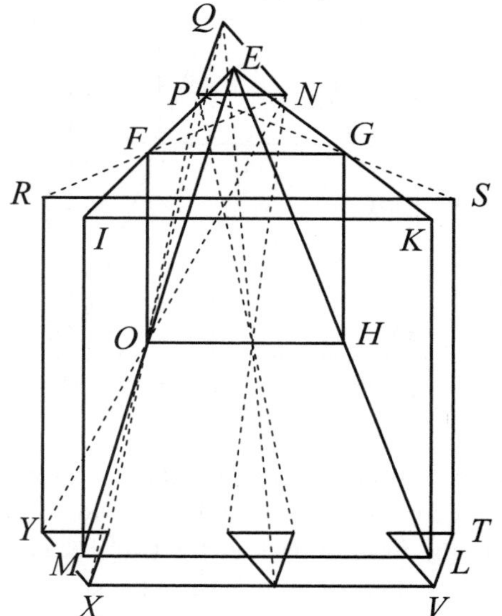

The quantity of every illumination on a wall is greater than the space of the window through which the illumination is sent in. For if, on the one hand, we pretend that it is a single point that shines, the rays transmitted through the boundaries of the window, since they meet at their origin, are proportionally farther apart as they go forth, and thus take up more space on a more remote wall than they do at the closer window, by 1 of this chapter. If, on the other hand, it be a surface that shines (as it always is), the proposition is all the more true.

Let PNQ be the luminous surface, whose center is E, and let
45 $FGHO$ be the window. Thus by the corollary of 1, the center E of the luminous surface will create the figure $IKLM$ on the wall similar to the window $FGHO$, and greater, by 1 of this chapter, or equal within the limits of sense perception, by 2. Now by the corollary to 3, through the individual points of the edges of the window, individual inverted images of the luminous surface are transmitted, such as you see at M and L, transmitted through the points O and H. And since EOM is the ray from the center of the luminous body, and the middle of all those

[14] Here Kepler used the Greek word ὁμόλογα.

that intersect each other at the point O, the remaining ones are therefore either beyond or this side of it, and the one that descends from the point Q, which is on the inside with respect to the window, is now made to be on the outside by the intersection that takes place at O. The same description can be applied to all the points. In this manner, a perimeter will be created that is greater than $IKLM$. But previously, this figure $IKLM$ was greater than the window $FGHO$. Therefore, this new expanded figure is much greater than the window $FGHO$, which was to be demonstrated.

Proposition 5

The shape of a ray on the wall is a mixture of the inverted shape of the luminous surface and the upright shape of the window, and it corresponds to them in position in this way. For by the corollary to 3, the inverted shape of the luminous surface, as if attached to the luminous surface, is drawn around following the boundaries of the window, with the result that by the individual points on the wall it describes lines corresponding to the sides of the window, and doing this on the same side. On the other hand, by the corollary to 1, the upright shape of the window, as if attached to the window, is drawn around on the wall, in an order contrary to the boundaries of the luminous surface (by the argument of 3), with the result that by the individual points on the wall it describes lines corresponding to the opposite sides of the luminous surface, on the opposite side. So the rays form the boundary of the figure that is made, where it has angles, and these rays are sent down from the extremities of the luminous body through the extreme angles of the window. Moreover, a geometrical motion is now assigned to the form of each of the two, because of the infinity of the multiplied form. Therefore, it is apparent that the boundaries of the luminous surface have a share in each of the shapes.

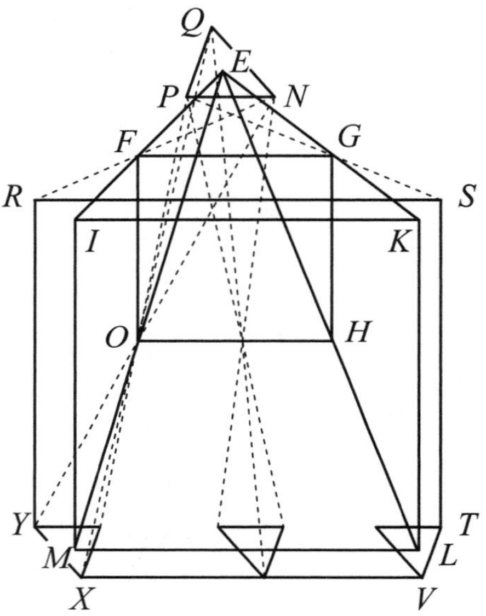

Let the above diagram serve as our example. The angle Q of the luminous surface, together with the whole triangular image, having descended past O to X, describes the line XV, corresponding to the line OH, when the point of intersection O is carried over to H. Similarly, the angle N, having descended past O to Y describes the line YR when O is carried over to F, and angle P, having descended past H to T, describes the line TS when the point of intersection is carried over from H to G. But PN, because it is raised up equally above FG, **46**
describes RS on the wall, after descending and being taken through FG. Thus the shape of the window is expressed approximately.

Again, suppose that the vertex E of the pyramid $EKLMI$ moves from the center to N while the pyramid is attached to $FGHO$. As a consequence, the angle IML of the image of the window will belong at Y. When the vertex is now moved from N to Q, the angle M will describe the line YX, corresponding to the line NQ on the opposite side. And if the vertex is at Q the angle MLK will be at V. When the vertex is transferred from Q to P, L will describe the line VT opposite the line PQ. And then IK will belong at RS, and K at S. Finally, when the vertex is moved from P to N, K will be separated from S until I belongs at R. The equidistance of the opposite sides PN and FG brings this about. Thus if RY, XV, and TS be extended to their common intersection,[15] the image of the shape of $FGHO$ will be completed. If, on the other hand, YX, VT, and SR meet when extended, this shape will be precisely similar to the luminous surface PQN, but inverted. Since neither of these things happens, it results that the two are to some degree mingled.

Corollary 1

It follows hence that if the sides of a pair of figures that are similar and on opposite sides shall be equidistant, the shape of the ray will perfectly imitate the common shape of the two, but in position it will imitate the window.

Corollary 2

But if the angle of one of the similar figures be placed opposite the side of the other, a shape with twice the number of sides will be created; e.g., a hexagon in place of two triangles, and an octagon in place of two squares.

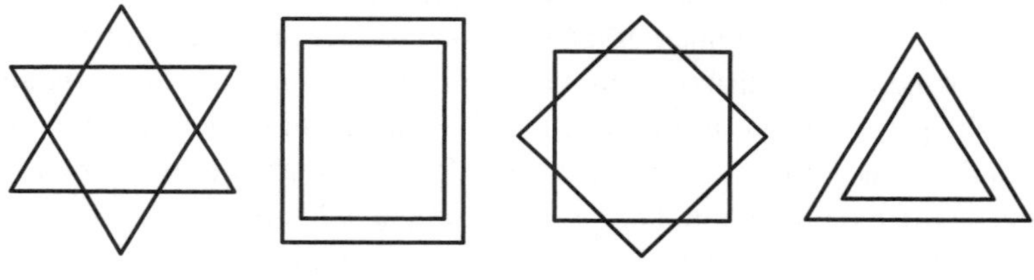

Corollary 3

Therefore, when the luminous surface and the window assume the shapes of a circle, the ray describes a perfect circle on the wall. For in a sense a circle has an infinity of both sides and angles, and thus the two circles can be considered to be equidistant, as in Corollary 1, and to have sides opposite sides and angles opposite angles, as in the second corollary. Therefore, whether you give the created ray a number of sides that is one times infinity or twice infinity, the result
47 is the same.

Lemma 1, for the following Proposition

Triangles whose sides cut off the same or equal portions of two equidistant lines are bounded by vertices on a third equidistant line. Let there be the straight line NEP, divided into equal parts at E, and let FOG be equidistant from it,

[15] Kepler uses the singular here, although the three lines obviously do not have a common intersection. He must have meant the common intersections of each of the nonparallel pairs.

likewise divided into equal parts at O. Let the straight lines PO, EF be drawn until they meet at K. In the same way, let EO and NF or PG be drawn until they meet at I, and let K and I be joined. I say that KI is equidistant from the bases NEP and FOG. For in the triangles NEI, FOI, the angles NEI, FOI are equal, by Euclid I 29. The same is true of ENI, OFI, and the angle at I is common. Therefore, the triangles have equal angles, and by Euclid VI 4 the sides are proportional. Therefore, as NE is to FO so is EI to OI. In the same way it is proved in EPK and FOK that as EP is to FO, so is PK to OK. But

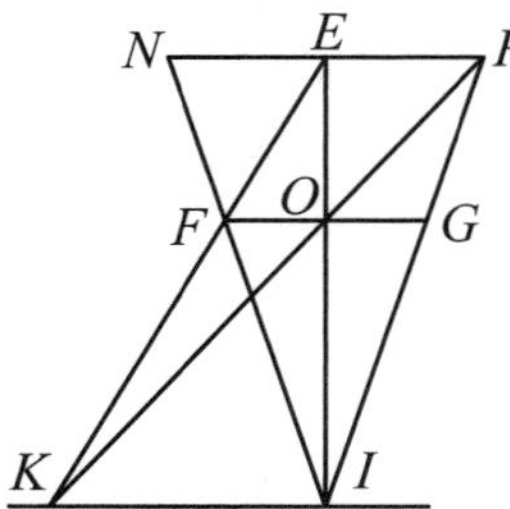

EP, EN are equal. Therefore, as NE is to FO, so is PK to OK. But previously EI was also to OI in the same ratio. As a result, as PK is to OK so is EI to OI. And by Euclid V 5 as PO is to OK, so is EO to OI, and alternately, as PO is to OE so is KO to OI. Further, EOP is equal to its vertical angle IOK. Therefore, by Euclid VI 6, the triangles EOP, IOK have equal angles, and OKI or PKI is equal to OPE or KPE. As a result, by Euclid I 28, EP and KI are equidistant. The same is also true of EPK and EPI, which in contrast have a common base but cut off equal equidistant portions of FG. The proposition is therefore evident.

Lemma 2. Problem

To find a point that is the same number of diameters of the luminous surface[16] *from the luminous surface as it is in diameters of the window from the window.* Let NEP be diameter of the luminous surface, and FOG be the diameter of the window, equidistant from it and lying perpendicularly below it, and let EO be the perpendicular to the two. From N and P let straight lines be drawn through the ends F and G, until they meet. Let the point of meeting be I. I say that this is the required point. For since in triangle NIE FO is equidistant from the side NE, IE will be to EN as IO is to OF, with the result that as IO (the
distance of I from O) is to FG (twice OF, which is the diameter of the window), **48**
so is IE (the distance of I from the luminous surface E) to NP (twice NE and the diameter of the luminous surface), which was to be accomplished. It is moreover clear from this that the diameter of the window must be less than that of the luminous surface.

Proposition 6

When a window is distant from the wall by the same number of its diameters as the luminous surface is in its diameters, the mixture of the shapes is most evident, and the shape of the ray partakes equally of both shapes. But when the window is distant from the wall by fewer of its diameters, the shape of the ray and its position approaches more nearly the shape of the window. And when on the other hand the luminous surface is distant from the wall by fewer of its

---

[16] Literally, "of the luminous." However, since Kepler uses "luminous surface" (*superficies lucens*) in Proposition 6, it seemed justifiable to supply "surface" here.

diameters, the shape of the ray more nearly imitates the shape of the luminous surface, in an inverted position, the more so as the former is true. Let $QRSPN$ be the luminous surface with center E and diameter NP, and let the window be FOG with diameter FG, and let it be equidistant from the entire figure of the luminous surface, lying perpendicularly below it. From E let a perpendicular be drawn through O, and let it be EI. And on line E let there be found, by Lemma 2, a point that is the same number of diameters of the window from the window as it is in diameters of the luminous surface from the luminous surface. Also, let EF and PO be joined, and likewise EG and NO, and let them be extended until they meet at K and L.

I say that if by this position of the window and the luminous surface a ray be created at I, it will be equally related both to the shape of the luminous surface E and to the shape of the window O. However, if it be created and it fall closer than I, towards O, the nearer it is to the center O the closer it will approach the shape of O, the shape of E being effaced. If on the contrary it fall beyond I, the farther it is from O the more precisely will it represent the shape of the luminous surface E, the shape of O being gradually effaced. For this to hold, note first, that the meeting points K, I, L, by Lemma 1 above, are in the same line, which is equidistant from the luminous surface NE. In addition, all
49 three are on the surface to be illuminated. Further, the rays FK and GL sent down from the center E of the luminous surface past the edges of the window FG carry the shape of the window down to the plane I, by the Corollary to 1 of this chapter, and carry that very midpoint of it, for all about it there stand as many other shapes as there are points lying about the center E. In the same way, the rays PK, NL sent down from the edges of the luminous surface N, P through the center O of the window, carry down the inverted shape of the luminous surface, by 3 of this chapter and its corollary, and likewise that very midpoint. For about this too there stand as many other shapes, similar and approximately equal, as there are points lying about the center of the window, since such an intersection takes place at each of them. Therefore, since these four rays intersect each other at the points K, L of the surface I, KL will be the common measure of both shapes, that of the luminous surface inverted, and that of the window upright. Now let the point be chosen, in accord with the preceding problem, that is distant from O by fewer diameters FG and it is from E in diameter NP. And let it be X, through which let a line equidistant from the said diameters, which will intersect the rays at H, M, T, V. Therefore, the part MT will represent quantitatively the diameter of the luminous shape inverted, as it falls above the position X. On the other hand, HV is the diameter of the shape that the window has. Thus the latter will be greater than the former. Consequently, more of the latter than of

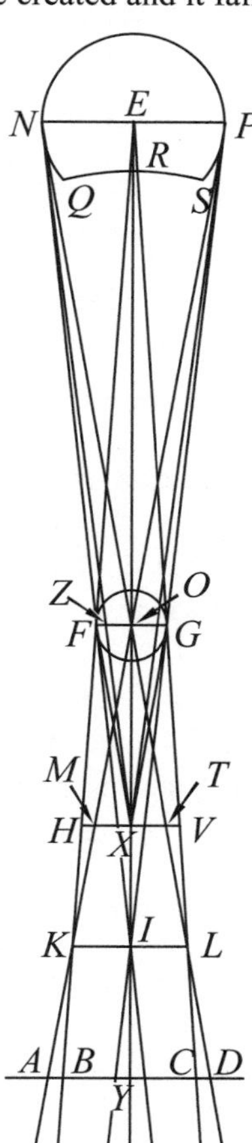

the former boundary enters into the eyes. Contrariwise, in accord with the same problem, let the point be chosen that is distant from *O* by a greater number of diameters of the window than it is from *E* in diameters of the luminous surface. Let this be *Y*, and through this, as before, let a line be drawn equidistant from the others, which will intersect the rays at *A*, *B*, *C*, *D*. But because the intersection of the rays already occurred at *K* and *L*, the ones which were on the inside are now on the outside, and *AD*, representing the inverted shape of the luminous surface, will be greater than *BC*, the diameter of the shape that the window has. As a result, the shape of the luminous surface enters into the eyes more here. In order that these three cases be more rightly understood, first let the shape of the window be described, just as it is produced upon the wall by any one point of the luminous surface, and let it be αβγδ, and let the inverted shape 50
of the luminous surface be εζη, of such size that it can be enclosed in the same circle [as that which encloses the shape of the window]. Now let the center of the shape εζη be moved along all the edges of αβγδ. Its position about α will be the triangle θιπ; about β, κλμ; and thus with the angle ι it has described the line ιλ equidistant from αβ. About γ, in turn, will be τνξ, and with the angle μ it has described the line μξ. Finally, about δ it will be σρο, and with the angle ο it makes the line οθ, while with the side νξ it makes the line ξο. Thus εζη has described 4 lines, bounding the figure ιλ, μξ, ξο, οθ, parallel to the four lines of the window.

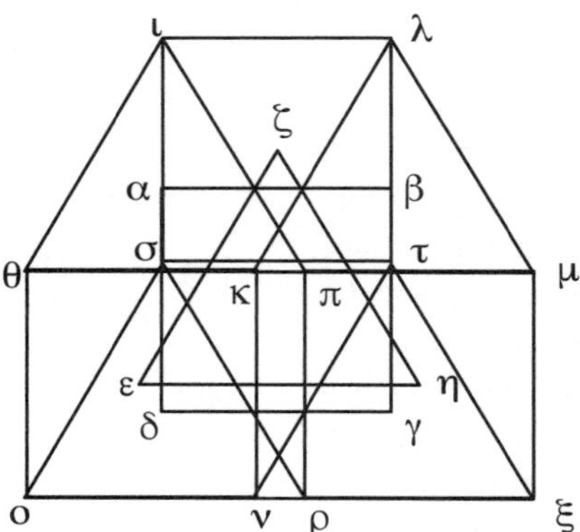

In the same way, let the center of the figure αβγδ be moved along all the edges of εζη. Its position about ε will be θκνο; about η, μπξρ; about ζ, ιλτσ; and with the line δγ it has described the line οξ, with the angle μ the line μλ, and with the angle θ the line θι.

Here you see both the diagonal sides of the triangle and the orthogonal sides of the square presented, on the perimeter of the entire figure, approximately equal, that is, of such a size that both are in the same circle. For those of the triangle are greater because they are fewer, and hence it can be said that the quadrangular sides are no more shortened than are the triangular ones.

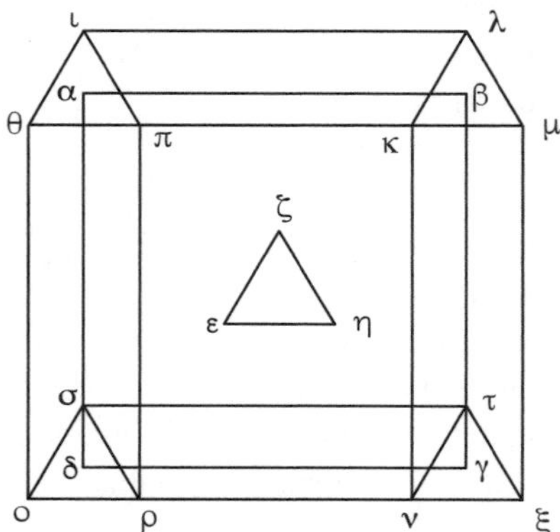

Now let the diameter of the figure of the window be greater, and let it be αβγδ, but let the inverted figure from the luminous surface be εζη, and let both be situated upon the same center, as before. Now let the center of the triangle be carried around on the edges of αβγδ: its positions will be θιπ, κλμ, ντξ, οσρ, and the intermediate points. You see that with the three angles it describes the three lines οθ, ιλ, μξ, while with the side it describes the line οξ. Let the center of the window be likewise carried around on the edges of εζη: its positions will be three, about ε, ζ, η, the points being these: θκνο, πρξμ, and ιλτσ. You see that with two angles it describes the two lines θι, λμ; with the remaining two and the side between them, the common line οξ. Thus this figure is lacking some least part, preventing it from being wholly similar to the window. The traces of the figure of the luminous surface, however, are very slight.

Contrariwise, let the diameter of the figure of the window be smaller, and let it be αβγδ, and about the same center let there be the inverted figure of the luminous surface εζη, and, keeping the same
51 twofold circular translation, let the lines just mentioned be drawn, only the proportion being changed. Thus some least part is wanting from the whole figure, preventing it from representing its original, namely the luminous surface, in an inverted position.

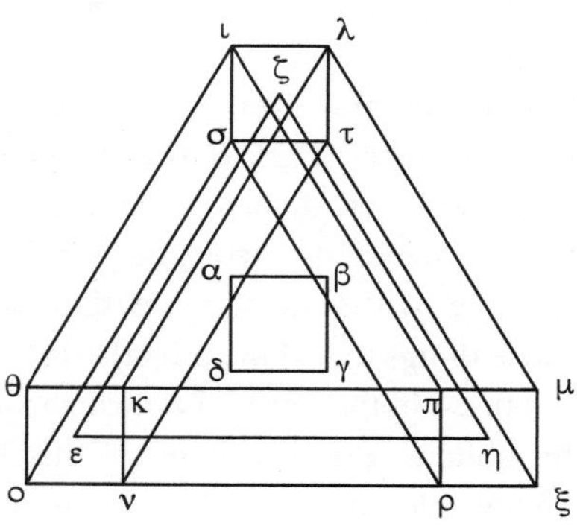

Proposition 7. Problem

In a closed chamber,[17] *and upon an appointed wall, to represent whatever either is or is done outside opposite the chamber; that is, that enters into the eyes.* This art was, as far as I know, first presented by J. Baptista Porta,[18] and constitutes not the least part of the *Magia naturalis*. But, being content with experiment, he did not add a demonstration. Yet it was in fact just this one experiment alone by which astronomers would have been able to make a determination concerning their picture of the solar eclipse.

So, let all the cracks of the chamber be closed, so that not the least particle of light can enter, while opposite the window, which has a view of what is to be represented, let there be a white wall, the other walls being black. Let an extremely narrow hole be made in the window, just as much as suffices the power of vision, in such a manner, however, that if the wall or the window should be thicker than the opening, the opaque parts surrounding the opening should be cut away, until your view through the opening may lie unobstructed to all things located outside that you wish to represent. And let the eyes of the viewer be shielded from the light of day for quarter to half an hour, until the images impressed by the spirits in the clear light of day might vanish, in accord with what was said in the conclusion to the first chapter. And let the things to be represented be located in the bright light, either of the sun, or of day, or of torches. I say that all things that are seen both to stand motionless and to happen outside, are going

---

17 *Camera clausa*. The term *camera obscura* had evidently not yet become standard; Porta called it a *cubiculum obscurum* (*Magia naturalis*, Naples 1589, XVII 7, p. 267).

18 *Magia naturalis* XV 6. This technique was also described, though obscurely, by several other sixteenth century authors.

to be represented inside on the opposite white wall, but in an inverted orientation.
For by Chapter 1 22, things outside illuminated by any light whatever tint the
communicated light, and scatter it in an orb. As a result, by 6 of this chapter,
the things outside will illuminate the opposite wall within, so that the shape of
the illuminated wall is indeed a mingling of the shape of the window, and that
of the things that are outside, but nonetheless, since the window is presupposed
to be very narrow with respect to the distance of the window from the wall, a
minimal part of the shape of the window will be mixed in with the shape of the
things standing outside. The only thing that will be lacking in this picture is, first,
that by the same Prop. 6 the things will appear inside in inverted orientation, and
second, that anything outside that falls within the reach of a cone having its vertex
at the wall and shaped or limited by the hole, cannot be portrayed upon the wall *52*
with its parts articulated, but whatever exceeds in breadth the boundaries of this
its cone, will be depicted within along with its parts. For the same reason also,
those things that shine directly through the hole shine more brightly and are more
confused in the image, because the hole is spread wider from the direct than from
the oblique. But the colors will also not be lacking in this picture, for by Chapter 1
25 the colored lights of cones coming together at the hole do not disturb or hinder
each other. And because the surface upon which these colors of exterior things
radiate is white, hence, by Chapter 1 29, it will receive these rays all the more
strongly, and will make them appear brighter. And since the chamber is closed,
and it is not the rays of the sun or of the whole sky, or rather of the glowing air,
that shine upon the individual points on the wall, but just whatever little part you
please that shines on the point opposite it, therefore, by the converse of Chapter
1 27 and 28, the senses distinguish between the individual lights, because they
are not at all tinted by a stronger light. This is especially so because we made
the other walls black, so that they might not be illuminated by the white wall
that was originally illuminated, and become bright themselves (by 1 29), and
in turn illuminate the representational wall, and thus confuse the colors coming
from outside (by 1 27), if these walls too had been white. Because if the chamber
were not so tightly closed, although the colored rays on any wall you please are
infinite, even without the representation of the hole, nevertheless, by Chapter 1 28
they could not be perceived because of the brightness of day. And because when
the configuration of the luminous hemisphere is changed, its image upon the wall
is itself also changed, while the things that are done, that is, motions, suddenly
change the appearance of the hemisphere, therefore, the image within will also
change, and thus the motions of exterior things are to be seen within.

However, in the representation of colors, there is lacking what was proved
by 29 and the converse of 30 of Chapter 1: that the distinctions of colors are not
well grasped by the eyes, because of the brilliance of the whites radiating upon
a white surface, and indeed, the radiations of the dark colors are apprehended no
differently than under the proportion of shadows. If you wish to change the color
of the surface, the radiation of colors will indeed be made distinct, by Ch. 1 30,
but with the polluted common color of that surface, and so weak that it can hardly
be grasped by the eyes.

The following cautions must also be added. First, if the hole be exceedingly *53*
small, things are indeed depicted distinctly and in detail, but just as very small

writing is hard to be read by a weak sense of vision, so here too the eyes, imbued with images seen in the brightest light of day, will have to be held in place for a very long time indeed before they may grasp so detailed a picture illuminated by such a bad light. If on the contrary you open up the hole, the picture will indeed be proportionately brighter and more apparent, but will also be proportionately coarser and more confused. Thus the hole should be of an intermediate size. Next, if the wall is at a great distance from the hole, the picture will indeed be more precisely detailed, as the ratio of the hole to the distance vanishes. But there is in turn this consequent disadvantage, that the luminous colors are more diluted, and, being thinned out, move the sense of vision more weakly; and that the air that is in the chamber, mingled with small dust particles, becomes luminous for a longer distance within, and effaces the picture on the wall with brightness. Thus colors come through badly to a wall that is too far, and from a nearby hole strike upon and color a white paper far more clearly. The same is to be held concerning the air and the distance outside. For even when one beholds remote objects directly, their color, diluted, moves the sense of vision more weakly, and is tinted by the blue color of the large quantity of intervening air, and is obscured: and it comes through to the chamber in the same way.

It is also helpful to place some sort of wall, like a brow, opposite the hole on the outside, to keep the sky or the air from enduing the part of the wall opposite with too much brightness, and to keep weak lights placed next to strong ones from becoming invisible, or to keep the air inside from becoming too bright and washing out the colors on the wall. The picture will be at its brightest if the sun should illuminate the things to be represented directly, when it is near the horizon.

You will avoid the inconvenience of a conical opening in a thicker wall, by making holes on both surfaces of the wall; thus the surfaces of the wall will remain nearly undamaged.[19]

The reader may expect me to follow G. B. Porta here through the rest of the devices that he has in Book 17 of the *Magic*, but this does not admit of method, nor does it further my plan.

Proposition 8

The shape of a ray of the sun or the full moon that passes in through a window with an angular shape, with the condition that the ratio of the window to its distance from the wall is less than the sun's ratio to its distance, approaches
54 *gradually more and more to roundness as it proceeds farther from the window.*
For since the shape for the luminaries, by which they pass into the eyes, is circular, the proposition is therefore evident through 6. And thus it is not true simply, that an angular ray that has passed through angular windows becomes perfectly round as it progresses. For the deficiency impinges on the sense of

---

[19] Instead of having one large conical opening in the wall, with the window and its hole on either the outer or the inner surface, one could make two smaller conical openings with their vertices meeting in the middle, and place the window in the middle. One could thus have the same field of view while removing only about one eighth as much material from the wall.

vision, if you look into it more carefully. This false persuasion completely racked the brains of the ancients. Further, it is clear from the same principles that *if a window be circular, the ray will be circular; if it be a good sized quadrilateral, the ray will not be a simple quadrilateral, but one of obtuse angles drawn back into a circle.*

Proposition 9

When the sun is in eclipse, the image on the illuminated wall will also be in eclipse, regardless of the sort of opening, when the image comes in high enough that the ratio of the window to its distance be less than that of the sun to its distance. For the eclipse of the sun consists of the motion of the moon beneath the sun's body, by which motion one and another image of the sun (but always horned or lunate) is constantly shown from beginning to end. The result will therefore be what we learned in Proposition 7, the demonstration of which, carried over to this case, is also valid. In the same way, eclipses of the moon, and its phases as seen by night, are depicted upon the wall, but less vividly.

Proposition 10

In the image or ray of the eclipsed sun sent in through a round window in the proper way, the horns appear, not sharp (as they are in the sky), but drawn back and blunted by the little circle of the window. Let $DACE$ be the true inverted image of the eclipsed sun, in shape and size such as passes in through the center of the window, by the corollary to 3, and on its edges D, E, C, A, let circles DH, EG, CF, AB, be drawn equal to the window, by 2.

By 5, the pointed shape of the luminous surface is necessarily mixed with the round shape of the window, while by Corollary 3 of that same proposition, this confusion with the circle, insofar as it is a circle, does not **55** at all detract from the likeness of the shape, but only moves the boundaries of the window D, E, C, A, out, so as to make the shape IGK, FBH. Therefore, only the pointed horns C, D, remain in question, for when they are drawn around following the circumference of the shape of the window, they themselves describe such a circumference HI, KF, just as was evinced by 6.

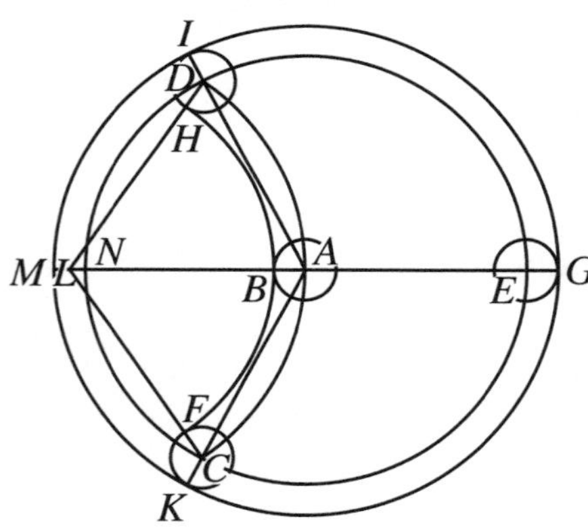

Proposition 11

The diameter of the moon in a ray of this kind appears less than it does outside in the sky. Keeping the previous diagram, let the sector DAC be extended, and let the center of this circle L be joined with C, A, D. Let the sector DEC also be extended through N, and let the center A of this circle be joined with the points C, E, D, and produced to K, G, I. Since by 3 Corollary, $DECA$ is the exact shape of the remaining part of the eclipsed sun, as it really appears in the sky, therefore $LA : AE$ or $LD : DA$ or $LC : CA$ is the true ratio of the diameters of the sun and moon.

But now since $IGKFBH$ is everywhere augmented by the semidiameter of the window EG or CK, two factors therefore cooperate to the same end. First, the semidiameters of the sun AD, AE, AC, are increased, and become AI, AG, AK. As a result, by Euclid V 8, the ratio of AI to AL is greater than AD to AL. Second, the semidiameters of the moon LD, LA, LC, are decreased, and become LH, LB, LF, likewise equal because equals are subtracted. As a result, by the same proposition, the ratio of AD to BL is greater than AD to AL. But AI to AL was also greater than AD to AL. Therefore, the ratio of AI, the semidiameter of the sun in the ray, to BL, the semidiameter of the moon in the ray, is much greater than that of AD, the semidiameter of the sun in the sky, to AL, the semidiameter of the moon in the sky. This inequality is quite perceptible, since a large enough window is usually chosen.

Proposition 12

The digits[20] *of eclipses appear fewer in the ray than in the sky.* For in the previous diagram let sector IGK be estended through M. Now since EN to NA is the ratio of the sun's diameter to that of the eclipsed part, while GM is greater than EN by two semidiameters EG and NM of the window, the ratio of NA to MG is therefore less than that of NA to NE. But NA is equal to MB, because AB and MN are equal. Therefore, the ratio of MB, the eclipsed part of the ray,
56 to MG, the diameter of the ray, is less than the ratio of NA, the eclipsed part of the sun, to NE, the diameter of the sun. The ratio is often less by a fourth or a third part, according to the breadth of the window.

20 In Ptolemy (*Almagest* VI 7, trs. Toomer p. 295), magnitudes of eclipses are estimated in twelfths of the sun's diameter, which are called δάκτυλοι, or "fingers". The Latin equivalent is "digiti," and the standard English translation has become "digits."

Chapter 3

On the Foundations of Catoptrics and the Place of the Image

1. Refutation of Euclid, Witelo, and Alhazen

At the very foundation of catoptrics, the demonstrations of the optical writers are still clouded, in that they require from sense perception the very thing that was to be demonstrated. Further, it is not true that no error arises from this procedure. In the present *Optics*, we examine refractions with more care, with a view to eclipses and stellar observation, but to get to refraction we had to pass over this gap. So this place, too, must be filled, the clouds must be driven off, so that the sun of truth may shine more clearly.

In theorems 16, 17, and 18 of the *Catoptrics*,[1] in order to prove any one of those things that are visible through the perpendicular of a mirror's surface, to appear upon the surface, Euclid makes a false assumption. Let CD be the mirror, B the observer, A the visible object, AC the perpendicular. That the place of the image of the object A is on AC, namely, at E, he proves thus: "For," he says, "when the position C of the mirror is taken, upon which the perpendicular falls, the visible object A is no longer seen." If by "taken" you understand "occupied" (that is, that the position C is covered), the axiom is false, even though Euclid brought this forward to the beginning of the book, placing it among his postulates (or, to put it differently, what he calls φαινομένον in this book) that are borrowed from experience. For even if C be covered, or taken away entirely, provided that D remain, A is nevertheless seen at E by an observer B.

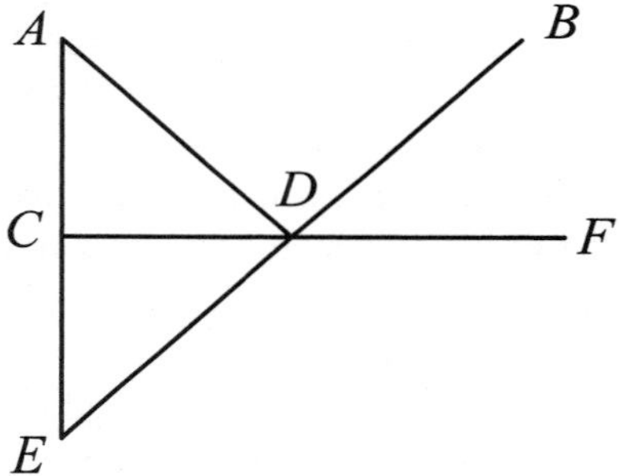

Furthermore, this axiom seems to smack of a false belief about the true and real ascent of the image on the line CE, which matters are indeed considerably at odds with the opinions of certain of the ancients concerning the emission of visual rays from the eye. For it is to this end that Euclid in the first postulate defines vision (ὄψιν, in Greek) to be a straight line, and uses it, taken thus, everywhere in [the proofs about] mirrors. Euclid thus appears to speak with

[1] Euclid's *Catoptrica* appears in *Euclidis opera omnia*, J. L. Heiberg and H. Menge, vol. VII, Leipzig 1895, pp. 288–343. There is an English translation of the medieval Latin version: *The medieval Latin traditions of Euclid's Catoptrica: a critical edition of De speculis with an introduction, English translation and commentary*, K. Takahashi trans., Kyushu University Press, 1992. There is also a translation from the Arabic: *The Arabic version of Euclid's Optics*, E. Kheirandish, ed. and trans., New York: Springer, 1999.

a certain *savoir faire*,[2] and assumes things which, though they are not easily granted, yet once admitted lead to no internal contradiction in what follows. And indeed, in this style of philosophizing, truth suffers violence: there exist false beliefs, which even seem to have stuck to Aristotle, when he speaks of his "visions",[3] from this Euclidean school,[4] which he did not well understand. At length, ignorance set up her tyranny under the guise of Art.[5] Let us now grant that Euclid's axiom is to be understood differently, so as to state that if the observer were situated at *A* and *C* were covered, then *A* would not be seen. Then the axiom is perfectly true, but the conclusion does not follow from it, except for perpendicular viewing. The argument does not carry over from a perpendicular to an oblique observer.

Alhazen and Witelo seem to have had a sense of this. For when they carefully brought forth what they had found in Euclid, they in fact omitted this, on the grounds that it was absurd. However, they say something similar in refractions.

Witelo tries to demonstrate the same thing in Book 5 prop. 36,[6] but this most careful author was harmed by the barbarism of his times, and by his close relationship with the Arabs, which makes him hard to understand today. Nevertheless, the obscurity of the subject also brought the delusion upon him.

First, I say that he does not do well to argue from the location of the thing seen to the location of the image, that is, out of fear that the image might cease to exist if the image should not correspond to the object in position. And indeed, in this way he would easily overturn all of catoptrics. For many things of this sort are different in the image than in the object. Next, for my part, I do not understand the postulate which he repeats from the beginning of the book, except for the amount of light that the Arab Alhazen reveals here in book V no. 9–10, from which Witelo transferred his own.[7]

Alhazen first gives a lengthy empirical proof that the position of the image is always on the perpendicular from the object above the surface of the mirror. Then, in no. 9–10, he tries to give the causes of this. However that may be, I even here do not understand anything any better, except this final statement: "The physical condition of natural things," he says, "looks to the site of their principles,

---

2 The word Kepler uses here is ἐντέχνως. Greek words were often scattered about in Renaissance Latin works in much the same way that French words and phrases are now introduced into English prose.

3 Although Kepler uses the plural, ὄψεσι, here, Aristotle seems always to have used the singular in *De anima*.

4 Kepler has the chronology of these two authors wrong: Aristotle died in 322 B. C. E., while Euclid flourished around the beginning of the third century B. C. E.

5 *Artificii*. This word corresponds to the Greek τέχνη, and has a considerable latitude of meaning. Here it evidently means "The rules or theory of an art, a prescribed method or system." (*Oxford Latin Dictionary*, Oxford 1982, p. 177, third definition.)

6 Witelo V 36 (*Thesaurus* II p. 207) and Alhazen V 9–10 (*Thesaurus* I pp. 130–1) state that the image falls upon the perpendicular for both flat and curved mirrors.

7 *Thesaurus* I pp. 130–1; II p. 207.

and the principles of natural things are hidden."[8] By these words he says two
58 things. First, he repeats the very thing that was proposed to prove (for they say nothing different), and second, he says by way of appending the cause, that it is hidden. But this is not demonstrating.

He nevertheless seems to be implying that this location of the image on the perpendicular was long ago thus established by God the Creator because it would be best so, and no more fitting place could be given to the image, which he proves by the sameness of the position, or from the contrary variation. Witelo, who also follows him [i. e., Alhazen] concerning the soul, which presides over vision, comes to suspect that the soul might assign the ratios of mirrors, by some particular scheme of its own: Book 5 Prop. 18.[9]

And in fact all these affects are consequences of vision by material necessity, where considerations of purpose or beauty have no place.

Furthermore, in the plane mirror, Alhazen does indeed allude more nearly *to the truth,*[10] but in such a manner that a certain instance can immediately refute him. He says that when an image is perceived on the perpendicular, it has the proper magnitude belonging to the thing itself.[11] But I object that it is not necessary that it have the proper magnitude: this is evident in curved mirrors, where the quantity is always changing. I therefore ask for the reason why it should have its true quantity in the plane mirror, rather than in the curved one.

But this fact, that they do not give the same cause of this matter in reflection[12] as in refraction, further strongly confutes the Optical writers.

In Book 10 Prop. 13, where Witelo is about to demonstrate that the image created by refraction likewise falls upon the perpendicular drawn from the object to the surface of the denser body, he faithfully repeats the words of Alhazen.[13] This he did, I believe, because he did not consider it prudent either to touch this sore, or to change rashly things that were not fully understood. They argue from the composition of motion on the diagonal, that it arises from motions parallel and perpendicular to the surface of the denser body. It is hard to see the connection, and even if you admit it, a mathematical deduction of what was proposed to be proved will not be forthcoming. And if it be established that the image plays about the perpendicular, it is not as a result immediately clear that it also falls upon the perpendicular itself.

---

8 No. 10, in *Thesaurus* I p. 131 line 7–8. This is not, however, "final" in the sense of being at or near the end of the section.

9 *Thesaurus* II p. 198.

10 These words italicized by Kepler are all adjacent in the Latin, though it was not possible to keep them so in English.

11 V 9, *Thesaurus* I p. 130. What he actually said was, "In this position, the truth, both of the point seen and of the image, will appear."

12 *Repercussio.*

13 *Thesaurus* II p. 415; cf. Alhazen VII no. 19, *Thesaurus* I pp. 255–6.

To Alhazen's opinion, Witelo appends the view that we had noted above as irrelevant and false in Euclid.[14] He says, "If on the surface of a transparent body 59
a point upon which there falls a perpendicular from the seen object, happens to be hidden by the interposition of something opaque between the seen object and the point, the object will not be seen."[15] I say that this is false. For provided that the point[16] be free, from which the ray from the seen object to the eye is refracted, the image of the radiating object in the depth[17] will perforce be seen. And indeed the first and more obvious image of an opaque object, created by reflection, will become visible, in that we have just said that for its own part it is submerged and set in the midst of the water, (a matter which perhaps deceived Witelo); while the second and less apparent image, that of the object radiating in the depth, which is established by refraction, will appear in the same place, commingled with the former image of the opaque object, or a little past it.[18]

This passage also provides a refutation of what we have just said that Alhazen used for the consideration of a plane mirror, namely that from the quantity of the image its position also follows. Thus through the subject of refraction a great instance arises. It is true that in convex mirrors the image appears to be both smaller and nearer, in plane mirrors it appears equal to the object itself in both quantity and remoteness, and in concave mirrors it appears both greater and more remote. Thus the remoteness approximately follows the quantity here. But in refraction, where the image is equal to or greater than its object, it gets closer, and when the quantity of the image is diminished, the image itself departs farther, as may be seen in lenses[19] diminishing the images of objects if you look through them from a distance.

2. True Demonstration

Next, in order to make evident the true cause of the place of the image, ignorance of which is a disgraceful stain in a most beautiful science, I should like to have placed before the eyes from the beginning the chief sinew of the demonstration: that in plane mirrors the angles under which objects are seen

---

[14] See above, p. 73.

[15] The quotation is from the end of the demonstration of Witelo X 13, *Thesaurus* II p. 415.

In the adjacent diagram, an opaque object at H will block the rays between C and E, and, according to Witelo, an observer at B will therefore not see the object at E.

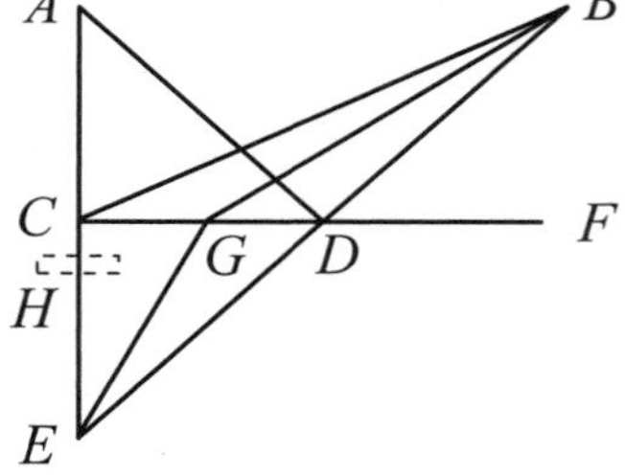

[16] Point G in the diagram.

[17] We might say, "in space".

[18] Kepler is comparing the reflected image of an object at A with the refracted image of an object at E. The latter will appear along the line BG (in the diagram in the preceding footnote), "a little past" the image of A, which appears on BD.

[19] *Perspicillae.*

are not changed by reflection, while they are entirely changed in convex and concave mirrors and in denser media. For when a convex surface receives rays that come together in a wide angle, it turns them back into a narrower angle, while a concave surface reflects into a shorter pyramid those that are hurrying towards a meeting with a moderate inclination. Those rays that are about to go into a denser medium, after being refracted, the surface alters almost not at all in breadth, but makes them come together slightly more acutely. In depth, however,
60 the surface acts in exactly the opposite way, bringing them together into a large angle. These general statements, I say, are set up before the eyes in the following demonstrations. Now I make ready the demonstration itself.

Definition 1

First, I place at the entrance the definition of the Image, taken from catoptrics, into which we are entering. *The Optical writers say it is an image, when the object itself is indeed perceived along with its colors and the parts of its figure, but in a position not its own, and occasionally endowed with quantities not its own, and with an inappropriate ratio of parts of its figure.* Briefly, an image is the vision of some object conjoined with an error of the faculties contributing to the sense of vision. Thus, the image is practically nothing in itself, and should rather be called imagination. The object is composed of the real form[20] of color or light and of intensional quantities.[21]

So, since the image is the work of the sense of vision, some preliminaries accordingly need to be said about vision. Now the image contains chiefly these four things: color, position or direction, distance, and quantity. For each of these it must be explained by what support of the visory apparatus, and by which reinforcements, they are comprehended, even though Witelo has explained this same thing in Book 3 and 4. Nevertheless, we have to row up a little closer to the demonstration that has been established.

Proposition 1

So, since seeing is a receiving,[22] and receiving occurs through contact, *therefore none of the things mentioned will be comprehended without contact, or something in some way of the nature of contact*. Contact is understood here as being between the surfaces of the eye and the image or rays which, by the foregoing, flow down from the objects.

---

20 *species.*

21 "Intensional quantities" are quantitative properties of things that can acquire different degrees of intensity. The theory of "configuration of quantities," as it was called, was developed in the fourteenth century, most notably by Nicole Oresme (ca. 1325–1382), and was familiar to Galileo, Kepler, and other prominent scientific thinkers of the seventeenth century. Cf. Oresme, *Configuration of Qualities*, in *A Source Book in Medieval Science*, Edward Grant, editor (Cambridge: Harvard University Press, 1974), pp. 243–253; Kepler, *Astronomia nova* ch. 33 and 36, Donahue trans. pp. 376 and 394.

22 *Passio.*

Proposition 2

Thus, *when something happens to the form of light and color in the middle of its path, such as either reflection or refraction by either polished surfaces or those of denser media,* this, on the grounds that it is entirely and in every respect removed from contact of the eye, *is not grasped by the eye or by the sensory faculties standing by to assist the eye.*

Proposition 3

Now let this be taken from the senses as generally admitted: that genuine vision occurs when the folding door[23] or pupil of the eye is exposed most closely to the arriving ray of light. Thence it follows that *vision from the direction whence the light approaches, is rendered more certain by this direction of the eye and of the entire face, which is like a support.* ***61***

Proposition 4

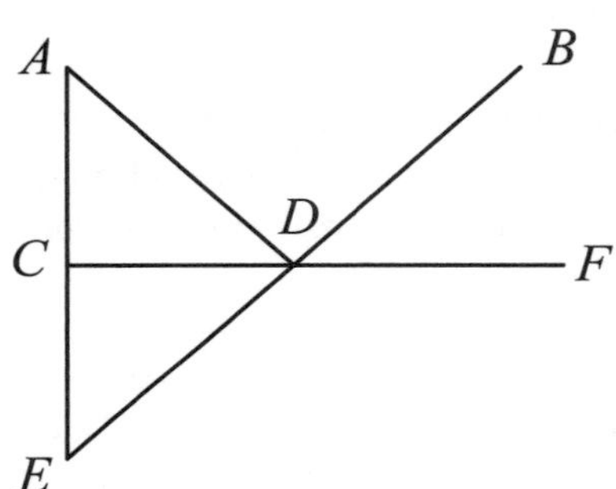

And now this is immediately clear as well, *that it is necessary that the sense of vision be in error in making judgements regarding the direction in the world or the position of an object, when the forms or rays proceeding down from objects, and falling upon a polished surface, are reflected to the sense of vision in the opposite direction.* For when, in the above diagram, the eye B is turned in the direction BD, it cannot become fully conscious of the reflection of the ray ADB that happens at D, by 2 of this chapter. Therefore, it imagines for itself the position of the object at A in the direction BD, namely, at E: thus the image is, as regards place, torn away from its object. The same judgement shall apply for refraction. For although here the directions are not exchanged for their opposites, they nonetheless differ perceptibly.

Proposition 5

Again, since vision is a receiving, and receiving is between contraries which are of the same kind, it is fitting that *whichever dispositions of visible things are to the eye common by virtue of their being of the same kind, all those are to a certain extent perceived by the eye with the assistance of the visual faculties and with the mediation of that disposition.*

Proposition 6

First, the *eye* consists of transparent humors, and so in this respect *it is capable of receiving light and colors.* See Witelo III 59.

---

23 The word Kepler uses here, *valuae*, was used in classical Latin to denote "A double or folding-door (esp. in a temple, palace, or sim.)" (*Oxford Latin Dictionary*, p. 2009), although by Kepler's time it was commonly used for the "valves" of the heart.

Proposition 7

Next, its shape is round, and it varies in the different coverings and humors, within and without. But this visible world is itself concave and round, and whatever we behold of the hemisphere or greater with a single fixed gaze, is a part of this roundness. It is therefore fitting that *the ratio of individual objects to the whole hemisphere be estimated by the sense of vision, in the ratio of the entering form to the hemisphere of the eye.* And this is what is commonly called the visual angle, or the vertex of the visual pyramid within the eye, whose base is in the object itself. For in any single gaze, the eye becomes the center of the visible hemisphere. Therefore, it is by means of this surface of its instrument (whether it be internal or external will be stated below in the consideration of the eye) that the strait of vision measures the angles of viewing. For although all the solid angles are in a point, which is the center either of the eye or of a particular one of its coverings, these nonetheless cannot be distinguished in a point: thus a surface is required, by which the eye may measure the solid angle, as is evident from the
62 geometrical writers. This round shape of the eye is therefore itself sufficient that among the principles we might know, by rough estimation,[24] that *the eye has a sense of the angles set up about it.*

Proposition 8

Thirdly, since to each animal a pair of eyes is given by nature, with a certain distance between them, by this support the sense of vision is most rightly used to judge the distances of Visibles, provided that that distance have a perceptible ratio to the distance of the eyes. For if the excess be enormous, that distance is apprehended no differently than any other you please, an unlimited number of times greater. For here it is simply the geometry of the triangle, as is more amply discussed below concerning parallaxes.

For, given two angles of a triangle, with the side between them, the remaining sides are given. In vision, the *sensus communis*[25] grasps the distance of its eyes through becoming accustomed to it, while it takes note of the angles at that distance from the perception of the turning of the eyes towards each other. Now when an object is so far removed that the distance of the two eyes vanishes in comparison with it, the axes of the eyes have a nearly parallel direction. But to the extent that the object is closer, the eyes will be more turned towards each other.

Now in catoptrics we are determining the principles of the place of the image, not only when we employ both eyes, but also when we are using only one. Nonetheless, this faculty of distinguishing distances, which first arises from this companionship of the two eyes, is later derived for single eyes also, through other dispositions of the eye.

---

24 *Crassa Minerva*, literally, "thick Minerva."

25 The power of perception common to all the senses. Aristotle writes, "There is also a common faculty associated with [all the senses], whereby one is conscious that one sees and hears . . . , and this is closely connected with the sense of touch." (*On Sleep and Waking*, Ch. 2, 455a 16).

That is, it is like what I said before (using rough and everyday imagery, which I shall refine below in ch. 5): that *the ratio of the object to the hemisphere* is grasped by the eye through the solid pyramid whose vertex is at a point of the eye, and whose base is in the seen object. Similarly, now, when we are estimating *the distance of the object*, the triangle becomes nearly isosceles: its point or vertex is (in contrast) in some point of the seen object, and its base is in the distance of the eyes. For the ratio applies to all dimensions of the surface viewed, while the distance is only considered in a straight line. Hence in the former instance a solid pyramid is required, and in the latter a plane triangle.

Note and Appraisal

The following propositions 9, 10, 11, 12, 13, 14, are a bonus: the main demonstration could stand without them. Experience bears witness that the image **63**
is on the perpendicular, even when we use one eye; with the necessary result that even with one eye the distances of points would be grasped. But although this ability does not lie in a single eye, for the following reasons, still it is consistent that it be there solely because of the motion of the head, by which motion a single eye stands in for two that are far apart.

Proposition 9

That distance-measuring triangle can also be considered in one eye: the vertex is in a point of the seen object, the base is in the breadth of the pupil, and in that diameter of the pupil which coincides with the line joining the points of the two pupils. Thus the one eye will learn this procedure for setting up the triangle from the two eyes, by becoming accustomed to it, but in this ratio, that if, because of the distance of the eyes, the sense of vision discerns the distance of those things that are a hundred paces away, by the breadth of the pupil it would approach to ten paces, or smaller distances. Although there are also other reasons why a single eye may measure distances, it does so through that diameter of breadth rather than through another diameter of depth. For although on the inside the eye is everywhere circular, on the ouside the folding doors of the eyelids are divided along the line of distance of the two eyes, parallel to the horizon. For Nature wished both the distance itself of the eyes, and also this opening of the folding doors as well, to be equidistant from the horizon. She did not at all wish one eye to be higher than the other, or one angle of the eyelids to be higher than the other (in animals of the most perfect vision, that is). This is because the plane of the horizon (coming into the eyes by means of its color) makes the greatest contribution to the imagining of the measuring triangle, that is, of its plane, to see whether the one is parallel to the other at its base. For this reason, when we gaze upon things with our head inverted or at an angle, we use the discriminating faculty in a more muddled way.

Proposition 10

For that reason also, *in convex mirrors, and in the flat surface of denser media, we strive, with nature as guide, to make the rays from both eyes strike upon the surface with equal angles (or vice versa, for geometrically it is the same thing).* For this faculty of setting up the measuring triangle is common to the

two eyes together and to each eye separately. There is a certain coordination of
64 motion, of such a kind that with the turning together of the axes of the two eyes there arises simultaneously some new disposition in the individual eyes (for the sake of the argument, and in a rather general sense, let it be granted for now that the *uvea tunica*[26] in the individual eyes is either wrinkled up or stretched out). Therefore, it happens that when the two eyes are directed at the same surface, whether reflective or refractive, and one of them is higher, the wrinkling of the uvea must necessarily differ, both from the turning together of the axes of the two eyes, and from the wrinkling of the other uvea, because of the difference of refraction, as will be said below.[27] However, if one tries this, first, he obtains it with difficulty, for he again and again perceives two images in place of one; and second, he does great injury to his eyesight, and induces headaches.

Proposition 11

However that may be, the external configuration of the folding doors of the eyelids is a sign rather than a cause of this faculty in the individual eye: one must reflect on the internal and authentic cause. Moreover, I have just given one (the wrinkling up of the uvea and the contrary dilation), which would be sufficient, but for the assertion of the Physicians that this motion in this tunic exists because of the abundance or weakness of light, for which see ch. 5 below.[28]

And so, fourth,[29] it should be considered that in the humors of the eye there is a certain ratio of density and tenuity, well known to the eye itself on its own. Thus, since both air, by whose mediation vision takes place, and light, which is seen, have established their ratios of density above, *it is fitting that there be in the eye the power of measuring either the density or rarity of both the air and the light.*

Proposition 12

But we shall steer into the light a number of arguments gathered from the air. For the eye seems to be nearly cut off from the function of seeing when the transparent medium matches the humor of the eye, as when the eye is submerged beneath water. Thus it is made completely albuminous in the kind of animals capable of breathing, for perceiving objects in air, not in water. For that reason, fish are endowed with a thicker and harder humor, almost bone rather than humor, in order to see through the waters. It is therefore reasonable to believe

---

[26] The "grape-like tunic." However, what Kepler means here is not the uvea as a whole, but the posterior part of the iris, which "is sometimes named *uvea*, from its resemblance in color to a ripe grape." (*Gray's Anatomy* (Philadelphia: Running Press, 1974), p. 831). Kepler's description of the uvea, derived from the anatomical studies of Felix Platter, is in Chapter 5 below, p. 177.

[27] The change in dilation of the pupil with convergence of the eyes, first noted by Kepler, is known as the "convergence reaction." It is described in standard physiology texts, e.g., Barbara R. Landau, *Essential Human Anatomy and Physiology*, p. 249.

[28] See p. 187

[29] Cf. propositions 6–8.

that animals of the highest rank[30] have an albuminous humor in the eye for the very reason that there is a difference in the transparent media of the air and the eye. For had she not kept an eye on this, Nature would have been able to send light in through the open hole in the uvea. The same thing, however, may also be brought to light from the shape of the eye. For since it is round, many things happen when the luminous rays enter refracted from the air into a denser round 65
body, which could not occur when the eye is flooded with waters, where, when the albuminous humor matches the surrounding medium, the round surface of the albuminous humor is nullified. At least, therefore, there is in the eye a perception of this airy thinness, in that way in which we have a deeply infixed perception of all the objects to which we have become accustomed. See Witelo book 3 prop. 3 and 4.[31] Above, however, in prop. 6 of the first chapter, we attributed to both light and the surface of the air the same condition of density: therefore, *the eye too will perceive the density of light.* And this happens all the more because, as was made plain in prop. 20 of the first chapter, in relation to this density, light receives something from all media that differ among themselves in density or transparency, and thus also receives something from the humors of the eye: why therefore should the eye or sense of vision in general not also receive something contrary from this density of light, likewise in relation to its own density?

Proposition 13

Fifth, if (by the preceding) the eye perceives the density of light, there must also be a rarefaction and condensation of it,[32] *so that it may sense the attenuation that in fact is going on inside the eye and in a way happens to it*, since the humors in addition are extended in depth, through which, by Ch. 1 Prop. 6, this rarefaction is carried out. For while light passes through this depth, it is spread out in a certain proportion (other things being equal, and ignoring the compaction that occurs through the configuration of the denser humors), and comes out more attenuated in the depth that it was when it first entered. Nevertheless, the eye might also be able to determine the rarefaction of light that occurs within the eye by using the evidence of the broader illuminated surface in the depth, since by Prop. 7 above we attributed to it a perception of the quantity of the surface touched by light.

Proposition 14

With these things thus proved by the four preceding propositions, I would now like to persuade you, with a geometrical demonstration of what was established only probably, *that by a single eye, with the assistance of the higher faculties, and through the actions upon the eye's humors by the density of light, as by an agent, the distances of points of a seen object, to which distances the diameter of the eye is perceptible, are perceived and compared.* For, first, any

---

[30] *Primas.*

[31] Witelo 3.3: "The organ of the visual power must necessarily be spherical." 3.4: "The eye is the spherical organ of the visual power, composed of three humors and four tunics, arising from the substance of the cerebrum, intersecting each other spherically."

[32] The antecedent of this pronoun is also uncertain in the Latin.

point whatever radiates in an orb, by 3 of the first chapter, and thus it also radiates in the breadth of the eye, which latitude, be it of the whole eye or of the
66 opening of the uvea, is by hypothesis not imperceptible. Next, the sense of vision perceives both the density and tenuity of the images or the light that has arrived from the point through the rays, by 12 of this chapter, and the quantity of the attenuation of these same things within the depth of the eye, by 13 of this chapter. Further, it grasps by habituation the depth of its own humors. Let αβ be the diameter of the pupil, γδ parallel to it, as far away as the depth of the eye allows. Next, from the boundaries α and β, let αγ and βδ be spread with a lesser inclination, αε and βζ with a greater one; and therefore, γα and δβ will meet at a more distant source θ, while εα and ζβ will meet at a nearer one η. For since the inner angles αγδ, βδγ are greater than the outer ones αεζ, βζε, when, by Euclid I 32, they are subtracted from two right angles, the remainder γθδ of the former will be less, while εηζ, of the latter, will be greater. Thus, by Euclid I 21, γα, δβ will meet at a more distant point θ, and αθ or βθ will be greater than αη or βη. Therefore, since the eye grasps the diameter αβ and the depth αγ, and since it is sufficient to observe the ratio of δγ to γε and that of each to αβ, either by the attenuation of light, or by the very small illuminated parts of the inner surface, it will consequently observe αη and αθ, not, indeed, by numbering, but by comparing the distances of the object through this habit, as it were, with the powers of its body, and the extension of hands and of paces.

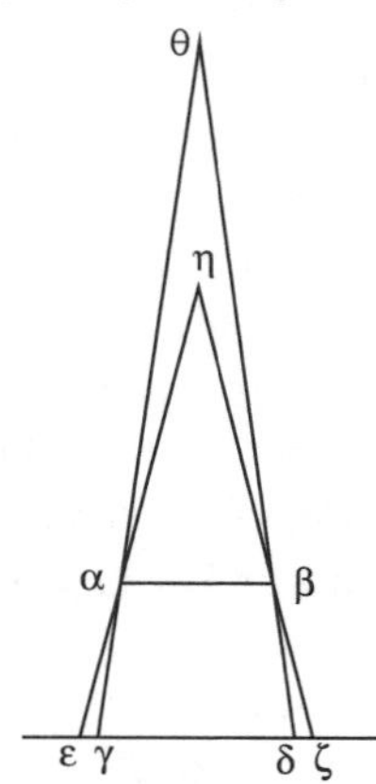

These things are sufficient to establish that those things, which we are now just about to demonstrate concerning the partnership of the two eyes, be understood as demonstrated also for the diameter of breadth of the single eye: which is what we had proposed above.

Proposition 15

And the comprehension of color, direction, and distance has been discussed: what remains is *the legitimate comprehension of quantity*, which I would account for in one word. For it follows upon the comprehension of the angle and of the distance, where the sense of vision, from the sides that come together at the eye and the angle set up between them, makes a judgement about the base of the pyramid, which is the quantity of the object seen. Now
67 let us make a nearer approach to the subject of the place of the image.

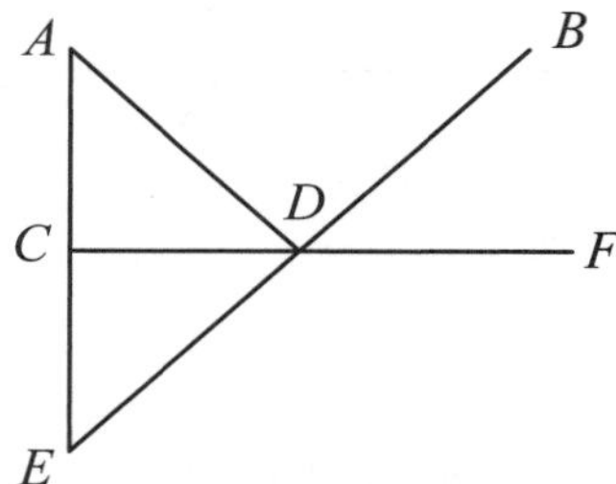

Definition 2

The surface of reflection or refraction, the Optical writers define as the surface determined by three points (for it is posited that a unique plane surface is always applied to three points not placed in the same straight line): *the center of vision, a point of the seen object, and the point of reflection (or of refraction).* In the above diagram in No. 1, it is determined by the points D, A, B.

Proposition 16

This kind of surface, they demonstrate, *is necessarily erected perpendicularly above the reflecting or refracting surface.* Witelo Bk. 5 prop. 25; Alhazen Bk. 4 no. 14.[33] The former carries out the demonstration using prop. 20, and the latter using no. 10, of the same books, namely, that the angles of incidence and reflection are equal, and they at length derive the cause from nature's acting through shorter lines, which they had absorbed from Euclid and Ptolemy.[34] And indeed, these operations are not those of a form that acts deliberately or keeps a goal in mind, but of matter bound to its geometrical necessities. Furthermore, we also perceive the images of stars in ponds by the same laws of reflections, where the comparison of lines completely vanishes.[35] Finally, if the demonstration were genuine, it would also follow in the consideration of refractions. However, Witelo, Bk. 10 prop. 1 and 2, and Alhazen Bk. 7 no. 9, change the form of the demonstration here, for it is ineffectual here to argue thus from least lines.[36] For us it is enough to establish that reflection takes place in the opposite direction, and refraction towards the perpendicular drawn from the radiating point to the refracting surface. The former is from Ch. 1 Prop. 19, and the latter from Prop. 20 of the first chapter. Each of these we deduced from its own proper cause, from which also followed at the same time the equality of the angles of incidence and reflection. Here we shall introduce in a general manner the argument of which Witelo makes use for refractions in particular, insofar as it can be gathered from that darkness.

Now, since directions are arranged in an orb, let there pass through the point of reflection A the pair of straight lines BC, DE in the reflecting surface, indicating the four directions.[37] Now let the surface of reflection $BFGC$ pass through
one of these lines BC. If this is inclined **68**
on the reflecting surface $DCEB$, it will as a result be inclined to one part of the second line DE. Let it be inclined towards D. Therefore, both the incident FA and the reflected AG will veer towards one direction D of the second line DE, which,

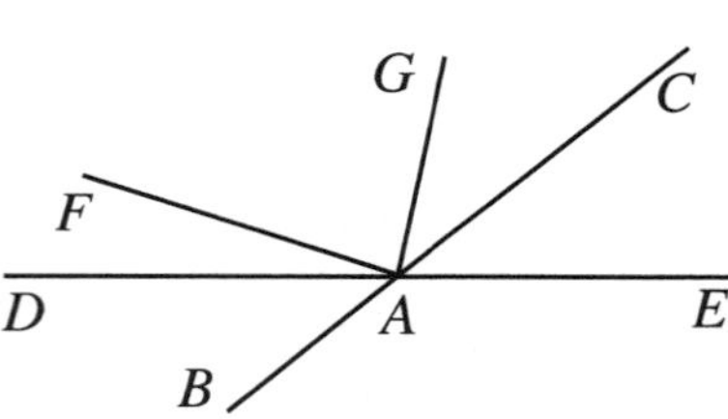

[33] *Thesaurus* II p. 202 and I p. 111, respectively.

[34] Kepler did not know of the *Optics* of Ptolemy. It was cited by Ambrosius Rhodius in 1611, and then lost until it was rediscovered by Laplace in Paris. (Cf. Hammer's note in *JKGW* II p. 440.)

[35] That is, the star is so far away that a change in the angles of incidence and reflection will make no difference in the distance travelled by the light.

[36] *Thesaurus* II pp. 405–6, and I pp. 242–3.

[37] Note that BC and DE define the surface of the reflecting medium; thus, BC should be imagined as coming out of the plane of the page perpendicular to DE.

as we said above, should not happen. For it is required that FAD and GAE be equal, no less than FAB and GAC. Therefore, it is necessary that the surface $BFGC$ be on the perpendicular to $DCEB$, so that the angle of incidence FAD veers to one direction of the second line DE to the same extent that the angle of reflection GAE, equal to it, veers towards the opposite part E of the second line DE.

In refractions, the same thing is evident more simply. For since refraction only occurs towards the perpendicular, the line CB, in the diagram of proposition 20 of the first chapter, which is drawn in the surface of the denser medium HC from the point of incidence C, perpendicular to AC, to the point of refraction B, will extend parallel to that which, coming forth at right angles from the extended perpendicular AC at E, at any depth whatever in the denser medium, intersects the refracted line BG. Unless this were so, the refracted ray BG would not approach the perpendicular BF or AE directly, but would at the same time veer in other directions. Therefore, by Euclid XI 7, AC, perpendicular to the refracting surface HC, is in the same plane surface in which the refracted line BG also is. And the plane surface in which the refracted line is, is called the "plane of refraction." Therefore, the plane of refraction, since it contains AC, perpendicular to the refracting plane HC, will be perpendicular to the same refracting plane, by Euclid XI 18.

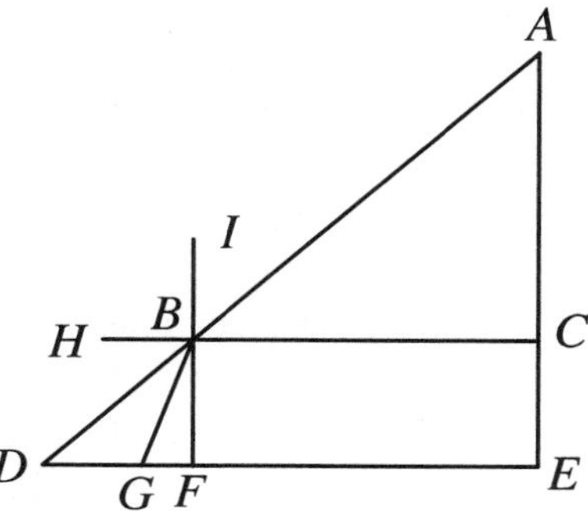

Proposition 17

There was a further difficulty in establishing principles. The demonstration itself is straightforward. First, the sense of vision errs in direction, as was said in prop. 14 of this third chapter: it imagines for itself an object in the same direction whence the refracted or reflected ray approached. Next, the sense of vision also errs in the angle. For it imagines for itself that the inclination by which the refracted or reflected rays proceed all the way to the centers of the
69 two eyes, is also the same as the inclination or angle by which proceed those rays which approach from the radiating point to the points of the reflections or refractions, corresponding to the eye, by 2 and 7 of this third chapter. And the genuine place of the image is that point in which the visual rays from the two eyes meet, extended through their respective points of refraction or reflection, by 8 of this third chapter. Now the visual ray of either eye (the luminous line drawn out by the imagination from the eye continuously through the point of reflection or refraction) is in the same surface with the surface of refraction or reflection, by def. 2. And either eye has its own surface of reflection or refraction, by the same. Therefore, where the visual rays of the two eyes meet, the surfaces of the refractions or reflections also meet, each one passing through its respective eye. Therefore, since the place of the image is in the meeting of the visual rays, by the first definition, it will be in the meeting of the surfaces of refraction or reflection of the two eyes. But those surfaces meet at a visible point [of the

object], by definition 2. And since these two surfaces of refraction or reflection are perpendicular to the refracting or reflecting surface, by 16 of this third chapter, and mutually intersect each other, their common intersection will first of all be a straight line, by Euclid XI 3, and consequently the object seen and all its images will be in the same straight line, since all are on the common intersection. Furthermore, by Euclid XI 19, this common intersection will be perpendicular to the same refracting or reflecting surface. *Therefore, again, all the images of the seen object will be on the perpendicular from the object to the surface, whether refracting or reflecting; and this will happen to such an extent that the distance of the points of the seen object is grasped in the manner described, whether by the two eyes, or by the diameter of the breadth of one eye.* And since (in particular) the surface cutting a sphere perpendicularly passes through its center, the common section of two such, that is, the line of the images and of the visible object, will therefore also pass through the center.

Therefore, it is not the occult nature of light, not the mind of universal nature, but the breadth of the sense of vision alone, that harmonizes with the causes for the sense of vision's placing the image on the perpendicular. **70**

Proposition 18

If you were to apply the sense of vision otherwise, so that (contrary to nature) the refractive or reflective[38] *surface were the same for the two eyes, an entirely different thing would happen. For in convex mirrors and in denser media the images will depart from the perpendicular and will approach the observer.* Which, as it is readily to be explored by anyone, and has also been proven to me through experience, I thus shall now demonstrate it in the cause of light.

First, let EG be a convex spherical surface with center L; a radiating point at D; the vision on CH, collinear with L so that the surface of refraction through the two eyes C and H is the same. Let the points of reflection be E and G. Let DE, EC, DG, GH, be joined, and let CE, HG be extended until they meet. Let the point of meeting be S. Also, let lines be drawn touching the circle at G and E, and let them be AO, OK; and let the points G and E be joined with a straight line, which shall be extended in both directions to M and N. Now, since DEK is equal to angle CEO, while DEN is less than angle DEK, therefore DEN is also less than angle CEO, and DEN will be doubly less than the greater angle CEM. And since DGO is equal to angle HGA (by the law of reflection), while DGE is **71**
greater than DGO, therefore DGE is greater than HGA, and doubly greater than the lesser angle HGM. But EDG is the difference between the greater reduced angle NED, and EGD, the lesser but made greater.[39] On the other hand, ESG is the difference between the greater CEG, made greater by the same measure, and

[38] *Refractoria vel repercussoria.* The terminology is becoming confusing, if not confused. This surface is clearly the same as the "surface of reflection or refraction" of Definition 2, as distinguished from the "refracting or reflecting surface" (Proposition 16, enunciation).

[39] That is, NED is less than the angle of incidence DEK (and hence is "reduced," while EGD is greater than the angle of incidence DGO (and hence is "made greater").

the lesser HGM, reduced again by the same quantity. Therefore, ESG is much greater than EDG. And ESG is the same angle as CSH in which the reflected rays CE, HG meet when extended, and S is the place of the image of point D, by 8 of this third chapter. But EDG is the angle with which the rays depart from the point D. Therefore, the angle at the image is greater than the one at the radiating point. Therefore, the image will appear to be nearer than the radiating point, by Euclid I 21. I say that S is between DL, the perpendicular from the object to the center, and the perpendiculars through E and G, the points of reflection.

For let ES be extended until it cuts DL; let it cut at T; and let the points T and G be joined and let the place of the image be at T. Further, let GO be extended until it cuts DL at V. Also, from the center L let straight lines be projected through the points E and G indefinitely out, and afterwards let a parallel to TE be drawn from D to LE, cutting it at X. Likewise, let a parallel to TG be drawn from D cutting LG at Y.

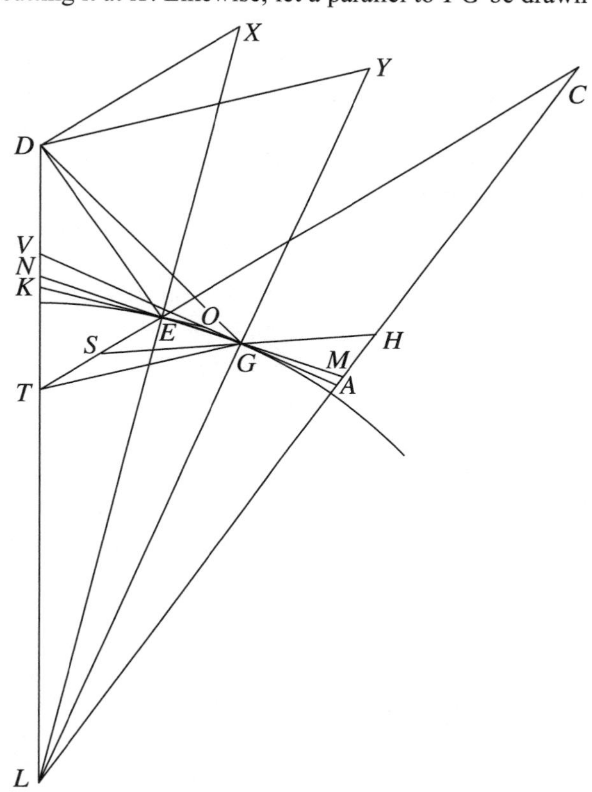

Therefore, if T is the place of the image, it will follow not only that the angle DET is cut in two by EK, the tangent to the circle at E, but also that the same thing is done to DGT by GV, the tangent to the circle at G, because DEK and KET are each equal to angle CEO. Likewise, DGV and VGT are each equal to the single third angle GHA. Consequently, since XEK and KEL are also equal, when equals are subtracted, the remainders XED and TEL will be equal. But TEL and DXE are equal, because DX and TE are parallel. Therefore, DXE and DEX are equal, and so also are DX and DE, opposite the equals. Therefore, in triangles DLX and TLE, the sides are proportional, by Euclid VI 4. Consequently, DL is to LT as DX (that is, DE) is to ET. But by Euclid VI 3, DK is to KT also as DE is to ET. Therefore, DK is to KT as DL is to LT. But by the same consideration it will be demonstrated that YDG is isosceles, and DL is to LT as DY (that is, DG) is to GT. And because, by hypothesis, DGV and VGT are equal, DV will be to VT as DG is to GT. Therefore, DV is to VT as DL is to LT. Previously, however,
72 DK was to KT as DL to LT. And, alternating, as the greater, DK, is to the less,
DV, so is the less, KT, to the greater, VT, which is obviously absurd. And so it

is false that the vertex S of the visual pyramid CSH, whose base at the observer is CH, is on the perpendicular DL, or that it goes all the way to T, as we had previously supposed, and from which this absurdity followed. And thus this is true of all inclinations that are between the perpendicular and the horizontal. But the perpendicular radiation itself, reflected to the sense of vision, moves it to think that the image is on the perpendicular, while the horizontal radiation moves the image away from the perpendicular DL more and more towards the points E and G, to such an extent that there is given a position where the line HG from the lower eye, extended through the point of reflection[40] G, cuts the circumference GE again, before it meets the line CE drawn from the other eye to the other point of reflection.

From these things it thus appears not to be universally true that the place of the image is on the perpendicular, unless this restriction also be added, that the sense of vision be so located with respect to the mirror as nature shows.

In concave mirrors the same is demonstrated in a contrary manner. For judgement concerning contraries is the same, other things being equal.

There is one further thing I should add. Whatever of this that can be gathered by a single eye, is either imperceptible, when the arc of the circle of reflections, varying the visual angle, is always less than the actual pupil of the eye, and thus the difference between the two reflections is extremely slight; or, if the two eyes were to come so near the mirror, and were to be so greatly inclined, that some large arc lie between the points of reflection, the sense of sight is hindered, as was said before, and impaired, because the proportion between the axes of the two eyes and the narrowing of the opening in the uvea has been disturbed.

Proposition 19

It must now be demonstrated that *the same thing also occurs in refractions which experience most plainly declares: that when both eyes are in the same surface of refraction, and are observing from a very oblique angle, the image departs from the perpendicular and approaches the eyes, provided that the space intervening between the eyes be moved near enough so that a perceptible difference is
interposed between the refractions at the two eyes.* **73**

Let the circle $EDAB$ cut the straight line AB at the points A and B, and join these to D and E, which are any points of the circumference lying on one side. These connections will be the lines DA, DB, EA, and EB, which shall be extended indefinitely in the directions of A and B, to F, G, K, and I. Thus DB and AE will cut each other: let the intersection be H. Therefore, by Euclid III 31, ADB and AEB will be equal. But DHA and EHB are also equal. Therefore, the remaining angle DAH will be equal to the remaining angle HBE. But DAH and FAK are equal, as well as HBE and GBI. Therefore, FAK and GBI are also equal.

Now, with these things being supposed, let the sense of vision be at F and G, and let all these points by supposition be on the same surface. Let the denser medium be the straight line AB; the point emitting the rays be at E; FAE and

[40] Reading *reflexionis* for *eflexionis*.

GBE be the refracted rays. Thus if the refractions at A and B were equal, as KAF and IBG are here made equal by construction, it is clear from the above that the image of the point E is going to be at D; that is, at the meeting of FA and GB. And since the points of the circumference towards A are always nearer to the perpendicular from A above AB (provided only that they are between A and the point of the circumference where the perpendicular to AB touches the circle), D will always be nearer the perpendicular than E. Therefore, on the supposition that the refractions at A and B are made equal, the image in the denser medium approaches the eye from the perpendicular drawn from the observed point E to the surface BA, doing so near the horizontal, since the angle FAB is small.

For the rest, because the obliquity of the incidence EA to AB is the cause of refraction, by 20 of the first chapter, when the cause increases, the effect will increase, and vice versa. Accordingly, the refraction will be less at A, because EA is less inclined away from the perpendicular, and at B the refraction will be greater, because EB is more inclined. Therefore, since KAF is less than IBG, DAH is also less than HBE. Therefore, ADH is greater than HEB. Therefore, ADH and the place of the image D locates itself inside the circumference, and approaches A nearer than the point D of the circumference, by Euclid I 21. Before, however, D itself approached nearer to the perpendicular from A than the visible
74 E did to its perpendicular on BA. Therefore, the image of E approaches much nearer the perpendicular at A than the visible E itself, receding from the point E. Which is what was to be proved.

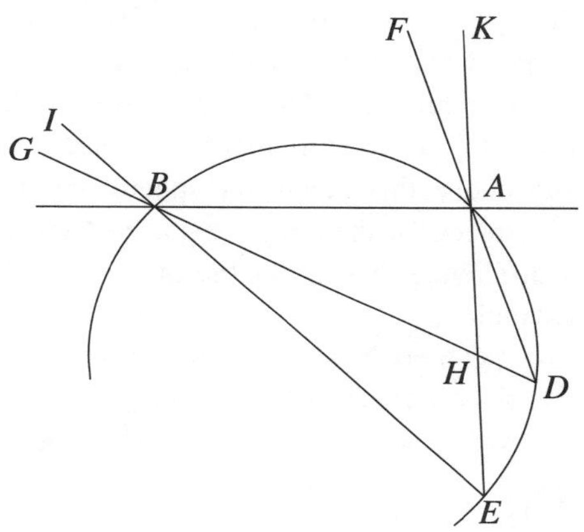

At points of the circumference more remote from A than is the point where the perpendicular to BA touches the circle, it does indeed happen that, on the supposition of equal angles of refraction, the perpendicular from E becomes closer to A than the perpendicular from D. But because the angles of refraction are unequal, D, where the angle is greater, again approaches and forsakes the circumference.

And in general, at this great a proximity to the perpendicular, the whole angles of refraction are slight, and thus the differences of these angles are much less perceptible. Accordingly, there is no departure of the image from the perpendicular that can be grasped by the senses, where the sense of vision perceives the object nearly perpendicularly beneath the surface of the denser medium. For other exceptions concerning the place of the image, see Chapter 5 below.

Proposition 20

Hitherto, then, it has been shown that the theorem of the optical writers is not completely general. Thus at the beginning it was more accurately determined, so that the predicate might be matched to the subject in breadth [of applicability];

and the instances were shown that were exceptions to the universal rule of the optical writers.

But so that you may see that there is nowhere that errors do not sprout up through ignorance, behold now also what sort of difference there is between my conclusions and those of the optical writers constructed upon this theorem. *They thought that the image put up a struggle to make its way by the direct path to the surface. Accordingly, even in conoidal mirrors they give guidance by the example (Alhazen Book 5, Witelo Book 7) of drawing the perpendicular from the object to the surface of the conoidal mirror, quite otherwise than what holds true.* For it makes no difference to the place of the image, what sort of mirror surface is placed opposite the object, since the proportions[41] of image formation are all taken from that part of the mirror upon which are the two points of reflection of light to the two eyes. So it is at this part of the mirror, not at the actual perpendicular from the object, that the cause of the image's place being on that perpendicular lies. And so one should understand mentally the continuation of the pattern[42] of curvature that had created the reflection on the whole circumference,
and above this imaginary sphere one should also draw a perpendicular from 75
the object for defining the place of the image. To make it apparent how great a difference there is in the outcomes of the two opinions, take the following example.

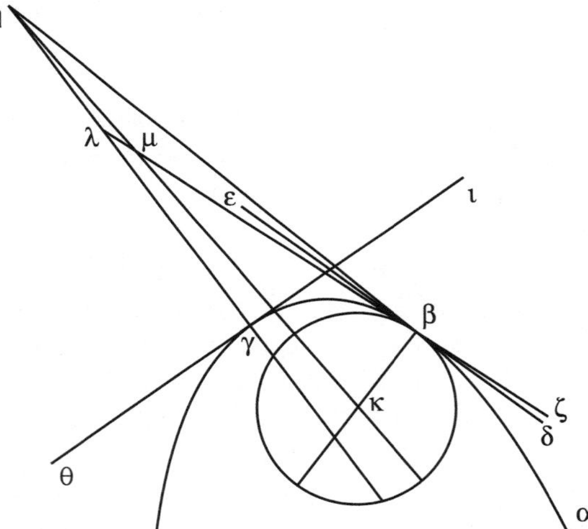

Let there be a parabolic section, the common section of a conical mirror and the plane surface of refraction (by the definition of a parabola in Apollonius), or even [the common section] of a rectangular conoidal mirror [and the surface of refraction] (by Archimedes, *On Conoids*, prop. 12). Let the resulting section be αβγ, and let the point β on it show the place where each eye (although in a pair of such sections) receives the refracted rays. Accordingly, let the section then be touched at β by the straight line εδ. And let the sense of vision be at ζ, and the visible at η, so that ζβδ and ηβε are equal. If you seek here the place of the image from the optical writers, they direct you to seek the point of the section upon which falls the perpendicular from η: let this be γ. After θι is drawn touching the section at γ, and the points η, γ are joined, they direct you to extend ζβ until it meets ηγ: let it meet at λ. They will then say that λ is the place of the
image of the point η. But a truer reason directs you to find the circle which 76

[41] *rationes*.

[42] *ratio*.

contains the pattern[43] of curvature that the section has at the point of reflection β (mixed lines have, moreover, one and another of this sort of circle[44]). Let its quantity be κβ, and, a perpendicular to εδ being drawn from β—call it βκ—the center of the circle will be placed upon the line βκ, ηκ will be joined, and the place of the image will be where ζβ extended cuts ηκ, namely, at μ. Here you easily see that even if β remain the place of reflection, and κβ also remain the pattern of curvature, and ζ the eye, while the visible[45] η moves farther away on line βη, the perpendicular from η is always going to fall upon a point farther away from γ, and the place of the image will be correspondingly farther from λ towards the outside. And indeed, this difference finally goes off to infinity. Now it very important for the account of mirrors[46] not to be kept suspended in doubt here.

But enough on the place of the image, which consideration was to be sure entirely necessary for what follows. Be indulgent, O reader, if we have anywhere been more concerned with the Paralipomena to Witelo than with astronomy. I know that the natural method requires that in the beginning the nature of light and its modifications, reflection and refraction, be treated; second, the eye, which depends upon refraction; and finally, catoptrics, or the image, which is the companion of the visual faculties. Had I followed this method, I would have been able to forego in this chapter many things that are unpolished and that are taken as known by anticipation, and even the coherence itself of the propositions would have been established with tighter bindings and a more geometrical form. But it seemed to me preferable for gaining the reader's trust to follow the actual series of my investigations:[47] at the same time to place these catoptrical paralipomena in subjection to the explaining of refractions, and so to drag these as well to some extent into astronomical servitude. Otherwise, contrary to the book's purpose, they would be set up on a royal throne, and would appear to embrace the full scope of the whole work, which is what would have happened had I followed the natural method.

---

43 *rationem.*

44 Lines may be categorized into straight, curved, and "mixed", which contain something of each of the other kinds: they are curved, but their curvature is not constant (cf. Kepler, *Mysterium cosmographicum* ch. 2 and Aristotle, *De Caelo* I 2). Hence, the circle of curvature (a concept which Kepler appears to have invented here) is different at different places.

45 Reading *visibile* for *visile*.

46 *Rationes specularias*.

47 This is similar to Kepler's professed method in the *New Astronomy*, on which he was working at the time of writing the *Optics*. There he wrote, " The scope of this work is not chiefly to explain the celestial motions, for this is done in the books on Spherics and on the theories of the planets. Nor yet is it to teach the reader, to lead him from self-evident beginnings to conclusions, as Ptolemy did as much as he could. There is a third way, . . . that is, an historical presentation of my discoveries. Here it is a question not only of leading the reader to an understanding of the subject matter in the easiest way, but also, chiefly, of the arguments, meanderings, or even chance occurrences by which I the author first came upon that understanding." (Donahue trs. 2015, p. 41).

Chapter 4
On the Measure of Refractions

It has been important to astronomy, for the attainment of certainty, to establish the angles by which the rays of stars are refracted from the straight path. It is now also important for the attainment of beauty, to know the causes of this incre-
77 ment of angles. Nor, indeed, is it without use to see whether there is something of certainty underlying this matter, so that we may dare to make pronouncements all the more confidently about whether the refractions are the same at all places. Now that this is attained, very important observations of the ancients must now be treated far differently than if no consideration were given to refractions.

1. On the Debate between Tycho and Rothmann upon the matter of refractions

A long time ago, Alhazen the Arabian,[1] and Witelo, following him, set out to explain the matter of refractions more carefully than the ancients had been accustomed to do. And since all our ideas derive originally from experience, they began by investigating with instruments the quantities of those angles with which the rays entering water from air are refracted, then both those from air to glass, and those from water to glass. And since the matter of the heavens, from the opinion of the ancients, was thought to be nearly glassy, that is, crystalline,[2] while air was akin to water, the authors, carried away with boldness, began with the aid of refractions to seek into the heavens' intimate secrets. Experience aided their attempts: a certain ratio of refraction was discerned even in the stars, and it was such that from it, by those experiments that were now confirmed in water and glass, it appeared to be possible to declare that aether is not denser than air, but much rarer than it. This carefulness was long neglected, but after several centuries Tycho Brahe arrived on the scene, who set out to measure with exceedingly accurate instruments the angles of refraction in air, a matter that Witelo had neglected. Contention about this as well as about many other discoveries arose between him whom I have mentioned (Tycho) and Rothmann, the Mathematician to the Landgrave of Hesse. There is much controversy on

---

1 Kepler is evidently thinking here of the author of the book whose Latin title is *Liber de crepusculis*, which was written, not by Alhazen, but by Abu 'Abd Allah Muhammad ibn Mu'adh. Cf. Lindberg, *Theories of Vision*, pp. 209 and 281, and the sources cited therein.

2 This may be the earliest published statement that the ancients believed the matter of the heavens to be crystalline. Although the Latin word *crystallinum* was used in earlier astronomical texts, it denoted specifically the ninth sphere, the *coelum crystallinum*, which represented the biblical "waters above the heavens," which, following Ezekiel 1.22, were thought to be congealed into a "terrible crystal." Aristotle, whom most medieval writers followed, believed the heavens to be made of a substance that has nothing in common with any ordinary material; cf. *De caelo* I.2–3 and Grant, *Planets, Stars, and Orbs* pp. 324–370.

refractions in the *Epistolae astronomicae* vol. 1, which Tycho published in 1597,[3] and anyone who wants may find it there. For the present, I shall add a summary. Tycho had previously given a notice about being aware of refractions when taking the sun's altitudes. In passing, he assigned the cause to the differences between air and aether, as he had read in Witelo. Rothmann seized the opportunity, and denied that there was any difference between air and aether by the very fact that no refraction occurred that could truly be assigned to this cause. For he said that
spheres intersect in a circle, even supposing they are different; refractions, as **78**
is in accord with this, do not occur throughout the entire circumference of the circle, but only near the horizon. Refractions therefore do not come from this difference of spheres, but from a lower cause, which does not exceed an altitude of twenty degrees from the horizon. The procedure of enquiring into twilight[4] (which accompanies the sunset up to a depth of 18 to 24 degrees) was helpful. He concluded, as a result, that the rays of the stars were bent in the matter that provides the occasion for twilight, which natural philosophers[5], with the geometrical art, prove not to exceed an altitude of twelve German miles.[6]

Tycho was not neglectful of his case. First, he stipulated[7] that the cause of the refractions appeared to him also to be twofold: the first cause arises from the difference between the denser medium of air and the rarer medium of aether, and the second, which drives refractions near the horizon so greatly downwards, does indeed reside in the vapors about the horizon. And he said that it is not sufficient, for blurring the distinction between air and aether, to demonstrate that no refractions of rays occur near the zenith, at least none that can enter into sense perception; for there can be other ones that are imperceptible, which would be rather small even at the horizon, were they not assisted by some other cause. But it grows suddenly when the stars approach the horizon, because of the vapors, under whose control the guiding principle of this cause is. Or indeed, from the very thing of which Rothmann made note, it appears that the transit for rays of the setting stars through the air 12 miles high, is six times as far as what it is for stars at an altitude of 30 degrees. So he thus far granted that

---

3 The *Epistolae astronomicae* was published in Uraniburg in 1596 (not 1597, as Kepler states). It is reprinted in *TBOO* vol. VI. Kepler's summary does not exactly follow the sequence of the published text; however, references to this edition for some of the more important points are provided below.

4 *Crepuscula.* Of this, Kepler writes in the *Epitome astronomiae Copernicanae* I (Linz, 1618) Part 3 p. 72 (*JKGW* VII p. 64), "For astronomers, this is that entire time that comes between the first lightness of the air noticeable by the sense and the actual sunrise; or, in turn, between sunset and the last trace of daylight in the air." The part of Rothmann's letter summarized here is in *TBOO* VI pp. 111–2.

5 *Physici.* To translate this as "Physicists" would be misleading, since Kepler is referring to those who, in the Aristotelian tradition, study φύσις, or nature.

6 The German mile was reckoned as five Roman miles, and hence as about 7.4 km. Tycho took the figure for the height of the atmosphere from the *Liber de crepusculis*, Prop. 6 (for this book, see the footnote on p. 93 above).

7 Letter of 17 August 1588, *TBOO* VI pp. 136–7.

for measuring the angles of refractions, the depth and the redoubled vapors, set obliquely to the stars, combine. This he repeated on pp. 92 and 95 of the *Progymnasmata*, with, however, an added correction.[8] Here he also opines that some very small refractions are removed from sight by the exceedingly great distance of the heavens, a view which is somewhat foreign to the way refractions work. Rothmann responded[9] that this is not at all the way that vapors become the cause of refractions. For he said that on the perpendicular they are 12 miles in depth; on the horizon, from the side, twelve times deeper; at an altitude of 30 degrees, six times. It will therefore happen that at an altitude of thirty degrees the
79 angle of refraction will still be one sixth of the horizontal refraction (which even Tycho himself would condemn, since he had measured these angles otherwise from experience), and furthermore, even right near the zenith some refraction will occur. And thus, following his own opinion, he presented another way that established vapors as the cause for refractions: there is a determinate space for the rays of the stars such that they can pass through the vapors unaffected, and so if they arrive at some place of the terrestrial surface through the vapors by a path that is shorter than is this space, they are minimally refracted. But so far as the passage from the side exceeds the established space, the rays are fully refracted. He hoped that in this way he had brought the matter to a satisfactory conclusion. Much was said pro and contra, and the dispute is so involved that I barely could disentangle myself. The opinion in which Tycho remained after this discussion with Rothmann, you have in the *Progymnasmata* Vol. 1 p. 92.[10] For the rest, as usually happens among the principles of things to be established, the channel is blocked for each. For if they had applied the genuine measure of refractions, it would not have been necessary for Tycho to introduce a twofold cause of refractions, by which I mean the two bodies, the one of air and the other of vapor; nor would Rothmann have denied that light is refracted by some imperceptible amount, even near the zenith.[11] Finally, it would have been apparent that the

---

[8] *TBOO* II pp. 77–8.

[9] Letter of 19 September 1588, *TBOO* VI p. 151

[10] *TBOO* II pp. 77–8.

[11] Here Kepler has the following note.

To p. 79 and p. 83, and to prop. 9 from p. 116 to 120. Vapor as a whole is of two kinds. One, when it arises from the bowels of the earth, conveyed upwards by a certain terrestrial heat; for then it both has the nature of ascending, bursting forth from mountainous lands like a fountain, and from the height of the bursting forth it is poured down in a circle upon the downward slopes as soon as it is affected by the mountain cold, and most of the time generates winds in this way. And thus in this state it is transparent, like water. For that reason, the lights of the stars then appear large and twinkle a lot: by this sign, winds, and the relaxing of cold in winter, are known to be at hand.

The other sort is of a cohering, cooled vapor, which stays inert in one place, mixed with thicker smokes, through which the lights of the stars appear dark, ruddy, and (as Virgil says [*Georgics* 4.183]) 'somber colored', which is by no means transparent.

The former kind is rare and ephemeral; the latter common and perennial. The former

surface that refracts the rays is not that of vapors blindly wandering about, nor that of some exalted body at the confines of the moon, but is just that of the air in which we humans draw breath in the way that fish draw water. Thus Tycho would have determined that there is no successive attenuation of air into aether and effacing of the density pertaining to air, but an obvous and evident division; and if one were to take a position above it, it would impinge upon his eyes no less than the surface that separates air from water now impinges upon the eyes. Rothmann, on the contrary, would not have collided with optical principles, that the surface of a denser medium is struck by light without suffering anything in return or being refracted; and that something is in the doubled parts that is not in the single ones; and the rays are refracted by the depth of the media, not by the surfaces; all of which are absurd.dummy **80**

The nature of our air.

2. Refutation of various authors' various ways of measuring refractions

The means and measure of refractions, even by itself, is established at a high price, and thus, reader, you may not be admitted without adverse consequences: not without first being led through the same briar patch[12] of enquiry that I myself have crept through, on the grounds that since you are going to partake of the common fruit, you should pour out labor as a first libation. This, however, turns out to be for your benefit, that, since there is not yet nothing left over that you might desire in the cause of refractions, you might nevertheless know that no other measure remains, since all crannies have been thoroughly gone over; and also, that you might have the method of seeking before your eyes, cognizance of which alone serves as the greatest argument that this way of measuring has not been assumed arbitrarily. For it is the sort of thing that, when striven for, cannot appear to be put forward from the nature of things, unless you were aware of it.

First, this in general is easily established from experience alone: that density is a cause of refractions. This we also demonstrated *a priori* above, in prop. 14 of the first chapter.

Next, this too is certain: that if light strikes perpendicularly upon a surface, it is not refracted, and that it is more evidently refracted as it strikes more obliquely. Therefore, the angle of incidence contributes to the cause, as I tried to deduce *a priori* above in prop. 20 of the first chapter.

It is therefore evident that the two causes are intermingled, so that density can accomplish nothing if you mentally remove the angle of incidence; for a type of incidence, namely, that which occurs at right angles, evidently strips the dense

---

most of the time sits upon the mountains, while the latter covers the entire face of the earth indiscriminately.

Thus when I deny that vapors have a part in forming refractions, I mean this last kind, since this used to be adduced as the permanent cause by the authors of this opinion, even when the heavens are very calm. The former kind of vapors, however, I too introduce, on p. 125 below, for creating certain more rare and unusual refractions.

[12] Cf. the "thorns" in *New Astronomy*, introductory remarks to the "Summaries of the individual chapters" (*JKNA* p. 79).

body of the refraction. A clear argument of this, which I deduced *a priori* above in prop. 10 and 14 of the first chapter, is that light is not affected by the body of the dense medium, but only by the surface. For incidence is the bounding of a motion, and motion is in a straight line; and the bounds of a straight line are points, and the bound of infinite contiguous lines is a surface (which has infinite points continuous in space), not corporeal bulk.[13]

Accordingly, the cause of refraction consists, not in the corporeal bulk of the dense body, but in the surface. Let these be for us the sure principles of judging the measures.

Thus, it is not a just measure of refractions, which follows from either of
81 these causes alone.

First to crash here is those people's opinion that seeks the measure in the bare length of passage through the denser medium. This is a multiple error. Let the surface of the dense body be AB, its bottom DC, the radiating light E, the rays EC, ED. This opinion states that the angles of refraction are to each other as are the parts of the direct rays beneath the dense [medium] from AC to BD. First, it will be declared that some refraction occurs in direct incidence, which is repugnant to sense and to prop. 20 of the first chapter. Next, the horizontal refraction in a dense body with a rectilinear surface will increase to infinity. For BD is to AC is as the secant of no angle, that is, the radius, to the secant of AEB, the complement of the angle of incidence.[14] It is, however, well known that the secant of 90 degrees is an infinite line. But it is known by experience that there is a certain specific limit to the horizontal refraction in water. For the circle, the measure of angles, is bounded by itself, and embraces four right angles. And since refraction is a striving towards the perpendicular, the angle with the perpendicular of the refraction that is parallel to the horizon is not greater than a right angle. Therefore, even the greatest refraction in the densest medium cannot be greater than a right angle.

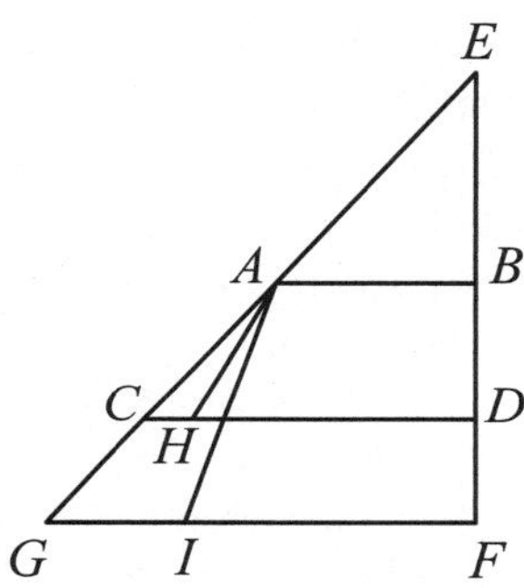

Third, the cause of refraction will become the corporeal bulk, which has been refuted above in prop. 10 of the first chapter.

Fourth, when the depth of the medium is increased, the refraction will increase. For let EC be extended to G. Thus the ratio of AG to BD, which stays

[13] *Corpulentia.* This is defined in the *OLD* as "the putting on of flesh, obesity." In later usage, according to du Cange's *Glossarium,* it can denote "the matter from which a body is composed." However, Tycho Brahe's use of the word shows it simply to mean "volume"; cf. *TBOO* II p. 416.

[14] The secant is the magnitude of the line from the center of a circle to a point of intersection with a tangent, in relation to the radius of the circle. At an angle of zero degrees, this point of intersection is the point of tangency itself, and hence the secant is equal to the radius.

the same, will become greater than that of AC to the same. Therefore, the refraction at A will be greater, while that at B stays the same. From this will follow the thing that Rothmann's arguments appear everywhere to smack of, that ACG is not a straight line, but an arc, which we denied in prop. 22 of the first chapter, and which is easily refuted by experience. Set up three points in air in the same straight line,[15] fill the basin with water, line up your sight so that you perceive the end points together, and the middle point will again clearly coincide with the end points, as before in air. And don't let the illusion of the rod bent in water, straight in air, sway you. For this is irrelevant to the experiment. For it is seen, not by the same, but by different refracted rays. But that what I have said follows, is shown thus. From A let the refracted ray AH descend to the bottom H. **82**
Now let the depth of the bottom be increased from that side; because the angle is increased at the same time, let a more refracted ray AI descend from A. It is obviously evident that the points A, H, I, are not going to be in a straight line. And even if, fifth, this opinion seems to connect density with the angle of incidence, it nevertheless separates them in fact, because it considers the angle of incidence in no other way than to the extent that it increases the length of the passage, which is also increased through another cause, as just now by the lowering of the bottom.

On the contrary, he will be held liable for the same fault who would say that the difference of the sun's rays of incidence upon the surface of the air and of the earth is the measure of refraction. This is because it is certain that the refractions that astronomers consider are brought about by the air poured around the earth. So let the surface of the earth be CB; let the straight line CF touch it at the point C; let the air's surface from the same center A be DE; and let CF cut this at D. Next, let some line touch the surface DE at the point of intersection D, and let this be GD. Thus if CF be made a ray of light, it will coincide with a tangent to the earth's surface, and will be set up at a certain angle to the tangent DG; the difference of the incident rays is GDC. This is diminished gradually, so that at the zenith it is nothing, because the ray from the zenith to either tangent is perpendicular, and descends at right angles. Therefore, if one should say that refractions are in proportional to these angles GDC, this procedure for measuring refractions will appear elegant, for the reason that for any particular altitude of the sphere of air there follows its specific maximum horizontal refraction, so that it seems to satisfy experience. In fact, this too is false, because, as stated, it separates the ratio of density from the figure or angle of incidence, which were to have been conjoined. For indeed, by this procedure, the refractions, or at least the increments of the refractions, from zero to the highest, would come out no different, whether the transparent sphere consisted of thin air or glass. But it is clear from experiments that different patterns of increments are also

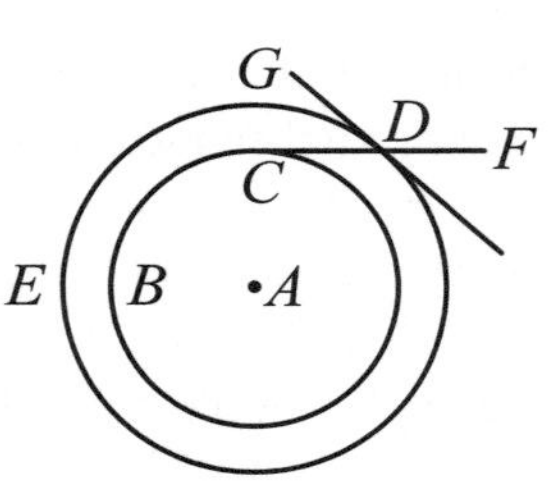

15 This is done in a basin which will be filled with water.

arrayed according to the variety of dense media. Besides, if the refractions were the result of shape alone, no refraction would occur when the surfaces of the
83 transparent body are planes. Thus, the same straight line has the same angles of incidence in water as it does at the bottom of the water, parallel to the surface. Here the difference in the angles of incidence is zero, while the refraction is not zero.

Let us proceed to those ways in which the two causes of density and angle of incidence are linked.

First to present itself is the one that was thrown up against Tycho by Rothmann, at the beginning.[16] Let the refraction at the horizon be only 3 minutes, he says, and it will have a refraction of a minute and a half at 45 degrees. The thought behind this will turn out to be that the maximum horizontal refraction is established by the density of the medium, and it is gradually spread out through all the degrees of the angles of incidence, so that to the extent that the illumination is raised on the arcs of a circle, the remaining refractions decrease proportionally from the maximum, until they become nothing at the zenith. Later, both Tycho and Rothmann acknowledged that this way is in disagreement with experience. For in the table of Witelo, Book 10 Prop. 8,[17] the refractions near the horizon are also driven down, nor do the ones at the zenith (or nearby) correspond to the horizontal ones, in this proportion of arcs of incidence. The refractions that take place in air drop off much more near the horizon, as is to be seen in the tables in Tycho's *Progymnasmata* pp. 79, 124, and 280. For the rest, neither Tycho nor Rothmann considered this sudden increment near the horizon to exist as a result of the actual form of the measure: it seemed to both likely either that the refractions are in proportion to the angles of inclination, or that a new cause—that is, a new body of vapors[18] —intervenes from elsewhere. I, however, say that a measure must be established from which these phenomena follow of necessity, without anything intervening. For the analogy from water and air makes us sufficiently assured to state that the surface of the air alone bears the full responsibility. In water, which is fairly dense, refractions perceptibly depart near the horizon, and at the horizon the maximum refraction is about 37 degrees, and nothing comes in here other than the surface of water. If this is so, much more so will the refractions decline towards the horizon where the density of the medium is small, and the horizontal refraction is insignificant and slight.

Moreover, by what has been said so far, that final opinion of Tycho Brahe (pp. 95–6 of the *Progymnasmata*)[19] has also been refuted, although it was the closest of all to the truth. It is approximately as follows. The standards of mea-
84 surement of the refractions are established by the length of the passage, and proportional parts of these are taken according to the quantity of the angles of inclination. That of water would become infinite at the horizon; it would arise out

[16] *TBOO* VI p. 111.

[17] *Thesaurus* II p. 412.

[18] Here Kepler inserts an asterisk that refers to the note included above on p. 95.

[19] *TBOO* I pp. 80–1.

of the corporeal bulk; it would be increased with the depth of the medium; and finally, the refraction in air is not in agreement. For in that place Tycho supposes that the horizontal thickness of the air is 142 miles, and 14 miles at an altitude of 60°. Therefore, 1/10 part of the horizontal refraction, namely, $3\frac{2}{5}'$, would be due to the altitude of 60°, of which 1 minute would be due to the inclination of 30°(because the altitude is given as 60°), which Brahe says does not happen.[20]

Nor did I leave this remaining case untried: whether, once the horizontal parallax is established from the density of the medium, the rest might correspond to the sines of the distances from the zenith. But computation did not give approval to this, nor indeed was there any need to make the enquiry. For refractions would increase in the same pattern in all media, which does not agree with experience.

On this score, the cause of refractions introduced by Alhazen and Witelo is censured.[21] They say that light seeks compensation for injury received from the oblique collision. For to the extent that it was weakened by the encounter with a denser medium, it gathers itself together again by approaching the perpendicular, so that it might strike the bottom of the denser medium with a more direct blow. For of blows, those which are direct are the strongest. And they add a subtle I know not what: the motion of light striking obliquely is composed of a motion perpendicular and a motion parallel to the surface of the dense medium, and this motion, thus composed, is not annulled by the encounter with the denser transparent medium, but only hindered. Therefore, the whole motion, as it was composed, fortifies itself again; that is, in the motion that has now been changed through the dense surface there resides the traces of its original composition, so that it becomes neither entirely perpendicular nor entirely parallel. However, it turns aside towards the perpendicular more than towards the parallel, because the perpendicular motion is stronger. They did not explain the matter much better than Macrobius, *Saturnalia* book 7, who attributed to the sense of vision a hesitation, and a retreat into itself from the encounter.[22] It is exactly as if the form of light were endowed with mind, by which it might reckon both the density of the medium and its own injury, and, using its own judgement and not an extrinsic force, acting and not being acted upon, might of itself perform its own refraction.

If this account were true, the measurement of refractions would be easy. For **85**
the refractions would increase with the sines of the distances from the zenith, because the impacts are diminished in the same ratio from the obliquity. Thus

---

20 The horizontal parallax is 34′, one tenth of which would be $3\frac{2}{5}'$. However, because the incident rays are inclined to the surface of the sphere at an angle of 30°, the refraction must be reduced by a factor of 1/3, making it about 1′.

21 Alhazen discusses the cause of refraction in VII no. 8, *Thesaurus* I pp. 240–242. Although Witelo devotes Book X to refraction, he considers its causes in Book II, especially prop. 47, *Thesaurus* II pp. 81–3.

22 Macrobius, *Saturnalia* VII 14.2, p. 448: "Water is denser in thickness than air, and therefore the sense of vision penetrates it more hesitantly; and, beaten back by the encounter with the water, the eyesight is broken and retreats into itself. But when it goes back after being broken, it does not do so by a direct impact, but flows in on all sides towards the outlines of the likeness, and it thus happens that the image appears larger than its archetype."

if you should ask how much more strongly the sun falls upon the earth from an altitude of 30 degrees than from an altitude of 45 degrees, the answer will rightly be given, that it is as much stronger as the side of a square is longer than the side of a hexagon. For near the horizon the force of the sun is suddenly accumulated, while about the zenith it changes little, and this also happens to the sines.

I devised another procedure for measuring, to combine both the density of the medium and the angle of incidence. For since the denser medium becomes the cause of refractions, it therefore seems to be exactly as if one were to extend the depth of that medium, in which the rays are refracted, to a size that the same amount of matter, in the form of the rarer medium, occupies. For then the rays entered in to the empty space of water towards the visible object without being turned back. When the object, through this kind of image-making, is sunk deeper on the perpendicular, the rays themselves, too, are understood to be sunk and thus lengthened at their other end, where they touch the visible[23] object. Let A be the light, BC the surface of the denser medium, DE the bottom. Let the oblique rays AF, AG, AB, descend, and let AB be extended to D, AG to I, AF to H, where they would have fallen had the medium been uniform. But because it is denser, imagine accordingly that the bottom DE is pushed down far enough that there is the same amount of matter in the depth CK under the form of the rarer medium as there is in the depth CE under the form of the denser. Therefore, the entire bottom DE being sunk to LK, the points D, I, H, E, will descend perpendicularly to L, M, N, K. So let the points B, G, F, be joined to L, M, N. The bottom DE will
86 be cut at the points O, P, Q, and the refracted rays ABO, AGP, AFQ will be made.

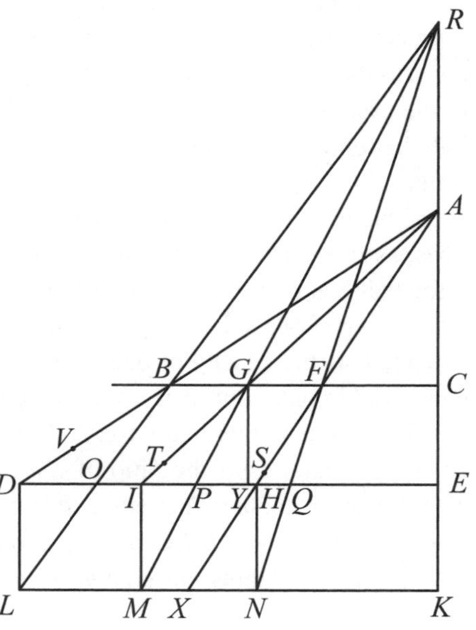

And it should be noted in this way of measuring, that if CK, FN, GM, BL be extended, they meet again in the same point. For let CA and NF be extended until they meet, and let the intersection be the point R, and let G and B be connected with R. I say that RGM is in turn a single straight line, and likewise RBL. For the triangles RAF, NHF, are between the parallel lines RA, HN, and therefore the angles RAF and FHN are equal, as are ARF and HNF; and the angle at F is common. Therefore, the whole triangles are similar. But side RA is common to triangle RAF and triangle RAG, and angle RAF is greater than angle RAG by the quantity of angle FAG. And points F and G stand in a single straight

---

[23] Reading *visibile* for *visile*.

line perpendicular to RAK. In this way, the three points of the two triangles are determined, so that they cannot be other than these particular ones. Triangles FHN and GIM, in turn, also have an equal side IM and HN, because these are parallels between the parallel lines DE and LK. And angle FHN is greater than the angle GIM, again by the quantity of angle FAG. And this occurs in similar triangles. For angles RAF and FHN are equal, as are RAG and GIM also. And, thirdly, angles G and F again stand upon the perpendicular above MI or the equal side NH, and HFN corresponds to angle RFA, at which the same thing had occurred before. Therefore, as triangle RAG is to triangle RAF, in lines and angles, so also will MIG be to NHF. But NHF and RAF are similar. Therefore, RAG and MIG are also similar. But they have the sides RA, IM parallel, and IG, GA in a single straight line, and the common vertex G. Therefore RGA and MGI are equal angles, and consequently MGR will be a single straight line. In the same way it will be proved that RAB and LDB are also similar, and that LBR is a straight line. What we had given notice of is therefore evident.

This way of measuring is refuted by experience. For the angles of refraction towards the perpendicular AC, such as HFQ, are made too great with respect to the horizontal refractions. And if you examine the angles of refraction from Witelo and Tycho, the directing point R of the refracted rays BO, GP, FQ, is not one, but near the horizon it is highest above A, coinciding with A at the
zenith. Whoever has the leisure may enquire into this either by computation or by **87**
a compass. Add that reason itself stumbles, and when it grasps at dimensions, it scarcely grasps or perceives itself.

I pass on to other ways of measuring. Since density is obviously a cause of refraction, and refraction itself appears to be a kind of compression of light (i.e., towards the perpendicular), it comes to mind to ask whether the ratio of the media in the case of densities is the same as the ratio of the bottom of the spaces that light has entered into and strikes, first in an empty vessel, and then one filled with water.

This measure is multiple. For, in one alternative, it is conceived in straight lines, as if one should say that the line EQ (in the above diagram), illuminated refractedly, is to EH, illuminated directly, as the density of one medium to the other. Or as if one should say that the refracted line FQ is to the part FH of AF extended, as the density of one medium to the other. In another alternative, it is conceived in plane surfaces, as if, the power[24] of EQ be to the power of EH (or the circle, or any other figure [on EQ] be to a similar figure [on EH]) in this ratio of dense media. Thus the ratio of EH to EI would be the duplicate ratio[25] of EQ to EP. In another alternative, this measure is conceived in the solidity of the truncated pyramids $FHEC$, $FQEC$, so that, as medium is to medium in density, so are these pyramids, an empty one to one full of fluid. Finally, because the

[24] That is, the "second power", or the square.

[25] "When three magnitudes are proportional, the first is said to have to the third the **duplicate ratio** of that which it has to the second." —Euclid, *Elements*, V definition 9.

proportion of the media has a threefold aspect, since they are subject to density in length, width, and depth, I also proceeded to seeking the cubic ratio between the lines EQ and EH. Indeed, I also considered other lines. From one of the points of refraction, such as G, let a perpendicular be dropped to the bottom GY. The question will be, whether the triangle GIY—that is, the base IY—would be divided by the refracted ray GP in the ratio of the density of the media. I linked all these measures together, because for all of them the procedure of enquiry is the same.

For whatever the procedure is by which the line, plane, or pyramid might keep the proportion EI to EP (or, shortened, YI to YP) the same everywhere—that is, in the ratio of the media—it is certain that EI, the tangent to the distance
88 of the point A from the zenith,[26] will become an infinite distance at the horizon, and therefore will make EP or YP infinite as well. Whence the angle of refraction IGP will become zero, and approaching the horizon will come out to be gradually smaller and smaller, which is refuted by experience. For it is greatest at the horizon. Nor do those pyramids brought into the scenario fit in well with prop. 7 and 8 of the first chapter, because the form of light has the dimensions of a surface, not of a body. Finally, this form is refracted right at the surface of the denser medium, and after this effect it is borne with a straight line motion, whatever sort of bottom encounters it. Therefore, the bottom is the lines EI, EP, and these contrived pyramids are clearly extraneous to the matter of refraction.

At last, therefore, one must go to the actual image of the object, whose place, by our *Optics* in ch. 3 above, we defined as the meeting of the visual ray with the perpendicular of the incidence. In the previous diagram, let the bottom now be LK, and the depth of the dense medium CK. Let the visible object be at M; erect a perpendicular from M to the surface of the water, that is, MI is perpendicular; let G be the point of refraction. The straight line AGI will be the visual ray, because the sense of vision does not perceive that the ray is refracted at G—indeed, it reckons that the entire ray comes from the place to which the vision, lying in the way, is directed by the viewer. Thus AGI meets with the perpendicular MI at the point I. Therefore, by definition 1 of the *Optics*,[27] it will be the place of the image. This image I first considered thus: whether, by always remaining on IE parallel to the surface BC, it provides a handle for measuring refractions. But that this is false, the sense of the eyes itself gave testimony. For the more obliquely you look into the water, the more the images rise to the surface. If you look directly down from A to C, no depth of the bottom K will be seen. Which is to say that this measure at length coincides with the one described above. For it has been proved that if D, I, H, E, (which are now the places of the images), be in the same line parallel to BC and LK, and DL, IM, HN, EK, are perpendiculars, as is supposed here as well, then KC, NF, MG, LB, meet at the same point R: thus this is one of the rejected measures.

Again, I asked whether the images were at an equal distance from the points

---

[26] If AE is taken as the radius, with magnitude 100,000 for Kepler, or unity for us, the length of EI is the tangent of the angle of incidence at G, which is equal to IAE.

[27] Chapter 3, Definition 1 of the present work.

of their refractions, and the ratio of the dense media were a measure of the minimum distance. As, if E be the image, C the surface of the water, K the bottom, and CE to CK as medium to medium with respect to density; and afterwards F, G, B be three other points of refraction, and the images at S, T, V, and CE, FS, GT, BV equal. But by this procedure, some altitude was set up 89
for the image E on the perpendicular AK, which is refuted by experience, not to mention other procedures of enquiry.

Third, I asked whether, as medium is to medium, so (if H be the place of the image) is FH to FX? By no means. For thus CE would be to CK in the same ratio, and consequently the depth of the image would be always the same, which we have now refuted.

Fourth, I asked whether, as CK is to FX, so is the depth of the image at K to the depthat H? By no means. For either the images would never begin to go upwards, or where they had once begun, they would go upwards to infinity, because FX at last becomes infinite.

Fifth, I asked whether the images go upwards in proportion to the sines of the inclinations? By no means. For the ratio of ascent would be the same in all media.

I therefore asked whether, sixth, they might at first be raised up in a perpendicular radiation in proportion to the media, and afterwards might more and more go upwards in proportion to the sines of the inclinations? For thus the ratio of upward motions would be compounded, and would be made different for different media. Nothing of the sort. For the computation disagreed with experience. And in general our consideration of the image, or of the place of the image, was in vain, for the very reason that it is an image. For what it is that happens to the sense of vision, from whose error the image results, has nothing to do with the density of the medium, nothing to do with that real affect[28] itself, the bending back[29] of light.

3. Preparation for the true measurement of refractions

Hitherto, we have followed an almost blind plan of enquiry, and have called upon luck. From now on let us open the other eye, proceeding with a sure method.

So, I had carefully assessed the fact that the image of an object viewed beneath water is so near to the real measure of refractions that it almost measures refractions: it is low when the object is viewed from the perpendicular; it becomes gradually higher when the eye bends down towards the horizon of the water. On the other hand, the reasons just stated deny that the measure is to be sought in the image, because the image does not exist purely from the nature of the object, but at the same time exists because of a deception of the sense of vision, which is accidental to the object itself. When I considered these 90
things, and compared the conflicting arguments, it finally occurred to me to look into the causes themselves of the image having being established in water, and in these causes to seek the measure of refractions. This opinion was all the

[28] πάθος.

[29] ἀνακλάσις.

more confirmed in me because I saw that the cause of the image as it appeared in both mirrors and in water, was not truly pointed out by the optical writers. And this is the origin of the labor that we took up above in ch. 3. Nor was it a trivial matter, in that among the principles in such a complex subject I follow other false opinions in place of the false traditions of the optical writers, in that thrice and four times I set out upon a new road, and repeat the whole procedure anew: and, as it happens, being rashly persuaded, I so many times embrace in my mind the very thing that had been sought after with such zeal, as already discovered.

And indeed, this very difficult Gordian knot of catoptrics I finally cut by analogy alone, in that manner which I have described above: when I consider what would happen in mirrors, and what fittingly should happen in water following this similitude. For in mirrors, it is not any matter, but only reflection from a polished surface, that places images outside of the place of the object seen. Therefore, it followed that in water, too, images ascend, and approach the surface, not according to the laws of density in water, greater or less according as the perception is more direct or more oblique, but only because of the refraction of the spark of light that has flowed from the object into the eye. When this was stated thus, whatever I tentatively proposed above for the measure of refraction through the image and its rising up, entirely collapsed. This was all the more so after I found the real cause for the image's being on the same perpendicular with the object itself, both in mirrors and in denser media.

So since the use of an analogy of this kind turned out happily in that most difficult demonstration of the place of the image, I began to pursue this analogy further, enticed by the desire to measure refractions. For I wished to lay hold of a measure of refractions, however blind it was, provided that it was something, in the confident hope that it would happen that once the measure were truly known, the cause would also become evident. I therefore proceeded as follows.

In the same manner that the image of an object is made smaller in convex
91 mirrors, it is also thus in rarer media; and as the image is made greater in concave mirrors, it is also thus in denser media. In convex mirrors, the middle parts of the image get closer than the surrounding parts, while in concave mirrors they get farther away. The same thing occurs in different media, so that in water the bottom seems lowered, the surrounding parts raised. Hence it was apparent that a denser medium corresponds to a concave mirror surface, and a rarer medium to a convex surface. It was therewith evident that the plane surface of water assumes a certain type of curvature. This required that causes be thought up to reconcile this effect of curvature with the plane surface, as for instance if a cause were given why the parts of water surrounding a perpendicular incident ray should represent a greater density than that of the water directly beneath the perpendicular. Thus the question kept returning to the contrivances described above, which, since they are refuted by reason and experience, were to be replaced by an enquiry into the cause itself. So I passed on to the measure. And since there are many kinds of media, differing in density, what was called for was to establish some kind of analogy with concave mirrors for this multitude as well. Thus, in order that the image be made greater in a concave mirror, the eye has to be inside the center, and the closer it is to the center, and farther from the surface, the greater will be

the image. And so, when denser media make a greater image in the same way, the different media corresponded to different places of the eye in the mirror's diameter.

I therefore accommodated the extremes: in a concave mirror, the place of

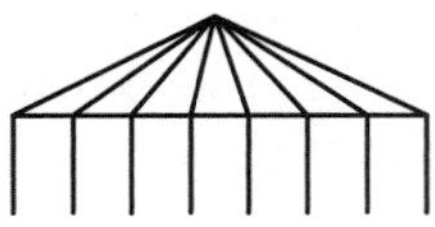

the eye, the surface, and the center; in media, the densest medium, and the medium that is plainly equal to that medium in which the eye is. If the eye were located at the center, the same things were required to happen that happen upon the object's being viewed in a medium of infinite density.

Here new forks in the road arose. For if you should consider what ought to happen given a medium that is quite the densest (or of infinite density), you will understand, by analogy with the other media, that if there were such a thing, all rays whatever coming falling upon such a surface from a single point must be completely refracted; that is, after refraction they coincide with the perpendiculars themselves, and thus are made parallel. For in the other media, the denser any one of them is, the closer the refracted rays approach their perpendiculars.

And yet in a spherical concave surface, the rays from the center flowing ***92***
in all directions to the surface are not parallel after reflection, but are gathered together again towards their origin. What had to be looked for, then, was a concave surface, and a point on its diameter, from which all rays going forth to the surface would be reflected into plain parallels. Had I not at this point had some kind of foretaste of conics, I never would have arrived whither I was striving. But I was mindful of those things that Witelo had written of the parabolic burning mirror, in Book 9 prop. 39, 40, 41, 42, 43, and 44.[30] For those things that Apollonius had demonstrated concerning the hyperbola and the ellipse in Book 3 prop. 48 and a few other neighboring ones, were omitted for the parabola; and Witelo to some extent supplies these in the places mentioned. In them he demonstrates a certain point from which any number of lines sent forth to the section or curved line make with the tangent angles equal to those angles which lines drawn from the same points parallel to the axis make with the same tangent. And it is this that we were seeking. However, because the consideration of the sections is difficult, because it is too little pursued, it would be good to say a few words about it in a mechanical, analogical, and popular vein. Geometers, be indulgent!

4. On the sections of a cone

Cones are of various kinds: right angled, acute, obtuse; again, right or regular cones and scalene or irregular or compressed cones; for which see Apollonius and the commentaries of Eutocius.[31] The sections[32] of all of these, regardless

---

30 *Thesaurus* II pp. 398–402.

31 Eutocius was born in Ascalon (now Ashqelon) in about 480 C. E., and wrote commentaries on works of Archimedes, Apollonius, and other mathematicians. His commentary was included by Commandino in the Latin translation of the *Conics* that Kepler used.

32 A "conic section" (from the Latin *sectio*, "a cutting") is created by the intersection of a

of kind, fall into five species. For the line on the surface of a cone established by a section is either straight, or a circle, or a parabola or hyperbola or ellipse. Speaking analogically rather than geometrically, there exists among these lines the following order, by reason of their properties: it passes from the straight line through an infinity of hyperbolas to the parabola, and thence through an infinity of ellipses to the circle.[33] For the most obtuse of all hyperbolas is a straight line; the most acute, a parabola. Likewise, the most acute of all ellipses is a parabola; the most obtuse, a circle. Thus the parabola has on one side two things infinite in nature—the hyperbola and the straight line—and on the other side two things
93 that are finite and return to themselves—the ellipse and the circle. It itself holds itself in the middle place, with a middle nature. For it is also infinite, but assumes a limitation from the other side, for the more it is extended, the more it becomes parallel to itself, and does not expand the arms (so to speak) like a hyperbola, but draws back from the embrace of the infinite, always seeking less although it always embraces more. With the hyperbola, the more is actually embraced between the arms, the more it also seeks. Therefore, the opposite limits are the circle and the straight line: the former is pure curvedness, the latter pure straightness. The hyperbola, parabola, and ellipse are placed in between, and participate in the straight and the curved, the parabola equally, the hyperbola in more of the straightness, and the ellipse in more of the curvedness. For that reason, as the hyperbola is extended farther, it becomes more similar to a straight line, i. e., to its asymptote. The farther the ellipse is continued beyond the center, the more it emulates circularity, and finally it again comes together with itself. The parabola, in the middle position, is always more curved than the hyperbola, if they be extended by equal intervals; and is always straighter than the ellipse. And since, just as the circle and the straight line bring the extremes to a close, and thus the parabola holds the middle, so also, as all straight lines are similar, as are all circles, so are all parabolas also similar, differing only in quantity.

Further, there are in these lines certain points to which special attention is given: these have a precise definition, but no name, unless you take the definition or some property in place of a name. For lines drawn from these points to lines touching the section, to their points of tangency, form angles equal to those that are made when the opposite points are joined with these same points of tangency. For the sake of light, and with an eye turned towards mechanics, we shall call these points "foci".[34] We might have called them "centers", since they are on

plane with a conic surface. The manner in which the construction is to be imagined is presented in detail by Apollonius, *Conics*, at the beginning of Book 1.

[33] In the diagram below, the "straight line" is the horizontal line through LL, the parabola is KG, and the infinity of hyperbolas lying between them is represented by LQ. The idea of an infinitely distant focus for the parabola, and the conception of conics as forming a continuum generated by a moving focus, are Kepler's original and substantial contributions to the theory of conics.

[34] This is the source of our present use of the word "focus." The foci had appeared in Apollonius, *Conics* III 45–52, but were named "the points arising out of the application" (τὰ ἐκ τῆς παραβολῆς γινόμενα σημεῖα), because of the way Apollonius had defined them. The word *focus* means "hearth", and refers to the properties of

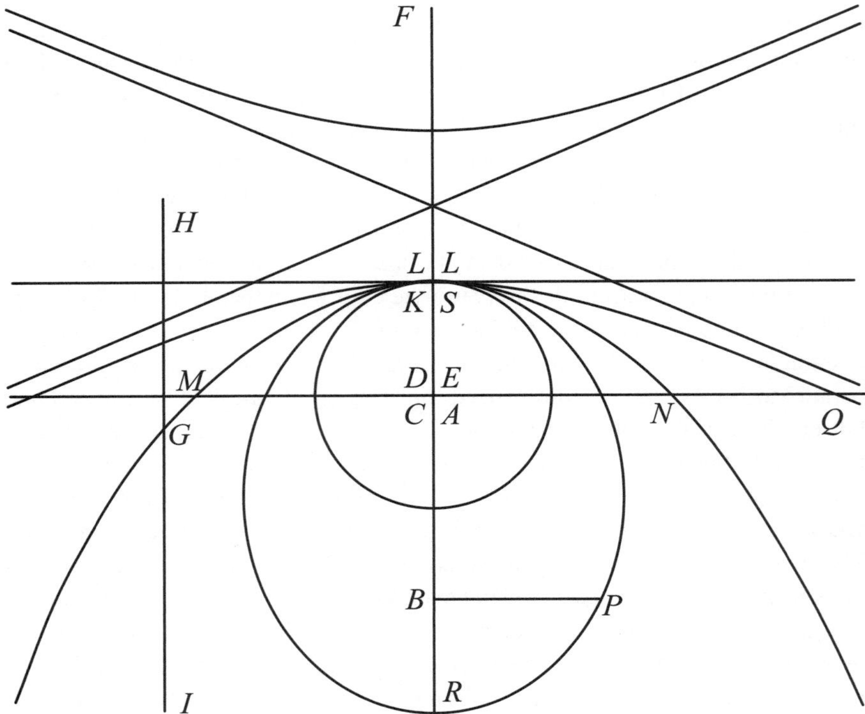

the axes of the sections, but in the hyperbola and the ellipse the writers on conic sections called another point the center. So in a circle, there is one focus A, and this is the same point as the center; in an ellipse there are two foci B and C, equally removed from the center of the figure, and more so in a more acute one. In a parabola one focus D is within the section, the other is to be supposed on the axis, either outside or inside the section, removed at an infinite distance from the former one, so that a line HG or IG drawn from that hidden focus to any point of the section G is parallel to the axis DK. In the hyperbola, the external focus **94**
F is nearer to the internal focus E to the extent that the hyperbola is more obtuse. And the one that is external to one of the opposite sections is internal to the other, and *vice versa*.

It therefore follows by analogy that in a straight line the pair of foci (we speak thus of the straight line, contrary to custom, only to fill out the analogy) coincides with the straight line itself, and is single, as in the circle. Thus in the circle, the focus is right at the center, receding as far as possible from the neighboring circumference; in the ellipse it now recedes less, and in the parabola much less; finally, in the straight line, it recedes from it by the least amount: that is, it falls upon it. And so in the limiting cases, the circle and the straight line, the foci come together, standing at the greatest distance in the former, and falling

---

conoidal mirrors of focussing light rays at one of these points when the rays originate at the other point (in the parabola, the originating "focus" is infinitely distant, and the rays coming from it are parallel).

right on the line in the latter. In the middle, the parabola, they are an infinite distance apart, while in the ellipse and hyperbola, which are at the sides, the the foci, paired in their function, are a measured distance apart, the other [focus]
95 being inside in the ellipse, and outside in the hyperbola. In all cases the accounts are opposite.

The line MN, which marks out the focus on the axis, standing perpendicularly to the axis, we call the chord;[35] and the line that shows the altitude of the focus from the vertex from the nearest part of the section, namely, the part of the axis BR or DK or ES,[36] we call the sagitta[37] or axis. Thus in the circle, the sagitta is equal to half the chord; in the ellipse the half-chord BP is greater than the sagitta BR, and the sagitta BR is also greater than half of the half-chord BP or the fourth part of the chord. In the parabola, as Witelo demonstrated,[38] the sagitta DK is exactly equal to one fourth of the chord MN; that is, DN is twice DK. In the hyperbola EQ is more than twice ES; that is, the sagitta ES is less than one fourth of the chord EQ, and is ever less and less, in all ratios, until it vanishes in the straight line, where, the focus falling directly upon the line, the altitude of the focus, or the sagitta, vanishes, and the chord at the same time is made infinite since it coincides with its own arc (to speak thus improperly, since the arc is a straight line). For geometrical terms ought to be at our service for analogy. I love analogies most of all: they are my most faithful teachers, aware of all the hidden secrets of nature. In geometry in particular they are to be taken up, since they restrict the infinity of cases between their respective extremes and the mean with however many absurd phrases, and place the whole essence of any subject vividly before the eyes.

Furthermore, in the description of the sections also, analogy has been of the greatest help to me. For from Apollonius III 51 and 52, the description of the hyperbola and the ellipse is accomplished very easily: it can be done even with a thread.[39] For the foci being given, and between them the vertex C, let pins be placed at foci A and B; and to A let a thread with length AC be tied, and to B a thread with length BC. Let each thread be lengthened with equal additions, so that if you grasp the doubled thread with the fingers, and as they move away from C you little by little pay out the two threads, and with the other hand you draw the path of the angle that the two threads make at the fingers, that drawing will be a hyperbola. The ellipse is described more easily. Let the foci be A, B; the vertex C. Put sturdy pins at A and B, to embrace the two with a thread in a single

[35] Note that Kepler's usage does not coincide with ours: his chord must pass through the focus and be perpendicular to the axis. Thus in a circle, Kepler's chord is a diameter.

[36] For the ellipse, parabola, and hyperbola, respectively.

[37] Literally, "arrow." Kepler's use of this term here is an extension of its previous application to the circle, where it denoted the versed sine.

[38] Witelo IX 42, *Thesaurus* II pp. 400–401. The proposition had been proved by Alhazen in his work on the parabolic burning mirror, which Witelo used.

[39] The following constructions for the hyperbola and the parabola were invented by Kepler. Although the ellipse construction had already been described, Kepler probably also developed his construction independently.

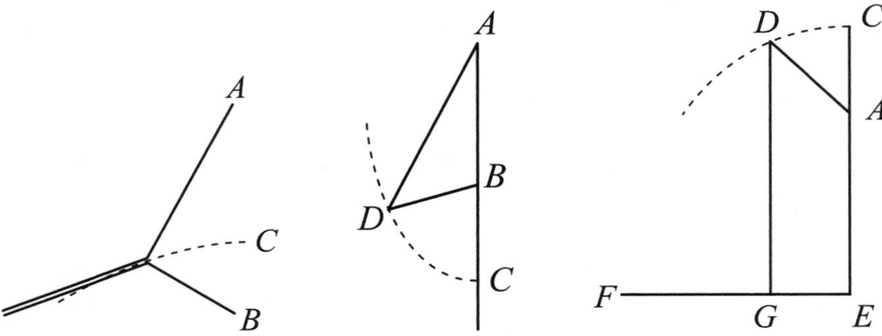

embrace, so that there be no thread between A and B. Let the length of the thread be twice AC, and let the ends of the thread be joined with a knot. Now insert the stylus D in the same loop of thread with A and B, and, with the thread stretched as tight as it allows, draw a line around A and B: this will be an ellipse. Since this description was so easy, not requiring those laborious compasses that some, by the forging of which, hunt for people's admiration, I was long disappointed that a parabola could not also be described thus. Finally, the analogy showed (and geometry confirms) how to draw this too, without much more labor. Let there be set out the focus A, the vertex C, so that AC should be the axis; let this be extended in the direction of A all the way to infinity, or as far as it is wished to draw the parabola. Let it be wished to draw it to E. Accordingly, let a pin be fixed at A, and from it let there be tied a thread with length AC, CE. With one hand, you should hold one end of the thread E, and with the other stretch the stylus with the thread all the way to C. Also, let EF be erected perpendicularly to CE, and then with the stylus C and the other hand E move equal distances away from the line AE, in such a way that the other hand and the end of the thread always remain on EF, and the thread DG always parallel to AE: the path CD which you will have drawn with the stylus will be a parabola. **96**

I have said these things about conic sections all the more willingly, because not only does the measuring of refractions require them here, but their use will also be apparent below in the anatomy of the eye. Then too, mention of them will be made in two places among the observational problems. Finally, a consideration of them is completely necessary for the most excellent optical devices, for setting up an image hanging in air, for increasing images proportionally, for igniting fires, for burning things at infinity.[40]

Porta's optical devices

5. What kind of quantity measures refractions?

I hit upon the conic sections by the best method, but not all reason for doubt

[40] This last remark is a reference to Porta's account of the parabolic burning mirror, in *Magia naturalis* XVII 17, p. 275, where he writes, "This mirror burns, not at a thousand paces, . . . but at infinity." The other "optical devices" mentioned by Kepler are presented by Porta in *Magia naturalis* XVII 8–20, pp. 267–278.

was yet removed. For we had proposed, for measuring refractions, infinitely many forms of media, differing in density, from that which is imagined to have
97 the ratio of infinite density to that which has no density at all. However, in the preceding it was demonstrated that the measure of the angles of refraction, through various angles of incidence of light upon denser media, is to be sought (at analogy's urging) from the angles of reflections through various angles of incidence of light upon concave mirrors. Now one extreme in the analogy was in good accord, namely, the analogy of the densest medium to the placing of light at the focus of the parabola. For, as the rays in the densest medium after refraction, so those which come from the focus of the parabola after reflection, are made parallel.

But concerning the media next in order, which are less dense, it was not yet established which way they were to be accommodated: whether by different positions of the light in the same parabola below the focus, so that the medium in which no refraction occurs corresponds to the position of the light right on the concave surface or the vertex of the parabola; or whether instead, since the parabola is the extreme of both hyperbolas and ellipses, the media different in degrees from the most dense are to be accommodated either to various hyperbolas or to various ellipses, so that the position of light might always remain at the focus of the section. In this way, that medium which lacked refraction, would, in the former case, call for the straight line or plane mirror, and in the latter, for the spherical concave. For this reason, I did not leave it untried to ask whether for any medium there might be its own hyperbola. For if you designate with points the places of the images in water through all the angles of inclination, a hyperbola will be approximately foreshadowed, which increases my confidence. Thus, for example, for the refraction of water, at an inclination of 80° let a refraction of 30° and a refracted angle of 50° be taken from Witelo.[41] And let B be the focus and A the opposite one; on the line BA at the point B let the angle 80° be set up, and let this be CBA. Likewise, on the line BA at the point A let the refracted angle 50° be set up, let this be CAB, and let AC and BC meet at C. Now since B is 80° and A is 50°, the remaining angle C will also be 50°, and AB and BC will be equal. Thus if AB be 100,000, AC will be 128,558. But BC is also 100,000. Therefore, the excess of AC over BC is 28,558. Therefore, by Apollonius III 51, 28,558 is the line between the vertices of the opposite sections, or the axis DE.[42] Subtract DE, 28,558, from AB, 100,000, and the remainder is 71,442, whose half, 35,721, is AD or EB. And so E is the vertex of a hyperbola, and D is the vertex of the opposite section. Now let us see whether the refractions

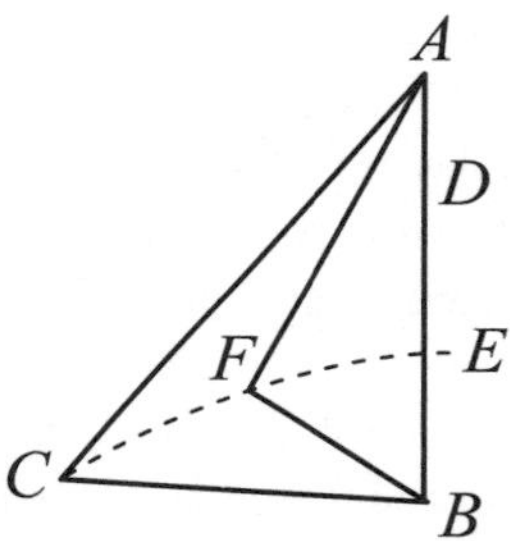

[41] Witelo's table for water is in Prop. 8 of this chapter, on page 128.

[42] As Kepler says below, D is the vertex of the opposite part of the hyperbola, which is not drawn in the diagram. Therefore, $AD = BE$.

of the other inclinations set forth by Witelo will follow. Let EBF be 70°, at which inclination Witelo places a refraction of 24° 30′ and a refracted angle of 45° 30′. Thus the question is whether, in this hyperbola just constructed, FAB is 45° 30′. Since AB is 100,000 and ABF is 70° and AF is longer than FB by the space DE, let us proceed by false position,[43] and let FAB be 45°30′. AFB will be 64° 30′. Consequently, as the sine of AFB is to AB, so is the sine of FAB to FB, and the sine of FBA to FA. Therefore, FB comes out to be 79,023, and FA 104,111, and the difference of these is 25,088. But it ought to have been 28,558. Therefore FB is too long with respect to FA. It is diminished, if angle FAB be diminished. This is one iteration. Now, for the second, let FAB be 44° 29′. AFB will be 65° 31′, and consequently FB will be 76,993 and FA 103,254. The difference, 26,261, ought to have been 28,558. You see that by the diminution of angle FAB by 1°1′ we gained 1,173 towards what is required. We were still off by 2,297. Therefore, that angle, which was to have represented the refraction for us, still has to be diminished by about two degrees. And in return, the refraction itself has to be increased by the same amount, about three degrees, so as to make it 27° 30′, closer to what it formerly was at 80°. Therefore, the hyperbolic mirror does not measure the angles of refraction at different inclinations. And in general, because BA represents the perpendicular in water, and the ray that is set in motion from B at right angles to BA, represents the ray parallel to the surface of the water or to the horizon, it must be noted that in the hyperbola, about the vertical angles of incidence, the angles of reflection suddenly increase, but do so hardly at all about the horizon, and reach a maximum quickly. In refractions it is otherwise, for near the horizon both the angles of refraction and the increments of those angles increase. Therefore, we seek the measure of refractions between the foci of hyperbolas in vain. ***98***

You could now guess immediately that because the hyperbola does the opposite of the refractions,[44] the ellipse, being the hyperbola's opposite, is going to do the same as the refractions, and will accommodate itself to the measure. This guess is made the more probable in that on this basis the analogy gives the concave spherical mirror to the medium lacking refraction, and in this mirror the rays going out from the center and those bent back coincide, and do not enclose any angles. Let B be the focus of the ellipse, A the opposite focus, $DABI$ the axis. ***99*** Again, let IBF be 80° and BAF 50°. Here, AFB will be 30°, representing the actual refraction; and let AB be 100,000. Accordingly, AF will be 196,962 and BF 153,208; and by Apollonius III 52 the sum of the two, 350,170, will be the quantity of the axis DI. Therefore, AD or BI will be 125,085. The ellipse being thus established, let IBC now be 70°. I know that BC, CA are equal to the axis,

---

43 That is, an iterative procedure. See e.g. *Mathematics Dictionary* s.v. 'False' and 'Regula Falsi'. In the triangle ABF, the side AB and the angle ABF are given, and the difference between sides AF and BF is known. Kepler did not know how to solve the problem directly. Henry Briggs, reading this passage, came up with a direct procedure, which he communicated to Kepler in a letter of 20 February 1625 (No. 1002 in *JKGW* XVIII, esp. pp. 225–8).

44 All editions have a period here, but the sense seems to require a comma.

350,170. Therefore, let the refraction, the angle BCA, be 24° 30′, as Witelo puts it. AC comes out to be 226,599, BC 171,995. The sum is 398,594. This is too great, for it was to have been only 350,170. Accordingly, BAC is too great, and BCA is less than it should be. So let it be greater, i.e. 27°, and let BAC be 43°. AC comes out to be 206,985, and BC 150,223. The sum is 357,208, still a little greater than it should be. Accordingly, the angle BCA still has to be increased a little. And so it will become greater than the Witelonian refraction. Therefore, the measure of refraction is also not to be sought in the focus of an ellipse. For in general this too increases the said angle with large increments from the vertex, but where a perpendicular to the axis goes forth from the focus, the increments of the angles are small, also nearly as in the hyperbola itself.

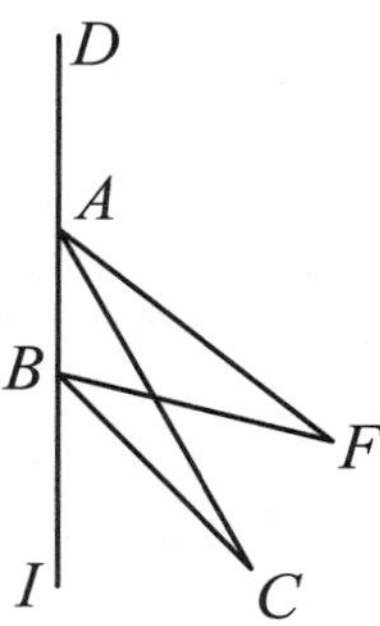

Therefore let another way be consulted, that for all media, in the matter of refractions, the measures are in the parabola alone, and let the measure of the densest medium be established by straight lines drawn from the focus, but let the measure of the most tenuous medium, or one lacking refraction, be taken (analogically speaking) by drawing straight lines from the point that is at the very bottom (or, geometrically, the vertex) of the parabola, situated at the very lowest level; let the remaining media lying in between have assigned to them the points on the axis likewise lying in between. For example, in the refractions of water, let A be the focus, in the following diagram, AB the axis, below A let a point C be taken, and let DC be inclined to AB at the angle DCB, 80°. And at D let FD be tangent to the section, cutting AB at E, and let DFG be made equal to angle EDC, and let GD be extended until it meets AB at I. The first question is, how high should the point C be from the bottom or the vertex B, in order that GIA become 50°, as great as Witelo puts the refracted angle at an inclination of 80°. Therefore, in CDI, DIC is 50°, ICD is 80°, and thus IDC is also 50°.

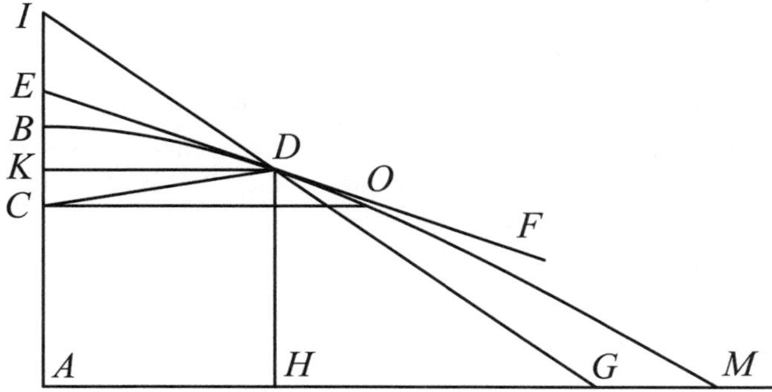

Consequently, CD and CI are equal. Therefore, where CI or CD is 100,000, DI is 128,558. And since EDC and FDG are set equal, and likewise IDE and FDG are vertical angles, therefore IDE and EDC are equal, and consequently, by Euclid VI 3, as CD is to DI, so is CE to EI. And so CE will be 43,753, EI 56,247. Now let a perpendicular be sent down from D to EC, and let this

be DK. KDC will be 10°, and thus DK will be 98,481 and KC 17,365, and the 100
remainder EK will be 26,388. And by Apollonius I 33, EB, and also BK, will be 13,194 and BC will be 30,559. But it was proved by Witelo that BA is half AM, and by Apollonius I 20, as BK is to BA, so is the square on KD to the square on AM. Therefore, where AB is 183,770, CB is 30,559.[45] Thus the parabola is found and set up, with its point, from the Witelonian refraction of the eightieth degree, and the height of the point is very nearly the sixth part of the height of the focus, or of the sagitta. And because as AB is to CB, so is the square on AM to the square on CO, therefore, when AB is multipled by CB, and the root is found, which is 74,940, AB will be to the root as AM is to CO. And since AM is 367,540, CO becomes 149,990.[46]

Let us now see how great the reflection will be at O. Let a straight line touch the section at O, and let it be OI (drawn in the imagination): in triangle OCI, two sides and the included right angle are given. For CO is at right angles to IC. And because OI is a tangent, CB and BI will be equal, and thus CI is 61,118. But CO is 149,990. Therefore, the angle OIC is 67° 50′, and IOC 22° 10′. If an equal angle be added to this,[47] or it be subtracted from OIC, 67° 50′, there remains 45° 40′, the angle which the reflected ray] makes with the axis, representing the horizontal refraction of water.

Now if it is desired to test the rest of the refractions in Witelo, we shall proceed thus. First, for the sake of an easier computation, let AM receive the round dimension of 100,000, and CB will become $8{,}314\frac{1}{2}$. Now let BC be 70°, at
which inclination Witelo puts a refraction of 24° 30′ and a refracted angle of 45° 101
30′: let this be represented by CID. KD must be taken large enough that together with KC it makes the angle KCD 70°, as it is supposed to be, and that at the

---

[45] In the previous sentence, Kepler says that

$$BA = \frac{1}{2}AM \qquad \text{and} \qquad BK : BA :: KD^2 : AM^2.$$

Combining these,

$$BK : BA :: KD^2 : 4BA^2,$$

and since BK and KD are known, BA can be found. One could proceed arithmetically, treating all these magnitudes as numbers and the proportion as an equation; or geometrically, by finding a mean proportional between BK and BA so as to remove the squares. Numerically, the result is the same.

[46] Let L be found on AB such that $AB : BL :: BL : CB$. Then since $AB : CB :: AM^2 : CO^2$,

$$AB : BL :: AM : CO.$$

But $BL^2 = AB \cdot CB$, or $BL = \sqrt{AB \cdot CB}$. This is the "root" that Kepler mentions. Thus

$$AB : \sqrt{AB \cdot CB} :: AM : CO.$$

[47] If IOC be doubled *and its complement be found*, this complement will be the angle sought. Kepler seems to have omitted a step here.

same time it be to AM in the subduplicate of the ratio that BK (the remainder of BC) has to BA. In order to pursue this artfully, always follow this rule: square the tangent of the angle KDC. To this square add the magnitude made of twice BC multipled by AM, which in these units is always 1,662,900,000. Again diminish the root of the sum by the same tangent of the angle KDC. The remainder will be KD, 18,262 at this inclination.[48] But since KD is always a mean proportional between EK and AM, and since KD is 18,262 where AM is 100,000, therefore where KD is 100,000, KE will be 18,262, which is the tangent of angle KDE, 10° 21′. KDE, 20°, added to this makes CDE 30° 21′. Twice this, CDI , is 60° 42′. And ICD is 70°. Therefore, the remainder CID to make two right angles is 49° 18′. Corresponding to this, Witelo puts 45° 30′. Thus a refraction of 20° 42′ would be shown here; for Witelo, it is 24° 30′. Therefore, this measure too is flawed, for it falls off too much near the horizon, the opposite of what happened above in the ellipse and the hyperbola. For it also shows a horizontal refraction of 44° 20′, of which, although Witelo omitted it, experience nonetheless testifies that it barely goes up to 37°. Nonetheless, it shows the required measure qualitatively, in that it makes the refractions grow with increasing additions near the horizon. For from 70° to 80° the refraction increases by 9° 18′, and from 80°to 90° by 14° 20′.

---

[48] As in footnote 46, let L be chosen on BA so that BL is a mean proportional between BK and BA. Because $BK : BA :: KD^2 : AM^2$ or $4BA^2$,

$$BL : BA :: KD : 2BA, \quad \text{and} \quad BL = \frac{1}{2}KD.$$

But since BL is a mean proportional, $BL^2 = BK \cdot BA$, and so $BK \cdot BA = \frac{1}{4}KD^2$; or, since $AM = 2BA$, $KD^2 = 2BK \cdot AM$. Further,

$$BK = BC - CK = BC - \frac{KD \tan KDC}{100,000}. \tag{4.1}$$

(The tangent must be divided by 100,000 because the radius (in this case, CD) is taken as 100,000 rather than as unity, as we would have it.) Combining these two equations,

$$KD^2 = 2\left(BC - \frac{KD \tan KDC}{100,000}\right) \cdot AM = 2BC \cdot AM - \frac{2KD \cdot AM \tan KDC}{100,000}.$$

Collecting the KD terms and completing the square,

$$KD^2 + \frac{2KD \cdot AM \tan KDC}{100,000} + \frac{AM^2 \tan^2 KDC}{100,000^2}$$
$$= 2BC \cdot AM + \frac{AM^2 \tan^2 KDC}{100,000^2}$$

or

$$\left(KD + \frac{AM \tan KDC}{100,000}\right)^2 = 2BC \cdot AM + \frac{AM^2 \tan^2 KDC}{100,000^2}.$$

Solved for KD, this becomes

$$KD = \sqrt{2BC \cdot AM + \frac{AM^2 \tan^2 KDC}{100,000^2}} - \frac{AM \tan KDC}{100,000}.$$

But since Kepler has made AM equal to 100,000, all the AM's but the first one drop out, along with the 100,000's, and this equation expresses Kepler's "rule" exactly.

Further, since the parabola is the most acute of all hyperbolas, it is easy for us to return to hyperbolas with this form of enquiry. It will thus remain at the focus as long as the medium in question is the densest, and with this only the parabola will square. Hyperbolas and ellipses will be at the service of other media, not at the focus as before, but at some point above or below the focus. For according to the various positions of light on the sagitta of the figure, there also occur various reflections, in the other sections no less than just now in the parabola. Moreover, the parabola will lead us by the hand. For in it, if we had remained at the focus, the reflection would have increased in proportion to the inclinations. Now, where a descent was made below the focus, the reflections were made to fall off near the horizon. From which it is given to understand that when the falling off is too great, the descent below the focus is **102**
too great. Again, if we had remained at the focus, the reflection of the horizon would have been greatest at 90 deg.; where we went down, we found 45° 30′. Therefore, if we go up again above the original position, we shall have more at the horizon (or at an inclination of 90 deg.)—more, I say, than 45° 30′. Thus there is more at 80 deg. than 30°. It is therefore too much. But hyperbolas diminish the horizontal reflections, even from the focus. They also decrease the falling off of the reflections of the horizon, so that there they increase with a lesser ratio than do the inclinations, although at the beginning, about the vertex, they increase in a greater ratio. The ellipses, too, do the same thing. So we shall have to descend below the focus of the hyperbola. There is no geometrical method of enquiry: luck has to be tried, a hyperbola a little more obtuse than a parabola must be taken, and two reflections investigated. If this does not do the job, another must be chosen, as if by a *regula falsi*, until we meet with the true one.

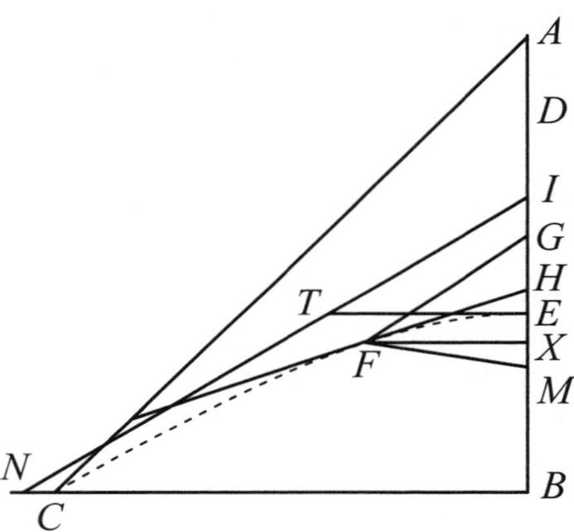

So, because in a parabola the half chord is twice the sagitta, let there be a hyperbola whose half chord BC is three times the sagitta, and on the sagitta EB let a point be chosen below the focus B, and let this be M, from which let the lines to be reflected be drawn. Let MF be drawn out such that FMG is 80°; and because FGM is 50° and GFM is likewise 50°, the tangent FH to the section at F will make FHM 75°, since it bisects GFM. Therefore, by Apollonius II 50, because the hyperbola is given, by the ratio EB to BC a tangent HF will be able to be drawn making a required angle with EB. For let A be the opposite focus: AC will be longer than CB by the space DE, the axis. And at the same time, the power[49] of AC is equal to both the squares on AB, BC. Therefore

[49] The second power, i. e., the square.

where AD is 1 and EB is 1, DE is 2,[50] or AB is 4 and BC 3 and AC 5. Next, bisect AB at I: I will be the center, and the meeting place of the asymptotes, by Apollonius II 1. But, in order to know the angle of the asymptotes, note that, by Apollonius II 10,[51] as rectangle DB, BE is to the square on BC, so is the
103 transverse side DE to the latus rectum. And conversely. Further, as DE is to the latus rectum, so the figure[52] is to the square of the latus rectum. Therefore, as the figure is to the square of the latus rectum, so is the rectangle DB, BE to the square on BC, and, permuted, as the figure is to rectangle DB, BE, so is the square of the latus rectum to the square on BC. But the figure is four times the rectangle DB, BE;[53] therefore, the square of the latus rectum is also four times the square on BC: that is, the latus rectum is twice BC. This is indeed always true, even in all three sections. Therefore, the latus rectum is 6 and the transverse side DE is 2. Therefore, the figure is 12, one fourth of which is 3, whose root is $\frac{173{,}205}{100{,}000}$. Therefore, by Apollonius II 1, the angle between the asymptote TI and BI is 60°.

Now, then, in triangle HFX (FX being sent down perpendicular to EM), because FHX is 75°, HFX will be 15°. Therefore, where FX is 100,000, XH is 26,795. But at the same time, by Apollonius I 37, as the transverse side of the figure is to the upright [i.e., the latus rectum], so is the rectangle IX, XH to the square on XF. So when one third of the square on FX (100,000) is divided by XH, 26795, the result is IX, 124,401, where FX is 100,000. But by Apollonius I 21, as the latus rectum of the figure is to the transverse side, so is the square on FX to the rectangle DX, EX. Therefore, the rectangle DX, EX is equal to the rectangle IX, HX, or one third of the square on FX. Accordingly, subtract one third of the square on XF from the square on XI, and the root of the remainder is the semiaxis IE. Therefore, the value of EX that comes out of this is 14, 209. And because XMF is 80°, XFM will be 10°. Consequently, where XF is 100,000, XM will be 17,633. Therefore, EM, the height of the point sought, is 31,842. But IE is 110,192, half the axis, to which the height of the focus EB was demonstrated equal at this place. Therefore, where EB is 100,000, EM is 28,897.

Now that a point has been found in the chosen hyperbola, from which a straight line going forth to the surface, making an angle of 80° with the sagitta, would be reflected into another line which makes an angle of 50° with the same axis, let us now again enquire how great the angle FGM becomes if FMG be 70°. The perpendicular FX should be made of such a size that, together with

---

50 In the equation $AC^2 = BC^2 + AB^2$, replace AC with $DE + BC$ or $DE + 3EB$, BC with $3EB$, and AB with $DE + AD + EB$ or $DE + 2EB$. The resulting equation in two unknowns reduces to $2EB = DE$.

51 The proposition that states what follows is I 21, not II 10.

52 Apollonius uses the term "the figure" to denote the rectangle formed by the transverse side and the latus rectum (i. e., the "upright side"). Thus, calling the latus rectum L, we have here $DE : L :: DE, L : L^2$, an application of Euclid VI 1.

53 In Apollonius III 45, the "point of application" (i.e., point B) is defined by this equality, and its focal properties are then deduced in this and the following propositions.

XM cut off, it includes the angle XMF, 70°; and that the remainder XE make with XD a rectangle equal to one third of the square on FX.

The procedure of the computation has difficulty. The line DE, 220,384, has to be added to the line EM, 31,842. The sum, 252,226, must be multiplied, both by EM, 31,842, and by the tangent of the angle 20°, namely, 36,397. Again, this 36,397 must be multiplied by EM, 31,842, and by itself. Next, the product of 36,397 and 252,226, and the product of the same 36,397 and 31,842, are to be joined, so as to make three factors in place of four; and let three times all of these be taken.[54] Now, from three times the square of 36,397 let the last five digits on the right be cut off, and let this abridged number be subtracted from 100,000.[55] Divide three times the composite[56] by the remainder,[57] and divide three times the product of 252,226 and 31,842 (increased by 5 zeroes at the right) by the same.[58] Once this is done, square half of the prior quotient, and to this squared number add the posterior quotient. You shall diminish the side or root of the sum by the quantity of half the prior squared quotient:[59] the remainder shows FX in this second establishing of it, as large as we are ordered to make it by the command of the hyperbola.[60] And by the procedure, FX will become 68,550, and XM 24,950 of the same units. For when FX was 100,000, XM was 36,397. Before, however, EM was 31,842. So when XM is subtracted from this, it leaves EX, 104

---

[54] The first product is $(DE + EM) \tan 20°$, and the second is $EM \tan 20°$. When the two are added, the sum is $(DE + 2EM) \tan 20°$. This whole expression is to be multiplied by 3.

[55] The number 36,397 is the tangent of 20° (in Kepler's units), and the expression developed here is

$$100,000 - \left(\frac{3 \tan^2 20°}{100,000}\right).$$

For future reference, we shall call this k.

[56] That is, three times the expression $(DE + 2EM) \tan 20°$.

[57] That is, by k.

[58] This expression is

$$\frac{3(DE + EM)EM}{k} \cdot 100,000.$$

[59] Kepler calls it the "squared quotient" only to identify it: half of the quantity itself is subtracted, not the square of its half.

[60] A careful reading of this nearly impenetrable description will show that Kepler has constructed the following equation:

$$FX = \sqrt{\left[\frac{3(DE + 2EM) \tan 20°}{2k}\right]^2 + \frac{3(DE + EM)EM}{k} \cdot 100,000} - \frac{3(DE + 2EM) \tan 20°}{2k}.$$

He undoubtedly derived this expression using what he describes below (p. 120) as "Geber's assurances"; that is, algebra.

6892; and when DE, 220,384, is added to this, it establishes DX as 227,276; and this, multiplied by EX (and this will serve us as a test), produces a quantity equal to that obtained if you should square the new value of FX, 68,550, and take the third part of its square. And you shall divide this third part of the square of FX by IX, 117,084 (that is, the sum of IE, previously known to be 110,192, and EX, now found to be 6,892). In this way the quantity HX, 13,379, will be established, whose ratio to FX, 68548, is as 19,518 to 100,000. It therefore shows the angle HFX to be $11^\circ\ 3'$. But XFM was assumed to be 20°, and therefore the sum HFM is $31^\circ\ 3'$. But from what was said before, GFM is twice this, and thus is $62^\circ\ 6'$. But GMF was by hypothesis established as 70°. Therefore, the remaining angle FGM comes out to be $47^\circ\ 54'$. But from Witelo's observations it should have been $45^\circ\ 30'$. In the parabola it came out $49^\circ\ 18'$. See how we have come down from the parabola to this obtuseness of the hyperbola: we have come one degree and $24'$ nearer to the target; and since we are still off, we have to carry on to more obtuse hyperbolas.

That also shows that we are on the way. In the parabola, the height of the point was one sixth of the sagitta; here it is a little less than one third. For IE (or
105 in this hyperbola, EB, which is equal to it) is 110,192 parts, while EM is 31842. We have therefore gone up. And this is what we said before ought to happen. Therefore, where we select another, more obtuse, hyperbola, we shall go farther up.

So, let there be a more obtuse hyperbola, the ratio of whose sagitta to half the chord is one to four. EB will be the sagitta and ED the transverse side of the figure, or the equal axes,[61] and the latus rectum of the figure is eight times the transverse (in the previous one it was three times). For that reason, where before you had tripled, you now multiply by eight. Briefly, so as to repeat the whole previous process in a word, the angle FGM comes out to be $44^\circ\ 58'$, which previously was $47^\circ\ 54'$. From Witelo, it ought to have been $45^\circ\ 30'$. And since EB is 60,868, the height EM of the required point becomes 33,850, more than half, which previously was one third, and in the parabola was one sixth. You now see that we have passed the limit in the obtuseness of the hyperbola, but just barely. Come now, if proportional computation delights you, or try new and intermediate hyperbolas, and the rest of Witelo's angles. I say that what will happen is that you find such a hyperbola, and in it such a point, from which the images of things positioned behind it will appear in exactly that form in which they appear under water; that is, that by this hyperbola and point is contained the measure of all the refractions of water; and besides, that by other hyperbolas are contained the measures of media that are different with respect to refractions, and these are by means of the angles of reflection that occur in concave hyperbolic mirrors.

---

[61] Substituting the appropriate magnitudes into the equation for the sides of the right triangle ABC, as before, one finds that DE and EB are equal in this hyperbola. These two are the "axes" in the sense that they are collinear parts of the axis AB. Kepler could not be referring to the conjugate axis here, since it would be $2\sqrt{2}$ times the transverse side DE.

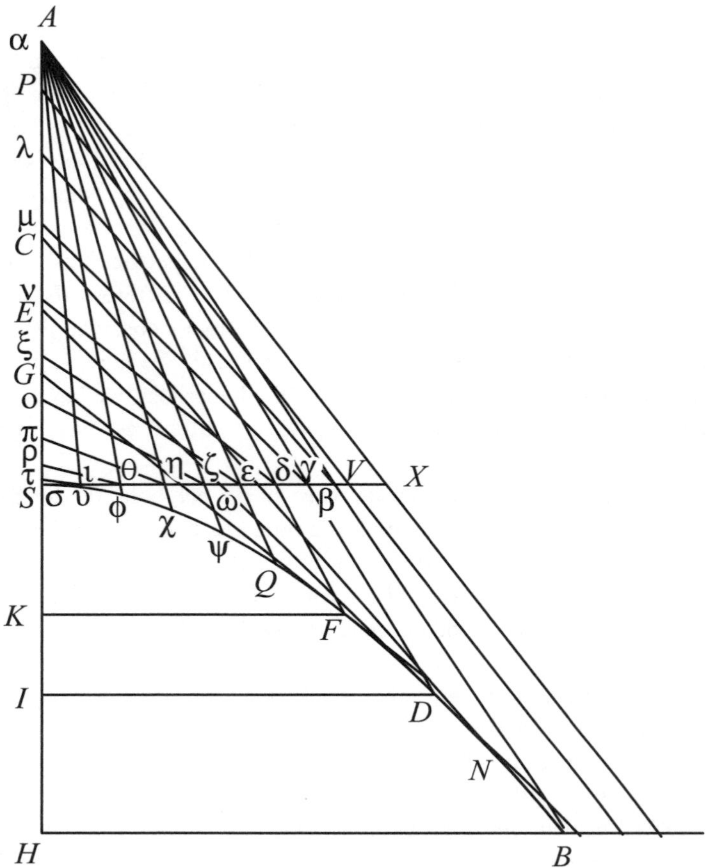

That was able to satisfy even the most enquiring mind, except that mechanics, and the consideration of the eye that follows in chapter 5, had drawn me back into a new labor. For we have hitherto used hyperbolas, and the mirror reflections from them, especially to represent the distorted images that water show us, when a refraction of rays of objects immersed in it has occurred. And indeed, those things, reflection and refraction, are almost entirely different in kind. The question therefore was raised, what sort of single, continuous surface of water it would be, that would prevent all the radiations received from some nearby point, and diverging in different directions, from diverging once refraction has occurred, but would send them onwards parallel. Whether it would be a parabola, or a hyperbola, or an ellipse, was for a long time in doubt. In favor of the parabola was the equidistance which the parabola shows in reflection. For the hyperbola, there spoke anatomical experience, which is treated below in the consideration of the eye.

God immortal, how much time have the assurances of Geber[62] lost me! I **107**

---

62 That is, Algebra (from al-Geber). However, Frisch (*JKOO* II 409) conjectures that this remark had something to do with an actual work by Geber that was published by Peter Apianus in 1534 as an appendix to his *Instrumentum primi mobilis*, and notes that

shall nonetheless add a diagram with a problem, if anyone be perchance aroused with desire for this cross. I shall prefix a geometrical demonstration that a surface similar to the hyperbola is required. From point A let radiations $\alpha\beta$, $\alpha\gamma$, $\alpha\delta$, $\alpha\varepsilon$, $\alpha\zeta$, $\alpha\eta$, $\alpha\theta$, $\alpha\iota$, $\alpha\sigma$, go forth, so as to make $SA\beta$ 30°, $SA\gamma$ 24° 30′, $SA\delta$ 19° 30′, $SA\varepsilon$ 15°, $SA\zeta$ 11°, $SA\eta$ 7° 30′, $SA\theta$ 4° 30′, $SA\iota$ 2° 15′, the magnitudes of Witelo's angles of refraction. Now let a perpendicular to AS be set up from S through ι, θ, and the remaining letters. And at the letters let there be added angles of such a magnitude as is the altitude given by Witelo to any angle of refraction (he himself gives the inclination, the complement of the altitude): $\alpha\beta\lambda$ will be 10°, $\alpha\gamma\mu$ 20°, $\alpha\delta\nu$ 30°, $\alpha\varepsilon\xi$ 40°, $\alpha\zeta o$ 50°, $\alpha\eta\pi$ 60°, $\alpha\theta\rho$ 70°, $\alpha\iota\tau$ 80°, $\alpha\sigma\kappa$ 90°.[63] What is required is to say what sort of surface it is upon which these radiations in this position coming forth from α strike, so that they strike just as do here in the lines $\beta\lambda$, $\mu\gamma$, and so on, so that these lines are either tangents to that surface, or lines equidistant to the tangents. For, such a surface of water being given, it is certain that the radiations diverging from A will turn out parallel after refraction. For a straight line striking upon parallels makes equal angles on the same side, and conversely. Therefore, since these are refracted at such an angle as those coming forth from A, they pass over into parallels after refraction. But they are refracted at an angle of such a magnitude, if they strike thus, as was said, which the experience of Witelo showed us in place of a hypothesis. Now, if $S\beta$ be tangent to the required surface at S, while the surface is to encounter the radiation $A\iota$ at obliquity $\tau\iota$, it therefore must be bent from S below ι to υ; and $\alpha\iota$ extended will encounter the surface beyond ι (at υ, say). But at the point of incidence υ there will touch some line parallel to $\tau\iota$. And whatever that line is going to be, it must not cut AS below S. For it would cut the surface extending all the way to S, to which it is nonetheless supposed to be tangent. Therefore, a tangent to this surface at υ cuts the tangent $S\iota$ at an intermediate place between S and ι. Let it be extended until it also cuts $\alpha\theta$ below θ. Therefore, again, because the required surface must meet $A\theta$ at the obliquity of the line $\rho\theta$, and has as a tangent at υ a line parallel to $\tau\iota$, it therefore must pass below the intersection to θ, and the line that is tangent to it at the place of meeting (ϕ, that is), will pass equidistant to $\rho\theta$ between ι and υ, and will cut $\alpha\eta$ below η; and there
108 again, having emulated the obliquity $o\zeta$, the surface will meet with $\alpha\chi$ below this intersection: that is, at χ. And a tangent drawn through χ equidistant to $\pi\eta$ will meet with $\alpha\zeta$ extended at ω. But the surface will pass below ω, so as to meet $\alpha\zeta$ at the point ψ, where a line parallel to $\xi\varepsilon$ is tangent. The same is to be understood of Q, F, D, and B. In this way, a pair of tangents to the surface always cut each other at places in between the points of tangency. And such an intersection is always given, since $\lambda\beta$, $\mu\gamma$, $\nu\delta$, and the remaining pairs always converge towards the direction $S\beta$, as is obvious from the angles. For the exterior angle $S\lambda\beta$ is equal to the interior and opposite angles, which are the angle of

Henry Briggs, in the letter cited above (p. 112), understood Kepler's remark in this way.

[63] Note that the points σ and κ coincide here, and are denoted by the point S. Kepler nevertheless treats them as though they contained an angle.

refraction $\lambda\alpha\beta$ and the angle of altitude $\lambda\beta\alpha$. Angle λ is 40°, μ 44° 30′, ν 49° 30, ξ 55°, o 61°, π 67° 30′, ρ 74° 30′, τ 82° 15′. So, since the lower is always greater, and goes to the same line AS, the remaining pairs will therefore meet on the right side. Thus since $\alpha\beta\lambda$ always becomes less in the direction of X, and finally vanishes at X, where SAX represents the horizontal refraction, and the altitude of AX is zero; and further, since between β and X there can be infinite radiations, each striking at a more inclined tangent; and since the tangents, too, the more they meet obliquely and at a smaller angle with the radiations and the surface, the more distantly they touch (as points BD are more distant than DF): therefore, the surface will be infinite, beginning, that is, from the magnitude $\sigma\upsilon$, accumulating $\upsilon\phi$ and so on, greater than $\sigma\upsilon$, an infinity of times; and will always situate itself more nearly to the straight line AX, because it always has longer parts with a smaller inclination—that is, straighter. These, moreover, are found only in the hyperbola, not in the parabola, which tends toward a straight line parallel to the axis, not one meeting the axis, as XA here. Consequently, if the surface is indeed a hyperbola, AX either is an asymptote or is equidistant from an asymptote PV that is farther in (which computation shows). Moreover, that this surface is not only similar to a hyperbola, but is also really a conic section, would immediately be proved if the converse of Apollonius II 29 and 30 were taken. For when a tangent $\tau\upsilon$ is drawn, it can either be drawn through S or through υ, and therefore through all the intermediate points; and therefore, both passing through that point through which the straight line has been drawn, and falling upon an intermediate point between S and υ, it is directed into the center of the figure. This is to be said of all other tangents. But because all these are also contained in other kindred surfaces turning slightly aside from the hyperbola, the theorem **109**
cannot be completely deduced.

Therefore, if it is wished to enquire whether this figure contain the rest of the properties of the hyperbola, I have gone off to Geber,[64] with the following things given. From B, D, F, let perpendiculars BH, DI, FK, fall to AS, applied ordinatewise,[65] and let BC be drawn equidistant from $\beta\lambda$. Likewise, let DE be drawn equidistant from $\gamma\mu$, and FG equidistant from $\delta\nu$, tangents to the surface at B, D, and F. Thus three angles of refraction from Witelo, with the same number of their inclinations or altitudes, give three classes of numbers, independent of each other and unconnected. If these plainly allowed of being so connected together that at the same time they should submit themselves to the rules of the hyperbola, the hyperbola will be given, and the hyperbolic surface will compute the refractions. For first, from the condition of Witelo's angles, where AH is 100,000, HB will be 57,735; CH, 68,806. Likewise, where AI is 100,000, ID will be 45,573; EI, 46,376. And where AK is 100,000, KF will be 35,412; KG, 30,249.

---

[64] Algebra.

[65] Apollonius I definition 4: "Of any curved line which is in one plane, I call that straight line the diameter which, drawn from the curved line, bisects all straight lines drawn to this curved line parallel to some straight line; . . . and I say that each of these parallels is drawn ordinatewise to the diameter."

Now we need to find the ratio of the sides of the figure, and the distance AP of the center P from A. And the semiaxis is PS, from these laws, because we are in the hyperbola, as HBH (which for the sake of brevity shall be the notation for the square of HB) is to IDI and KFK, so is the rectangle CHP to EIP and GHP: that is, as the latus rectum of the figure is to the transverse side. And at the same time, CPH, EPI, GPK, and PSP are equal, by Apollonius I 37. Also among the beginnings, for assistance, are Apollonius II 29 and 30, because by its power you will be able to establish for yourself the bounds within which the unknown HI lies. If you prevail in showing that through three angles of this kind what is ordered is more than sufficient, and that the case of a certain hyperbola can be conclusively proved by only two, the operation will be all the more clear, and it will immediately become evident in another combination of angles whether a surface of this kind is a hyperbola. Mechanics shows that it has no better resemblance to anything else; nevertheless, it is a little more acute than the hyperbola itself near the vertex. When you shall have acquired perfect knowledge of this surface by some procedure, know that you have achieved something great in mechanics. Optical mechanisms using crystals.

6. Causes of the quantity of refractions

Forsooth, reader, I have kept you and myself hanging long enough now,
110 while I tried to gather the measures of different refractions in a single packet, meanwhile acknowledging that the cause is not in this measure. For what do refractions, which we have established to be fundamentally in the plane surfaces of transparent media, have in common with conic sections, which are mixed lines? For that reason—may God look kindly upon us—we shall now also busy ourselves with the causes of this measure. For even if we shall perhaps still stray somewhat from the goal, it is nonethless preferable to show our industry in looking around, rather than our lassitude in inaction. If among the optical propositions above we have explained the cause of refractions correctly in general, the specifics must also be correctly derived from the same source. But in prop. 20 above, we proposed as a cause the resistance of the medium, by which the spreading of light is hindered, by material necessity. Now it must be seen whither we are able to arrive by following these tracks.

Proposition 1

Where light strikes more obliquely, it is refracted in a greater angle. This is Witelo X 14, but faultily and obscurely demonstrated. If one were to deny this, he would indeed be taking on a great deal of toil in order to demonstrate it legitimately. Therefore, I must make an attempt at another demonstration, which goes like this. Unless the refractory angle were to increase continually with the obliquity of incidence, there would be no universal cause of refraction. For since the oblique incidence is the basis for the spreading of rays, when the former increases, the spreading will increase. But if the angle of refraction were to stop increasing at some incidence (80 degrees, say), at incidences of both 80° and 82° it would be 30 degrees; therefore, the medium would resist the spreading up to an inclination of 80°, from which, over the 2 degrees to 82°, it would no longer resist. But Ch. 1 Prop. 20 of this book demonstrates that it always resists; therefore, the angle of refraction always increases as the obliquity of incidence increases.

Hence, there is a corollary.

If the medium itself were to be considered in isolation, in respect to its density, the angles of refraction would become proportional to the angles of incidence.

Proposition 2

When light strikes more obliquely, the resistance from the same medium becomes greater than at a more direct incidence, even with respect to the same medium. For since refraction is a product [*Affectus*] of motion when light strikes upon the surface of a denser medium, and the surface, because of its infinite points, 111
terminates, or rather affects here, the infinite motions of the infinite points of light, while it takes on a consideration of density in this regard, not less than the corporeal bulk itself, it will accordingly produce a greater effect if in a certain respect a denser medium encounters the light. But a denser medium encounters light obliquely. Let A be the light, BC the denser medium, AB, KM parallels, or nearly so, from the sun, whose perpendicular distance is ML. So, since BLM is right, and LBM is by supposition an acute oblique angle, LBM will therefore be less than BLM; and the side LM opposite the smaller angle B, will be less than the side BM which is opposite the greater angle L. But LM is the measure of the breadth of the medium encountering the light coming down directly, because BLM is right, while BM measures the breadth of the medium encountering the light obliquely. Therefore, there is more density in BM than in LM. Therefore, the resistance, in this respect, is greater.

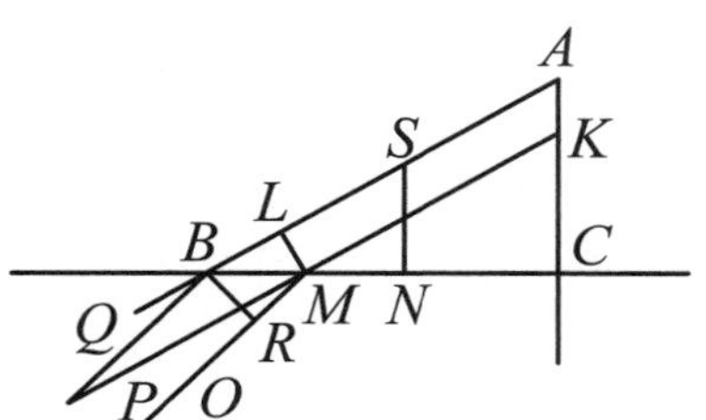

Proposition 3

The angles of refractions increase with greater incremental proportions than does the obliquity of incidence. For by the corollary to the first, if even the density alone be considered, the angles of refractions will be proportional to the incidences. Now, however, by the second, the ratio of LM to MB will also mix itself in. Therefore, the angle of refraction is compounded of one ratio, which is proportional to the incidences, and another, which is proportional to the lines BM. But the lines BM initially increase very little, at a low incidence increase greatly, as the table of secants shows, where ever greater and greater secants correspond to equal degrees. Therefore, part of the angle of refractions is proportional to the incidences, and part increases with greater increments of proportion. Thus the whole angle increases with greater increments.

Proposition 4

There is no difference between the refraction of weak light and denser light, other things being equal. For if the light that is weaker were more overcome by

the denser medium, the refraction would be a characteristic of light itself, owing to its essence, and not of motion. But refraction is only the diminution or inhibition of the motion of spreading, and this motion does not accept any modification from the density or dilution of light, i.e., from its strength or weakness,
112 nearness or remoteness, since all motion of light, without exception, is perfectly swift, and simply instantaneous. Therefore, refraction is not at all varied by the weakness of light. The force of the proposition is more evident in an example. In the preceding diagram, let the radiating point be at A: thus, the surface of the medium enclosed between the perpendicular AC and the oblique radiation AB is represented by BC. Now let A descend on the same radiation AB to S, and, a perpendicular [SN] being dropped, BN will be less than BC; but BN represents the surface contained between the perpendicular SN and the oblique radiation SB. And there is the same amount of light on BN as there was before on BC, which is greater, when the point A was more distant. Therefore, if as a consequence of the light of A on BC being more dilute than the light of S on BN, it were that much more refracted, there would as a result occur different refractions at the same incidence AB by the same medium, according to the approach or departure of A on the line AB. This, however, is in conflict with experience. For the rays of all illuminations, whether approaching or receding, that fall upon the surface of the same medium at the same angle are refracted at the same angle. This is most readily experienced in water, whose refractions are most evident. In air, Tycho Brahe put it to a test, in finding the same refractions of the sun and moon, the former of which is twenty times as far away as the latter. And although in the fixed stars a very few minutes are wanting, no sane person would assign the cause to their distance, if he were to consider that Tycho ascribes the same refractions to the fixed stars and to the planets, some of which are nearer than the sun. I shall accordingly give consideration to the cause of this below. It is therefore in vain that Rothmann, in Tycho's *Epistolae astronomicae*, vol. 1 p. 124,[66] opines that unless the cause of refraction be foist upon an adventitious vapor, which should not stand above the observers' zeniths, it would happen that the refractions of the fixed stars would come out different from the sun and the planets. And although Brahe, in p. 280 of the *Progymnasmata*, suspected that the moon's refractions are greater because of its proximity, he nonetheless found, on p. 124 (which he wrote later), the same refractions as the solar ones, as I have just said.[67]

Greater distance of illumination does not increase refraction.

Nor will refraction show the moon in conjunction with a star, if they would have appeared separately without refraction.

Proposition 5

Near the horizon, refractions increase in suddenly decreasing proportion to
113 *the increments.* For since two causes alone combine, there is no third (by the preceding).[68] Therefore, one cause is proportional to the inclinations, and the

---

[66] *TBOO* VI p. 152–3. The passage cited by Kepler actually starts on p. 123 of the 1596 edition.

[67] *TBOO* II pp.287 and p. 136. However, despite Kepler's claim, the lunar refractions in the table on p. 136 are slightly different from the solar refractions on p. 64 (p. 79 of the 1602 edition).

[68] Proposition 3.

other increases with the secants of the inclinations. For because BLM, above, is right, and BCA also; and LBM is common to the two triangles, the triangles will be similar. Therefore, as CA is to AB so is LM to MB. But AB are the secants of the inclinations ABC, which at the end of the quadrant suddenly increase. Therefore, BM do so too. But we have said[69] that it is in the ratio of BM to ML, or approximately so, that the refractions increase, apart from their increasing in proportion to the inclinations themselves. Consequently, the refractions also increase suddenly.

Proposition 6

BM, however,[70] is taken into consideration twice,[71] according as, when it makes one or another set of angles with the rays undergo refraction in either a rarer or a denser medium. Therefore, it must be known that *it is the secants of those angles of incidence that are set up at the surface in the denser medium that contribute to the measure of the refractions.* This is clear from a [reduction to] absurdity. For since near the horizon the secant becomes infinite, if the rarer medium be considered, its effect should also be infinite, and all horizontal radiations should be refracted at an infinite angle. Which is absurd to state. Therefore, it is not the secants of the angles in the rarer medium that are to be taken as a measure; that is, it is not the ratio BM to ML but the ratio BM to BR. Besides, even if it were only a whole right angle (not more) that is set up by an infinite secant, a single ray from the depth of water or any other medium, exiting perpendicular at a single point of the surface, would be scatterd over the whole hemisphere in the free air. For the path of the entering and departing ray is the same. And here there would be two paths supposed for the entering ray, one from the perpendicular, the other parallel to the surface of the water, and this in all directions.[72] In demonstrating the true cause of this directly and *a priori*, I am stuck. Perhaps it is this, that since the angles of refraction are in the dense surface, and so is the cause resisting the scattering of light, it is appropriate also to seek the measure within.

Tell in which lands, and for me you will be the great Apollonius.

Proposition 7

In the densest medium, all refractions take place on the perpendiculars themselves, and are equal to the inclinations. In less dense media, those that have

---

[69] In proposition 3, Kepler states that "the angle of refraction is compounded of one ratio, which is proportional to the incidences, and another, which is proportional to the lines BM." He does not specify that it is the ratio BM to LM that is to be taken, and for good reason: in Proposition 6 he will show that one must use the secant of the angle that the refracted ray MO makes with the perpendicular. This is equal to angle RBM, whose secant is the ratio of BM to BR. It is not the same ratio as BM to ML, but the two ratios are approximately the same; hence, Kepler adds "or approximately so" below.

[70] The printed text has only the letter "*a.*" here, undoubtedly an abbreviation for *autem*, which is what Frisch, *JKOO*, vol. II p. 198, gives.

[71] Cf. Proposition 3 above.

[72] The marginal note is a parody of Vergil, *Eclogue* 3, 104: *Dic quibus in terris, et eris mihi magnus Apollo.*

a smaller refraction make it suddenly diminish near the horizon, and begin to be
114 *perceived later.* For the highest density, by reason of its infinity, causes so great a refraction as to put a stop to all dispersion of light. Therefore, it makes light descend perpendicularly, and appropriates every angle of incidence, leaving as a result nothing for the ratio of the secants. For, by the premise, the angles beneath the water will be forever right, and therefore the secant forever the same. All that remains, then, is the ratio of the incidence. But where the refraction is small, by the premise, there is also too little removed from the quadrant for finding the last secants, at a time when they now increase vigorously. On the other hand, where the density is great, the horizontal refraction is great, and too much is removed from the quadrant for extracting the secants of the last refractions. Thus in the former instance, the last refractions increase in a great proportion, while in the latter, in a lesser proportion. Finally, if the greatest refraction of a medium of low density, such as air, is so many times as great as the inclinations of the refraction, and as the smaller secants of the refraction, it is in fact, as supposed, very small indeed; and therefore, any fraction of it that corresponds to the inclinations that are not yet completely horizontal, will be much less perceptible. And so it will begin to be perceived slowly and only near the end.

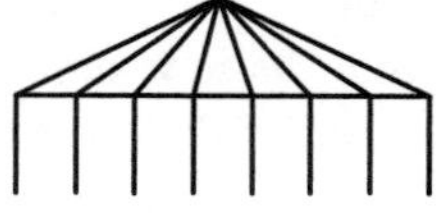

Proposition 8

Problem 1. *From a known composite refraction*[73] *of any inclination, to hunt the elements of refraction, and the composite or whole refractions of the remaining inclinations.* Let the medium be water, the inclination 80°. The refraction, from Witelo, is 30°. I shall take this since it is larger, and less subject to experimental error. So, that which is under water is inclined 50 degrees from the perpendicular, the secant of which angle is 155,572. As this is to the secant of the angle 0°, that is, to the right sine,[74] so is the composite refraction, 30°, to the proportional refraction of the inclination 80°. For this has been demonstrated in the preceding. Thus the refraction that is simple or proportional to the inclination, for an inclination of 80°, is 19° 17′, to which is added 10° 43′ because under water, in refraction, it encounters a medium that is denser than in straight motion, by the ratio of 155,572 to 100,000. Once the simple refraction of an inclination of 80° is obtained, let there be a distribution to the other inclinations, since the simple refraction is proportional to the inclinations, angle for angle. Next, let any one be multiplied by the secant of the refracted ray, which is not yet fully known. And

[73] From the preceding propositions, the refraction is the combined effect of the angle of incidence and what might be called the apparent density, which is measured by the lines BM (or, more accurately, by the secants of the angles of the refracted rays); hence, Kepler calls it the "composite refraction."

[74] The sine of 90°, or 100,000.

let this search for the secant be iterated sufficient times that there be no remaining discrepancy. This could be done algebraically,[75] if there were also a way of going from straight lines to curves in algebraic operations. For example: for an inclination of 50°, the rays entering through the air, from the distribution of the simple refraction, are assigned 12° 4′ for a simple refraction. But this is not yet the total refraction. I nonetheless shall proceed as if it were the total, and subtract it from 50°. The remainder is 37° 56′, whose secant, 124,787, or the excess over the whole sine, 26,787, multiplied by 12° 4′ and divided by the whole sine, gives 3° 14′ for the part of the refraction that comes from the secants, which previously likewise had to be subtracted from 50°. Let this now accordingly be subtracted from the now diminished 37° 56′. The remainder is 34° 42′. The excess of the secant of this over the whole sine is 21,633, which, multiplied by 12° 4′, gives 2° 37′. Therefore, when we subtracted 3° 14′ before, it was too much. So subtract now this 2° 37′ from 37° 56′: the remainder is 35° 19′, the excess of whose secant is 22,554, which gives 2° 43′. This still differs from 2° 37′. So subtract this 2° 43′ from 37° 56′: the remainder is 35° 13′, the excess of whose secant, 22,402, multiplied by 12° 4′, gives 2° 42′. So, because this 2° 42′ differs imperceptibly from the former 2° 43′, I rest here, and declare that the refraction due to the secants at the inclination of 50° is 2° 42′, and therefore, the total, by adding the proportional part, 12° 4′, is 14° 46′. I accordingly introduce below the entire table of the refractions of water, and add the refractions published by Witelo which he found out with his instrument, so as to show the agreement.[76] **115**

Distance of Radiant from Zenith, in Rare Medium	*Part of Refr. Proportional to the Inclinations Deg. Min.*	*Addition because of Secants Deg. Min.*	*Whole Demonstrative Refraction Deg. Min.*	*Witelo's Experimental Results*	*Difference*
10	2. 25.	0. 1.	2. 26.	2. 15.	0. 11. –
20	4. 49.	0. 10.	4. 59.	4. 30.	0. 29. –
30	7. 14.	0. 35.	7. 49.	7. 30.	0. 19. –
40	9. 39.	1. 23.	11. 2.	11. 0.	0. 2. –
50	12. 4.	2. 42.	14. 46.	15. 0.	0. 14. +
60	14. 28.	4. 40.	19. 8.	19. 30.	0. 22. +
70	16. 52.	7. 19.	24. 11.	24. 30.	0. 19. +
80	19. 17.	10. 43.	30. 0.	30. 0.	0. 0.
90	21. 43.	14. 47.	36. 30.		

This tiny discrepancy should not move you; believe me: below such a degree of precision, experience does not go in this not very well-fitted business. You see that there is a large inequality in the differences of my figures and Witelo's. But my refractions progress from uniformity and in order. Therefore, the fault **116**

[75] *Per Cossam. Cossa* (from the Italian *cosa*, "thing") originally denoted the unknown in an equation, and later came to refer to the theory of equations as a whole.

[76] Witelo 10 8.

lies in Witelo's refractions. You will believe this all the more, if you look to the increments of the increments in Witelo. For they increase through 30 minutes. It is therefore certain that Witelo laid his hand upon his refractions gathered from experience so as to bring them into order through an equality of the second increments. For the straight lines that are taken from a circle, or from any sort of thing of a circular nature, of which kind refractions are, have increments that vary infinitely, and never become equal.

Proposition 9

Problem 2. *Given two refractions of the stars arising from the air in a certain altitude above the ground, to seek out the refractions of the other inclinations.*[77] The business is veiled in many perplexities. Let A be the center of the earth's orb BC, and likewise the center of the sphere of air DE. If both the surfaces BC and DE were plane and parallel, from a given angle DBE and a given refraction at that angle upon the surface of the earth (let this be GEF), the inclination above the surface of the air of the free ray EF in the aether, would also be given. Now, because BC and DE are spheres, let BEG be the tangent to the earth at B, and IEH the tangent to the air at E. So, from the given inclination DBE of BE above the surface BC, it does not follow that either the inclination GEH of the same GEB, or the inclination FEH of the refracted ray FE, above the surface of the air, is immediately given. But if only the ratio AE to AC, or the altitude of the airy surface EC that introduces the refraction, were known, the angle too would immediately be known. For since EB is tangent to BC at B, the line AB drawn from the center A to the point of tangency B, will be perpendicular to BE, and likewise AE will be perpendicular to EH. Thus EBA is right, and IEA is also right. And BEA, BAE together are equal to a right angle. Therefore, the common angle BEA being subtracted, IEB or GEH and BAE will be equal. But from the altitude of the air CE being given in proportion to BA, EA and BA and EBA would also be given; and consequently
117 also EAB and hence also GEH, equal to the latter. And, when the refraction GEF is subtracted, there would remain FEH, the inclination of the refracted ray FE above the tangent EH. But in fact, EC, the altitude of the air making the refraction is unknown. For as for the optical writers' demonstration that the

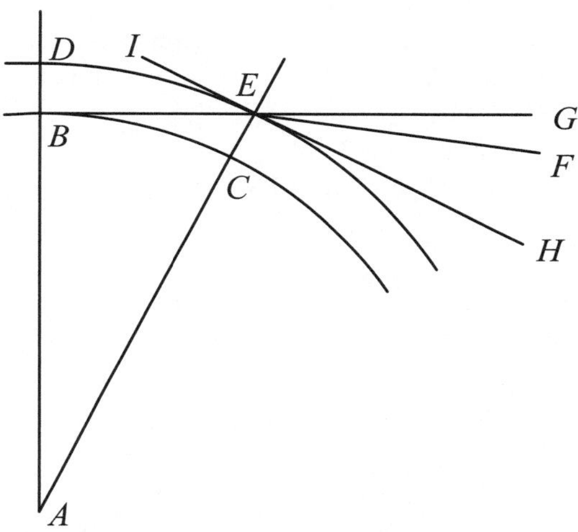

[77] Here Kepler refers to the note introduced on his p. 79, on the effects of vapors. The note is on p. 95 of this edition.

altitude of the matter of twilight is equal to twelve miles, it does not immediately follow that it is the same matter that makes the twilight and the refractions. We might now suppose that what refracts the rays of the stars is a liquid matter, moist and heavy, distinct from the matter of water only in degrees. It will thus follow that if there exists here on earth something dry, smoky, and fiery, it is going to force its way above this humid body no differently from the way air forces its way up out of water. On this account, the smoky matter will occupy places that are higher than this refractive matter; and, being there, will be illuminated (since it is material) to the brightness of twilight, while not bringing about any refraction. For the definition of the transparent, by ch. 1 prop. 11, does not belong to smoky matter, and hence neither does refraction. Matter of twilight

Surounded by these difficulties, let us try approaches, not the geometrical ones that we would like, but rather *gauche*[78] ones, comparing pairs of refractions of pairs of inclinations. And at the beginning let us establish reliable inclinations of the same ray in air and on earth, whose discrepancy is imperceptible. For to the extent that the ratio EC to CA is smaller, this discrepancy of angles is smaller, and is perceived more slowly. So, for example, let EC be 95 where AC is 100,000. The incidences will be as follows.[79]

On Earth		In Air
90°	. . . .	87° 30′
89°	. . . .	87° 19′
88°	. . . .	86° 48′
87°	. . . .	86° 6′
86°	. . . .	85° 17′
85°	. . . .	84° 25′
60°	. . . .	59° $54\frac{1}{2}$′

Let us therefore suppose that at an inclination of 60° or an altitude of 30° the rays do not differ perceptibly. At this altitude, however, Tycho, *Progymnasmata* book 1 fol. 79, supposes a refraction of the sun of 1′ 25″.[80] And the secant of the angle 60° is twice the radius. Therefore, the simple refraction proportional to the angles is 43″. Therefore, to one degree correspond 43‴. Now let us pass on to **118**
consider the remaining inclinations. At an altitude of 1° or an inclination of 89°, the sun is refracted by 26′, as is stated on the same page. It is required to find

78 ἀτέχνους. It was the fashion of the day to use the occasional Greek word, in much the same way that we might use French. Where possible, I adopt this analogy in the translation.

79 It should be kept in mind that these are angles of incidence of the *same* ray BEG to the respective verticals BA and EA: that is, refraction is ignored for the moment. Kepler assumes, as a first approximation, that the altitude of the surface of the air is 95 where the earth's radius is 100,000. From this, he calculates the angle of incidence of the ray GE at the surface of the air, which is the same as the angle BEA. He does this for seven values of the angle DBE, which is now no longer required to be right. The results are in the following table.

80 *TBOO* II p. 64.

how much the altitude of the sun is refracted above the air.[81] It is required to find the number of degrees which, multiplied by 43‴, and the product multiplied by the secant of that number of degrees, with the last five digits dropped, amounts to 26′.[82] Let it be 84° 16′. 1` 24° 16′ multiplied by 43‴ gives a proportional refraction of 1′ 0″ 23‴ 28⁗.[83] The secant is ten times the radius.[84] Therefore, the refraction is 10′ 3″.[85] It should have been 26′. Therefore let the trial angle be 87° 8′. Multiply 1` 27° 8′ by 43‴: the simple refraction comes out to be 1′ 3″.[86] The secant is twenty times the radius. Therefore, the refraction is 20′.[87] It should have been 26′. Again, let the trial angle be 87° 48′. Multiply 1` 27° 48′ by 43‴: the simple refraction comes out to be 1′ 3″. The secant is twenty-six times the radius. Therefore the refraction is 27′ 20″. It should have been 26′. Therefore, in air that ray is inclined in refraction by 87° 47′.[88]

If the two refractions were correctly determined here, we would now have reached the goal. But because the very small refraction of 1′ 25″ at 30°, which we have taken over from Tycho, is very easily subject to observational error, not large in itself, but intolerable after subsequent multiplication, it must now be seen whether the rest also correspond. So in the diagram, since it is supposed that EBA is 91°, because DBE is 89°, and the angle in air corresponding to it was found to be 87° 47′, the difference BAE is therefore 1° 13′. Consequently, BEA is 87° 47′, and as the sine of BEA 87° 47′, 99925, is to AB, so is the sine of EBA, 99985, to AE, 100,060. So if we now use this to seek the horizontal

---

81 This is a confusing way to state the problem, so a clarification is in order. It is known that when a star appears to be 89° from the zenith (angle DBE, in the diagram), it is actually 26′ lower, at 89° 26′. (That is, in the diagram, the angle FEG is 26′.) It is also known that for each degree, the part of the refraction that is proportional to the angle of *refraction* (that is, proportional to angle BEA) is 43‴. From these it is possible to compute (through an iteration) a value for the angle BEA, and this in turn will yield a new altitude of the air EC, which can be used to compute refractions at other incidences.

82 What Kepler needs is the ratio of the secant to the radius. Since the radius was by convention taken to be 100,000, the required ratio is obtained by dropping the last five digits.

83 The unit indicated by the symbol ` is a unit of 60°, the next larger unit in sexagesimal arithmetic. The multiplication then becomes

$$(1 \cdot 60 + 24 + 16 \cdot \frac{1}{60})(43 \cdot \frac{1}{60^3}),$$

which comes out as Kepler says.

84 This is the secant of the trial angle of 84° 16′.

85 As in Proposition 8 above, the part of the refraction that is proportional to the inclination is multiplied by the secant of the refracted angle to obtain the total refraction.

86 This should be 1′ $2\frac{1}{2}$″; however, Kepler rounds it off to 1′ 3″, knowing that the main determinant of the result is the magnitude of the secant.

87 This obviously should have been 21′, but the error has no effect upon the final result.

88 This is the inclination of the refracted ray EB from the vertical EA; that is, the angle BEA.

refraction on earth, and EBA becomes 90°, then the secant 100,060 results in the angle BAI of 2°.[89] So the secant of the angle 88° is 2,865,370. And the simple refraction of the inclination is 63″.[90] Therefore the total refraction comes out to be 30′ 5″; from Tycho's inclination it should have been 34′. So, although we have not quite hit the target this time, the light is nonetheless open to us, so that we may tell whether it has to be diminished or increased. The increments of the refractions are in fact not large enough. So one should resort to a sharper drop in the secants. But the moment we do this, the refraction at 89°will be increased too much. Therefore, the simple refraction, which we have derived from the altitude of 30°, has to be decreased.

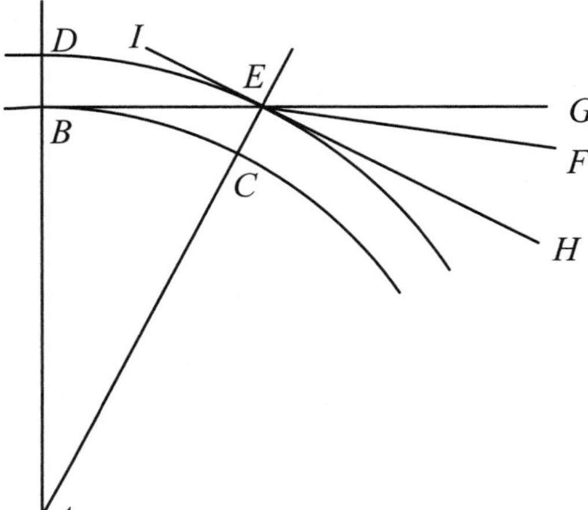

In sum, it should be noted that the proportional refraction from 89° to 90° is going to be extremely small in the difference, because, while the inclination on earth varies 119
by one degree, in air it varies by barely 13 minutes.[91] And to one degree there belong 43‴; therefore, to 13′ there belong barely 10‴. Therefore, the proportional refraction, being so small, remains perceptibly the same. And it must be multiplied by same. And it must be multiplied by two secants, so that one makes 26′ and the other 34′. And since the ratio of equimultiples is the same, look for two secants, about 13′ apart from each other, that are to each other as 26 to 34. Such, approximately, there is from 88° 54′ to 89° 10′, by 15 or 16 minutes. Now see whether the inclinations in air corresponding to the inclinations on earth of 90° and 89° differ by 16′. If BAE is 0° 50′, AE will be 10,001,058, where B is right.[92] But if EBA is 91°, BA will be to sine BEA as EA is to sine EBA; BEA comes out to be 88° 45′; before it was 89° 10′; the difference is 25′. They are 25 minutes apart, and the secant of 88° 45′ now shows a smaller refraction than 26′ at 89°. So we have already passed the mark. The truth is between the secant of 88° and 89° 10′ for a refraction of 34′, and likewise between 87° 47′ and 88° 45′ for a refraction of 26′. I shall try 89° for air, corresponding to 90° for earth. Thus BAE is 1°, AE 10,001,524. As this is to 9,998,477,[93] so is 10,000,000 to the sine of BEA. BEA comes out to be 88° 35′. The secants are 573, etc., for the former,

---

[89] The angle here should be 1° 59′, which makes the total refraction 30′ 19″ instead of 30′ 5″.

[90] That is, 43‴ × 88°.

[92] Without comment, Kepler has made AB 10,000,000 units in length, to allow greater precision in the evaluation of the secants.

[93] This is the sine of 91°. Having established a magnitude for angle BAE, Kepler raises the ray BEG to an altitude of 1°, and calculates the corresponding angle BEA.

and $404\frac{1}{2}$ for the latter. If 34 gives 573, what does 26 give? It comes out to 438; it should have been $404\frac{1}{2}$. We have come closer, but the secants nevertheless still drop off too fast, as you see.

We shall go down another 10 minutes, making *BAE* 1° 10′. *AE* will be 10,002,074. As this is to 9,998,477, so is the whole sine to the sine of *BEA*, 88° 28′. But the secant of the complement of angle *BAE* is $49\frac{1}{10}$. The secant of *BEA* is $37\frac{1}{2}$.[94] The ratio of 34 to 26 required that it be $37\frac{1}{2}$. We have therefore come very close; and nonetheless we are still a little off, which will be overcome by going down another 5 minutes, an amount that suffices for the refractions of 26′ and 34′.

In fact, since the refractions presented by Tycho could not attain such precision from observation as is here required, you accordingly see them rounded off to exactly 34 minutes at altitude 0° and 26 minutes at altitude 1°. Tycho himself, on pp. 79 and 124[95] does not try to hide this; thus, before the measure is established, others that are a little higher and more easily perceptible have to be compared.

Above, for a refraction of 1′ 25″ at altitude 30° and a refraction of 26′ at altitude 1°, a 2° difference of angles on earth and in air sufficed. Again, for refractions of 26′ and 34′ at altitude 1° and 0°, a 1° 15′ difference of angles
120 sufficed. It must be seen whether a refraction of 8′, which Tycho gives for an altitude of 14°, likewise results from this hypothesis. So, because the refracted ray inclined in air by 88° 45′ makes a refraction of 34′, so that the one in the aether is inclined 89° 19′, while the secant of 88° 45′ is 4,584,023, as this is to 100,000, so is the composite refraction of 34′ to the simple refraction of the inclination, which is $44\frac{1}{2}$″. When a distribution of this into $98\frac{1}{3}$ is made, to one degree there comes 30‴. Now let *BAE* be 1° 15′. Therefore *AE* is 10,002,381. Again, let *EBA* be 104°. Its complement, 76°, makes *BEA* 75° 57′. To this there should correspond a refraction of 8′, so as to make the inclination of the unrefracted ray in air 76° 5′. The difference of this and 89°19′ is 13° 14′, to which a simple refraction of 6″ 38‴ corresponds, which is to be subtracted from $44\frac{1}{2}$″, so that there remains 38″ proportional to the inclination of 76° 5′. This, multiplied by the secant of 75° 57′, which is 411,915, gives 2′ 35″. From Tycho's indication, this ought to have been 8′. Therefore, this latter hypothesis does not suffice for the intermediate refractions, which are perceptible enough. For the secants have dropped off too much: the refraction that is proportional to the inclinations has been made too small.

We shall, however, also try the former hypothesis for the refraction at 14° of altitude. There, to one degree corresponded a simple refraction of 43‴. And, that we not demand excessive precision, the secant of the angle in air, and the angle in aether, and likewise also the angle on earth and at the air's surface, hardly differ. Let it accordingly be 76°, or 1ˋ 16°: multiply by 43 thirds, and the product is the simple refraction, $54\frac{1}{2}$ seconds. But the secant here is four times the radius, plus

---

94 Should be $37\frac{1}{3}$, which is probably what Kepler intended here: this would explain why Kepler says "we are still a little off", although the two numbers in the text are identical.

95 Of the *Progymnasmata* (*TBOO* II pp. 64 and 136).

one tenth of it, so the product here is about 4 minutes. It ought to have been 8, and thus these secants also drop off too much.

Although this dissonance contains much that is disturbing, you still do not yet have cause to suspect a flaw in the general hypothesis, as long as we can suspect something unequal in the Tychonic refractions. That there is an inequality in them is proved first by reasoning. For because the height of air as a result of the circularity of air and earth, and its density, combine in altering refractions, the two certainly change daily, as Tycho Brahe attests on p. 79 of the *Progymnasmata*, and as will be shown below in its own chapter.[96] Tycho, however, investigated refractions at different times, and indeed a variation in altitude did occur that was perceptible in this matter. For the surface of air that brings about refractions is very low. Next, consider Tycho's differences of refractions: there
immediately appears from them something unequal, namely, that it cannot have **121**
its locus in any alignment, whether it imitates a circle or any other linear configuration whatever. See Table 3 below.

Someone might therefore wish to see how much the refractions at these horizontal inclinations vary in one and the same day. This may be sought in many places from Tycho's observations themselves.[97] I shall give one example. In 1587, 16 January, while the sun was setting, within the 32 minutes from 3:26 to 3:58 the sun's declination was observed nine times, from altitude 3° 50′ to
0° 35′. It is an easy task to get the refractions for the altitude from these. For **123**
the triangle between the sun, the zenith, and the pole, is given, in which four elements are known; therefore, the angle at the sun cannot remain hidden, from which is usually obtained the ratio of the observed refraction in declination to that refraction which belongs to the altitude. Now the side between the zenith and the pole is 34° 5′ 15″, the complement of the altitude of the pole at Hven.[98] The observed declination, added to a quadrant, constitutes the side between the pole and the sun; the observed altitude constitutes the side between the zenith and the sun. Finally, the time constitutes the angle at the pole. The place of the sun at the fourth hour of that day[99] is 6° 19′ Aquarius. The true declination of this is 18° 45′ 10″, and half an hour before is only 20″ greater. But something must be added to these declinations for the sun's parallax, which is 3 minutes at such a small altitude,[100] so that it may appear how large the declinations of the sun were

96 *TBOO* II p. 64. The text has 29 instead of 79, evidently a typographical error.

97 The observations for the following example may be found in *TBOO* XI p. 193.

98 Hven, an island near Copenhagen, was the site of most of Tycho's observations.

99 The convention of astronomers of that time was to begin the day at noon; hence, the fourth hour is 4 pm.

100 Kepler followed astronomical tradition in setting the sun's horizontal parallax at 3′, which is too large by a factor of twenty. This error had a drastic effect upon Kepler's determination of the eccentricity of the earth's orbit, which in turn created problems for all of his planetary theories. (On this subject see Curtis Wilson, "The Error in Kepler's Acronychal Data for Mars," *Centaurus* **13** (1969), 263–8, reprinted in Wilson's *Astronomy from Kepler to Newton* (London: Variorum, 1989).) Here, its effect was to make all the refractions in the following table 3′ too large.

Tycho Brahe's refraction table

	Degr. of Altitude	*In the Sun*		*In the Moon*		*In the Fixed Stars*	
		Refraction *M. S.*	*Decrements* *M. S.*	*Refraction* *M. S.*	*Decrements* *M. S.*	*Refraction* *M. S.*	*Decrements* *M. S.*
	0	34. 0	8. 0	33. 0	8. 0	30. 0	8. 30
	1	26. 0	6. 0	25. 0	5. 0	21. 30	6. 0
	2	20. 0	3. 0	20. 0	3. 0	15. 30	3. 0
	3	17. 0	1. 30	17. 0	1. 40	12. 30	1. 30
	4	15. 30	1. 0	15. 20	1. 0	11. 0	1. 0
	5	14. 30	1. 0	14. 20	0. 30	10. 0	1. 0
	6	13. 30	0. 45	13. 50	1. 5	9. 0	0. 45
	7	12. 45	1. 30	12. 45	0. 45	8. 15	1. 30
	8	11. 15	0. 45	12. 0	0. 40	6. 45	0. 45
	9	10. 30	0. 30	11. 20	0. 35	6. 0	0. 30
	10	10. 0	0. 30	10. 45	0. 35	5. 30	0. 30
	11	9. 30	0. 30	10. 10	0. 35	5. 0	0. 30
	12	9. 0	0. 30	9. 35	0. 35	4. 30	0. 30
	13	8. 30	0. 30	9. 0	0. 30	4. 0	0. 30
	14	8. 0	0. 30	8. 30	0. 30	3. 30	0. 30
	15	7. 30	0. 30	8. 0	0. 30	3. 0	0. 30
	16	7. 0	0. 30	7. 30	0. 30	2. 30	0. 30
	17	6. 30	0. 45	7. 0	0. 30	2. 0	0. 45
	18	5. 45	0. 45	6. 30	0. 30	1. 15	0. 15
	19	5. 0	0. 30	6. 0	0. 30	1. 0	1. 0
	20	4. 30	0. 30	5. 30	0. 30	0. 0	
122	21	4. 0	0. 30	5. 0	0. 25		
	22	3. 30	0. 20	4. 35	0. 25		
	23	3. 10	0. 20	4. 10	0. 25		
	24	2. 50	0. 20	3. 45	0. 25		
	25	2. 30	0. 15	3. 20	0. 20		
	26	2. 15	0. 15	3. 0	0. 20		
	27	2. 0	0. 15	2. 40	0. 20		
	28	1. 45	0. 10	2. 20	0. 20		
	29	1. 35	0. 10	2. 0	0. 15		
	30	1. 20	0. 10	1. 45	0. 15		
	31	1. 15	0. 10	1. 30	0. 10		
	32	1. 5	0. 10	1. 20	0. 10		
	33	0. 55	0. 10	1. 10	0. 10		
	34	0. 45	0. 10	1. 0	0. 10		
	35	0. 35	0. 5	0. 50	0. 5		
	36	0. 30	0. 5	0. 45	0. 5		
	37	0. 25	0. 5	0. 40	0. 5		
	38	0. 20	0. 5	0. 35	0. 5		
	39	0. 15	0. 5	0. 30	0. 5		
	40	0. 10	0. 1	0. 25	0. 5		
	41	0. 9	0. 1	0. 20	0. 5		
	42	0. 8	0. 1	0. 15	0. 5		
	43	0. 7	0. 1	0. 10	0. 5		
	44	0. 6	0. 1	0. 5	0. 5		
	45	0. 5		0. 0			

going to be if the air had not caused any refraction. See the following table of the entire operation.[101]

Series of Observations at Altitude of	*Observed Declination*	*Parallax of Declination*	*Declination Without Parallax*	*True Declination*	*Refraction of Declination*	*To Fill Out the Refraction of Altitude*	*Refraction of Altitude Sought*
3°. 50′	18°. 35′. 30″	2′. 42″	18°. 32′. 48″	18°. 45′. 30″	12′. 52″	1′. 30″	14′. 22″
3. 30	18. 34. 0	2. 42	18. 31. 18	18. 45. 28	14. 10	1. 35	15. 45
3. 10	18. 32. 30	2. 41	18. 29. 49	18. 45. 26	15. 37	1. 37	17. 14 +
2. 50	18. 31. 45	2. 40	18. 29. 5	18. 45. 24	16. 19	2. 2	18. 21 −
2. 40	18. 30. 30	2. 40	18. 27. 50	18. 45. 22	17. 32	2. 12	19. 44
2. 5	18. 29. 0	2. 39	18. 26. 21	18. 45. 20	19. 0	2. 23	21. 23 −
1. 15	18. 23. 30	2. 38	18. 20. 52	18. 45. 16	24. 24	3. 29	27. 53 −
1. 0	18. 22. 30	2. 38	18. 19. 52	18. 45. 14	25. 22	3. 37	29. 0
0. 35	18. 20. 0	2. 38	18. 17. 52	18. 45. 12	27. 20	3. 54	31. 10

Now to these refractions, which were taken on the same day by the method just now investigated above, we shall compare the measurement, hunting it from the first and last refraction, and comparing the rest. In the diagram, let EBA be 35′ above 90°, at which altitude we have here found the refraction to be 31′ 10″. **124**
Now let the corresponding angle in air, BEA, be 87° 30′,[102] so that, when GEF, 31′ 10″, is subtracted from GEH, 2° 30′, the remainder FEH is 1° 58′ 50″, or its complement, 88° 1′ 10″. The secant of 87° 30′ is 2,292,558. Divide this into 31′ 10″, or 1870″, and the result is $81\frac{1}{2}''$.[103] This portion, divided into 88°, establishes the simple refraction for one degree as $55\frac{1}{2}'''$. And because as sin BEA is to BA, so is sin EBA to EA, EA consequently becomes 10,009,000. Now let EBA be 3° 50′ above 90°, at which altitude we find the refraction here to be 14′ 22″. As EA is to sin EBA, so is BA to sin BEA, which becomes 86° 59′.[104] This differs from the previous value by half a degree, and consequently the inclination of FE will be 87° 14′, less than the previous value by three fourths, to which belongs 42‴ of simple refraction. There therefore remains a simple refraction of 1′ 22″. But the secant of 86° 59′ multiplies this little sum by 19. Therefore, it will make

101 Since the full refraction occurs along a great circle perpendicular to the horizon, the amounts of the refractions measured in the declinations, which are at an angle to the horizon, need to be increased by a factor of the cosecant of the "angle at the sun" (i.e., the angle zenith–sun–pole).

102 Kepler knew that this would be too small. As before, it is the starting point for the iterative process that will determine values for AE and the rest that will correspond to this particular series of observations.

103 The quotient must, of course, be multiplied by the radius, which is 100,000, to obtain this result. Kepler is applying the refraction computation in reverse, dividing the total refraction (measured) by the secant of the angle from the zenith in order to find the "simple refraction."

104 This should have been 85° 28′, an error first pointed out by Delambre, *Histoire*, I pp. 365–6.

the refraction too large, and the ratio of secants is too close.[105] One must carry on towards the horizon. Thus the supposition that *BEA* is 87° 30′ is false.

Therefore, since *EBA* is 35′ above 90°, let *BEA* be 89°, so that the inclination *FE* becomes 89° 31′ 10″. Thus, if the secant of arc 89°, 5,729,871, be made 100,000, what would 31′ 10″, or 1870″, be? The simple refraction comes out to be $32\frac{1}{2}''$, and for one degree, 22‴. Hence, *EA* becomes 10,001,005. Now let *EBA* be 3° 50′ above 90°: *BEA* will become 86° 5′, and *FEA* 86° 20′, differing from the previous [value] by 2° 40′. That many times 22‴, subtracted from $32\frac{1}{2}''$, leaves $31\frac{1}{2}''$. But the secant of 86° 5′ is 1,464,011, [which], multiplied by $31\frac{1}{2}'$, produces 7′ 40″. It ought to have been 14′ 22″.

Therefore, the setting of *BEA* at 89° was false, and the secants drop off too sharply. And so we have now gone past the true value. Accordingly, let *BEA* be 88°. *EA* will become 10,005,578. The simple refraction, divided into $88\frac{1}{2}$, produces a simple refraction of 44 for one degree. And where *EBA* is 93° 50′, *BEA* will become 85° 43′, 86° without refraction; previously it was $88\frac{1}{2}$°. Therefore, subtract two and one half times 44‴, or 1″ 50‴, from 65″: the remainder is 63″ 10‴. But the secant of 85° 43′ is $13\frac{39}{100}$ times the radius, and therefore the refraction becomes 14′ 49″. Therefore, by setting *BEA* at 88°, we have hit the target well enough. The simple refraction should be investigated, when the inclination in air is 90°. To a simple refraction of 1′ 5″ for $88\frac{1}{2}$° must be added about 1 second for $1\frac{1}{2}$ degrees, so that it becomes 1′ 6″. From this, the simple refractions for all degrees of air are easily derived.

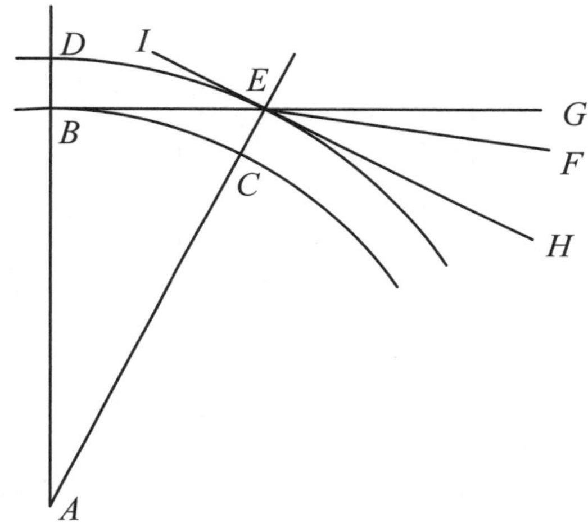

126 By the simple refractions accommodated to the refraction-free inclinations in air, the table is carried through thus. The whole refraction of the preceding degree is subtracted from the following degree; secants are taken of the remainder, casting off the last six digits, since the whole sine has 7 zeroes; and they are multiplied by the seconds expressing the simple refraction of that degree; and the product is divided by 600. If something large comes out, differing from the previous refraction, the operation will have to be repeated twice, sometimes thrice, and so on.

At 60° let the whole refraction be 1′ 28″; subtract from 61°, and there remains 60° 58′ 32″. The secant, 206 and some, multiplied by the simple refraction, 44″, gives a product of 9,064, which, divided by 600, makes 1′ 31″, the whole

[105] That is, the secants now do not drop off too quickly: successive secants are too close to each other.

125

Inclination of Free Rays in Aether	*Simple Refraction*	*Composite*	*Inclination of Free Rays in Aether*	*Simple Refraction*	*Composite*	*Inclination of Refracted Rays in Air*	*Inclination of Refracted Rays on Earth*	*Tycho's Refraction, Approx.*
Deg. 1	0″	0″	45	33″	47″	44.° 59.′ 13″	45.° 0′	0.′ 5″
2	1	1	46	33	49	45. 59. 11	46. 1	0. 6
3	2	2	47	34	51	46. 59. 9	47. 1	0. 7
4	3	3	48	35	54	47. 59. 6	48. 1	0. 8
5	3	3	49	36	57	48. 59. 3	49. 1	0. 9
6	4	4	50	36	1.′ 0	49. 59. 0	50. 1	0. 10
7	5	5	51	37	1. 2	50. 58. 58	51. 1	0. 15
8	6	6	52	38	1. 5	51. 58. 55	52. 1	0. 20
9	6	6	53	38	1. 8	52. 58. 52	53. 1	0. 25
10	7	7	54	39	1. 10	53. 58. 50	54. 1	0. 30
11	8	8	55	40	1. 13	54. 58. 47	55. 1	0. 35
12	9	9	56	41	1. 16	55. 58. 44	56. 1	0. 45
13	9	9	57	41	1. 19	56. 58. 41	57. 1	0. 55
14	10	10	58	42	1. 22	57. 58. 38	58. 2	1. 5
15	11	11	59	43	1. 25	58. 58. 35	59. 2	1. 15
16	11	11	60	44	1. 28	59. 58. 32	60. 2	1. 25
17	12	12	61	44	1. 31	60. 58. 28	61. 2	1. 35
18	13	13	62	45	1. 35	61. 58. 25	62. 2	1. 45
19	14	14	63	46	1. 41	62. 58. 19	63. 2	2. 0
20	14	15	64	47	1. 49	63. 58. 11	64. 2	2. 15
21	15	16	65	47	1. 53	64. 58. 7	65. 2	2. 30
22	16	17	66	48	1. 58	65. 58. 2	66. 3	3.
23	17	18	67	49	2. 4	66. 57. 56	67. 3	3.
24	17	19	68	50	2. 15	67. 57. 45	68. 3	4.
25	18	20	69	51	2. 23	68. 57. 37	69. 3	4.
26	19	21	70	51	2. 31	69. 57. 29	70. 3	5.
27	19	22	71	52	2. 41	70. 57. 19	71. 4	5.
28	20	23	72	53	2. 50	71. 57. 10	72. 4	6.
29	21	24	73	54	3. 4	72. 56. 56	73. 4	7.
30	22	25	74	55	3. 18	73. 56. 42	74. 4	7.
31	22	26	75	55	3. 32	74. 56. 28	75. 5	8.
32	23	27	76	56	3. 50	75. 56. 10	76. 4	8.
33	24	28	77	57	4. 22	76. 55. 38	77. 5	9.
34	25	30	78	57	4. 44	77. 55. 16	78. 5	9.
35	25	31	79	58	5. 2	78. 54. 58	79. 5	10.
36	26	32	80	59	5. 36	79. 54. 24	80. 5	10.
37	27	33	81	1. 0	6. 18	80. 53. 42	81. 6	11.
38	27	34	82	1. 0	7. 6	81. 52. 54	82. 7	11.
39	28	36	83	1. 1	8. 14	82. 51. 46	83. 7	13.
40	29	37	84	1. 2	9. 43	83. 50. 17	84. 8	13. 30
41	30	39	85	1. 3	11. 36	84. 48. 24	85. 10	14. 40
42	30	40	86	1. 3	14. 51	85. 45. 49	86. 13	16.
43	31	42	87	1. 4	18. 27	86. 41. 33	87. 20	18.
44	32	45	88	1. 5	25. 34	87. 34. 26	88. 31	23.
					33.	88. 5.	90. 0	34.
			89	1. 6	38. 30	88. 21. 30	A little below.	
			90	1. 6	1° 1. 30	88. 58. 30	Below the earth.	

for degree 61°. When this is subtracted from 62°, the remainder is 61° 58′ 29″, whose secant is 21 and some. This multiplied by 45″, the refractions of degree 62, leaves a refraction of 1′ 35″ for degree 62°. Again, at 84° let the refraction be 9′ 43″. Subtract from 85°. The secant of the remainder, 111 and some, multiplied by the simple refraction 63″, and divided by 600, gives 11′ 39″, which differs in position from the previous one, 9′ 43″, by nearly 2 minutes, where the secants are now increasing vigorously. Therefore, repeat the operation, and in place of 9′ 43″ of the previous degree, now subtract 11′ 39″ from its degree of 85°. The secant, $110\frac{1}{2}$, multiplied by 63″, gives 11′ 36″, the correct refraction.

Moreover, in order that the inclinations of the refracted [rays] be accommodated to earth, always multiply EA, 10,005,578, by the sine of BEA or the inclination of the refracted lines[106] in air.

I conclude this tiresome investigation with the setting out of this table, in which we come very close to the Tychonic solar refractions, differing by only four minutes near a solar altitude of 15°, and at the same time we satisfy those that were observed by Tycho within a single day. They would perhaps not differ even by that much if what was said in proposition 6 of this chapter had been correctly carried out. And so I beg the reader not to misuse this very scrupulous investigation of mine into the refractions provided by Tycho, for undermining the whole treatment of refractions, for the confirming of which I undertook it. And if I judge well of these matters, I have fully confirmed it, while that which Tycho considered to result from the unequal thickness of the air in its varying
127 depth, I myself demonstrate this, or something not much different, from the ratio of the circle and of secants, from principles that are clearly harmonious. Let it be enough for me to have given notice. I know how today many blind persons would argue about colors, and how they desire to bring forth someone who would in some way help their rash insults upon Tycho and even upon this very matter of refractions.[107] Had they held their childish errors and their sheer private ignorance, they would have been without blame, since that happens to many great men; but because they publish, and make treacherous attacks with thick books and fine sounding titles for the profit of the ignorant (so that today there is more danger from the abundance of bad books than there was once from the scarcity of good ones), let them know accordingly that room is available to them to emend publicly their own public errors, in that, where they have postponed doing so for a very long time, it will be permissible for me or anyone else to do the same for them (having unfortunately entered upon geometrical matters) as they have taken upon themselves to do against the most prominent men. Although this task is disdained, and is sure to be involved in a worthless topic of absurdities, it will, however, be as much more necessary than what they themselves had taken up

Excuse.

Against the Tycho whippers.

[106] The noun is supplied conjecturally. One would expect "rays" here, but *refractarum* is feminine, so the best plausible candidate is *linearum*.

[107] It is evident from the correspondence of Kepler and David Fabricius that the present remarks were directed against Jacob Christmann, a Heidelberg professor who wrote a book entitled *Observationum solarium libri III*, Basel, 1601. He argued that the vapors that refract the solar rays and make the solar body bigger have no effect upon the measurement of the true solar altitude: the solar parallax alone suffices for this.

against others, in proportion to the greater public harm that is done by one who endeavors to undermine good and necessary discoveries of others, in defraud of the truth, than by one who persuades himself that he has found something that cannot be found. Let them meanwhile cease, in the silence of others, to boast in their own obscurity.[108]

Proposition 10

Problem III. *From the quantity of refractions to investigate the ratio of media to each other, such as that of air to water, as concerns their density.* In Proposition 6 of this chapter it was pointed out that the simple refraction of that inclination that belongs to the ray in the rarer medium above the common surface, is multiplied by the secant of that inclination that belongs to the refracted ray in the denser medium above the common surface. And so let the simple refractions of the same inclinations be compared, so that one might not place different positions in the two media upon the same incident ray. As, in Prop. 8 of this chapter, at an inclination of 80°, the simple refraction of a ray from air to water is 19° 17′. At the same inclination of 80° in Prop. 9 of this chapter, the simple refraction of a ray from aether to air is 59″. Therefore, from aether to water there would be a simple refraction of 19° 18′. Thus the ratio is the same
as that of unity to $1177\frac{2}{3}$.[109] And as a cause for refractions of this kind, there **128**
was alleged only that density which is considered on the straight line, because the surface of refraction, as named by the optical writers, intersecting the surface of the denser medium, establishes an intersection line. And although the same surface intersecting the denser body establishes an intersection surface, it was nonetheless demonstrated in the first chapter, that no refraction occurs in the dense body, but it all happens on the surface. Thus if you were to make a cube from the two terms of the ratio, you will find that ratio of the bodies as concerns their densities, which is between 1 and 1,633,304,460.[110] Nor is there any doubt that if one were to stand in pure aether and were from here to pour out one dipper of water, and then sixteen myriads of myriads of dippers of air, that these would weigh the same.[111] And in a chamber or cube 12 feet long, wide, and high, there is contained no more matter, when it is filled with that purest air that is contiguous with the aether, than there is in a cubicle that extends for the eighth part of an inch[112] in all dimensions. He should, however, doubt whether it might not be enough to square the terms of the ratio, or rather to multiply them by their

---

108 About this syntactically labyrinthine passage, Frisch remarks (*JKOO* II p. 433), "Kepler appears to have imitated Christmann's mode of expression at the end of the letter quoted above." (This was a letter from Fabricius to Kepler of 14 March 1603, *JKGW* XIV no. 252; the Christmann letter is quoted on pp. 385–6).

109 This is the ratio of the simple refraction of aether to water, to the simple refraction of aether to air, or 19° 18′ to 59″.

110 This is the cube of $1177\frac{2}{3}$, which is the ratio of refractions found above.

111 A myriad is ten thousand, or 1×10^4, so "sixteen myriads of myriads" amounts to sixteen times 10^8, approximately the same as Kepler's cubed ratio.

112 Literally, "thumb" (pollex). There are twelve "thumbs" in a foot; thus the ratio of

roots,[113] because the density of corporeal bulk is itself represented in the surface of the medium, by prop. 13 of chapter 1. But this ratio, too, is not perpetually fixed, not unless it is to be ascribed to that part of the air which occupies the highest spaces in a sky clear of impurities. But the vapor that bubbles from boiling pots, as well as that which, providing material for sudden rains, comes forth from caverns of the high mountains and covers the sky with darkness, is much closer to water, into which it is immediately turned.

Source of rain.

I am not unaware, believe me, that I am going to be incurring the censure of the natural philosophers[114] in stating both here and previously that air is heavy or has weight. But the contemplation of all of nature teaches me this; indeed, along with Cardano and many other very prominent philosophers of this age, I also ascribe coldness to it.[115] I also do not see what is left for Aristotle with which to defend the absolutely light and hot in its own nature, if you take one combination of qualities away from him. Neither medicine nor physiology is brought to ruin, provided that air remain comparatively both light and more congenial to being heated. But there will be a more suitable place for speaking of these things elsewhere, and something is said in my astrological theses.[116]

Air cold and heavy by nature

Proposition 11

Problem 4. *From the refractions to investigate the altitude of the air from*
129 *earth.* Now because the matter of the air is fluid, as is argued by the evidence of winds, which are, in my opinion, nothing but an abundant boiling up from the highest mountains, from where, by its own nature, matter, in again seeking the low places (which is not out of keeping with Aristotle, *Problems*, Section 25 Probl. 13), is poured around the mountain in a circle; and from the initial impulse, one circle arises from another, as also occurs in pools of water. What follows from this is what was used in Proposition 9 of this chapter, that the matter of air embraces the orb of the earth circularly. Therefore, for investigating the measure of the refractions of air, there was obviously need of supposing an altitude for the sphere of the air. The altitude of the air will accordingly be most nearly that by supposition of which we most nearly approached the measure of refractions. Now EA, above, was 10,005,578, where the distance from the center of the earth A to its surface was 10,000,000. Therefore, by the rule of proportions, if

---

spaces is $(12^2)^3 : {\frac{1}{8}}^3$, or a bit more than fifteen myriads of myriads, nearly the ratio of densities given above.

113 That is, to take the $\frac{3}{2}$ power.

114 *Physici.* These are not "physicists" as we use the word today; rather, they are those who profess one or another variant of the natural philosophy of Aristotle.

115 Aristotle considered air to embody heat and wetness, water being cold and wet and fire being hot and dry. Cf. *On Generation and Corruption* II 3, 330b 4, trs. Forster pp. 274–5.

116 *De fundamentis astrologiae certioribus*, Prague [1601]; English translation in J. Bruce Brackenridge, ed., and Mary Ann Rossi, trans., "Johannes Kepler's On the More Certain Foundations of Astrology, Prague, 1601," *Proceedings of the American Philosophical Society* **123** (April 1979).

10,000,000 becomes 860 German miles, according to the received tradition of the geographers,[117] the altitude of the air, 5578, will give $\frac{4,797,080}{10,000,000}$ miles. That is, in the arc of Uraniburg,[118] where the observations of the heavens were made, the altitude of the air was half a German mile, no more.

Here the reader might also be advised of the controversy which Tycho had with Rothmann concerning the substance of the air and the aether, in volume I of the *Epistolae astronomicae*, which Tycho concluded on p. 92 of the *Progymnasmata*.[119] Rothmann and Pena had said that from earth to heaven there is nothing but air, with the exception of a little vapor.[120] Tycho joined them in eliminating the Aristotelian fire, and admitted vapors around earth different from air, but extended the air to the boundaries of the moon, and said that there it gradually comes to an end in aether. I, moved by the present experiments, hold the middle. First, what they called vapors, I call air, and I end it with the peaks of the mountains; above that, the smoky exhalations, lamps of the twilight, are set, and the aether follows immediately.

What is in the space between the earth and the moon.

7. Consideration of those things that Witelo advised were necessary for astronomy.

This would now be time for me to insert Witelo XI 49 and the following propositions, had they not already been well enough brought home by 130
Witelo himself: in Prop. 49 he shows by experiments that refraction of light occurs among observations of the stars.[121] Tycho Brahe proved the same thing with instruments of the most exquisite accuracy. See the various ways on p. 15 and p. 93 of Tycho's *Progymnasmata*.[122] Others have proved the same, for which see below. And indeed, by armillary spheres, when the pole is high, the stars appear closer near the horizon than on the meridian; and this experimental procedure is a good one, used by Brahe along with the others. However, he was unable to believe that it was properly carried out by

Lights of the stars are bent in air.

[117] The German mile was five Roman miles, or about 7.4 kilometers. Kepler describes how to measure the radius of the earth in the *Epitome astronomiae Copernicanae* I (Linz, 1618), *JKGW* VII p. 39, where he comes up with a slightly greater number.

[118] Tycho Brahe's observatory on the island of Hven (*q. v.*).

[119] For this debate, see Section 1 of this chapter (p. 93 above. The relevant part of the *Epistolae* begins on p. 134 of *TBOO* VI. The passage from the *Progymnasmata* is on p. 77 of *TBOO* I.

[120] In his introduction to his 1557 edition and translation of Euclid's *Optics* (*Euclidis optica et catoptrica, nunquam antehac graece aedita. Eadem latine reddita,* Paris 1557), Joannes Pena argued on the basis of lack of refractions that the heavens cannot consist of "spheres" of different substances. (Frisch includes extensive quotations from Pena's work in *JKOO* II p. 573.) In the traditional cosmology, there was a sphere of fire between the air and the heavens.

[121] *Thesaurus* II p. 444.

[122] *TBOO* II pp. 16 and 78–82.

Witelo in that age, or by Alhazen (p. 91 of the *Progymnasmata*).[123] But that Witelo added another way, that the place of the moon be computed and compared with observation, is a splendid fiction, really from Witelo's time. There is, of course, no doubt that the same thing happens to the moon that happens to the other stars. But there is so much uncertainty in the parallaxes and in the mean motions of the moon, that the refractions would often have ended up in the wrong direction, even right at the horizon, where the refraction was greatest.

Aether is rarer than air.

Thus Witelo's X 50, that aether is rarer than air is also pertinent here, as Rothmann protests in vain, denying the difference and placing the blame for refraction upon fortuitous vapors which would not cover our heads.[124] But he was in error, not considering that the actual cause of refractions makes them imperceptible at high altitude. This proposition, moreover, Witelo demonstrates legitimately, from the fact that the altitude of the stars above the horizon appears not less, but greater than it should.

The distances of stars at the zenith are imperceptibly less than they should be.

Now, Prop. 51,[125] that the distances of stars at the zenith overhead appear less than they should, is in fact true, but whatever there is of this effect is imperceptible. For, by 9 of this chapter, the refractions at the zenith are also imperceptible.

The distances of stars equally near the horizon are imperceptibly less than they should be.

Thus also, in Proposition 52,[126] that the distances of a pair of stars, or their diameters, appear smaller than they are in actual fact when they are parallel to the horizon and near it, is again true, but completely imperceptible, unless the stars are separated by a semicircle or a little less. Let AB be a portion of the horizon, C the zenith, CA, CB quadrants, and let A appear
131 by refraction at D, and B at E, in such a way that the distance AB should appear to be DE: indeed, DE will be less than AB. But because the arcs AD, BE, are not greater than 34 minutes, AB and DE will be perceptibly equal. For let ACB be a full quadrant, a distance that is seldom actually taken with instruments, never, of course, by the Tychonic sextants, never by astronomical radii or staffs, to keep the eyes from having to turn around too much and make the observation untrustworthy. Therefore, perpendicular CF being dropped, CFE will be right, FCE 45° because DCE is isosceles,[127] and CE is 89° 26′. And as the sine of CFE is to the sine of CE, so is the sine of FCE to the sine of FE, whose double is DE. Thus in the case in which AB is 90°, the distance DE involved in refractions will be

[123] *TBOO* II p. 76.

[124] *Thesaurus* II pp. 444–5; cf. *Epistolae astronomicae*, *TBOO* VI pp. 150–1.

[125] *Thesaurus* II pp. 445–6.

[126] *Thesaurus* II pp. 446–7.

[127] Note that arc DE is a great circle arc, and is not parallel to AB. Therefore, the angles at D and E are not right, but they must be equal. Under these circumstances, a perpendicular dropped from C to DE must bisect DE; hence, angle FCE is 45°.

$89^\circ\ 59'\ 43''$: the difference is $17''$. It will be even smaller where a smaller distance AB is taken.

The distances of stars set up from the zenith to the horizon are noticeably smaller than they should be.

Against those who deny the refractions of the heavenly bodies.

Those things that are said in Proposition 53,[128] however, that the distance of the stars or the diameters of the heavenly bodies set up from the zenith towards the horizon, are less than they should be, especially if one end is near the horizon—these things, I say, are both true and evident to the senses, and are in the highest degree necessary for the astronomer to know. So much so that I do not believe that this proposition has been read by those who accuse the Brahean refractions, concerning whom see the end of Prop. 9 of this chapter. I therefore refer them to this proposition.

The sun near the horizon is not round.

Against the Tycho-whippers who deny refractions.

Therefore, what Witelo presents in the following Proposition 54,[129] that all stars near the horizon appear smaller than they should, is true, but his contention that they also appear round, is false. For it is not true that all of their diameters appear uniformly smaller: one that is set up towards the zenith comes out perceptibly smaller, while one that is equidistant from the horizon is diminished by an imperceptible amount. Thus the refracted figures acquire an oval figure, like that of a coin thrown into water, if viewed from a very oblique angle. And this is the sole proposition on which those who deny refractions founded their apprentice-work in optics: that there is no force to the refractions, except that of expanding or contracting bodies seen under refraction. What they subsequently build upon it, concerning the directing of vision to the center of the heavenly body, cannot be any more solid than this proposition of Witelo's. Let it be that the lower edge of the sun graze the horizon: its refraction will be $34'$. But the refraction of something higher, such as has an altitude of half a degree, will be $29'$. Thus, in place of the sun's diameter, $30'$, there will be seen a quantity of 25 minutes, while the transverse diameter, or that parallel to the horizon, will be $30'$. The semidiameter of altitude $12\frac{1}{2}$, when the sight is directed to it, besides erring from the whole body, will also err from the center of the body seen under refraction by $2\frac{1}{2}$ minutes. **132**

Whether refractions shows the diameters of the heavenly bodies as greater?

In the same Proposition 54, Witelo proposes two other causes that alter the diameters of the luminaries. The first, he says, is a thick vapor bounded on this side and that by two surfaces, one of which is opposite the star, and the other opposite our vision, like a convex lens[130] Of this, certain books of spherics have imitations, even using the same words about a coin thrown into water. Therefore, reader of these matters, be studiously careful not to get confused. The action of ordinary refraction is in itself such that a decrease of diameters follows upon it, not an increase—so far, Witelo has it right. Or rather, if on the contrary the diameters should increase anywhere because of refraction, it is necessary that the distances also increase there, as in certain examining glasses, as well as in the coin thrown into water, if it be viewed from the perpendicular. But here there occurs exactly the opposite to that case that exists in which the coin is thrown out into water. For you are viewing from a rarer medium the coin that is lying in a

[128] *Thesaurus* II pp. 447–8.

[129] *Thesaurus* II pp. 448–9.

[130] *Perspicillum.*

denser medium, but we see, from the denser medium of the air, the stars existing in the rarer medium of the aether. Thus, when Witelo, following Cleomedes,[131] and others following them, explain the reason why the stars sometimes appear larger, he adduces, not the air, in which we observers exist, as is done in the usual explanation of refraction, but a thick vapor, which stays in the middle of the air like a cloud, so that the ray of a star viewed thus has a transition, first from the aether into the air, from the air into the convex vapor, again from the convex vapor into the air, and finally from the air into the eye, something that could delude you.

The cause of the delusion I believe many find in a typographical error, which all copies have, even the corrected ones published by Frideric Risner. For instead of what we read, that all stars appear round, greater, and so on, "smaller" should be read.[132]

Why the lights of stars sometimes are greater than usual.

However that may be, if I may unveil my own opinion here, this cause adduced by Witelo, following Cleomedes, is hardly a sound one for this event (even though I am going to introduce it below in demonstrating another phenomenon). The causes are these. It is a very common event, that when winds, or warmth after a hard freeze, arise, the lights of the stars appear enormous to us. Aristotle plainly
133 asserts this in Sect. 26 Q. 53,[133] saying that when Euro blows, everything looks bigger. And so since this happens so often, it could not exist because of a chance intervention of vapor between us and the stars. Besides, all the stars of the whole hemisphere are seen in this way at once, but if some pellucid vapor separated from the rest in one place were a cause, this would be seen to occur sometimes in one part of the heaven. Then I ask, by what force does a pellucid vapor, much heavier than air, stay suspended in air, in order that it, like lenses made of glass or crystal, could magnify the stars? The instance of clouds has no force. For clouds hanging in air, and mists, and things of this sort, introduce darkness, and are not pellucid; and since they appear to be of a somewhat dryer nature after they rain (for when they are heavy with humidity they do not hang, but fall down in drops), it is no marvel that they float upon air, as wood upon water. But how will vapor that is humid, pellucid, fluid, thick, and heavy, be held suspended in air, which is much lighter than itself? Why when this happens, is it not rather evidence that humid matter ascends from the lowest viscera of the earth, and that the parts that are lower and contiguous to both the earth and to our eyes are filled up first: the result is the same for us as for those who weep, for in both cases the eye is in physical contact with the humid. Furthermore, the same things usually happen in refraction, to both the diameters and the distances of the luminaries. But this accident

Why clouds hang in air.

---

131 Cleomedes, *De motu circulari* II 6 pp. 222–4, trs. Goulet pp. 171–3. Witelo himself does not cite any source.

132 Risner was the editor of the *Thesaurus*. It should perhaps be remarked that the change proposed by Kepler would require the substitution of "smaller" for "greater" (or some such) in at least four places, and a substantial emendation of Witelo's geometrical argument.

133 *Problems* 946a 34–5, trs. Hett Vol. II pp. 106–7.

does not affect the distances. For it has never been observed that the distances appear greater than they ought to: they are always either as they should be, or less than they should be. Finally, this thing that happens to the stars in a sense contrary to that of the refractions, happens at the zenith no less than at the horizon. Therefore, it is necessary that this does not belong to that class of refractions of which we have spoken so far. It will thus not be an accident of quantity, or of the visual angles, but of light alone, roughly the same sort of thing as haloes, rainbows, scintillation, and the like. Of these meteorological phenomena, the causes do not hitherto appear well enough explained, but since they are hardly necessary to astronomical considerations, I have not hitherto given them special attention.

Haloes, rainbows, scintillation: see the end of the chapter.

A third cause by which Witelo says the apparent diameters and distances of the stars vary, is truly optical, and does no harm to the light, as does the second, nor to the visual angle and the actual quantity of the image, as does
the first. Nor does it make the ones near the horizon smaller, but it does harm ***134***
to the estimation of quantity, and deceives this faculty of vision into imagining for itself a larger object than ought to appear by virtue of the visual angle. For when the eyes are turned upwards, nothing in between encounters them by which they may estimate the distances of the heavenly bodies. We therefore reckon those stars overhead to be extremely close, and consequently also smaller, the angle remaining the same. The opposite happens at the horizon. For then, if the interposed tracts of the earth on one horizon are comprehended in a single view, they instruct the vision to some extent about the enormous distance, from which the quantity of the object viewed (whether it be the distance of a pair of heavenly bodies or the diameter of one heavenly body) appears rather large, the angle remaining the same. For of those things that are perceived under the same angle, those that are farther away are greater, and thos that are less [far], smaller. This cause does not affect astronomers much, because the observation that is carried out with instruments does not introduce any error from this source. The only important thing is this: when we read that the ancients carried out their observations, not with instruments, but by estimation of the distances, we know that they might have been in error in this matter, since the faculty of estimation itself is egregiously in error through this cause. This was taught by Tycho Brahe, and by Ptolemy himself in Book 9 Ch. 2.

Why the stars appear greater about the horizon.

8. Whether the refractions are the same in all times and places.

The refractions in maritime locations are more constant; in those inland they are sometimes nearly nothing, and sometimes prodigious. For in maritime locations the air keeps about the same altitude, and it is deep enough that even if some altitude were added to it, that would not be so perceptible. In inland locations, however, the air is lower, so much so that on certain mountains one could not even stay alive unless one put a sponge to the nostrils, as Aristotle attests of Olympus in the *Meteorologica*.[134] The Peruvians relate something similar of

---

[134] Kepler is evidently citing a passage from memory, but his source has yet to be identified.

mountains that one cannot approach because the human spirit fails. And Bodin[135] relates from an Indian history that a great many of the Spaniards died of cold when they moved across the highest ridges of the mountains beneath the equator, while the plains were burning with heat. For the highest mountains project out
135 of the air, like rocks out of the sea, and the inland regions, near to the sources of the rivers, because of the altitude above the shores of the sea, which provokes the downwards flow of the rivers, correspond to some extent to the shallows in the sea. Hence the common opinion that the air in the Alps is more healthful, because it is thinner and more clear of impurities, the thicker air settling down into the valleys. And so in some places there will be no refractions, and in other places they will be very slight, and all nearly at the horizon itself. For it is in accord with this that there are many places that are elevated half a mile above the surface of the sea; and this, by 11 of this chapter, is the elevation of the surface of the air in the Danish strait. Vitruvius,[136] in Book 8 Ch. 7, specifies the two hundredth part of the course for the gradient in aquaducts. There is no doubt that such a height will be ruinous for navigation. And so if you allow a hundred miles for the meanderings of the Elbe, there will be much less than half a mile left for the elevation of the Austrian plain, arising from which it flows into the Ocean, because it flows gently and forms ponds. But the Danube, carried down through almost another 100 miles, again delimits the plain of Swabia, higher than that of Austria. So it will be brought about, that the Vosges and the continuous Alps of Raetia[137] nearly surpass the altitude of the air, which brings about the refractions. The perpetual snows confirm this, for they indicate that their summits are not clothed in the vaporous air. Therefore, to the greatest extent there will be no refractions there.[138] But if storms threaten, and sudden vapors arise out of the neighboring mountains, before they flow down and distribute themselves into the uniformity of the airy sphere, it is fitting that they perform their function in the matter of refractions, and that this would vary very greatly because of the surface

Gradient of rivers.

Altitude of the Vosges.

[135] Jean Bodin, *Methodus ad facilem historiarum cognitionem*, many eds. It is not certain which passage Kepler is referring to; possibly he had confused this book with another. Cf. *Epitome* I 3 (*JKGW* VII p. 63), where he presents the same example and, in the Errata, says he doesn't know the author.

[136] Marcus Vitruvius Pollio, fl. 1st C. B.C.E., known entirely for his *De architectura*, which centers upon architecture but contains excursions into clocks and timekeeping, mensuration, and astronomy.

[137] A former Roman province including much of the Tirol, Switzerland, and Bavaria.

[138] Here Kepler refers to the following endnote:

To p. 135 l. 20. A perfectly clear example: On 1601 May 10/20, the sun set from a fair number of degrees of altitude without any rays, with a dull face, as if through water, the sky being to the sight entirely clear. I was in the part of Styria which they call by the German word that means "The Hills"; to the west are the highest mountains. Therefore, the sun's distinct appearance, and the uniform color of the air, were arguing for a pellucid matter, but denser, as might be proportionate to sunlight that is very much weakened, its light being drunk up, so that the sun could not draw a shadow around anything. In three days there followed a tremendous rain, and a ruinous overflowing of those regions.

that, in the boiling up of the vapors, is uneven. In that case, then, a place sees greater refractions in proportion as it is nearer to the boundary of the air,[139] because air is circular, and (as a consequence) it refracts at the greatest angle the rays of the sun falling upon it most obliquely—indeed, by $61\frac{1}{2}'$, as was just now clear in Prop. 9. But those rays, thus refracted, traverse only the highest places of the air; those, on the other hand, that penetrate to lower places, are refracted with a smaller angle, because they also strike the air more directly. See the diagram in Ch. 7 No. 5 [p. 290 below].

This same thing is confirmed by the testimony of learned men and by experence. Certainly, Rothmann in Hesse, which is near the sources of the Weser, constantly affirms that the refractions of the stars commonly appear less[140] than those that Tycho Brahe observed in Denmark (pp. 29 and 85–6 of Tycho's *Epis-* **136**
tolae astronomicae),[141] and he urged Tycho to reckon the refractions to be diminished at places to the south, which is indeed true of Hesse, but not in proportion to its removal southward from Denmark, but in proportion to its greater elevation above the center of the earth and its greater nearness to the surface of the air. See vol. 1 of Tycho Brahe's *Epistolae*, pp. 63, 64, 65, and 112, where Tycho, refuting Rothmann's argument, confirms my argument.[142] Indeed, his having experienced

---

139 Here Kepler refers to the following endnote:

On line 25 following: When I attribute greater refractions to high places, you should understand "horizontal" refractions. For those that occur at some altitude of the heavens are proportionally smaller, and drop off proportionally more rapidly, as was demonstrated above.

140 Here Kepler refers to the following endnote:

Below this on line 34: In Hesse, which is high, the refractions are said to be smaller, that is, they are rather high, and are those at which the stars can be observed. For elsewhere they observe the completely horizontal refraction elsewhere to be enormous, and right at the horizon they rarely captured the stars with instruments. For these two things must be carefully distinguished: First, that in some region or time, the refractions can be smaller because of the thinness of the air; and here, upon the diminution of those that occur at some altitude of the heavenly bodies, there does indeed also follow the diminution of the whole refraction, that is, of the refraction that occurs in the ray that is tangent to the sphere of the air. But, second, in that same region and time, the horizontal refraction can nonetheless appear greater, for the reason that the observer is nearer the surface of the air because of the altitude of the region, and is blessed with the ray tangent to the sphere of the air, most greatly refracted, which ray would have passed far above his head, if the observer had been standing in a lower place. For it does not happen in all places that the greatest amount that the refraction attains is also that which the setting sun has, as I have taken pains to impress upon you.

And it is clear *per se* that there is some part of a smaller quantity, nearer to the whole, which can be greater than some part of a greater quantity, more deficient from its whole. Therefore, even though the whole refraction of a tangent ray is sometimes smaller in one place than in another, nevertheless, that part of it which the setting sun has, can be greater than the part at another place.

141 *TBOO* VI pp. 57 and 114.

142 *TBOO* VI pp. 92–3 and 140–1. The text here has 85 instead of 65, which cannot be correct as p. 85 is in a letter of Rothmann; 65 is a conjectural emendation.

colder winters in Bavaria (which he afterwards also asserted of Bohemia) clearly argues that the depth of the air is less in the mountains than on the coasts. For the coasts of the Atlantic Ocean, they bring in another cause, that warm winds are stirred up by the daily tides of the sea, which spread over the coasts, so that snow rarely lasts long. In the Baltic, however, there are no tides, or very slight ones, and the cause I proposed stands intact; not, however, that which Rothmann brings in, that the air in Denmark is thicker because of the slightly increased elevation of the pole; but this, that the air is deeper in that part of the sea in which the Weser is poured forth than it is in the plain whence that river arises (for the level of its Ocean is the same as that of the Baltic, which flow together in a nearby strait). I meanwhile do not deny that it can also be thicker, but this is in the highest North. On the same p. 112, Tycho opines that even in different regions of the same horizon, the refractions can change because of different vapors; much more so, therefore, in different places. The Landgrave of Hesse, in turn (p. 22 of Tycho's *Epistolae*), asserts that on a certain night the star Venus was carefully observed by him to remain stationary on the horizon for about a quarter of an hour, as if it were not being carried at all by the motion of the *primum mobile*, while it would now have set a further two degrees.[143] Afterwards, it suddenly vanished. This unusual sight arose from no other source than a gradually rising thick vapor, that was immediately again diffused in the level of the air.

A not dissimilar experience was related by my teacher Maestlin in the *Theses de eclipsibus* which he published in 1596. These are his words in Thesis 55:[144]

> In the year 1590, July 7, as the center of the sun was emerging above the
> **137** horizon, we here at Tübingen saw the moon, already two digits[145] eclipsed from the south, at an elevation of almost two degrees; and on the other hand, as the center of the moon was descending beneath the setting point, we noted that the sun's altitude was two degrees above the rising point. Moreover, the moon set before the eclipse reached its maximum darkness.

From these it is deduced that the horizontal refraction on that day was much greater than two degrees, half of which was due to the sun, and the rest to the moon. For, with the center of the sun rising, it was fitting that the center of the shadow should be setting, and since the moon had now at that time begun to enter the shadow, while it is usual for it to enter from the west side, the center of the moon was therefore more westerly than the center of the shadow; and since this was setting, the center of the moon had already set. And yet to the sight it was still raised by two degrees. Therefore, there was more than two degrees in the aggregate refraction of the sun and the moon together, not to mention that the horizon of Tübingen does not lack mountains, whose level is a little higher than the gradient of water, and so when the sun was halved, if the mountains had not

---

143 *TBOO* VI p. 50.

144 Maestlin, Michael, *Disputatio de eclipsibus solis et lunae. (Def. Marcus ab Hohenfeld).* Tübingen 1596, p. 19. (Citation from *JKGW* II p. 447).

145 See the note to Ch. 2 Prop. 12, p. 71 above.

Mount Hoeberg of Swabia.

been there, it would have been entirely clear of the horizon. Thus the refraction of one star was greater than a whole degree. Moreover, that part of Swabia is not far removed from the sources of the Neckar and the Danube, and Mount Hoeberg[146] near Horba[147] is notorious for sorcerers and gatherings of witches, since the credulity of the people treats the wit of Nature and the closely spaced eruptions of storms, as portents.

But that at different times, refractions are different even in the same place, Rothmann argues stubbornly, on pp. 85–6 of Tycho's *Epistolae astronomicae*, and Tycho too grants it without difficulty in the *Progymnasmata* pp. 79 and 280.[148] And in vol. 1 of the *Epistolae* p. 64, he requires that the air be rather pure and calm when he makes a test of refractions.[149] Thus in the Observations, I have found an annotation for January 1587, "refractions near the solstice appear to be greater."[150] Nevertheless, in my opinion, the cause of the increase refraction on the day just mentioned is not to be assigned to the winter solstice, but to the warmer air created by Jupiter and Mars being in quadrature, as they were then.

Why are the refractions of the fixed stars smaller?

For that Tycho found a smaller refraction in the fixed stars than in the sun, there seems to be no other cause than this, that the refractions of the sun are observed in summer and those of the fixed stars in winter, most correctly, but in summer the air is more humid and higher than in winter. For it appears on p. 64 of Book 1 **138**
of the *Epistolae*, and on p. 94 of the *Progymnasmata* it is openly taught, that refractions must above all be observed when the sun both passes through the meridian fairly high and does not change the declination perceptibly within the same day, which happens at the summer solstice.[151] For in the wintertime the sun does not surpass the altitude of refractions. On the other hand, the refractions of the fixed stars are difficult to observe in summer, partly because of the brightness of the air all night in Denmark, and partly because the thick air near the horizon intercepts the view of the stars. In winter, then, and on a very calm night, when the air is lower. And why would the refractions not increase in east winds, since, on p. 122 of the Tychonic *Epistolae*, Rothmann states that when he stood in the middle of a heated observatory he himself had very often seen the stars refracted through the vapor of the hypocaust?[152]

And so, to finish up this section, let this be certain, that in different places and weather the refractions are different, and that on extraordinary occasions they are extraordinary. For example, if the altitude of the place be high, the refraction will be none; if there is an exceptional releasing of vapors, the refraction will be prodigious. If, on the other hand, the places and weather keep themselves moderate, the refractions will be approximately the same.

---

146 Oberhohenberg, in the western part of the Swabian Alps, near Rottweil.

147 At Kepler's time, this was the seat of the county of Hohenberg.

148 *TBOO* VI p. 114; II pp. 64 and 287.

149 *TBOO* VI p. 92.

150 Observation of Jupiter, 15 January 1587, *TBOO* XI p. 173.

151 *TBOO* VI p. 92 (where the passage referred to appears to be on p. 63 of the first edition); and II p. 79.

152 *TBOO* VI p. 152.

9. On the observation of the Dutch in the far North

The story is familiar to everybody, of the journey of the Netherlanders, that is, the description of the sea voyage through the northern Ocean to the desert regions called Nova Zembla, in search of the strait by which a passage might be made to the Scythian and Eastern Ocean.[153] In this book, among other memorable things, they relate that when night overtook them, stuck in the ice, and they saw the sun for the last time on 1596 3 November, New Style, although from the altitude of the pole, which they reckoned to be 76 degrees, they considered it certain from astronomical principles that the sun would not return before 1597 11 February, it actually happened that they saw the sun again, its highest edge right at the meridian point, on 24 January, seventeen days before the proper time. A few hours after this time, they noted the conjunction of Jupiter and the Moon at 2 degrees of Taurus, lest anyone think that because of the continuous darkness they
139 had failed to observe the correct intervals of days and nights. Moreover, to remove all doubt, they saw the whole sun standing clear on 27 January. Therefore, on 25 January, the center had risen. Taken by wonder at this fact, many people everywhere consulted many mathematicians, of whom the others answered something else,[154] while I answered thus. Since the account of the Netherlanders appears to be trustworthy, the entire cause must not be assigned through suspicion to an error of the seamen about investigating the altitude of the pole, which the others tried to do. For if you deny the seamen this, that they cannot have any certitude about the altitude of the pole within five degrees (which is in fact about the size of the arc by which the sun was still below the horizon on the stated day, in the truth of the matter), you would be overturning nearly all the nautical knowledge of this age; nor will these Palinuruses bear without indignation this being said of them.[155] Besides, as their account states, the setting of the sun's center between November 2 and 3 does not at all correspond to this its rising. Therefore, that whole error cannot be in the altitude of the pole. On 2 November the sun was at 11° 37′ Scorpio, when they did not see the whole, and on 3 November it was at 12° 38′ Scorpio, where they could barely see the top edge. Therefore, the center set at 12° 7′ Scorpio, whose declination is 15° 27′, which is the same as what the sun has when it enters 17° 53′ Aquarius on February 6, not when the sun was at 5° 28′ Aquarius on January 25. But it is also not true that the place or shore in which they were held captive by chains of ice was floating in the Ocean (to reply here to Bodin's followers), so that they

---

[153] This was Wilhelm Barents's expedition of 1596–97. The description of the voyage mentioned by Kepler is Gerrit de Veer's *Diarium nauticum seu descriptio trium navigationum admirandum*, Amsterdam, 1598 (and other eds.), a book which gained a wide readership.

[154] Kepler had corresponded with a number of colleagues concerning the problem raised by this observation, most notably David Fabricius and Michael Maestlin. See Letters 132 and 214, *JKGW* XIV pp. 55 and 229.

[155] Palinurus was a pilot of Aeneas's ship, chiefly known for having been lost overboard after being overcome by sleep at the helm. Cf. Vergil, *Aeneid*, V 827–871 and VI 337–383.

might have been carried through sixty miles from north to south during those three months. For when, shortly thereafter, the altitude of the pole as well as that of the sun were taken again, they again corresponded closely to the previous observations. And in returning home, they traversed the same course that they had made going out. What is left, therefore, is that refraction alone bears the responsibility for this phenomenon.[156] But to make such a prodigious refraction probable, of so many degrees, I prescribed first a look at the examples just presented of Tübingen and Hesse, where trustworthy authors had noted a horizontal refraction greater in the former case than one degree, in the latter than two. For although the Netherlanders' place was maritime, where according to the nature of humidity the altitude of air would be about the same as that in Denmark, nevertheless, another of the causes, namely the density of that air, 140
was able to boost the refractions. If it is true that air is condensed in darkness and expanded in light, the darkness in those places was indeed of sufficient duration for so great a refraction, being of about three months. I also added another example, of the drops in the clouds of Greenland, of an entirely incredible size, as someone or other left attested. Before I add anything to these at present, let their altitude of the pole first be examined. Let EF be the horizon; CD the equator, whose poles are A, B; the sun at F on day $2\frac{1}{2}$ November, with declination 15° 27′. But on April 30 let the sun be at E, on which day (I mean at 12 hours before noon) they relate that they first saw the whole sun above the horizon. It was at 9° 20′ Taurus with declination 14° 39′, which is CE. Subtract the semidiameter, 15′. Therefore, at declination 14° 24′, the center of the sun could have been on the horizon. So AE is 75° 36′, and DF 15° 27′; the sum ought to have been 90°, but it surpasses 90° by 1° 3′, which is the standard of measurement of the two refractions at E and F together. Indeed, rather small. But if the observation just cited had been on 30 April at 12 hours after noon, this aggregate will become less, barely 44 minutes. Unless perhaps the sun at E had attained some altitude above the horizon, which they do not add. But if this aggregate refraction be bisected, and the half be subtracted from AE, and the remainder from DF, there remains, in the former case, the altitude of the pole, 75°, and in the latter, that of the equator, 15°, more or less by a few minutes. Or, more probably, let the refraction in the former be 20′, because of the length of the day, and in the latter 43′, because of its shortness, the altitude of the pole will become 75° 16′; of the equator, 14° 44′. And the declination of center of the sun when first seen is 18° 58′. Therefore, the refraction is 4° 14′; by bisection of the sum of the refractions it would be 4° 3′. Let me now supply the previous computation of the

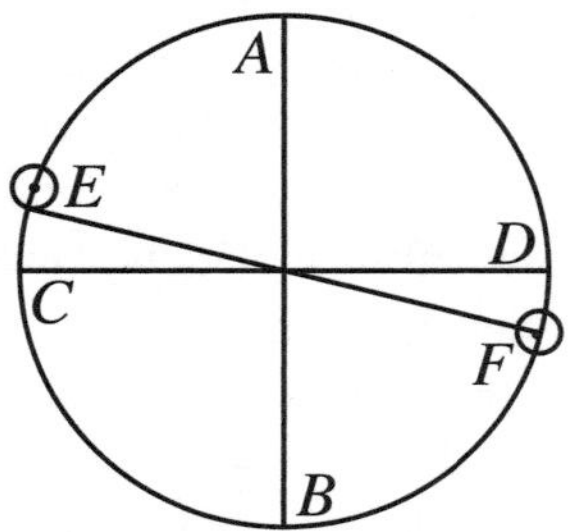

This opinion was expressed very recently by Miverius, in a book written against Christmann.

156 [To the marginal note]: In his *Observationum solarium libri III*, Basel, 1601, mentioned above (see page 139), Christmann had argued that certain trigonometric theorems of Phillip Lansberg (in *Triangulorum geometriae libri IV*, Amsterdam, 1591 and 1631) were false. Against Christmann, Daniel Miverius (English by birth, though he practiced medicine in various places in the Netherlands) wrote a book entitled *Apologia pro Lansbergio ad J. Christmannum*, Middelberg, 1602, which Kepler cites here.

altitude and density of the air, what is the least it could be to bring about such a refraction. And we would be assuming a minimum density, if we should stipulate that this refraction of 4° 14′ is entirely horizontal; that is, that the ray of the sun from the aether that is tangent to the surface of the air, being refracted in air, is again tangent to the surface of the earth. Then within the air there will be set up an angle of 85° 46′ with the surface of the air, whose secant is 1,354,677. Hence,
141 the simple refraction of 90° is 19′. But above, air was normally refracting from 90° through its density with a simple refraction of 1′ 6″.[157] Therefore, this density that it has is nearly seventeen times as great, but the sphere of the air is nearly four times higher than before, almost two miles.[158] It is in fact not credible even to me that this altitude should be so great. But you cannot assume a smaller one. About center A let the greatest circle of the surface of the earth BC, and of the air E, be described; and at some point E let a ray of the sun FE be tangent to the surface of the air E, and let it be refracted there into EB, so that EB is tangent to the surface of the earth at B; and let the angle of refraction be 4° 14′. By what was demonstrated above [in Prop. 8], BAE will be equal to this, and the altitude CE will become fixed. If you now should wish to depict the surface of the air as lower, then let a circle inside E be described about center A: it will cut the refracted ray EB; let it cut it at D; and from the point D parallel to EF let DC be drawn out, so that the angle of refraction again be 4° 14′, i.e., so that BEF is equal to BDC. I say that DC is no longer in the aether, but is going to be in the air, that is, that it is going to follow the circle through D in the direction EF. For let points D, A be joined. Then, because EF, DC are parallel, DCA, FEA will be equal and right, because EF is the tangent to the circle at E. Therefore, in the right triangle DCA, the right angle DCA is equal to the remaining angles CDA, CAD, together. Thus DCA is greater than CDA alone. Thus DA is greater than CA, by Euclid I 18. And since DA is the semidiameter of the most recently described circle of the air through D, the circumference D is therefore farther away from A than is point C. Therefore, DC cuts the circumference in the direction C no less than at the point D. So it will happen that CD is beneath the air, not in the free aether. Consequently, either this phenomenon will not be of the class of genuine refractions, or we have to believe that the altitude of the air was 2 miles.

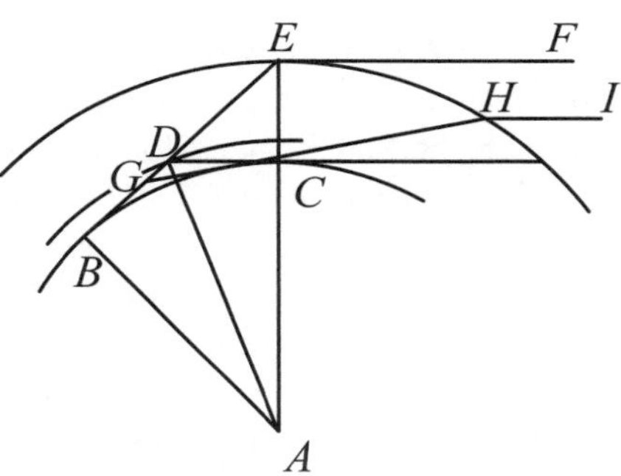

I shall therefore show two other ways by which such a phenomenon might come to be possible, both drawn out from Cleomedes, if his words at the end

157 This is 90 times the simple refraction of 44‴ per degree. See the refraction table on p. 138 above.

158 This is proved immediately below. Since the angle of refraction is about four times as great as that in the table, the angle BAE in the following diagram must also be four times as great, and the height of the air CE must increase accordingly.

of Book 2 be well compared with these.[159] The first consists in this, that if the observation occurs at point B, DC be drawn tangent to the earth's surface, and **142** containing with DB the required angle, so that, because the refraction is to be 4° 14′, BDC is 175° 46′. Let it be tangent to the earth at C, and cut BE at D. And suppose now that there is no air at C. Therefore, in places lying beneath D, there is some boiling up of thick vapors with an irregular surface, which nevertheless is in such a state that through the mediation of the density of pellucid matter it might refract the ray CD of the sun approaching from the free aether through 4° 14′. I believe this is what Cleomedes intended when he compares vaporous air near the horizon to water into which a coin is thrown. For just as, in the former, the eye is placed outside the denser medium, so also, in the latter, Cleomedes locates the eye of the observer in the air outside that thicker vapor. Therefore, since DBA, DCA are right, and DAB, DAC are equal, each being $2\frac{1}{9}$ degrees, we have now come closer here to the previous measure, for in the refractions that usually are seen, the angle BAD was 2°. If it should appear to anyone that the altitude of the point D or of the vapors is still too high, the vapor can still be lowered by doubling the refraction. For let a lower point G be chosen on EB, and from G let GH be drawn tangent to the earth, so that it may cut the air at H and be refracted there. Finally, from H let HI be drawn parallel to EF. There will be the usual refraction at H, and afterwards the extraordinary refraction at G. However, if you consider that in turning their faces to the south, the Netherlanders looked towards mountainous, wooded, and very high Tartary, whence flows the river Ob, so that BE is raised up by about sixty miles in that inland region, it will probably not be incredible to you that the point E on the surface of the air was so greatly raised above BC, the level of the Netherlanders' quarters at that moment. There will be much on this inequality below.

The other way is based upon reflection caused either by a dense and uniform cloud (as Cleomedes maintains), or, as I maintain, by the upper surface of the air surrounding us, such as if IH were refracted into HG, and at point G of the concave surface passing through G it were reflected to B. This can happen through the laws of reflection, if CD, DB are tangent to the same spherical surface. Nor is there fear that the surface may transmit rays, for the reason that after that convex surface it encounters a rarer medium. For we see in glass that both surfaces, both the **143** exterior one at the top and the interior one at the bottom, reflect rays, so much so that when a glass mirror is set facing the sun, it comes to a third or a fourth reflection. Let there be a glass, whose upper surface is ABC, lower surface ED, by which the mirror is bounded by an application of white lead. And let the sun be at F, the incidence be FC, the consequent reflection being CG, at equal angles.

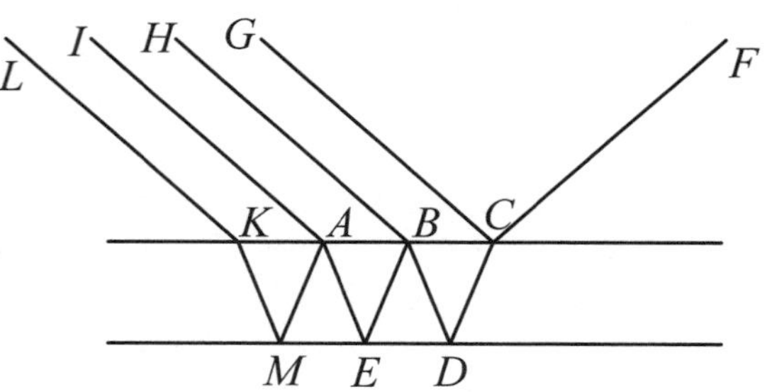

[159] Cleomedes, *De motu circulari* II 6 pp. 222–4, French trs. Goulet p. 173.

But because glass is transparent, the most powerful part of the solar ray passes through the surface at C and is refracted into CD, and thus the remaining part, which is scattered through CG, is rather faint. Next, CD is reflected from the point D strongly (because the mirror is bounded at D), and at equal angles, into DB; and, going out into the air there, that is, at B, it is refracted into BH, and here the ray BH is strongest. Again, because DB encounters the polished surface B, it is partly (but weakly) reflected at equal angles into BE, notwithstanding the presence of air above B; and at E it is again strongly reflected into EA. And, going out into the air at A, it is refracted into AI, but here it is very weak, because part of it went away at C, but much more of it at B: the former of these departures was scattered into CG, the latter into BH, so that at AI it is very much weakened. And nonetheless, even at A some part of EA is reflected into AM, and from M into MK, and is refracted into KL. But now this fourth ray can seldom be perceived, only when cast forth into dark places, because of its excessive rarefaction. Therefore, in that way in which the rays are reflected in glass by the upper surface, even though it is not bounded, they can also be reflected in air by its highest surface, so that, in place of the sun, its image[160] could have been seen by the Netherlanders in Nova Zembla. The learned should see whether the phenomena of twilight, too, may be justified in this way, so as to make it unnecessary for the matter that is to be illuminated to ascend to a height of 12 miles: it would suffice that once the sun's ray has entered the surface of the air, from which it again departs, it be reflected at equal internal angles, and that this occur twice, thrice, or four times, until the illumination of the air made by the reflected rays is completely obliterated and no longer enters into the eyes.

Phenomena of twilight.

10. Conjectures from antiquity concerning refractions

Much to astronomy's loss, it happened that in the investigation of the sun's
144 motion and the equinoxes, refractions were ignored by the ancients. If it could be proved by reliable arguments that the rays of the heavenly bodies are refracted in all places and times, that loss might partly be made good by us. I shall make a try at accomplishing this, as much as I can. First, that refractions are different in different places, we have just now proved clearly enough. Further, from the same argument it was evident that one or another magnitude of refraction resulted from taking one or another density of air and altitude of its sphere above the earth. Next, Egypt and Rhodes, homelands of Ptolemy and Hipparchus, respectively, are maritime places, and are therefore immersed in the depth of the air. On the other hand, while cold condenses, heat expands, and there is a greater heat in those climates, so the air will also be thinner, which, when denser and heavier air comes in underneath from the North, is driven upwards, and places itself above, by which position it is made to slope, and in turn seeks the north, pouring itself

Could this be the origin of the etesian winds?

---

160 *Idolum.* This is the Latinized form of the Greek εἴδωλον, a somewhat unusual word in classical Latin. Although it can mean "ghost, apparition", it was also used by the Epicureans to denote the small images given off by illuminated bodies. These images enter the eye of an observer and provide the information about their source that constitutes vision.

down onto a lower surface of the air, unless it is driven back by a stronger force (such as if the entire level of the air were set in motion by the initial impulse), and, being driven back, overflows and ponds, as it were. And in sum, as regards density, there is a perpetual intermingling of our air with the southern air, and in turn of that air with ours, and this is because of the continual blowing of winds. And so with respect to density, there will not be a very great difference between us Europeans and Egypt or Rhodes, but in what follows, in Chapter 7, we shall make use of the liberty to seek further whether in all places the altitude of the air be the same, and the density the same. And so, once the cause is set in place, it is necessary that the effect follow, and since it is not just today that the air is poured around the earth in a circle, but this law of nature, as is fitting, endures from the beginning of things all the way to us, it should be thought reasonable and fitting that there was never a time when there were no refractions. These supports deduced by reasoning are to be bolstered by testimony and conjecture fetched from antiquity, from which it may appear that, over the continuing succession of ages, refractions of the stars in air were either noticed or at least perceived by the sense.

Pliny provides clear testimony in the *Natural History* II 13.[161] His words are as follows: "For what reason is it that when, as the sun rises, that shadow that dulls [sc., that causes lunar eclipses] ought to be beneath the earth, it happened once already that the moon was eclipsed as it set, both bodies being visible above the earth?" Here you have at the same time the sun and the moon, although **145**
they are at opposite positions, distant by a semicircle, nevertheless appearing by refraction above the horizon, and at a distance that is less than a semicircle. This is in agreement with the theory of refractions as hitherto presented.

Cleomedes relates in Book 2 that certain more ancient mathematicians thought they could untie this knot by saying that the earth, because of its roundness, is like a sort of mountain, from which the same observer easily looks upon what is being done in two valleys.[162] However, Cleomedes refutes them by showing the dissimilarity. For if the place in the mountains from which the valleys are viewed is a rather high place, it is also necessary that the mountain be everywhere precipitous, conical like a top. But we are settled upon the surface of the earth, so much so that it appears flat to us. To Cleomedes' refutation, I add that if we look down from some mountain, that happens along lines that descend below the level of water. But if this phenomenon just described should happen to us, the lines of vision on both sides are necessarily raised above the level of water (to which the visible horizon appears parallel). For in the similar eclipse mentioned above, and in explaining this very passage of Pliny, Maestlin testifies that each was alternately seen to shine at an elevation of two degrees.[163]

A clear passage in the Greek text was wonderfully distorted by the Latin translator.

Therefore, at the start, Cleomedes calls this account of Pliny's into doubt. He says, "What if we should say that this sort of account was fabricated by certain people whose aim was to throw into doubt those astronomers and philosophers

---

161 *Natural History* II.13.30, trs. Rackham Vol. I pp. 206–7.

162 Cleomedes, *De motu circulari* II 6 pp. 218–222, trs. Goulet (French) pp. 171–3.

163 See p. 149 above.

who devoted themselves to this investigation?"[164] But more than once, Tycho and Maestlin strongly restrained the incredulous Cleomedes by observations. Not unjustly, Maestlin is amazed that Cleomedes was able to assert that "no one who professes to be a mathematician has ever left written testimony that this had been seen by him, even though all eclipses, from the Chaldean, Egyptians, and the rest all the way to the time of Cleomedes, had been noted." Even so, Cleomedes also labored over this: how Pliny could have abandoned his trust and transferred the cause of the phenomenon to the sense of vision; and how this sense, from a reflection from a cloud or refraction by thicker air, might have erred, and might have looked upon the sun's image[165] in place of the sun. We have aired out both ways above, and have not rejected them completely, in refraction of unusual mag-
146 nitude. But the legitimate and commonly encountered cause for this phenomenon is sufficient, namely, the refraction of the light of the two luminaries in the surface of the air that is poured around the earth. Therefore, there were once refractions in Italy, and this testimony does not allow any exception. There follow weaker conjectures. In Book III of the *Great Work*,[166] Ptolemy asserts that he had often observed the equinox twice on the same day on the Alexandrine armillaries.[167] He places the blame upon the instrument's position having settled to that extent from the beginning of its installation. Hipparchus, however, who was closer to the first installation, and who affirmed that he had observed the equinoxes on the same armillaries—Ptolemy, I say, draws him into the same accusation. "For in year 32 of the second Calippic period, on the 27th day of the month of Mechir,"[168] (Ptolemy speaking), "he observed the equinox in the morning, but the Alexandrine armillaries were equally illuminated on both sides on the fifth hour of the day." Ptolemy therefore concludes, "the two observations do not agree within five hours".[169] And indeed, this is driven home by Hipparchus: "that observation is the best which is made using the Alexandrine armillaries, for that is the hour of the true equinox in which the two surfaces of the armillary are equally

---

164 Oddly, Cleomedes's Greek text, quoted in *JKOO* II p. 415, says, "But first one must confront those who say that this account is fabricated by certain astronomers and philosophers wishing to create confusion for those concerned with these matters." Although Kepler's version is better than Valla's (which Frisch also included), it is still not what Cleomedes wrote.

165 *Idolum*. See the footnote on p. 155.

166 This is a loose translation of μεγιστὴ συντάξις, one of the titles by which the *Almagest* was known.

167 Ptolemy, *Almagest*, III 1, trs. Toomer, p. 134. Ptolemy describes the construction of an armillary in the *Almagest* I 12, trs. Toomer pp. 61–2.

168 −145 March 24.

169 From Ptolemy's wording, Toomer concludes (note 9, p. 134) that the former observation was made by Hipparchus, presumably at Rhodes, while the latter was by another observer. Evidently, Kepler believes both observations to have been made by Hipparchus, and takes it as evidence of multiple apparent equinoxes on the same day.

illuminated."[170] So even if it could indeed happen that this inconsistent observation of the equinoxes originated from a fault in the instrument, and Ptolemy's critique might be true, it is nonetheless also true that this same phenomenon (i.e., the equinox appearing twice on the same day) was very often observed in our age by Tycho Brahe using highly accurate and very precisely installed instruments. For in the morning, before the equinoctial moment has arrived, the rising sun appears higher and closer to the pole of the world because of refraction, so that it may already be seen on the equator while it is still to the south. After noon, when the sun is free from refraction, suppose that it has now completed the winter semicircle: thus it will then at last be both truly and apparently on the equator. The opposite must be said of the autumnal equinox. It is therefore uncertain whether the cause for the authors' having persuaded themselves that they saw the equinox twice is to be derived from a fault of the instrument or from refraction.[171]

Whether the eccentricity was once really larger.

I should like to say something that is not really proper to this place, and should instead be deferred to the debate on the determination of the year. For what if this intricate matter of refractions alone were the cause for Hipparchus's once finding a greater eccentricity of the sun than Tycho finds today, and Al- **147**
bategnius[172] eight hundred years ago; while Ptolemy had carefully adjusted his observations in such a subtle matter to Hipparchus's demonstrations (which he fully kept)? If there is agreement on the former, then the latter is also decided, for to observe such a close consensus between the two of them in numbering the length of the year's quadrants was completely impossible using the Ptolemaic and Hipparchan instruments. Now the cause for my voicing this suspicion of Hipparchus is this. It is known from Ptolemy and Proclus Diadochus[173] that Hipparchus and others usually measured the diameters of the luminaries with clepsydras.[174] For while the highest edge of the sun was rising, water was flowing

170 This is not an accurate translation of Ptolemy's quotation of Hipparchus in the *Almagest* III 1, trs. Toomer p. 133, although there is some resemblance between the two passages.

171 This explanation was proposed by Karl Manitius in his German translation of the *Almagest* (Ptolemaeus, *Handbuch der Astronomie*, Leipzig, 1912, Vol. I p. 427 note 21), apparently unaware that Kepler had offered the same explanation more than three hundred years earlier. See Toomer's note in his translation of the *Almagest*, p. 135, for further discussion and sources regarding Manitius's explanation.

172 Al-Battānī (Latinized Albategnius), Arab astronomer of the late ninth and early tenth centuries, known chiefly for his work on the motion of the sun and the length of the year. His chief work was published in 1527 and other editions under the title *De motu stellarum*.

173 Proclus (410–485), Neoplatonist philosopher and astronomical theorist, was greatly admired by Kepler, who included a quotation from his commentary on Euclid's *Elements* on the title page of Book I of the *Harmonice mundi*. Cf. Max Caspar, *Kepler*, pp. 92 and 266–273.

174 On the use of the clepsydra, or water clock, for such measurements, see Ptolemy, *Almagest* V 14, trs. Toomer p. 252, and Toomer's note referring to Pappus, Heron, and Proclus. On the clepsydra in general, see W. K. C. Guthrie's note in his translation of Aristotle, *On the Heavens*, p. 226–8.

from the open hole in the clepsydra into a separate vessel, and once the whole sun stood clear, the remaining water was received by another jar, throughout a whole day and night, until the edge of the sun was again seen to rise in the same place. Since they used this observational procedure to measure the sun's diameter, what is more likely to be believed than this, that Hipparchus prescribed that day for the equinox upon which the sun should touch two diametrically opposite indicators on the visible horizon, one in rising, the other in setting; and that once those indicators had been established by observation, they would forever after be kept as markers of the equinoxes, if it should stay in the same place. If I have guessed right here, I shall obtain the rest without difficulty. For up to the time when the sun is in the southern semicircle of the zodiac, it is impossible to accomplish what we have just said was seen by Hipparchus, that in rising, it should touch the point of the horizon diametrically opposite that which it touches in setting. For in both cases it still verges towards the south, and therefore, on the horizon, where we have presupposed that the observation occurs, the sun, appearing higher than is correct because of refractions, in the spring will appear to get to the equator more quickly than is correct, and in the autumn will appear to have sunk down to it more slowly than the truth. And so the time of the summer semicircle will appear to be more lengthy, since both ends are extended. Thus the eccentricity will be greater than is correct. And so one who wishes to believe with a clear mind that the sun's eccentricity is the same both in the past and today, has the perceptible argument of refractions that Hipparchus had observed.

148 You would not be entirely ridiculous to add at this point what Proclus of Lycia presents in the *Sphaera*,[175] where the antarctic[176] is described by the forward foot of Ursa Major; that is, where the star in that foot appears to touch the horizon daily. In the same place, the summer tropic is cut by the horizon into 5 and 3. These two teachings are not consistent with each other. Suspicion can be plausibly cast upon refractions, that either the foot of Ursa appears to touch the horizon, which in fact it goes beneath; or that the sun, in rising more quickly than is correct and setting more slowly than is correct, indicates on the clepsydra a diurnal space that is more drawn out than it really is, whence a portion of the tropic will be thought to stand clear that is greater than is correct, and the altitude of the pole will be considered greater than is correct. I realize that those teachings are not written particularly accurately, and can be taken more roughly. But since this way of taking the altitude of the pole is rather easy, I wanted to advise the learned (as Tycho does on p. 95 of the *Progymnasmata*) to consider which places on earth it was, at which Ptolemy established the altitude of the pole by this observation of the longest day.[177] This may very well be why, 100

---

[175] The Greek text, with a Latin translation by Thomas Linacre, was published in Basel in 1547. The passages cited by Kepler are on pp. 3 and 11, according to Frisch (*JKOO* II p. 415).

[176] Or, "antarctic circle." In either case, the use of the word is puzzling, and I have been unable to check Proclus's text, other than the brief passage quoted by Frisch.

[177] Tycho does not appear to have urged any such consideration. He did advise great care in determining the polar altitude, and discussed the effects of refraction, on p. 15 of

Whether the altitude of the pole is changed.

years ago, Antonius Maria,[178] teacher of Copernicus, thought that the altitudes of the pole had decreased in all places in Italy, namely, if these altitudes belonging to the places of which Ptolemy had once correctly noted the latitudes, were excessively increased in later times, by estimation of the longest day, because of the refractions exceeding their measure. Or, contrariwise, that Ptolemy had investigated them from observation on the shortest day, which, since it is greater than the truth because of refractions, creates the illusion of a greater portion standing above the horizon, and of a lowering of the sphere. William Gilbert warns of this same thing in that abstruse study of his on the magnet, in Book 6 ch. 2, in concise words.[179] A man of such excellence, in whose divine discoveries it is fitting for all students of Nature greatly to delight. I would hope to be able without much difficulty (since it is not a haughty philosophy) to earn his friendship through my eagerness to learn, unless that ostentatious Tethys begrudge me this.[180]

Those things that I have brought in from Proclus, things similar to them can be added from Cleomedes, who in turn gets them from Posidonius, but are also referred to by the same Proclus at the end of the *Sphaera*, and likewise by Pliny Book II, that in Egypt, Canopus at noon shows an altitude of $7\frac{1}{2}°$; at Rhodes, it **149**
grazes the horizon, on which island the ratio of night and day in summer is that of 19 to 29. For since from this the latitude[181] of the city of the Rhodians is 36°, while that of Alexandria is 31°, how is it that Canopus, which at Rhodes is barely visible from the highest mountains, can, on Proclus's authority, stand $7\frac{1}{2}°$ out, or the fourth part of a sign, in Egypt? For this altitude is given consistently and in the same words by Proclus and Cleomedes and Pliny, with Ptolemy also supporting, who gives that star a latitude of 75°. Therefore, either all the authors with a single

---

the *Progymnasmata* (*TBOO* II pp. 16–17); perhaps a 1 was substituted for a 9 in the typesetting.

178 This was Dominicus Maria of Ferrara (not Antonius Maria), of whom J. G. Rheticus said (in the *Narratio prima*) that Copernicus had been his "assistant and witness of observations." See Edward Rosen, *Three Copernican Treatises*, Third Edition (New York: Octagon Books, 1971), p. 111. Maria's views on the decreasing altitude of the pole were published by Antonio Magini, *Tabulae secundorum mobilium*, Venice, 1585, Canon 8, p. 29.

179 Gilbert, *De Magnete*, trs. Mottelay pp. 316–7. Gilbert considers Maria's argument at some length and rejects it on the grounds that the observations upon which it is based were not sufficiently accurate. At the very end of the chapter, he remarks that "observations of latitudes cannot be made with exactitude save by experts, with the help of large instruments, and by taking account of refraction of lights." Evidently, these are the "concise words" Kepler means.

180 Here, "pomposa illa Tethys" is the North Sea; this remark shows that Kepler was already contemplating a journey to England. Later, in 1606, Kepler sent a copy of his newly published *De stella nova* to King James I, and in 1619, he dedicated the *Harmonice mundi* to that monarch. In 1620, Kepler was visited by Sir Henry Wotton, the English Ambassador, who invited him to come to England (see Wotton's letter to Francis Bacon, *JKGW* XVIII, no. 892). However, he never succeeded in visiting England, and by the time he expressed this hope, Gilbert had died (in 1603).

181 Reading *latitudo* for *altitudo*.

voice must be hallucinating, or in Egypt, towards the sources of the Nile, towards the Mountains of the Moon, and those regions that are molested by continuous rains, it is necessary that there be interposed for observers in Alexandria an air that causes a refraction of some prodigious number of degrees, and that this air is not thus precisely to the south of the island of Rhodes, as Rhodes is indeed a few degrees west of Alexandria.

What the same Cleomedes confidently alleges, probably possesses less soundness: that there are two stars, distant by the diameter, the Eye of Taurus and the Heart of Scorpius,[182] when either of which rises, the other sets, and they are seen on the horizon at the same time. If anyone ever really saw this, the two must have been elevated above the horizon, since in reality it could not happen otherwise, unless one or both were always beneath the horizon, in this northern part of the world. For even though they are exactly a diameter apart, with respect to longitude, nonetheless both are in the southern hemisphere, one having a latitude of $5^\circ\ 31'$, the other $4^\circ\ 27'$.

To this point there can be related something, though it is obscure, from Proclus Diadochus's book *Hypotyposes astronomicae*[183] and Question 8 of his *Mirabilia*, even though this too appears: that he speaks of the excess of the year, as then accepted, above its true measure. However, his words, in Valla's translation, sound as if he were amazed at something that appears in the sun's motion upon the occurrence of refractions : "That the sun is fully seen, and that it accomplishes its rising as if taking a seat." Now William, Landgrave of Hesse used nearly the same words in his first letter on astronomical matters to Tycho, explaining the occasions of his taking note of refractions.[184] And although
150 those things at which Proclus wondered there can have occurred because of the computation in general use at that time, nevertheless, in giving reasons for them, Proclus sweated so, that you might suspect, not unjustly, that in tangling himself in refractions, he made the undertaking more bewildering, and that without them it would have been easy for him to untangle it.

Martianus Capella[185] is somewhat clearer in the *Encyclopaedia*, Book 8, whose words concerning the latitude of the zodiac are these: "For the sun, nowhere leaving [the zodiac], is carried in a central equipoise, outside of the proper boundary of Libra. For it deflects itself there either to the south or the

---

[182] Aldebaran and Antares, respectively.

[183] Giorgio Valla's translation of Proclus's *Hypotyposes*, to which Kepler refers here, is included in *Claudii Ptolemaei Pelusiensis Alexandrini Omnia quae extant opera, . . . ,* ed. E. O. Schreckenfuchs, Basel 1551.

[184] The Landgrave's letter is in *Epistolae astronomicae* pp. 21–23, *TBOO* II pp. 48–51. He discusses refractions and the measurement of the pole's elevation on p. 22 (*TBOO* II p. 50).

[185] Martianus Capella lived in the first half of the fifth century C. E. in northern Africa. His encyclopaedia, entitled *Satyricon*, or *On the Marriage of Philology and Mercury and on the Seven Liberal Arts*, covered Astronomy in Book VIII. Capella is noted for having placed the orbits of Mercury and Venus around the sun, rather than beneath it, as Ptolemy had done.

north at approximately the half moment."[186] He expressed the horizontal refraction carefully enough; and therefore the absurdity and strangeness of the statement that the sun, at a single place on the zodiac, declines in zodiacal latitude, instils me with a suspicion, as follows. Since the equinoxes, chiefly the autumnal one, have been carefully observed, and particularly on the horizon, at sunrise or sunset, because the visible horizon, as was said before, is like an instrument, it happened, from frequent repetition, that the mathematicians finally noticed the refraction that occurred at the horizon, and since they were ignorant of the cause itself, they attributed this little space, the amount by which the sun is moved closer to the pole by refraction, to the motion of the sun itself. If anyone be less pleased with this interpretation of Capella, let him present a more appropriate one to defend the author.

For the rest, let us leave out the guesswork and turn to the surer testimony of later authors. The Arab Alhazen, to whom Risner, in a conjecture not lacking in judgement, attributed an age of fifty years, and, active about two centuries after Alhazen, our Witelo, assert fully the same thing in their *Optics*, bringing in astronomical experience, which nowadays Tycho Brahe, using the most precise instruments, has brought fully out into the light, that stars are seen near the horizon under refraction, on which there is enough above.[187]

Bernhard Walther, in the book of observations that is appended to Regiomontanus's *Torquetum*,[188] gives fully the same testimony, that he had learned from experience that stars often appear above the horizon when in fact they are beneath it. That was about 120 years ago. But the passage deserves to have light shed on it, in order to have that man's discoveries compared with those of our Tycho. The author's words are these.

> In the year 1489, on March 6, about sunset, while, to be exact, 25 Gemini **151**
> was at mid heaven, the sun was at 25° 15′ Pisces, with the armillaries. Venus was found to be at 27° 15′ Aries through another circle, by the ecliptic dividing the sun. But by the circle of latitude mediating the sun, as is usual near the horizon, it was found at another place, namely, at 25° 30′ Aries, the cause for which I shall append later.

These things are explained a little later. There follows:

---

186 I have not found this passage in the *Marriage of Philology and Mercury*, though there is something rather like it in par. 849: "although the equipoised sun moves along the middle line of the zodiac belt, the obliquity of its course causes it to be depressed or elevated." (trs. Stahl and Johnson p. 330).

187 Alhazen discussed stellar refraction in general in VII 15–16, *Thesaurus* I pp. 251–2, and in detail in VII 51–55, pp. 278–282. Witelo's treatment of stellar refraction is in X 49–53, *Thesaurus* II pp. 444-8.

188 Joannes Regiomontanus, *De torqueto, astrolabio armillari etc.*, Nuremberg, 1544. Pages 36–60 contains "*Joannis de Monteregio, Georgii Peurbachii, Bernardi Waltheri ac aliorum, eclipsium, cometarum, planetarum ac fixarum observationes*." The observation cited here is on fol. 52^v–53^r.

> On March 7, the sun was 26° 15′ Aries by observation of the armillaries. Venus, from the ecliptic, was 28° 15′ Aries. From the circle of latitude, 27° 38′ Aries.

To these observations, he immediately appended this:

> Further, so that it might not be hidden too long from those who are going to read this, how I could proceed, since the position of Venus was found so variously in practically the same instant, it must be noted that, because of broken rays, stars appear above the horizon when according to the truth they are beneath it.

And this was clearly required by the explanation and theory of refractions, hitherto handed down, but not its contrary. For when the rays of the sun that strike obliquely upon the surface of the water immerse themselves within the water, they proceed at an angle more inclined to that surface of the water, exactly as if the sun were higher. In clearly the same way, if the same rays of the sun striking upon the surface of the air, which completely covers us humans like a flood, were to follow that path with which they strike upon the surface of the air, they would never strike upon our eyes, but would be carried off far above our heads. But, being refracted at the surface of our air, they now descend more steeply to us, and appear to us to be more elevated. And in fact, if we were dealing with water, the sun could not ever be beneath the surface of the water, always above, because the portion of the water surface in which we create the opportunity for this refraction is narrow, so much so that it seems to us perfectly flat. However, since air is round, the part of it that places itself between our vision and the sun is not narrow, and the part is indeed great enough that it now begins to curve towards the sun. And so, although some particular ray of the sun is tangent to the surface of the air at some point that is overhead at our zenith (when the sun shall be fixed right at the horizon), or at a point that is now nearer the sun than that which is above us (when the sun shall have already set), nevertheless, the parts of the airy sphere that tend towards the sun after that point of tangency, because of the air's being massed together into a globe, will send themselves down beneath that tangent, so that what is now setting for us is still elevated
152 for those points. In this way, the same thing happens again that also happens in water, that the sun is never below that point of air in which its ray is refracted, and nonetheless, for us it has in truth already set. But Walther proceeds: "Which very often appeared to me perceptibly with the instrument of the armillaries, before I saw the Perspectives of Alhazen and of Witelo of Thuringia, in which I later found this stated within a hair's breadth." The same thing happened to Tycho Brahe, that he noticed the refractions of the luminaries before he had read these optical writers or the discoveries of Walther, as he states in the *Progymnasmata*, Book 1 fol. 91. For the occasions upon which he discovered them, see fol. 15 of the same book, and the same fol. 91.[189] For this reason, both deserve all the more trust, in that in ignorance, and not caught by any prejudices, they realized that this was so. For it is characteristic of truth, to come together by diverse ways.

[189] *TBOO* II pp. 16 and 76.

There follows, in Walther, the occasions upon which he himself came to notice this fact. "But in order to escape the differences in the moon's aspect, I even examined Ptolemy's way, in the second chapter of the seventh book." That is, the occasion that showed me the refractions, was this. I was about to investigate with instruments whither on the ecliptic the fixed stars had proceeded in my time. The easiest way to inquire into this is if a perfectly correct position of the moon from the sun, or from the equinox, could be obtained for any desired moment by computation; for then one may see by inspection which stars it is that the moon has approached at that moment. Or, conversely, if at some particular hour of the night the moon should cover some fixed star, there may be computed, by a completely reliable calculation, the moon's true position from the equinox. For that will also be the star's true position from the equinox. Or if anyone should mistrust the calculation, he should carefully observe the middle of some lunar eclipse, and not the distances of the moon from the neighboring fixed stars. For he will thus have the distance of the sun from those fixed stars, and from the sun's meridian altitude and the latitude of the place he will also have the sun's distance from the equinox, through its declination. Thus the distance of those fixed stars from the equinox will also be given.

The procedure could easily be carried out in these ways if the moon's true position could be had, either in eclipse or outside of it. But, among other things, it happens that although we may have sure knowledge of the moon's true position, which one might see it holding if he were looking out from the center of the earth, nonetheless, parallax, whose variety is incredible, places the moon in one or another place. And so the moon's distance from the fixed stars is far different **153**
than if the moon were perceived in its true place. Therefore, so that the moon's parallax not bring a case against me, I used that method that Ptolemy pioneered in book 7 ch. 2. He, of course, used it for the moon, while I imitated it for Venus, whose parallax is less, and whose diurnal motion is less. "And I used an armillary instrument," (for a description of which see Tycho Brahe's *Mechanica*),[190] "to take the distance of the sun and Venus about midday, or afterwards," (a procedure which Tycho Brahe, unaware of Walther, imitated many times with the greatest care, as you see in the *Progymnasmata* Vol. 1 fol. 152),[191] "the sun's position being previously acquired by rules, or armillaries, or by both instruments." Both methods are described by Regiomontanus in that same book from which this is transscribed.[192] Moreover, the order is this. First, from observation of the fixed stars, the altitude of the pole and of the equator are obtained. Next, the meridian altitude of the sun is observed, which, compared with the altitude of the equator, gives the declination. And to each declination there corresponds a certain distance of the sun from the cardinal points, or the position on the ecliptic, the maximum declination being presupposed. "But since I had ostensibly been examining the positions of the two (the sun and Venus) at sunset, by turning the

---

190 *TBOO* V pp. 52–67, where Brahe describes four different armillaries.

191 *TBOO* II p. 159.

192 Regiomontanus, *Scripta clarissimi mathematici Ioannis Regiomontani de torqueto . . .* (Nürnberg 1544).

ecliptic armillary, and to be examining the sun's place on it," (by turning that part of the ecliptic in which the sun was perceived to remain by the meridian observation), "towards itself," the sun; not, however, by accurately locating the center of the armillary, the found position of the sun, and the center of the true visible sun, on the same straight line, but by turning that position of the sun to one side and the other of the sun's body, "until both parts of the surface," both edges of that circle of the inside of the ecliptic armillary, because the material circles extend over the latitude, "that is, the lower and upper," or south and north, "were equally illuminated by the sun," already setting, "also by moving the circle of latitude, until the two parts of its interior surface, that is, the right and left," or east and west, "received equal illumination from the sun. By this procedure, if the solar rays had come to the sense of vision," or the instrument, " without refraction, the circle of latitude," by its intersection with the ecliptic, "would accordingly have shown the the position of the sun," which was previously found on the meridian, and was corrected by the addition of the required hourly motion.
154 "But I found a noticeable difference." The sun was found in one position by the meridian observation, and in another position by the evening one, "and further, this," the difference, "was different," not always the same: "it was greater when the sun's declination was increasing or decreasing moderately, but smaller when the declination was changing noticeably, so as to be greatest at the solstices, and least at the equinoxes." This is not because there was that great a variety in the refractions themselves, but because even if the sun's declination were equally altered in both cases, nevertheless, since the longitude is obtained from the declination, and a false longitude from a false declination, this discrepancy in the declination therefore causes a greater error, where to one degree of longitude there corresponds a single minute of declination, which happens near the solstice, than where to one degree of longitude there corresponds 24′ of declination, as at the equinoxes.

"So, when I saw both circles," the ecliptic and the circle of latitude, "illuminated by the sun at sunset, as was said before, I by no means had either the true place of the sun or, from it, of Venus." For it was just said, that if this should happen, the true place and this refracted one would be seen to differ. "If, however, with the circle of latitude placed over the sun's position as found by the meridian altitude and restored by the addition of its intermediate motion, I showed the same circle towards the sun at sunset," that is, if I found the place that the sun has on the meridian by a different and reliable method, and if I have corrected it by the addition of the part that corresponds to the time from noon to sunset, if, to that place (I say) I were to apply the circle of latitude, and were to turn them, thus affixed to each other, to the setting sun, "until its inner surface," that of the circle of latitude, "was illuminated in the manner previously described," it being ignored that the same would happen on the ecliptic, "I have come closer to the truth," in investigating the place of the sun's longitude. For the circle of latitude nearly coincided with the vertical, especially at the spring equinox with the sun setting, and so the refraction, which as a rule happens on the vertical, nearly all went over to the latitude, the longitude being hardly affected. However, "this way, too, was insufficient," because the circle of latitude and the vertical or refractory circle would not have entirely coincided. "On account of this fact, from

the opening at the eye," which represented the true place opposite the sun, found by the meridian observation, and reduced to the time of this evening observation, "I added a perpendicular to the pinnule,[193] through a special contrivance," which 155
perpendicular represented the vertical circle, "in considering that the point whose form is refracted," whose function in this place is performed by the intersection of the circle of latitude with the ecliptic, "the center of vision," represented by the opening mentioned, "the point"[194] on the surface of the air "by which the refraction is generated, and the perpendicular from the point of refraction, are in one surface," which we here imagine to be the plane of the vertical surface. "And I saw, when the intersection of the ecliptic and the circle of latitude" representing the sun's true position "had come into contact with the horizon," which position corresponded to the sun's true setting, "that the sun, coming through the opening of the other pinnule," the one towards the sun, which I elsewhere used only for the stars, "still illuminated the thread of the perpendicular." I took care, he wishes to say, that the sun illuminate uniformly, not the circle of latitude, not the ecliptic, but the thread of the perpendicular suspended from the opening of one pinnacidium[195] through the opening of the diametrically opposite pinnacidium; "so that the previously described things are in one surface, namely, the center of the true sun, that is, the point whose form is refracted, the center of vision," in this place, the opening from which the thread is suspended, or rather, the lower place of the thread, in which it is illuminated by the sun, "the point of refraction in the air and of the perpendicular; likewise also the sun's true place on the ecliptic" of the instrument, and finally, the position of the image, or the visible sun.[196]

Now because the openings are diametrically opposite, and the thread is suspended from one opening, the plane passing through the thread and the openings bisects the instrument; and since the visible sun illuminates the thread through one opening, the visible sun is therefore also in the same surface, which, because of the thread, is perpendicular to the horizon, therefore in a vertical plane, therefore the sun's true position is also in the same plane, since the true place of an object and the place of the image are always in the same surface. When this occurs in an instrument (if in fact the one opening or the intersection of the ecliptic and the circle of latitude represents the perfectly true place of the sun, as is supposed), it serves as proof that the instrument was so directed that the center, the

---

[193] The pinnule, or sight, usually had a hole in it through which the object was sighted. From what follows, it appears that Regiomontanus added a perpendicularly oriented thread across the hole.

[194] Reading *punctum* for *punctus*.

[195] *Pinnacidium* is the term used by Tycho for the entire assembly of a sight, which might consist of one or more pinnules, with slits and adjusting screws. The construction and use of the Tychonic sights is described in *TBOO* V p. 154.

[196] Kepler has considerably altered the words he claims to be quoting. The original reads, ". . .*scilicet centrum Solis, idem punctus cuius forma refrangitur, centrum visus punctus refractionis et perpendicularis similiter et locus Solis in ecliptica.*" Or, in English, ". . .namely, the center of the sun, the same point whose form is refracted, the center of vision, the point of refraction and of the perpendicular; likewise also the sun's position on the ecliptic."

openings or the intersection mentioned, and the perfectly true position of the sun in the heaven fall upon one straight line; and that they cannot be arranged differently in such a way that this may occur. It is indeed an ingenious short cut to eliminate the refractions instrumentally by observation, without calculation from
156 longitude and latitude, and to establish a sound longitude and latitude, indeed also to measure the refractions themselves by all three sides of a right angled triangle. It could also be carried over to the parallaxes. I, at any rate, have somewhat less doubt than did Tycho Brahe, on fol. 91 of the *Progymnasmata*,[197] concerning Walther's carefulness, after I have helped myself out with an explanation of the passage. Besides, if anyone is so careful that he does not want to trust blindly in instruments in such an exacting matter, he has the way of distributing the parallaxes and refractions, which affect the altitude, into the longitude and latitude, described in Tycho Brahe's *Progymnasmata*, chiefly vol. 1 fol. 93, 94, and 96, where he also shows how to investigate the whole refraction or the altitudes.[198] In chapter 9, below, you will find a most useful and easy method for this *entreprise*.[199]

In going through the examples, I think it is sufficiently established that astronomers of all ages took note of refractions. I therefore anticipate how those who deny the whole matter of refractions, except those that consist only in the enlarging of diameters, would want to excuse their rashness: they would say that the phenomenon that I showed above is very different from the usual matter of refractions. Now it is true that all people consider themselves permitted to rise up against those who appear to profess something new to them. Where this happens without reason, it degenerates into disparagement. It is, however, much more shameful that at the same time, people of this sort openly betray their own ignorance, and the smug contempt of others. They should but read only the very recently published disputation of my teacher Maestlin,[200] whose words (from Thesis 58), as the most recent writer, even now (by the grace of God) alive, are fittingly appended to this chapter in place of a colophon. I append them so that this man's discoveries may be compared with the Tychonic ones, and the authority of this theory may be all the greater, being propped up by the votes of many, by which the mouths of the naysayers may thus be finally shut. After the testimony of Witelo and Walther is presented, he speaks thus:

If our observations made with the astronomical radius are to be trusted, we

---

[197] *TBOO* II p. 76. However, Tycho does not mention Walther on this page, though he did consider the observations of Regiomontanus and Walther on pp. 59–61.

[198] *TBOO* II pp. 78–82.

[199] Chapter 9 Section 4, pp. 326. The word Kepler uses here is πραγματείᾳ, which is the term he uses in the Mars Notebook (*JKGW* XX.2) for the working out of orbital parameters. I have used a French word because of the analogy between Renaissance humanists' use of Greek and modern English writers' use of French.

[200] Maestlin, Michael, *Disputatio de eclipsiblus solis et lunae* (def. Marcus ab Hohenfeld), Tübingen 1596.

> have not seldom found, [sometimes in my presence],[201] that the distance of Venus, higher above the horizon, from the sun, placed next to the horizon, is noticeably less than if, on the same day, the distance of the same planet from the sun were taken higher up and more free from vapors [concerning
> the opinion of the cause, there is enough above]. Therefore, through vapors ***157***
> [as well as through the surface of the air], it appeared higher than is correct. Therefore, that both it and other stars can likewise appear above the horizon while they are still beneath it, we do not hold as impossible, but conclude as certain.

Further, this is one of the ways by which Tycho very frequently investigated refractions, as will some day be evident, God willing, from his books of observations. Furthermore, Maestlin too, because of his unique keenness of vision, is very experienced at observing Venus by day.

Note to Section 7

Of haloes, of whose circles parhelia are certain parts, as well as parselenae, I wished to inform those who are interested in these things with this note. First, the colors of haloes around the sun and of parhelia are the same as those in rainbows, but darker, because of the brightness of the sun shining nearby, and rather weak in haloes around the moon and parselenae. Therefore, they are the offspring of refractions, as are rainbows as well. See the experiments through watery globes and glass triangles, in ch. 5 below.[202]

Next, the diameter of the rainbow, when the sun is setting, is always 90°; that of a halo, 45°. Therefore, you, whoever you are, who wants to know something beyond Aristotle here,[203] *know that the cause of this fact is to be demonstrated to you.*

Third, haloes are perceived in clouds that are closest, which is evident from their swift flight. And nevertheless, by absolutely all people, in whatever place they are located, they are perceived to be $22\frac{1}{2}°$ *from the luminary, as from its center. Therefore, in clouds that are so nearby, everyone sees his own halo, no less than the rainbow, because for a different spectator the heavenly body is placed among different clouds.*

A halo and a paraselene have often appeared to me now, while remaining in the same place, when a cloud was suitably located, and have now vanished, when the cloud departed, and have again appeared, when another cloud followed.

It is therefore necessary that the rays of the heavens be refracted, and thus be sent down refracted, in a much higher place, and nonetheless in a matter that is fluid and bounded by a distinct surface, and is always of the same density and depth. Nevertheless, these rays are not seen, unless they be thus received from

---

201 This is Kepler's parenthetical insertion, referring to his own presence. Other bracketed words, below, are also Kepler's.

202 Ch. 5 Sect. 3, beginning on p. 191.

203 Aristotle considers these phenomena in the *Metorologica* III 3–7.

above by a pellucid cloud, in order that this picture might appear from below; that is, through the body of the pellucid cloud.

Fourth, parhelia and parselenae are very near neighbors to haloes themselves, because nearly always there are parhelia in a halo. And whenever these appear, they are perceived at an equal altitude with the sun or moon.

Fifth, in order that a rainbow appear, it is necessary that the place where it can appear be both clothed in matter and shadowy. For this reason, whenever there is rain with the sun shining in between, with a cloud appearing immediately above the sun, that is, where the shadow of the cloud has barely gone away, it is then that rainbows are perceived most clearly. Therefore, the watery matter between the spectator and the sun, whether it be rain, or pellucid clouds, or mist (for I have also seen rainbows in mist)—this watery matter, I say, forms and shapes the refractions of the sun's rays, while the matter which is beyond the
158 *spectator, receives these refracted rays of the sun. But a cloud standing overhead puts the place in shadow, and turns aside the direct rays of the sun, in order that the colored rays be capable of being perceived.*

Thus it is not true that the rays, whether of the sun or of the vision, are reflected or refracted at the exact place of the cloud in which the rainbow appears.

Sixth, it must not be thought unimportant that the centers of vision, of the halo, and of the sun or moon are in the same straight line, no less than the centers of vision, of the rainbow, and of the sun.

Chapter 5
On the Means of Vision

While the diameters of the luminaries and the quantities of eclipses of the sun are noted as fundamental by astronomers, there arises a certain deception of the sense of vision, arising partly from the technique of observation, which we discussed in ch. 2, above, and partly from the simple sense of vision itself, which, insofar as it is not removed, creates a great deal of work for the practitioners, and detracts from the reputation of the art. And so the occasion of looking into error in vision must be sought in the formation and functions of the eye itself. If the optical writers Alhazen or Witelo, or after them the Anatomists, had treated these subjects clearly, lucidly, and without risk of incertitude, they would have freed me from this task of continuing the *Paralipomena to Witelo* in this additional chapter. But, to proceed with this matter methodically, I shall first gather together, as if in the role of principles, the descriptions of the parts of the eye that will come under consideration, from the testimony of the most reputable anatomists, for Witelo's description is untrustworthy and confused. Second, I shall foreshadow in a summary manner the way that vision occurs. Third, I shall demonstrate each and every thing. Fourth, I will uncover that which escaped the reasonings of the optical writers and the physicians concerning this function. Finally, I shall explain the deceptions of vision arising from the construction of the instrument, and shall accommodate them to astronomical use.

1. Anatomy of the Eye

It is conducive to the winning of trust in the demonstration which I am about to bring in, to introduce, not my own experiences, but the public experiments of the most eminent physicians concerning the eye. For what if one were to charge me either with bad faith, as if I were intent upon establishing my own opinion,
159 or with inexperience in the dissections, of which I never before had been either spectator or performer? Therefore, let men of accepted authority speak for me on the subject that is best known to them, up to the point where the undertaking shall have reverted to the limits of my profession. For then they too will ungrudgingly hand the torch over to me at the point where I am going to carry it forward legitimately into mathematics, concerning which the judgement will belong to the expert.

I have consulted chiefly Felix Platter's plates concerning the structure and use of the human body, which, published in 1583, were deservedly reprinted in this year 1603.[1] With these I compared the *Anatomia Pragensis* of my friend Mr. Johannes Jessenius of Jessen,[2] for the reason that he not only professed

---

1 Felix Platter, *De partium corporis humani structura et usu libri III*, Basel, 1583 and 1603. Platter (1536–1614) was Professor of Medicine at Basel from 1560 to his death.

2 Johannes Jessenius a Jessen, *Anatomiae Pragae anno 1600 ab se solemniter administratae historia*, Wittenberg, 1601. Jessenius (1566–1621), Professor of Medicine at Wittenberg from 1596–1601, and thereafter at Prague, where Kepler came to know him.

chiefly to follow Aquapendente,[3] but on his own prowess devoted himself chiefly to anatomical labors. If I, being myself chiefly occupied in the mathematical profession, have passed over any of greater merit in this succession, they will grant me pardon.

The eye [oculus] is from the Greek ὀπή, whence come ὄπτεσθαι, ὄψις, ὄμμα, ὀφθαλμός, because these are the cracks or uncovered openings reaching from the shadowy head into the bright air, just as, by the opposite reasoning, lupus (wolf) appears to have been said from λύκος; it is not from 'occulo' [cover up], for when they are closed they are of no use. They are paired in animals, and are parallel, and at a suitable distance apart, as is said above in ch. 3[4] and below in ch. 9,[5] because they have to note the distances of things, and not to have another in case one is lost. For nature does not destine anything for loss. Their location is in the highest place, so that the vision may reach to places that are that much more distant. For since the earth and the sea are most of all spherical, an eye that is not placed at some height would not be able to perceive anything at all of the surface, hindered by the curvature of the globe in all directions. This is shown to us by storks, and especially by giraffes, capturing the sight and sound of distant objects all the more distinctly by an erect neck; and above all by eagles, who, with the intention of surveying a whole region for the sake of predation, seek out the highest places. For this reason Jessenius is clearer than Platter, in saying that the eyes are high up in order to see more things, while the latter says it is in order that they might better see individual things from on high. For as regards individuals, they are better inspected from nearby. And when we are going to scrutinize a coin thrown upon the ground, or a low plant, we bring our eyes downwards. Ovid ethically and wisely says of humans,[6]

> While prone the other animals behold the earth,
> A high mouth he gave to humans; to view the heavens
> He ordered them, and to raise to the stars **160**
> their upright countenance.

But if you should consider physically, the eyes are not at the top, nor is the eye socket upwards, but they lie open more to the sides and downwards, following the nose. And so, for the sake of changing the story, they seem to be so located in humans because the human being is a social and political animal. Accordingly, the human bodies themselves invite them to association, more than those of animals. So this opinion of Platter's (that they obtained their anterior place in order to be directly opposite their objects), insofar as it pertains to humans, you may take to apply beautifully to one person opposite another. Another cause is derived from human dignity. Since pasturage is required for other animals, they have eyes

---

[3] Hieronymus Fabricius ab Aquapendente (1537–1619), *De visione, voce, et auditu*, Venice 1600.

[4] Proposition 8, p. 79.

[5] Section 2, p. 321.

[6] Ovid, *Metamorphoses* I 84–6.

that are directed downwards to things closest to them. The human, master of creatures, has his face so directed that he should be invited continually to contemplate how far flung are the limits of his possession: they are the heaven itself, contiguous to the mountains, as it appears.

But the cause for the ears and eyes being directed to the same side is most properly derived from motion; and since this was required to take place in a straight line parallel to the horizon, in every respect the bend of the members and position of the necessary senses were required to be directed to the same region of the circle of the horizon. Further, Galen said that it is for the sake of the eyes that the head is elevated, since the eyes could not be far from the cerebrum, and it is fitting that the neck also be made flexible so that the senses could be turned to see and hear those things that are behind; and hence this line of reasoning also appears to be constructed not badly, and, as it were, from the elements. Thus the eyes are close to the cerebrum because the objects of sight and hearing are the final causes of local motion, and in fact all local motion is placed in the authority of the cerebrum. Therefore, in order that the limbs of the body, as a result of some sort of sudden emergency[7] might get ready for a more swift pursuit of those things whose ghosts might be carried through the senses to the cerebrum, it was fitting that the eyes be set in the very embrace, as it were, of the cerebrum.

Each eye is enclosed in an individual chamber, for safety's sake. However, they protrude, lest, hindered by the sides of the chamber, they be unable to look out upon the whole hemisphere in a single gaze. But since this total view could not be distinct, except in the middle, but would be confused at the sides, it was appropriate that the eyes have a very fast motion, by which the view might better be carried across to the particular parts of the hemisphere.

161 Further, since there are two eyes, in order to distinguish the distances of things, there was a need for them to have the power of direction, or turning closely together, which is likewise motion. It was not, however, useful that the eyes be moved from their center in this motion. Therefore, in order that they move about their center, they had to be spherical on the outside, except to the extent that they are clothed with oblong muscles within the head, and reach into the cerebrum by means of a certain long process or canal. Moreover, the need for eyelids seemed to require that they be spherical even on the anterior part, on which they are not rubbed by motion with respect to the parts of the face. For because the eye had to be transparent, that could hardly be maintained in a member that would always be hot along with the entire body, unless tears, and air reduced into water and sticking to the outside, and small dust particles flying in, were repeatedly wiped away, and unless the eye were also moistened, wherever it is inclined to dry out and get rough. The eyelid appears to accomplish all this, as is manifest in those who weep. The same eyelid also at the same time acquired the function of covering, when danger should threaten or when sleep creeps up, nature also turning aside from the sleeping person the chances of danger that were a threat when the eyes were open. Moreover, to this end of wiping off, the interior of the eyelid is a membranous and soft covering, of the same substance as that

---

[7] Literally, lightning (fulmen).

covering which it wipes off, while there is an arch at the edge, called the tarsus, smooth and round and of a rather hard substance inside, as are all cartileginous parts, yielding to some sensible extent at its highest surface, while resisting by the very mass of its body and the condensed hardness. Thus it happens that the upper membrane desires to be wrinkled up, and is stretched by the globe of the eye; and when the tarsus is lifted up, it folds into itself, while when the tarsus is lowered, it never falls away from the globe of the eye, and is not stretched in one place more than in another, since the whole tarsus moves as one. So it was hardly possible for this tarsus to open and close without getting tired, and to unite with its companion the lower tarsus, otherwise than if the globe of the eye were spherical, while the arches themselves of the tarsi were semicircular, attached on either side as if at opposite poles of the globe. From this mechanism there also arise a pair of angles in each eye, which they call the "canthi" or "hirqui".[8] Why it is that the line between these coincides with the line between the two eyes, is said in chapter 3 above,[9] namely, so that the eyes, singly as well as jointly, might imitate the **162**
trace of the visible horizon. And so that this might be possible, and the breadth of vision might follow this line, the chamber of the eye, between the brow and the jaw, had to be somewhat depressed, so that nothing stand in the way there. At the nose this was indeed not necessary; rather, a palisade had to be made of the nose, so that the vision of the two eyes might not be confused. This palisade seems in fact to accomplish two things: one, that it remove the view of one eye from the other, lest they be blinded by the borrowed brilliance arising from reflections; the other, that by its height it prescribes a boundary of the approach of visible things. But that which has been removed from one eye by the nose is compensated for by the other eye. For nature established the perfection of vision in two eyes. Two other palisades of this sort go off downwards from the chamber of the eye between the jaws and the nose, so that those things that are nearest the mouth and the feet may lie open to the eyes. This the necessity of motion and eating requires. The arch of the brow, however, has the greatest protuberance, in order to close off the light of the heaven, by which visible things would be overpowered. For this reason, those troubled by the brightness of light wrinkle the brow, and draw that palisade down closer to the eyes. The eyelashes and eyebrows are both for nearly the same purpose. For light that is sent down through them does not strike suddenly upon the eyes with its full ray, but is partly obscured by the eyelashes, and the eyes create the risk for the light approaching through the small gaps in the eyelashes, that it might not be able to be carried along. And when the light is coming from a greater distance, we defend ourselves with the eyebrows and the wrinkling of the brow, while when it is directly opposite, we also make use of the eyelids' lashes and their squinting. This Aristotle, in sect. 31 prob. 16,[10] thought he had to dispel, but he used an absurd emission of rays, which elsewhere he himself rejected. This is, to be sure, the cause for the eyebrows' being curved

Jessenius says, however, that this is is denied by those who have had a mutilated nose.

[8] Literally, "tires" or "billy goats," respectively.

[9] Proposition 9, on p. 80

[10] *Problems* 959a 3–8, trs. Hett Vol. II pp. 192–3; Cf. *On the Parts of Animals* II 13, 657a 25 ff.

into an arch, so that they might enclose the circular eyes circularly. Further, as something extra, is that, because they hold up small dust particles, the eyelashes seem to play the role of fans, to sweep out the small dust particles that fly against them; and finally, that here too, as in the whole body, the moist and soft parts appear to be defended by hairs against abrasion. Particularly, when the eyelids are wiped off, and dirt must be pushed out through the double doors by being rubbed off, the hairs function as channels, to divert the dirt through the doors, so that they not be driven into the eye. In fact, all hairlike things, being arid and dry,
163 repel moisture by their roughness and by a certain sort of stiffness, to the extent that that is not overcome. And from these the cause is apparent why the hairs are erect, not lying forward or bent back, not more sparse, not more dense, not longer, not shorter, and moderately curved. It is these which, standing about the eye itself, assist in the function of seeing.

In surveying the globe of the eye itself, just this one thing seems to me to need my care, that I understand the anatomists correctly, and that I have a good command of the nomenclature. So, therefore, the anatomists appear to me to depict and describe the eye as if the overall form of the eye, as it arises from the cerebrum, were to represent the image of an onion:[11] just as this consists of a bulb and a stalk, so the eye consists of the visible globe and of the hidden process reaching in to the cerebrum. And what is more, in onions there are found nuclei, lying entirely hidden in the bulb, which the coverings common to the bulb and the stalk enclose and conceal. To these nuclei you might not unaptly compare the humors that are in the sphere of the eye, and that are enclosed in nearly the same way by coverings, which are common to the eye and to the process going to the cerebrum, as well as to the cerebrum itself. For the actual substance of the cerebrum is surrounded by two coverings, the inner being thin, the exterior thick, which they call "meninges."[12] Thus there are in the cerebrum three parts, from each of which is extended its own covering into the process of the eye, and thence into the globe of the eye itself. The thick meninx produces the exterior covering of the eye, the thin membrane produces the middle covering, while the actual substance of the cerebrum produces the optic nerve, which becomes the inner covering, which is itself not simple. Platter, however, asserts that this is called a covering (*tunica*) improperly, no doubt because he holds that the optic nerves, which the others assert to be entirely hollow, are similar to the other nerves, and denies that they are hollow.

---

[11] Kepler refers here to the following note.

To p. 163 and 177. Forgive the crude analogy, reader. For when I wrote this, I had not yet hit upon the idea of copying Platter's drawing, which depicts the parts of the eye true to life, in print, and including it in the work, as friends later persuaded me to do, so that it could be inserted at p. 177, below. And so I wanted to accomplish here with words and comparisons, what I did not then intend to include in a picture.

You should therefore disregard this comparison and repeatedly look at the drawing referred to, and carefully compare what has been said in this entire argument with that drawing.

[12] These are the arachnoid membrane and the dura mater, respectively.

Let us now divide the globe of the eye into the apparent and hidden hemispheres; and let us further subdivide the one that appears, into the white and the sun or iris, to speak more clearly. Accordingly, the anatomists say that the exterior covering of the ocular globe, which alone enfolds the whole eye, develops from the thick meninx, but according to its different parts it acquires different names and modifications. For the part from the end of the optic nerve through the whole hidden hemisphere, and through the white of the eye, has the name σκληρώδης tunic,[13] and it is hard, thick, opaque, white, and nearly cartilaginous, but that part which covers the iris is pellucid and protrudes, as **164**
Jessenius points out, and ex-perience confirms by a careful inspection of the living eye. Thus the cornea is a portion of a smaller sphere, the sclerodes that of a greater: this must be carefully kept in the memory. However, in such a narrow space it is not possible to tell whether it is a portion of a sphere rather than of a spheroid. Further, before I pass on to the other two tunics, more has to be said of this exterior tunic. For even though it is the outer with respect to the other principal tunics, it is still not the outermost of the whole eye. For in the hidden hemisphere, it is completely covered with muscles on all parts, both those on the globe and those on the optical process; and these muscles also send tendonlike ends or thin parts out even to that part of the tunic of the sclera that is in the white of the eye, and these enfold it everywhere. Not only this: there is also a certain membrane that arises from the pericranium, called the side-swimming (*adnata*), adhering, or ἐπιπεφυκυία,[14] to which the globe of the eye is attached at that circle by which the hidden hemi-sphere is divided from the apparent one. This membrane is made doubled there: on the one side it surrounds the eyelids, and on the other it covers the tunic of the sclera and its overlying tendonlike muscle ends. It is itself transparent, so that the white tunic of the sclera may nonetheless be seen through it by those who ex-amine it. If I understand Jessenius rightly, his sense is that not just the white of the eye, but the iris itself as well, is covered by this pellucid and extremely thin membrane. This seems in fact to require a mode of attachment, since it is above all by this bond that the eye is kept in its socket. However that may be, Platter openly asserts that the corneal tunic, which is part of the sclerodes, stretched over the iris, is bare, chiefly because those little reddish veins that are seen to run about over the white of the eye for the benefit of this adhering membrane, do not go in over the iris or the cornea.

These things are added for the sake of distinguishing the parts, although they do not contribute to the forming or perceiving of the likenesses of objects, and only pertain to motion and fastening, perhaps also to the nutrition of the eye. The whiteness of the sclerodes and the different colors of what is called the "iris" [rainbow], might give occasion for suspicion that the exterior parts contribute in

[13] Evidently the term "sclera," which we use today, was not then common. In the Platter diagram, below (p. 188), it is called the *crassa tunica* (thick tunic). I have therefore kept Kepler's Greek term in the translation.

[14] "Clinging close." This term was used by Galen to denote the membrane we call the conjunctiva, which covers the anterior surface of the cornea and the inner surface of the eyelid. Cf. *De symptomatum causis libri III*, in *CGOO* VII p. 101.

a way to vision, since the inner parts are perfectly black, while the outside is white to the same degree. However that may be, I do not understand how they
165 could carry anything inside, since those parts beneath which they are carried do not allow them a passage within, because of their darkest black color and opacity. And so if the cause of these colors of the eye is something higher than the necessity of a material that is nearly cartilaginous for supporting the round chamber of the eye, it is not discordant with the truth that the reason why what is called the "sun" [i. e., the iris], surrounded by the white, and black in the center, is itself of one color, while the other parts are of another, is so that the sun or the iris in the white, and the black in the iris, might be all the better seen from without, and so that someone might be able to judge at a glance whether someone else were casting his eyes upon him. But consider also Prop. 38 of the first chapter, whether the reason for the globe of the eye's being surrounded by a white tunic is so that it may be more mildly struck by the sun's light, and that it not be destroyed, had it been made black.

Now the second tunic of the eye, which lies beneath the sclerodes, arising from the thin meninx of the cerebrum, is in its posterior part called the choroid, in the anterior, the uvea. These parts do not differ only in location, but also in thickness: the uvea is regarded as twice as thick as the choroid. Finally, they differ also in the application to the exterior tunic of the sclerodes, for where it is called the choroid it adheres to the sclerodes by fibers spread out everywhere, and originates in the same circle by which the sclerodes goes over into the cornea. But the uvea that arises thence does not originate in the remaining part of the sclerodes, which is called the cornea, but diverges from it towards the interior of the eye, the aqueous humor filling up the gap. And while the sclerodes with its cornea constituted an integral tunic, here the uvea is perforated in the middle, the aqueous humor filling the opening, and also wetting it from the inside, to the extent that this whole part that is called the uvea floats in the aqueous humor, with the sole exception of that circle by which I said it was bonded to the sclerodes, and its posterior part, which is called the choroid. The inner part of the choroid is stated to be black—not like the black of the uvea, but more blue. The inner part of the uvea is rough and very black; the outer part, by which it displays what is called the "sun" [the iris] in the eye through the translucent cornea, is light, and along with the exterior surface of the choroid is of various colors, now dark, now grey, now light brown, now brilliant. Thus displaying a broad circle (in the middle of which is the black of the eye or the opening), it is called the Iris. It is by the rubbing of this, since it is light and rigid, that those sparklings which we perceive from time to time in the eyes, appear to be aroused, exactly like the sparkling of the furry
166 backs of cats, and also of the recently cut hair of humans, when rubbed. In the humors themselves, there can be no light without the vision at the same time being impaired.

However, the persistent lights that are in the eyes of cats, not the transitory ones, like the sparklings, are thought by Jessenius likewise to reside in this iris.

The third quasi-tunic, arising from the nerve spreading itself out into the hollow hemisphere, is likened, in Platter's opinion (not in the configuration of its substance, but of its shape), to a net bounded by a sac, or better, to a funnel.

Hence, it has received the name "retiform" [net-shaped].[15] It is said to resemble the substance of the cerebrum, but to be more mucous and reddish (bluish, according to Jessenius), whence one concludes that it seems to be above all a diluted white tinged with redness or blueness. Jessenius also divides this into the substance of the nerve and the tunics.

Here, Jessenius, along with Witelo, and Platter go off in different directions. Jessenius, from Witelo,[16] says that the retiform, where it has embraced the vitreous humor, having progressed, is attached to the greatest circle of the crystalline humor, which is the middle of the humors; and that as a benefit of this it happens that the crystalline is established as the middle in the vitreous humor. Platter, on the other hand, says that beneath that circle of the concave surface, by which circle the second tunic is joined to the first, and begins to be called the "uvea"—that is, from the uvea, and not at all from the retina—there arise certain processes all around, which by their resemblance to hairs are called "ciliary", that end in the greatest circle of the crystalline. Thus they join the crystalline to the uvea rather than to the retiform. Those processes are black, and are divided like the teeth of a comb. He states, moreover, that the retina ends below these ciliary processes, nor is it entirely of the same origin as either the choroid or the vitreous humor, which it surrounds, or with these ciliary processes. Thus, in fact, when the vitreous humor is poured out (by whose stuffing alone the retina is brought into contact with the choroid), the retina wrinkles up and does not adhere to anything except its own nerve. I tend to agree with Platter, seeing that he also treated these things in greater detail. For so far as it appears, Witelo ascribes the faculty of discerning visible things to the crystalline humor, with the result that he sought a passage for this faculty from a nerve (or the retina, offspring of a nerve) into the crystalline through these ciliary processes. Platter, on the other hand, leaves the faculty of discerning in the retina, which is more in agreement with the truth. The crystalline, however, he includes in the listing of instruments, concerning which see **167**
below. Further evidence for this is that the ciliary processes, by which the greatest circle of the crystalline is held in place, resemble in their blackness, not the retina, but the uvea.

And Jessenius himself inclined in this direction in his "second thoughts".

Now for the humors: there are three in the eye, distinguished by their respective seats. The vitreous is in the posterior part of the eye, the aqueous is in the anterior, and the crystalline is in their midst, it and the ciliary processes making a wall between the vitreous and the aqueous. And the vitreous has its very thin and pellucid skin, called the hyaloid tunic, of the same name as the humor; and it contains the vitreous, even when that humor has been freed from the tunics of the sclerodes, choroid, and retina. The crystalline, in turn, also has its tunic (called the arachnoid or spider's web because of its fineness), likewise pellucid, by which the crystalline is contained, also when freed from the surrounding humors. But the anatomists should consider the cause of the name. For

[15] This is, of course, the retina. I shall, however, follow Kepler here in calling it the "retiform," except where Kepler himself calls it the retina.

[16] Witelo, *Optica* III prop. 4, in *Thesaurus* II p. 86.

this name, spider's web,[17] appears to belong not just to the skin of the crystalline, but to the ciliary processes taken together. For truly, just as the spider stays suspended in the center of her web, so the crystalline humor is suspended in the center of the ciliary processes, by means of threads drawn inwards to the center on all sides. The aqueous humor, in turn, has no skin, but is held in place before the cornea, behind the spider's web, by means of the ciliary processes of the uvea, and the underlying hyaloid tunic of the vitreous humor. It is, however, constricted in the middle, and nearly divided in two by the uvea, which it bathes within and outside the opening. The vitreous has the greatest mass, the crystalline the least, its axis being hardly the fourth part of the axis of the vitreous.

The vitreous has most nearly the shape of more than half of a sphere. The aqueous does consist of spherical surfaces, but it has been driven back in the middle, as I have said, into a narrow neck. The crystalline, on that face that is bounded by the ciliary processes and immersed in the aqueous, acquires either a spherical figure, or a portion of a lenticular spheroid, a rotated ellipse, divided by the axis, the side remaining straight. On the posterior side, which is bounded by the same ciliary processes, it is immersed in the vitreous. Its figure is a hyperbolic conoid, a hyperbola rotated around the axis. For Jessenius thus relates, that it is not spherical, as Platter said, but that it protrudes markedly, and is made oblong, stretching up almost into a cone; and that on its anterior face it is of a flattened roundness, the anterior roundness of the cornea and the crystalline humor being
168 perceptibly similar, their dimensions being in the ratio of 4 to 3. The aqueous is least dense, the crystalline densest, both are perfectly transparent. The vitreous is slightly more tinted. If an axis is drawn through the center of the opening of the uvea to the optic nerve, the centers of the circles of the crystalline lie upon it. Thus the crystalline is rightly said to be fixed in the center of the eye, if you understand the eye to be cut by a plane perpendicular to the axis, and the center of the eye to be the center of that circle. But as regards the overall globular shape of the eye, the crystalline verges more to the front of the axis described. The crystalline is at the greatest distance from the end of the optic nerve, less from the opening of the uvea, and the distance of the opening of the uvea from the surface of the cornea is least.

2. The Means of Vision

Although the brow, eyebrows, eyelids, sockets, also themselves contribute to the means of vision, they do not meet with the chief tasks. Therefore, I have now prefixed a description of their use, and joined it with an account of them. But the parts that I have appended, relating to the actual globe of the eye, all contribute to the principal function. Therefore, in order that the use of each may be apparent, I shall describe the means of vision, which no one at all to my knowledge has yet examined and understood in such detail. I therefore beg the mathematicians to consider these carefully, so that thereby at last there might exist in philosophy something certain concerning this most noble function.

---

17 *Aranea*, which can mean either spider or spider's web.

I say that vision occurs when an image[18] of the whole hemisphere of the world that is before the eye, and a little more, is set up at the white wall, tinged with red, of the concave surface of the retina. How this image or picture is joined together with the visual spirits that reside in the retina and in the nerve, and whether it is arraigned within by the spirits into the caverns of the cerebrum to the tribunal of the soul or of the visual faculty; whether the visual faculty, like a magistrate given by the soul, descending from the headquarters of the cerebrum outside to the visual nerve itself and the retina, as to lower courts, might go forth to meet this image—this, I say, I leave to the natural philosophers[19] to argue about. For the arsenal of the optical writers does not extend beyond this opaque wall, which in fact occurs first in the eye. I do not think that Witelo should be heeded in regard to Book 3 Prop. 20, in thinking that this image of light originates ***169***
further along through the nerve, as far as where in mid course the nerves of the two eyes come together in a kind of joint, and again diverge into their individual cavities of the cerebrum. For what can be pronounced by optical laws* about this hidden confluence, which, since it goes through opaque, and therefore dark, parts, and is administered by spirits, which differ entirely in kind from humors and other transparent objects, has already completely removed itself from optical laws. And so, when Witelo argues, in Book 3 prop. 20, that the images[20] need to be united; in prop. 21, that refraction therefore occurs at the back of the vitreous humor; and, in prop. 22, that spirits are transparent; I invert the sequence: spirits are not an optical body, and the narrow cavern of their nerve is not optically straight, and if it were so it would immediately become bent when the eyes are turned, and it would place its opaque parts against a very slight opening or mouth of a passageway. For this reason, light does not pass through the back surface of the vitreous humor, nor is it refracted there, but is driven against it. And how can an image be refracted if it enters perpendicularly? It is remarkable that Witelo was not struck by this in writing proposition 31 of the third book.[21] As a result, he was thrown into difficulties that were not trivial in prop. 73 of the third book, on account of this conjoining of images at the confluence of the nerves. If anything is to be said about this conjoining of the nerves in mid course, it should be according to physical principles. For it is more certain than certain that no optical image[22] carries through this far. It appears, accordingly, that if each nerve had carried through freely and directly to its seat in the cerebrum, the result would have been that when we used two eyes we would think we were seeing two objects instead of one. Or this conjoining has been brought about so that when one eye is

* Nevertheless, through this nerve, vision, from which the word "optics" is derived, occurs; and it is therefore wrongfully cast out of optics, because, owing to the poverty of our science, it cannot be tolerated in optics.

---

18 *Idolum.*

19 *Physici.*

20 *Species.* The propositions from Witelo are on pp. 94–5 of Vol. II of the *Thesaurus*.

21 Witelo III Prop. 31: "When a single point of the object seen strikes perpendicularly upon the surfaces of the two eyes, it is necessary that the radial axes be refracted angularly in the centers of the openings of rotation of the concave nerves." *Thesaurus* II p. 99.

22 *Imago.*

closed, that hollow of the cerebrum might not also cease its function of judging. Or perhaps that doubling of the hollows is not only for the sake of the eyes, but in itself facilitates the making of correct decisions, just as the association of the eyes does for distances. Accordingly, in order correctly to distinguish visibles as well, those that are perceived with one eye as well as with two, this confluence of the passageways had to come to pass. This one thing must be said here from the conclusion of chapter 1, from an optical perspective, that spirits are acted upon by colors and lights, and this action is, so to speak, a coloring and an illumination. For the images of strong colors remain in the vision after perception of them has
170 occurred, and mingle with colors impressed by a new perception, and a confusion of the two colors occurs. This image, which has an existence separable from the presence of the object seen, is not in the humors or the tunics, as was proved above. Therefore, it is in the spirits, and vision occurs through this impression of images upon the spirits. However, the impression itself is not optical, but physical and mysterious. But this is a digression.[23] I return to explaining the means of vision.

Vision thus occurs through a picture of the visible object at the white of the retina and the concave wall; and those things that are on the right outside, are depicted at the left side of the wall, the left at the right, the top at the bottom, the bottom at the top.[24] Further, green things are depicted in the color green, and in general any object whatever is pictured in its own color within. The result of this is that if it were possible for this picture on the retina to remain while the retina was taken out into the light, while those things out in front that were giving it form were removed, and if some person were to possess sufficient keenness of vision, that person would recognize the exact configuration of the hemisphere in the compass of the retina, small as it is. Moreover, that proportion is preserved, such that when straight lines are drawn from the individual points of objects that can be viewed to some particular point within the compass of the eye, the individual parts within are depicted at very nearly the same angle at which the lines will have come together, so much so that not even the smallest points are

---

[23] Here Kepler inserts an asterisk referring the reader to the following endnote:

To p. 170. It is most worthy of consideration, what Master D. Matthaeus Wagger, the Emperor's Councillor of the Imperial Hall, told me once happened at Heidelberg. Someone who had lost one eye and who covered the other completely with the palm of his hand observed that, if something shining were to be set beneath his nostrils, he would be aware of its shining and would also to some extent differentiate it. It is therefore in doubt whether some opening might have been set up in the empty eye socket upwards through that line into the head and to the ends of the visual nerve, the seat of the spirits, or whether the same spirit might also be poured forth into the instruments of smell.

What is chiefly confirmed by this example is that the spirits are set up to have a capacity for light and colors.

[24] For contrasting discussions of this dramatic and paradoxical discovery, see Lindberg, *Theories of Vision*, pp. 193–208, and Straker, *Kepler's Optics*, pp. 450–470.

left out. And to such an extent that the fineness of this picture within the eye of any person you please is as great as the acuteness of vision in that person.

Now, in order to approach closer to the way this picturing happens, and to prepare myself gradually for the demonstration, I say that this picture consists of as many pairs of cones as there are points in the object seen, the pairs always being on the same base, the breadth of the crystalline humor, or making use of a small part of it, so that one of the cones is set up with its vertex at the point seen and its base at the crystalline (though it is altered somewhat by refraction in entering the cornea), the other, with base at the crystalline, common with the former, the vertex at some point of the picture, reaches to the surface of the retina, this too undergoing refraction in departing from the crystalline. And all the outsides of the cones come together at the opening of the uvea,[25] at which space the intersection of the cones takes place, and right becomes left.

See the diagram of Proposition 23, following.

So that these things may better be grasped, I shall repeat a step at a time. Let there be some visible point, directly opposite the eye, so that a line through the orifice of the nerve and the center of the opening of the uvea falls upon this visible point. Since any point you please radiates in an orb, it will therefore also 171
radiate to the parts of the orb, with the result that it will radiate to the whole of the small portion of the little sphere of the cornea, and will illuminate the iris and its black center, or the opening of the uvea. And since the iris is opaque and black, it turns away and keeps out the rays that are all around the edges, admitting within only those in the middle, so far as it is spread out at the opening. And since the cornea, and the aqueous humor that is beneath it (both of which I take as the same medium with respect to density), are a medium denser than air, the rays sent down to the inclined surface of the little sphere are accordingly refracted towards the perpendicular. Thus those rays which previously were spreading out in their progress through the air, are gathered together now that they have entered in to the cornea, so much so that any great circle described by those rays upon the cornea, which in their descent touch the edges of the opening, is wider than the circle of the opening of the uvea; however, these rays, all the way to the opening of the uvea, are so strongly gathered together through such a small depth of the aqueous humor, that now the edges of that opening are trimmed off by the extremes, and by the descent they have made they illuminate a portion on the surface of the crystalline that is smaller than the opening of the uvea. Now, having entered into the front surface of the crystalline humor, if indeed they all have first arisen from a point at a certain and proportionate distance (which is peculiar to each eye, and not just the same for all), they all fall approximately perpendicularly, because of the similar convexity of the cornea and the crystalline humor. As a result, hardly any new refraction of rays of a point directly in front of the vision and at a suitable distance occurs at the front surface of the crystalline, even though the medium of the crystalline is denser than that of the aqueous humor. Again, however, I here ascribe to the spider web and to its humor, i.e., the crystalline, the same density with regard to refraction, as I do below to the vitreous humor and to the hyaloid tunic. Thus these rays of one point, in the

[25] The pupil.

amount that was admitted through the opening of the uvea, descend through the entire depth of the crystalline, always coming closer and closer together, until they come to the hyperbolic posterior surface of the crystalline. So if it were possible for one to make a likeness of this series of rays in a cross section of the eye, he would make one and the same conic surface from the heights of the cornea all the way to the last parts of the crystalline, whose breadth is, of course, that of the opening of the uvea at that place, while its vertex would stop at some point beyond the eye. But now, when these rays go out through the conic posterior
172 surface of the crystalline into the vitreous humor, whose medium is rarer than the medium of the crystalline, they are refracted from the perpendiculars drawn to the surface through the points of refraction. As a result, it happens that the refracted rays converge towards the axis, and so are bounded by a shorter and more obtuse cone than that by which they had previously come through. Finally, then, all these rays from a single visible point come together in exactly one point, which is the very center and end point of the optic nerve, at the place where it is joined to the retina.[26] For corresponding to this density of the crystalline and the magnitude that these refractions are going to have, Nature has measured out the space of the vitreous humor between the crystalline and the retina.

And this vision, finally, is the most distinct, when all the light of the same point, howsoever much it is spread over the breadth of the cone admitted through the opening of the uvea, is brought together by two refractions, one at the cornea, the other at the posterior surface of the crystalline humor, and illuminates most strongly a single point of the retina, namely, the orifice itself of the nerve bearing the visual faculty or spirit; and no other rays from any other lucid point can fall upon that point, because of the beneficial action of the blackness and opacity of the uvea, of the narrowness of the opening, of the ciliary processes, and of the rest, which will be described shortly.

For the rest, what we have so far set up as a visible object was a point, not a body, and accordingly it had no parts, nor was right distinct from left, above from below. And it was not really visible, but was an element, or rather a boundary, of a visible object; and thus the vision itself of the point that has so far been explained must not be taken as the sum total of vision, but as a certain element of vision. For just as there are many points in the visible object, so there are, as it were, many elements of the vision of that object. Witelo III 19 nonetheless stands, that nothing is seen unless it has some proportionable magnitude.[27] Therefore, let there be another point next to and directly opposite the previous one, inclining off to the right of it. This point will therefore also itself illuminate the cornea, and will face the underlying iris and its opening obliquely. Thus the rays sent in through the circle of the opening will display the form of a scalene cone, which will cut the right cone of the previous point at the opening of the uvea, and after the intersection will pass over to the left side within the uvea, illuminating
173 a part of the surface of the crystalline also illuminated by the previous one, but also a part not illuminated by the previous point, but more to the left; and

26 Reading *quo in loco* for *quo is loco*.

27 *Thesaurus* II pp. 93–4.

nearly the same thing happens that we proved in ch. 2 above happens in a closed chamber.[28] For the pupil takes the place of the window, the crystalline takes the place of the panel opposite, except that, because of the proximity of the pupil and the crystalline, the complete intersection has not yet been brought about here, with the result that everything is still confused. Next, when this cone constructed on the left strikes upon the anterior surface of the crystalline, it will be refracted towards the right cone, and nonetheless takes an oblique path through the crystalline, and to that extent strikes more directly upon the hyperbolic surface of the crystalline, where it will again be refracted towards the previous right cone, but slightly so, with the result that it departs less from the previous right cone in the vitreous humor than in the crystalline, but it does nevertheless depart, and thus will fall upon the left wall of the retina. But the ratio of the hemisphere might be perturbed if the points outside in the air, opposite each other from the center of the eye, were deflected from the opposition by the triple refraction made by the surfaces of the cornea and the crystalline, and were to flow into the depth of the eye as if from an angle, and were thus to gather themselves into a portion of the hemisphere that is less than the hemisphere. To keep this from happening, Nature has found an excellent proportion, by having established the center of the retina, not at the intersection of the axes of the cones which go all through the vitreous humor, but far within, and by having moved the edge of the retina from the sides, so that thereby the longer cones, which are more divergent, would intercept the straight (and therefore short) parts of the retina, but those that are shorter, and diverge less to the sides of the retina, would mark out large parts of the retina, presented obliquely, with a narrow angle: and those cones belonging to points radiating from opposite sides, though they are not opposite when refracted, nonetheless would fall upon opposite points of the retiform, and thus compensation would be made. And thus, finally, if straight lines were drawn from the points of the hemisphere through the center of the retina and of the vitreous humor, those lines will mark the points of their own respective picture on the retina opposite. If this were not to happen, the quantity of things seen indistinctly from the side would always vary with the turning of the eye, as happens when lenses are moved towards the eyes. For these, even though they adhere to the eye without motion, if they are carried around with it, represent those things that are at rest with some motion, because of the variation of the apparent quantity of the hemisphere from the sides.

Further, it is worthwhile here to consider the difference between direct and **174**
lateral vision. First, the cone of direct vision is enclosed by the uvea alone, so that the whole of it is in the cornea; some oblique cones are left outside of the cornea itself. For they could be wider in proportion to the uvea, and so they measure out light grudgingly to the retina. The direct cone is circular or right; the oblique cones are flattened or scalene. The axis of the direct cone is not refracted at the cornea, the axes of the oblique cones are refracted at it. All the lines of the direct cone are approximately perpendicular to the crystalline, none of those of the oblique cones are. The direct cone is cut equally by the anterior surface

---

[28] *Camera.*

of the crystalline; the oblique cones are cut very unequally, because where the anterior surface of the crystalline is more inclined, it cuts the oblique cone more deeply. The direct cone cuts the hyperbolic surface of the crystalline, or the boss, circularly and equally; the oblique cones cut it unequally. All the rays of the direct cone are gathered together at a single point of the retina, which is the chief thing in this process; the lines of the oblique cones cannot quite all be gathered together, because of the causes previously mentioned here, and as a result, the picture is more confused. The direct cone aims the middle ray at center of the retina; the oblique cones aim the rays to the sides. The direct cone falls directly upon the retina; the ones at the sides, obliquely, because it was just now said that the center of the retina is below the intersection that the axes of the cone, which are in the vitreous humor, will have made. Finally, the sensory power, or spirit poured in through the nerve, is more concentrated and stronger, with respect to both the source and the orb, in that place where the retina is presented to the direct cones; it is spread out from that point with the spherical surface itself of the retina, and withdraws from the source, on which account it is also weakened. And, as in a funnel, and in a fishing net with a small bag, which above were analogies for the retina, all the sides turn either the liquid or the fish aside into the channel or the little bag, so the sides of the retina use their measure of sense not for its own sake, but whatever they can do they carry over to the perfection of the direct vision. That is, we see an object perfectly when at last we perceive it with all the surroundings of the hemisphere. On this account, oblique vision is least satisfying to the soul, but only invites one to turn the eyes thither, so that they may be seen directly. And so Witelo III Proposition 17 needs interpretation.[29] For by the perpendicular alone nothing is perceived distinctly. But by all the radiations of the same point (from which a perpendicular can be drawn through the centers of the opening and of the humors), gathered together at the center of the opening of the nerve (by Witelo III 29), vision that is completely direct and detailed, or distinct, occurs.[30]

175 Now The color of the retina is not dark or black, so as not to tint the colors of objects, nor is it quite as bright as possible, so as not to pour excessive brightness into the vitreous humor, and so that the things represented as white and shining upon it might not obscure the more colored things in an immoderate proportion. See the corollary to Prop. 30 and 31 of ch. 1. The figure of the retina is more than a hemisphere. First, because it was fitting that it should be a hemisphere, for receiving a picture that is proportioned to the objects, as has already been said. But the remaining edge is extended all the way the the ciliary processes, so that the retina might be stretched out by the stuffing of the globe of the vitreous humor, the neck now being made narrower than the belly. For it was impossible

---

[29] Witelo III 17 states, "Distinct vision occurs only following perpendicular lines from the points of the object seen extended to the surface of the eye. From this it is evident that every shape seen is oriented on the surface of the eye just as it is oriented on the surface of the seen object." (*Thesaurus* II p. 92).

[30] Witelo III 29: "Every point, belonging to a form, that is incident upon the surfaces of the eyes along the radial axes, must necessarily reach the center of the opening of rotation of the concave nerve." (*Thesaurus* II p. 99.)

to attach it anywhere because of the tenderness and fineness of the visual spirit, which required a channel in the nerve, which, contrary to the nature of the rest of the nerves, does not even support the substance of the nerve. For if the retina had not occupied more than a hemisphere, it could easily have gotten wrinkled up and have run back to the junction with the nerve. And further, the hollow had to be filled up by the hemisphere all the way to the ciliary processes, and thus this would have had to be accomplished either by the choroid or by the vitreous humor. This, however, is done more beautifully by the retina. But this edge also does not appear to cease its function of seeing. For even though none of the cones formed through the crystalline should touch it, the fissures of the ciliary processes nonetheless appear to be molded into this edge itself, so that light might to some extent come in from the sides through the ciliary processes and might be received by this extended edge of the retina. For a straight line drawn from the extremity of the cornea through the nearer edge of the opening of the uvea nearly falls upon the juncture of the crystalline with the ciliary processes; drawn through the opposite edge of the opening, it nearly touches the origin of the ciliary processes from the uvea. And it is by this device that nature seems to have secured our ability to perceive more than a hemisphere without moving our eyes, fully as much as is admitted through the corner of the eye, with the eye turned back ever so slightly. It is indeed little short of your being able to perceive the ears themselves with the neighboring eye, especially if they are rather long. I have often seen, to my amazement, both the sun and my shadow, as if the two were out in front and not opposite each other. This precaution nature seems to have provided for the guardianship of the eyes, so that things that were not turned away from the chamber, where it ought to be open, might in approaching come immediately into view, wheresoever the eye might be looking. For why should that which takes care of the whole animal, not also take good care of itself, and give this function to the care of its body?

The vitreous humor, in turn, has acquired a skin, so that it might not weaken **176**
the nerve or the retina, and make them grow limp, by wetting them; and so that it not be mixed in with the aqueous humor on its anterior side through the fissures of the ciliary processes. It had to differ in density from the crystalline, as well as from the aqueous, for the sake of refraction. For unless it were rarer than the crystalline, the rays would not be gathered together from their perpendiculars to the axis of any of the cones. And if the vitreous humor is denser than the aqueous, the rays that come in through the ciliary fissures will be able to descend deeper in proportion to the density, and will approach by way of the outside cones that are given their form by the crystalline.

Further, the crystalline has acquired a small tunic, so that it, being soft, should not become mixed together with the aqueous humor. It is, moreover, too small to touch the sides of the retina, in order that through this gap the rays of the cones may be gathered together at a point. Because it had to be both supported and connected, for the sake of nutriment, it was bound to the choroid by the ciliary processes. These are black all over, so as not to create glare when illuminated, and closely spaced, so that the lower chamber of the vitreous humor might be dark, as it should be, and so that the vitreous humor might not dilute the images by being illuminated. It is no doubt for this application that they [i.e., the ciliary processes] will acquire the motion of puffing themselves up, of becoming cloudy in bright light, and of becoming rare in bad light, like the uvea. And a hyperbola, or a figure akin to it, is at the posterior part of the crystalline, so that the rays of the same right cone that tend downwards within the hyperbola and converge towards the same point, should be gathered together at the same point, but at a shorter distance. That this cannot happen in another figure, will be demonstrated below.[31] But the anterior surface of the crystalline is gibbous, so that the more obliquely the opening of the uvea is presented to the radiating point, the more on the slope this surface is cut by the scalene cone, and thus the equality of the quantity intercepted by the cone is preserved, as much as possible; and so that it might lie perpendicularly beneath all rays refracted at the cornea (coming from the same particular point) and tending to the same point after refraction. I maintain that this anterior surface of the crystalline is entirely circular, or spherical.

The whole uvea with the ciliary processes is present only to produce darkness, so that the light not be excessively troublesome. It brings nothing to the
177 forming of the image or picture, and if it had brought something, it never would complete it and bring it to perfection, because the opening is too wide in proportion to the narrowness of the eye. In fact, when it is established in darkness[32] it is dilated three times more widely than in the light of the sun, so that in darkness it might uncover a greater part of the surface of the crystalline, so that thereby more of the light that is so weak, gathered at the same point through the crystalline [humor] (which gathering at the same point is accomplished by the crystalline without the help of the opening), may move the sense all the more obviously. In the light, on the other hand, it is narrower, so as to keep out more of the light, lest so strong a light injure the sense. Thus the location of this opening is the place where the rays intersect, existing as an effect of the crystalline itself. However, this intersection does not occur in a point, but is spread out into a very long cone, because of the circular surface of the crystalline. Therefore, the location of the opening becomes the base of this cone of intersections. For between this opening and the crystalline there is no intersection, and if something were set out there before the sense of vision, it would appear inverted and confused. On the inside, the uvea is rough, so as not to cause a reflection of a ray reflected at it from the surface of the crystalline, as it would if it were smooth. It is all black, even where it encloses the retina, with a likeness of substance. For the nutriment, by which the anterior parts had to be made black, has its passage through the choroid. Unless perhaps the retina too has transparency, and this black is the boundary of that transparency.

The aqueous humor was needed to fill up the chambers, and to continue the refraction created at the cornea, so that it should be the same all the way to the crystalline.

---

[31] See Prop. 24 of this chapter, p. 214.

[32] Reading *constituta* for *constitutis*.

The tunic of the cornea also seems itself to be a small portion of a spheroid, so that the rays that fall perpendicularly upon the anterior surface of the crystalline might come together in one point. There is, however, nothing to prevent the cornea's being perfectly round, as will be said below.

*   *

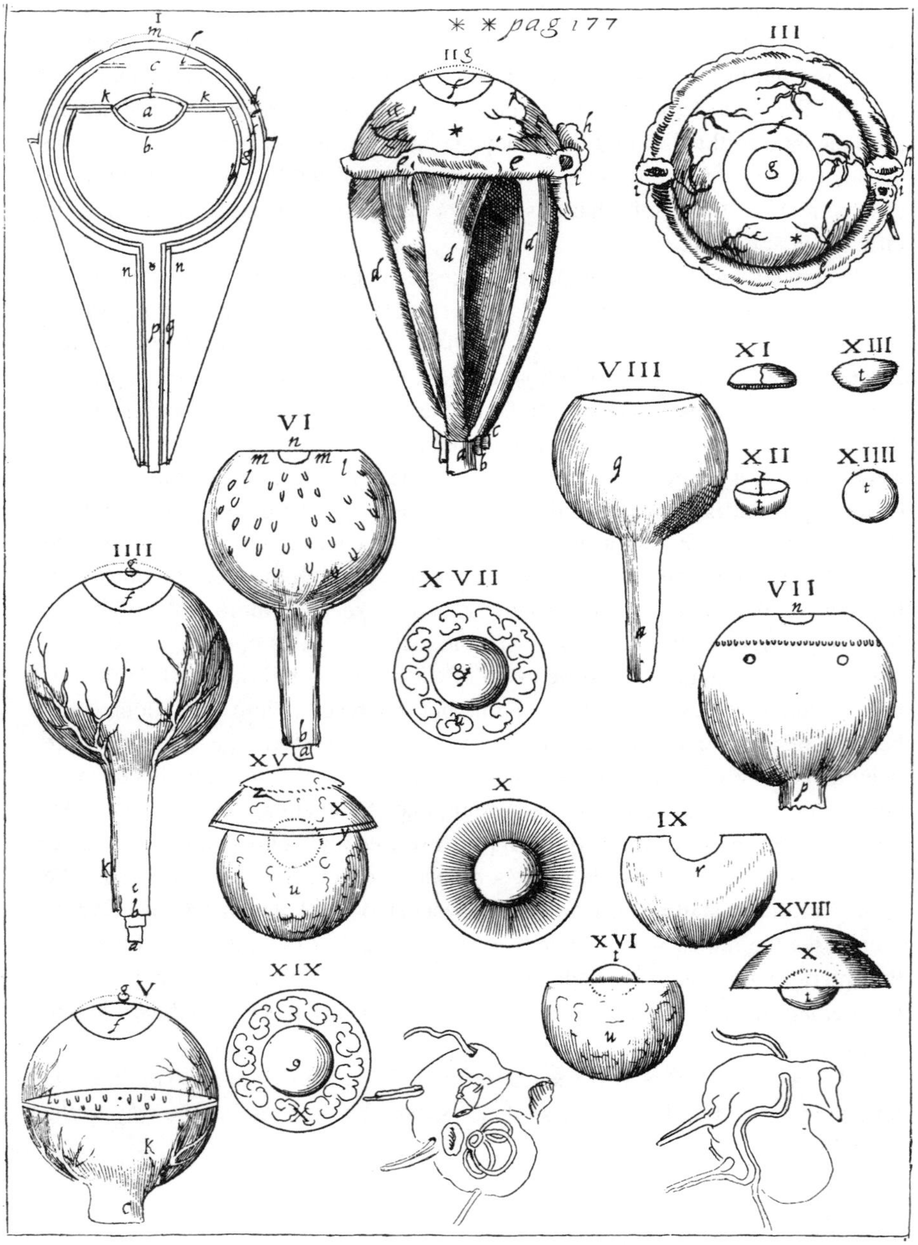

In order to counsel the reader for whom anatomical plates are not at hand, I thought it advisable to place here a copy of Plate 49 of the eminent man Felix Platter. I would that this plan had presented itself right away at the start of work, for I would then have accommodated the text to it. No explanation of the letters and notes had to be added other than those of Platter himself; I have added only the references to my text.[33]

There are 19 figures, for the last two (the organs of hearing) the engraver added unbidden.

I. Portrayal, through a line drawing, of the membranes and humors of the eye, in imitation of the real eye. In which *A* is the crystalline, *B* the vitreous, *C* the aqueous, humors; *D* the adnata tunica [i.e., the conjunctiva]; *E* the opaque part of the thick tunic [crassa tunica, i.e., the sclera]; *F* the uvea, *G* the retinal, *H* the hyaloid, *I* the crystalloid, tunics; *K* the ciliary processes of the uvea tunic; *L* the indentation of the uvea separating off from the thick tunic; *M* the corneal part of the thick tunica, whose protruding convexity, noted by others, is indicated by dots; *N*, *N* the muscles of the eye; *O* the visual nerve; *P* the thin membrane of the nerve; *Q* the thick membrane of the nerve. p. 163.

II. The whole eye with the muscles, removed from the skull, separated only from the eyelids.

III. The anterior seat of the globe of the eye. p. 163.

IV. The thick tunic [i. e., the sclera] of the eye with a part of the optic nerve. pp. 161, 163.

V. The thick tunic [i. e., the sclera] of the eye, divided by a transverse section.

VI. The uvea tunica with part of the optic nerve. pp. 165, 176.

VII. The everted inner surface of the same. pp. 165, 177.

VIII. The retinal tunic with the substance of the visual nerve. pp. 163, 166, 168, 175.

IX. The hyaloid tunic. p. 167, 176.

X. The ciliary processes, spread radially through the anterior [parts] of the hyaloid tunic. pp. 166, 167, 175, 176.

XI. The crystalloid tunic. pp. 167, 176.

XII. The crystalline humor, still covered by the tunic.

XIII. The crystalline humor bare, placed so as to be seen from the side.

XIV. The anterior seat of the crystalline humor. p. 167.

XV. The three humors of the eye at the same time together, the aqueous, the vitreous, and the crystalline, only shadowed in incidentally. p. 167.

XVI. The vitreous humor containing the crystalline. p. 167.

---

[33] These are the page numbers following each item in the list. They refer to Kepler's page numbers (the numbers in the margin of this edition). The list, along with the plate, were inserted between signatures Y and Z on two unnumbered sheets.

XVII.	The anterior seat of the vitreous humor alone.
XVIII.	The aqueous humor poured above the crystalline. pp. 167, 177.
XIX.	The anterior seat of the aqueous humor alone.

A in II, IV, VI, VIII.	The visual or optic nerve. p. 172.
B in II, IV, VI.	The thin tunic surrounding the nerve.
C II, IV, V.	The thick tunic, wrapping the nerve.
DDD II.	The muscles of the eyes of one side. pp. 161, 164.
EE II, III.	Part of the adnata tunica [i.e., the conjunctiva], stretched beneath the eyelids. p. 164.
* II, III.	Part of it expanded, with veins interwoven. p. 164.[34]
F II, III, IV, V.	The dark or iris of the eye, which the white surrounds. pp. 163, 165.
G II, III, IV, V.	The dark or center of the eye, in the middle of the iris. p. 165.
Note to II, IV, V	The arc at the letter *G*, delineated by dots, arising from the extremities of the iris, and a portion of a smaller circle than is the globe of the eye, was added by me from the observation of others, and indicates the protuberance of the cornea, by which it protrudes from the white. p. 164, 177.
H II, III.	The tissue at the inner angle of the eye. p. 161.
I II, III.	The tear openings.
K IV, V.	Ducts spread through the thick membrane.[35] p. 164.
L, *L* V, VI.	The fibers by which the uvea tunica is joined to the thick [i. e., the sclera]. p. 165.
M, *M*, VI.	The indentation of the uvea, where it separates off from the cornea. p. 165.
N VI, VII.	The opening of the uvea, or the pupil. pp. 165, 177.
O, *O* VII.	The beginnings of the ciliary processes. p. 166.
P VII.	The beginning of the uvea, expanded out of the thin tunic. p. 165.
Q VIII.	The extent of the retinal tunic, stretched forward beyond the middle of the eye. pp. 166, 173, 175.
R IX.	The hollow of the hyaloid tunic, supporting the crystalline humor.
S XI, XII.	The breadth of the crystalloid tunic. p. 167, 168.
T in XII, XIII, XVIII.	The spherical posterior part of the crystalline humor (protruding in a cone, as some explain it; hyperbolic, as I explain it). pp. 167, 176.

[34] See Kepler's note on p. 175.

[35] *Crassa membranea*, or *crassa tunica*, i. e., the sclera.

in XIV, XVI.	The flattened anterior part of the same. pp. 167, 176.
V XV, XVI.	The extent of the vitreous humor. pp. 167, 169.
X XV, XVIII, XIX.	Extent of the aqueous humor. p. 167.
Y XV.	The seat where the vitrious humor is distinguished from the aqueous, by the interposition of the hyaloid tunic. p. 167.
Z XV, XVIII.	The seat where the uvea floats upon the aqueous humor. p. 167.
& XVII.	The cavity in the vitreous humor that remained when the crystalline [humor] was removed.
§ XIX.	The cavity in the aqueous humor, from the same cause.

3. Demonstration of those things that have been said about the crystalline in regard to the means of vision

Common experiences of nearly all the things that have so far been said of the crystalline may be perceived in crystalline balls and in urinary vessels filled with clear water. For if one were to stand with a crystalline or aqueous globe of
178 this kind in some room next to a glazed window, and provide a white piece of paper behind the globe, distant from the edge of the globe by a semidiameter of the globe, the glazed window with the channels overlaid with wood and lead, enclosing the edges of the windows, are depicted with perfect clarity upon the paper, but in an inverted position. The rest of the objects do the same thing, if the place be darkened a little more, to such an extent that if the globe be brought into the chamber which we described above in ch. 2 prop. 7, and set opposite the little window, whatever things are able to reach through the breadth of the little window or opening to the globe are all depicted with perfect clarity and most pleasingly through the crystalline upon the paper opposite. And while the picture appears at this distance uniquely (that is, a semidiameter from the globe to the paper), and nearer and farther there is confusion, nevertheless, exactly the opposite happens when the eye is applied. For if the eye be set at a semidiameter of the globe behind the glass, where formerly the picture was most distinct, there now appears the greatest confusion of the objects represented through the glass. For the glass appears either entirely bright, or entirely red, or entirely dark, and so on. If the eye comes to be nearer to the globe, it perceives the objects opposite erect and large, where they are poured together on the paper; if it on the other hand recedes farther from the globe than a semidiameter of the globe, it grasps the objects with distinct images, inverted in situation, and small, and clinging right to the nearest surface of the globe. Before, however, when the paper was placed there, the picture had entirely vanished. All these things happen with regard to an aqueous globe, because of the refractions and the shape, as a result of there being some convexity in the shape. And since the crystalline is made of convex surfaces, and is also denser than the surrounding humors, just as water in the glass is denser than air, therefore, whatever we shall have demonstrated concerning the aqueous globe in this way, and using these media, have also been proved concerning the crystalline, with privileges reserved to it because of the particular

convexity of shape, inconsistent with the convexity of the globe. Let us proceed, then, to the demonstration of those things that happen in relation to a crystalline or glass globe.

Proposition 1. Problem

To find the place of the image by the commonly known path, when the object is observed with both eyes through a globe of a denser medium. Let E, F, H, G be points on a globe of a denser medium, A a visible point, B, C the eyes, and I the center of the sphere in an intermediate place. And let F, H be the points of the refractions towards the eyes; E, G the points of the refractions towards the visible object at A. First, if E is the point of refraction, and EF be the **179**
ray refracted from AE, then A, E, F will be one surface, by ch. 3 definition 2. But by 16 of that chapter, the surface AEF is perpendicular above the globe; that is, it passes through the center I. Therefore, $AEFI$ are in one surface. Now if F is the point in which EF is refracted into FC, and through FC it comes to one eye at C, by the same proposition, $EFCI$ will therefore be one plane surface. And since surfaces $AEFI$ and $EFCI$ have a common part, the surface EFI, therefore $AEFCI$ are in one surface; and AI extended will meet with FC at the intersection D. In exactly the same way it will be proved that the other eye B and $DHGAI$ are in one surface. Therefore, D is on both surfaces, and consequently BH and CF cut each other mutually at D. Thus D is the place at which the image of the object or of the viewed point A is seen with the two eyes, by chapter 3 17.

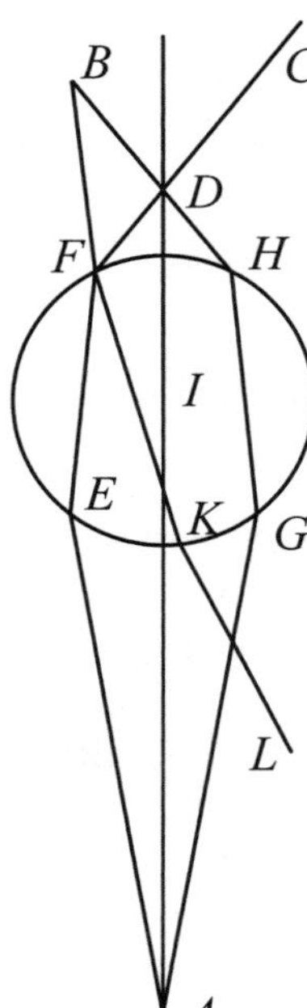

Proposition 2

The sense of vision looks upon things that are very close with greater difficulty than upon things that are more distant. For it was said that in the viewing of nearby things the eyes have to be turned towards each other.[36] Turning towards each other is at variance with nature, which has given the eyes a parallel situation. Fatigue follows, as a result, and less fatigue from a lesser turning. Hence, those who are given to thought are easily distinguished from others, although no one understands the indication. It is just this: that they relax the muscles of the eyes, so that they are less turned towards each other for nearby objects; for thus they return to their parallel situation.

Behavior of those given to thought.

Proposition 3

The sense of vision is drawn to the obvious, poorly attracted by the faint or vanishing. What experience testifies, the property of vision confirms. For it has been given in order to be moved by light, and it will therefore be moved strongly

---

36 Ch. 3 prop. 8, p. 79.

by a strong light. But to be moved by light is to see. Therefore, one who had first been looking at a weak light will follow a stronger light when one such arises from the same direction, and will drop the former.

Or, by chapter 1 prop. 28, when strong lights illuminate the eye excessively, a weaker illumination will remain hidden beneath it. But the sense of vision
180 follows the manner of illumination.

Proposition 4

Darkness accommodates an image, but when a stronger light arises from the direction of the image, the image passes away. For an image is in part a creature of intention,[37] the work of the sense of vision. Thus, when the act of vision passes away, the image passes away.[38] But the act of vision passes away with these conditions, by the preceding. On the other hand, when the place or matter is dark in the direction of the image, the stronger light of the image, by the preceding, consequently causes the act of vision.

Proposition 5

In front of an aqueous ball or globe there is no place for the image of an object hiding behind the ball. Experience confirms this. Causes therefore had to be sought. For this proposition is not only a means of correcting what J. B. Porta proposes in Book 17 ch. 13, "with a crystal ball, to make an image appear to be hanging in air,"[39] but also detracts from proposition 1 of this chapter. So, since the place of the image, by 1 of this chapter, is nearer the eyes than is the aqueous globe, therefore, by 2 of this chapter, the eyes will be turned towards each other to the place of the image with more difficulty than to the globe. And by 3 of this chapter, the eyes will be drawn to the globe, whose light is stronger and more obvious than the rays of the object reaching through the ball to the point or place of the image. Therefore, by 4 of this chapter, the light from the aqueous globe, illuminating the intermediate place of the image, drives away the image, when a new act of its seeing is aroused. Hence Porta, vacillating, says, "If a visible be of the greatest visibility, such as fire or a candle, the object will be seen without difficulty, and more clearly."[40] Therefore, those that are not fire, by Porta's admission, are perceived with greatest difficulty in the image's usual place. What need of more? These are the conclusions of the optical writers originating from an insufficient and non-universal demonstration, concerning the place of the image. Pertinent to this is what Porta had taught in chapter 10 preceding, "with a convex

---

[37] *Intentionale ens.*

[38] Here Kepler refers to the following endnote.

To p. 180, prop. 4. This is the reason why a bluish and nearly colorless material is laid beneath glass mirrors, while metallic ones are mostly made of iron or another black material. Hence also, when darkness outside surrounds glass windows, i.e., at night by candlelight, the windowpanes serve us as mirrors.

[39] J. B. Porta, *Magia naturalis* (Naples, 1589) p. 270.

[40] Porta, *Magia naturalis*, p. 271.

crystalline lens, to see an image hanging in air."[41] For the reason is very nearly the same. For this reason, he adds, "If you will place a piece of paper in the way, you will see clearly that a lighted candle appears to be burning upon the paper." That is, the image will be seen weakly and hardly at all in the bare air itself, by Porta's admission. But if you put a piece of paper in the way—if, I say, you interpose a piece of paper between the lens and the sense of vision (for, with me, Porta here is still speaking about the image, not yet about the picture, of which this is true, as will be clear below), the image will now appear, not hanging in air, but fixed on the paper. For the paper, striking the eyes more **181**
obviously, steadies them on the place of the image, so that they may be turned towards each other in that direction. And nonetheless, because the paper is then brighter than the image, the paper will be seen primarily, the image secondarily. For it is not mathematical dimensions alone that create the image, but also, and much more, the colors and lights and physical causes, with which prop. 2, 3, and 4 of this chapter are concerned. If you should focus the eyesight upon one place, namely, upon the place of the image previously investigated, as it has been described in prop. 1 of this chapter, when a clearly visible object is placed nearby, then the eyes, coming together upon this object, will also see the required image secondarily. Another more tricky caution, from 4 preceding, can be applied, an experiment of which I saw at Dresden in the Elector's theater of artifices. But what I, steeped in demonstrations, stated that I had seen, the others denied. I therefore attribute it, not to the overseer's intent, but to chance. A disk thicker in the middle, or a crystalline lens, a foot in diameter, was standing at the entrance of a closed chamber against a little window, which was the only thing that was open, slanted a little to the right. Thus when the eyesight travelled through the dark emptiness, it also, fortuitously, hit upon the place of the image, nearer, in fact, than the lens. And so since the lens was weakly illuminated, it did not particularly attract the eyes. But the walls were also not particularly conspicuous through the lens, because they were in deep darkness. But the little window and the objects standing about it, which had the benefit of much light, lying hidden beyond the lens, set up a bright image of themselves in the air (between me and the lens). And so at first glance I perceived this aerial image, but with repeated gazing, gradually less and less. The games can be made more elaborate. We will here propose things that are more obvious and ready at hand, that is, suited to our purpose.

Proposition 6

Images of objects seen through an aqueous globe with one eye adhere to the nearer surface of the aqueous globe. Since this is confirmed by experience, it strongly confirms the trustworthiness of my demonstration of the place of the image, proposed in ch. 3 above. For vision is deprived of the association of the eyes when only one is used. Therefore, the place on the perpendicular will result entirely from this association of the eyes. Here I prove what is proposed to be proved from 5 and 3, preceding. For since between the eye and the globe there **182**

[41] Porta, *Magia naturalis*, p. 269.

occurs no place suitable for the image, the globe is the first thing that holds the eye by its light, and it consists of the same light of which the image consists. Consequently, by 3 preceding, both the globe and the image will be seen in the same act of seeing.

Proposition 7

Images of objects seen through an aqueous globe with both eyes most often appear confused and double. This would never come to be confirmed by experience, if the experiments of Porta just introduced were to stand without further caution, or if Witelo 10 47 were true without any restriction.[42] I prove what is proposed thus. By 5, the usual place of the image in relation to the use of both eyes is most often made inapplicable. Therefore, by 6, for either eye the image appears to inhere in the surface of the globe. Now if the surface of the globe were the usual place of the image for the two eyes (as can happen in a number of media), then one and the same image would be perceivedby both eyes. But here it is supposed that the place in which the same image is perceived with the two eyes is nearer to the sense of vision than the globe is. It is therefore necessary that the image be double, and the vision be confused, for the reason that in the same glance it perceived the same glass directly, and at the same time and in the same light in which it perceives the glass, it also perceives the image directly with one eye, indirectly with the other. In the above diagram, while *D* ought to have been the place of the image of point *A* when the two eyes *B*, *C* are observing, it is also impossible for the vision to be fixed at *D* because the object *FH* is more obvious. It therefore happens that when the vision is carried over to *F*, the other point of refraction, the point *F* is indeed seen properly, but with two lights, one coming from *A* to the eye *C* along the lines *AE*, *EF*, *FC*, the other to the eye *B* along the lines *LK*, *KF*, *FB*. For that reason, the sense of vision, unaccustomed to sighting two different objects *A*, *L* directly (by refraction), when the two eyes are turned towards *F*, is indeed correctly informed about *F*, but is confused about the images *AL*. For if the vision remains the same, and the eye *B* is directed along *BF*, the point *A* comes into the eye *B* laterally by the lines *AG*, *GH*, *HB*, while it had come directly into the eye *C*. Therefore, the sense of vision, seeing the point *A* with one eye directly, with the other indirectly, rightly considers that it sees two objects. If the vision be directed neither towards *F* nor towards *H*, but towards some intermediate place,
183 then this place of the globe will be seen directly, but some image on the two sides indirectly.

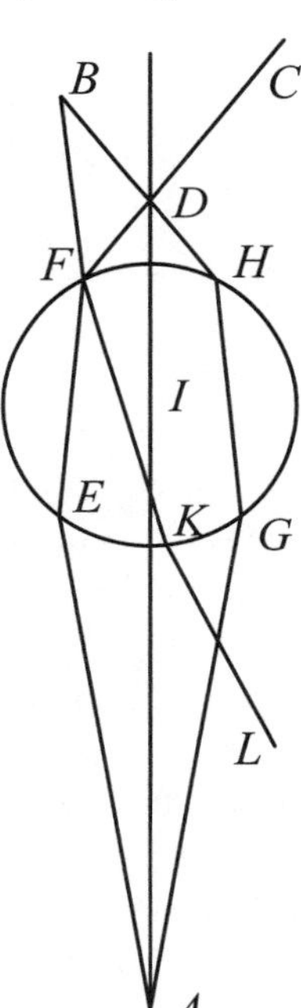

[42] Witelo 10, 47: "A single refracted image occurs through the same observer's seeing with both eyes." (*Thesaurus* p. 443).

How it happens that an object may appear double.

Here, because the diagram provides the opportunity, although it is off the track,[43] I append a demonstration of what Aristotle seeks in section 31 problems 11 and 17.[44] How is it that, although each eye is brought into use by the same principle, there is still in the eyes a faculty for taking one object as double, even outside the consideration of the aqueous globe or concave mirror? I reply that the duality of the eyes, and the distinction of direct vision from oblique, or of oblique visions from each other through different parts of a surface, are in agreement. In the preceding figure, let FH be the wall, the eyes B, C, whose powers of sight are directed to one point: let it be F. Let the visible object, whose form is to be doubled, be located in an intermediate place: let it be D. Now, since FC is the axis of the eye C extended, D will appear to the eye C on the line of the axis. And since FB is the axis of the eye B extended, while BD is a line other than the axis, D will appear to the eye B outside the line of the axis at another part of the surface. Thus, according to different locations in the eyes, different forms will appear. Therefore, this happens to the inebriated and the sick, and to boys and to the old, and those whose voluntary motions are impeded.

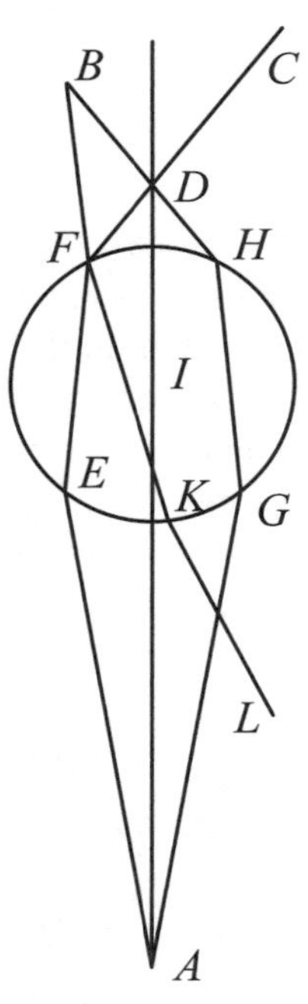

Proposition 8

Rays from some point at a distance beyond comparison that flow to any points of an aqueous globe, meet the axis after a double refraction, that is, they meet the line that is drawn from the radiating point through the center of the globe. Let A be the center of the aqueous globe BC, DAF the axis. Let the radiating point be towards D, at a distance beyond comparison from the globe. Therefore, the lines of a ray drawn from that point to any point of the sphere will differ imperceptibly from lines equidistant from the axis DAF. Let points B, G **184**
be lines parallel to the axis KB, LG. I say that they will meet the axis DAF. For because KB, LG strike obliquely upon a denser sphere, they will therefore be refracted towards the perpendiculars BA, GA, and KBC, LGH become a pair of lines making angles at B, G. Therefore, CB cuts BK, and consequently also, when extended, cuts its parallel DA. Likewise, HG cuts GL, and consequently also its parallel DA. And because BC, GH, which are going to meet AF, go out into a rarer medium, they will therefore be refracted from the perpendiculars AC, AH in the opposite way, but with the same angles, so that KBC, BCF are equal, and also LGH, GHI, because the angles of incidence B and C are equal, as are

[43] *Extra oleas*, literally, "outside the olive trees", a Latin translation of the Greek ἐκτός τῶν ἐλαιῶν, an idiom found e.g. in Aristophanes, *Frogs* 995. Olive trees stood at the end of the Athenian racecourse; hence, to go "extra oleas" is to go outside the racecourse.

[44] Aristotle, *Problems* 958b 11–13 and 959a 9–19, trs. Hett II pp. 188–191 and 192–3.

G, H as well. See Witelo X prop. 9.[45] But BC itself, as well as GH, were going to cut AF. So much more will CF, HI cut AF at F and I, because they are more inclined to that side.

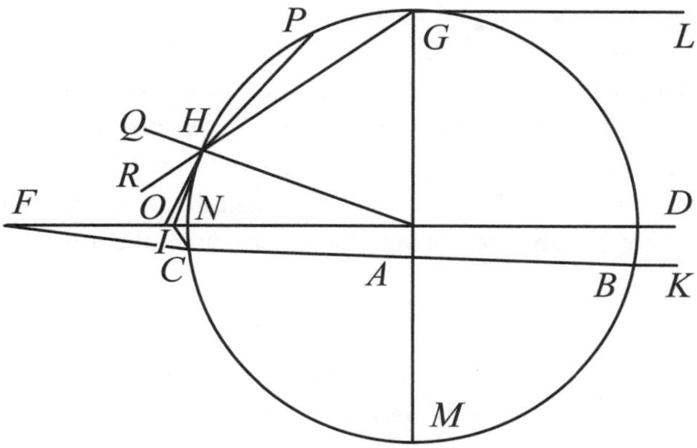

Lemma for Proposition 9, following

When straight lines are drawn from equal arcs of a circle to a point outside the circle, the angles subtended by the equal arcs at the point are unequal, and those whose arcs are closer to the diameter through the point are greater.

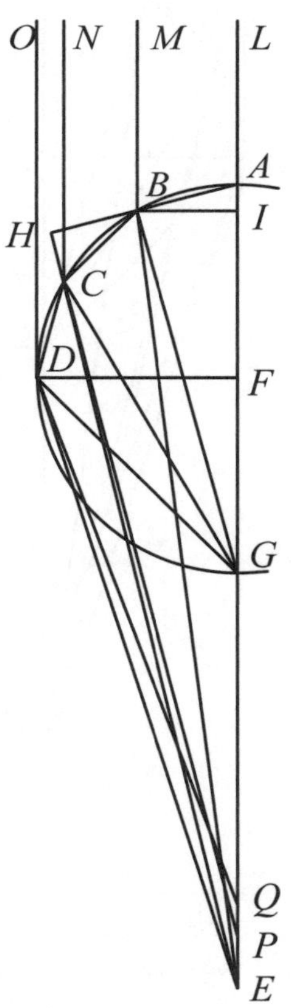

Let AB, BC, CD be equal arcs of a circle about center F. And on diameter AF let E be a point outside the circumference of the circle. Let the ends A, B, C, D be connected with E and with the center F, and with point G of the surface; and let the neighboring ends be connected among themselves, by the lines AB, BC, CD. I say that AEB is greater than BEC, and that this is greater then CED.

What is proposed is readily apparent. For at F all the angles are equal; from there towards G, those that are farther from EA become greater; until at G they again become equal, by Euclid 3, 21. Therefore, beyond G towards E the ones that are farther from EA become smaller. The cause is the inclination of the equal lines AB, BC, CD, greater in the more distant ones. For let EC be extended to H, and let perpendiculars
185 BI, BH fall from B to EA, EH. CHB, BIA will be right triangles upon equal bases CB, BA. I say first that EAB is greater than HCB. I prove it thus. GAB and GCB make the sum of 2 right angles, by Euclid 3, 22. Therefore, GAB and the supplement of GCB are equal.

---

45 *Thesaurus* pp. 413–4. The proposition shows that if the ray were to go from the vision to the object, rather than from the object to the vision, its path would be the same.

But ECB is greater than the part GCB. Therefore, HCB, the supplement of ECB, is less than the supplement of GCB, that is, less than EAB. Therefore, since EAB is greater than HCB, while the bases of the right triangles are equal, and thus the right angles fit into the same semicircle, BI will be greater than BH. Therefore, the right triangles BIE, BHE, upon the common base BE, will again fit into the same semicircle, and therefore, BEA will be greater than BEH or BEC, because for the former the longer line BI is the subtense; for the latter, the shorter BH. In the same way, DEC will also be proved to be less than CEB, provided that you keep GBE much smaller than GCE, because GE is more directly, and more closely, presented to the latter than to the former.

Proposition 9

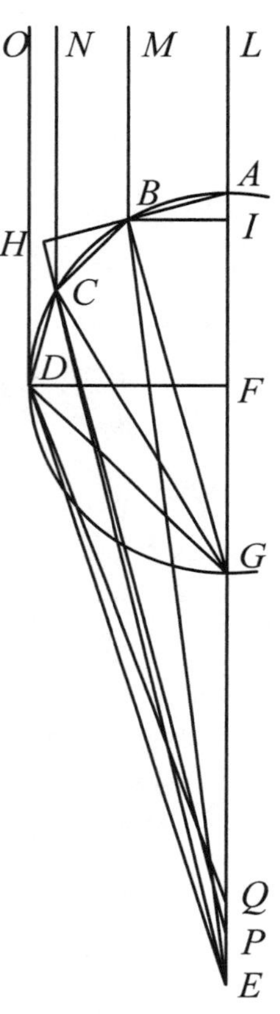

Parallel rays that have entered an aqueous sphere meet the axis sooner according as they are farther from the axis. In the preceding diagram, let lines parallel to the axis AE fall at points A, B, C, D, and let them be LA, MB, NC, OD. And first, let the refractions be proportional to the inclinations. The inclinations here are LAF, nil; then MBF, NCF, and ODF horizontal or maximum. And let MB, by one refraction occurring at B, have met with the axis at E beyond the globe, the second refraction, which occurs in the back part of the globe's surface, being neglected for now. Now, since by the preceding lemma, the angles at E are unequal, while the excesses of the inclinations are equal, the angles at E will not be able to correspond to the angles of incidence proportionally. That is, if BEA (the supplement of MBE) is the correct angle of refraction of MB, then CEA will not be able to be the angle of refraction of NC, because the inclination NCF is twice MBF, while CEA is less than twice BEA.[46] So if the angle of refraction of NC should also be twice that of MB, it should be increased. But it will be increased by drawing CP so that it cuts the axis EA before E at P. For the interior angle CPA is equal to the exterior angles CEP, ECP.[47] And thus it is greater than CEP or CEA. Therefore it is evident, if the angles of refraction were to be proportional to the angles of

---

46 In ch. 4 sect. 6 prop. 6 and 8, Kepler establishes that the angle of refraction—that is, the angle through which the ray is deflected in passing through the refracting surface—increases in a composite ratio which is greater than the ratio of increase of the inclination. Here, however, the angles at E increase in a ratio that is less than than of the inclinations. Therefore, the angles at E cannot represent the angles of refraction.

47 If this is intended to apply to the triangle CPE, using Euclid I, 32, angle CPA is the exterior angle and CEP, ECP are interior angles.

186 inclination after one refraction, then, of parallel rays, those that are nearer the perpendicular are going to meet with it farther out, while those that are farther from the perpendicular are going to make an intersection nearer the sphere.

Now, however, if the refraction should happen twice—that is, if another one should be added at the opposite part of the surface, which happens in those media whose density is so adjusted that all the radiations finally meet with the axis beyond the sphere—then this ratio is doubled, since, by Witelo X 9,[48] the angles of refraction are equal whether the light pass from water to air, or from air to water by the same line.

Finally, because the refractions are not proportional to the angles of incidence, but the angles of refraction are much greater in rays that are more inclined than the size of the angles of incidence bears, therefore, a new cause is again added, to bring the radiations that are more distant from the perpendicular together with the axis closer than before, while making the radiations that are nearer the axis come together with the axis farther from the sphere than before.

Proposition 10. Problem

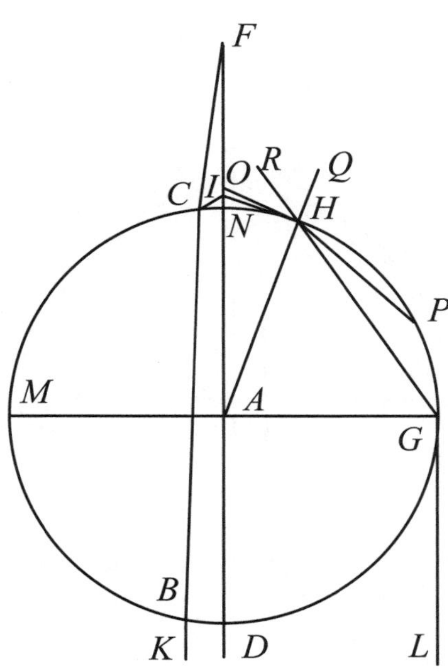

At a given point of a sphere, to apply a ray of some point of the lucid body so that when refracted it should not meet with the axis drawn from the radiant point through the sphere. Let the point H, in the next to last diagram (Prop. 8), be taken, and from H let any straight line be drawn, not through the center A, cutting the sphere, and let it be HG, to which let HI be inclined at the correct angle that the angle of incidence of the refracted ray GH requires at the surface of the sphere, by prop. 9 ch. 4. Let it be IHG, and let HGL be made equal to it, and parallel to GL let there be drawn through the center A the straight line DA cutting HI at I. I say that the point I projects the ray to the given point H, that is,
187 IH, which, refracted at H and G does not meet AD.[49] The demonstration is evident from the preceding, of which this is the converse.

---

48 See the note on this proposition above.

49 Here Kepler refers to the following endnote.

The demonstration of this is very easy. For GL, AD [reading AD for AK] are parallels by construction. They therefore do not meet, by the definition of parallels. And so there is no point in the whole arc NH that does not have its own non-cutting line [ἀσύμπτωτος]. This is the meaning of line 32 p. 188 [i.e., the statement, "such an arrangement is possible at any point," near the end of Kepler's p. 188. —trans.].

Proposition 11

It is impossible that the same ray, refracted in a medium not perfectly dense,[50] *should descend from many conjoined radiations out of a rarer medium coinciding at the same point of the surface of the denser medium.* For by Prop. 8 of the fourth chapter, the refracted angles at all inclinations are established as follows: first, some portion of the angle of refraction increases with the inclinations; then, when this portion is subtracted from the inclination, the secant of the remainder multiplies the portion again. Therefore, since the elements differ quantitatively at different inclinations, the composites, or the angles of refraction, must necessarily differ also. As a result, those of different inclinations, that is, those of different radiations to the same point of the surface of the denser medium, are different: but the refracted rays are by no means the same.

Proposition 12

Rays coming from different directions to the same point of the surface of a denser medium cut each other at that point, and the refracted ray of the higher radiation becomes lower. If not, let the radiation OH be higher than IH, and its refracted ray be HP, which cannot coincide with HG by the preceding. Accordingly, let HP also be higher than HG, so that IH is refracted into HG and OH into HP. Let the perpendicular also be extended to some point Q. For it is certain that QH, when it goes beneath the water, is not refracted, but proceeds along HA. Therefore, when QH is made to incline and arrives at OH, HA, which is beneath the water, arrives at HP. Therefore, when the former lies somewhere between Q and O, the latter will lie between A and P, and will coincide with HP. Previously, however, HG was the refracted ray of IH, lower than OH, while now it is that of some ray higher than OH, which is impossible, by the preceding. Therefore, OH is not refracted into HP above HG, but into some line below HG, and the intersection takes place at point H.

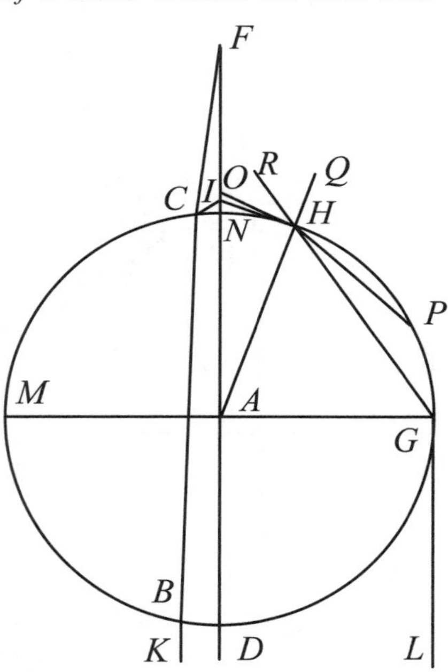

Alternately. If OH does not intersect IHG and descend below it, then all the higher ones will have refracted rays that are higher than IH, because the ratio

---

[50] Here Kepler refers to the following endnote.

To Prop. 11, p. 187 (in a medium that is not perfectly dense): For in a perfectly dense medium, all the radiations from the whole hemisphere entering at the same point are refracted to the center along a single line, as was pointed out above in ch. 4 p. 91, demonstrated in prop. 7 p. 113, and as you will find in use below in ch. 6 p. 225.

of all is the same, since it has been proved that the refractions increase with the inclinations; and where any one is higher, its refracted ray will also thereby be
188 higher. Let GH be extended some distance to R. Now, since RH is higher than OH, its refracted ray will be higher than HG. Let it be HP. Now, since RHG is one straight line, RH, upon entering the denser medium at H, will be refracted away from the perpendicular HA into HP, which is absurd; for it should have been refracted towards the perpendicular in the denser medium. Therefore, the refracted rays of rays having a higher angle of incidence are not higher. But, by the preceding, they are also not the same. Therefore, they are lower. Accordingly, an intersection occurs at H, which is what was to have been proved.

Proposition 13. Problem* [51]

To find a point on the axis[52] *outside the sphere which forms the nearest boundary of the radiations into the sphere, meeting with the axis.* In the diagram of prop. 8, let FAD, through the center A, be the axis of the sphere BC, to which let AG be erected at right angles from A, and let GL be tangent to the circle at G. This incidence will be "horizontal", as we named it above in ch. 4; thus, LG would be refracted with the angle of the horizontal refraction, in the amount that the density of the medium requires. And let LGH be the correct angle. Now let GHI be set up equal to it; that is, let another line be tangent to the circle at H, and let this be HI, cutting the axis AF at I. I say that this is the boundary of the radiations meeting the axis; that is, that I projects no further ray upon the sphere which after the refractions meets the axis; much less, if the point which is taken is nearer to A than is I. But if a point is taken farther from A than I is from it, some radiations upon the sphere certainly do meet the axis. First, because IH, GL are tangent to the circle, IH is accordingly the horizontal radiation, and there cannot go forth from I to the sphere a line that is more inclined. And because GL is parallel to FA, this last radiation of the point I will therefore not meet the axis. Now let another line be drawn out from I to a nearer point of the circumference (let it be C), and by the converse of 10 of this chapter,[53] let the straight line CB cut the sphere at C in such a way that the angles of refraction in both places, or their supplements CBK, BCF, accommodated to this standard of measure of the inclination CBA,[54] make BK go forth parallel to FAD. For such an arrangement

---

[51] Here Kepler refers to a long endnote, which has been included below as an appendix to this proposition.

[52] Here Kepler refers to the following endnote.

To 188. The axis is always taken as in prop. 8, to mean the straight line drawn through the center and the radiating point (in prop. 17 and 18 below, through the point of the eye also).

[53] In Prop. 10, we are given a point, a ray is chosen arbitrarily, and the axis is constructed. Here, the axis and the point are given, and the final direction of the ray (BK) must be parallel to the axis. These conditions will determine the position of the line CB.

[54] The radius BA is the normal to the surface of the sphere, while CB is the path of the ray. Thus angle CBA is the angle of incidence or the angle of the refracted ray (depending upon which direction the light is considered to be travelling). This angle

is possible at any point. Let there be, I say, such an arrangement. Therefore, by 9 preceding, CF will cut DF at F, more remote than HI, for the reason that we have supposed C to be nearer the perpendicular FA than H, whence B was also nearer then G. Therefore, from two points in air, which are F and I, there **189**
exist two radiations FC and IC to the same point of the aqueous sphere, with the result that, by 12 preceding, a cutting of the rays occurs at C. And IC, which was inside, after the refraction and the cutting at C, becomes outside. Moreover, since CB, BK have just ceased to meet FAD on that side, much less will IC meet after making two refractions.

It has therefore been demonstrated that no radiation made from I meets IAD.

But also, no meeting radiation will proceed from any point nearer than I.[55] For again, that which is farther in than IC, or any whatever from I, after making an intersection at C, will stray farther out than IC. And since the latter does not meet, much less will the former meet.

Now let some point be taken that is higher than I. And let it be O. I say that some radiations from O do in fact meet the axis. For again, let the ray OH be drawn to the point H, which is the boundary of the meeting point of the radiation from any point nearer than itself. Therefore, an intersection occurs at H. And since $IHGL$ is the boundary of the meeting point from I, and GL is parallel to AD, when OH is made to go farther in, it will by all means be inclined, and as a consequence will meet the axis. This will happen to all radiations from O from the radiation tangent to the sphere to that radiation which is drawn from O following the rules of 10 above. However, those which fall within CN from the point F do not meet beyond the globe. Or, what is the same thing, no parallel between KB and DA can meet with AF this side of F. Hence it appears that Witelo X 43 is not universal.[56] Which he indeed does not conceal in concluding as follows: "But there is great variety in these, which we leave to the study of the **190**
curious." But these studies languished after your time, Witelo, and those who are trying to stir them up are nearly wasting their time. For I fear that you, and now I as well, have used up more time and effort in plowing up, sowing, and carving out these propositions, than everyone else together are going to use up in reading them.

*Kepler's endnote to this proposition:

[The note begins near the top of p. 440 of the 1604 edition.]

To Prop. 13 p. 188–189. For the sake of clarity, let me place before the eyes all the steps of this procedure (even though the proposition says a certain amount

---

is equal to BCA, and so the refractions are equal at C and B, as Kepler noted in Proposition 8.

55 Here Kepler refers to the following endnote.

To 189. The words "nearer" and "farther" are with respect to the globe.

56 Witelo X 43 (*Thesaurus* II pp. 440–1) argues that the image of something seen through a diaphanous sphere "is seen as being in the shape of a ring, much greater than the seen object."

about some things, while 16, following, speaks of the rest), and all the more so in that the demonstration on p. 189 line 30 itself also meanders rather long.

There are two points on the axis AN extended, one I and the other a little above F, which divide the radiations into three classes. The first is that of the points between I and N, the second that of the points between I and the other point slightly above F. The third is that of all the remaining points above that point at F along the infinite line.

The conditions of the first and third are contrary. For the radiations from the line IN not only do not meet with AD, but all, without exception, also diverge from it. But the radiations above the sign at F on to infinity all (without exception) meet the axis AD.

So the conditions of the bounds of the first and third class are again contraries. For the radiations from I no less than those from the line IN all diverge from the axis, except, obviously, for the last IH ending at a lesser circle of the globe, at which circle those radiations are tangent to the globe. For these are the first and the only ones for IN that neither diverge nor meet, but become parallel to the axis AD, as GL going forth from I.

Further, the radiations from the sign at F all meet the axis, all the way to the innermost FN which coincides with the axis. All the rest above F do this same thing, but this is the first among them, of which all the rays meet the axis.

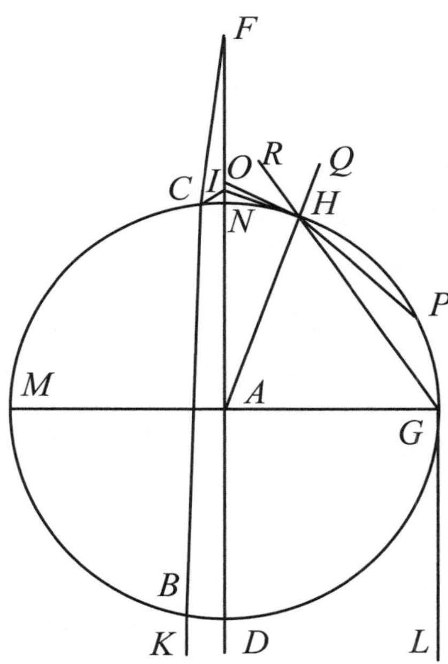

But the condition of the second class, that is, of points between I and F, is intermediate. For the radiations of (for example) point O, which are close to the axis OA, diverge from the axis, all the way to some point of the surface at which the laws of proposition 10 are expressed. For at the point, or rather, at the circle of the sphere about the pole N, the radiations become parallel to the axis. And from that point all the way to the one at which the radiation from the point O is tangent to the sphere, all the
441 interposed radiations coming out from O meet the axis. And thus there is a broad limb or belt in the globe, situated uniformly about the axis, at whose points all the incident radiations from O meet; within that limb the surface is like a cap or like the frigid zone, or the polar land on the sphere of the world, the radiations of whose points do not meet the axis.

All the radiations of the first class remain within the cap NH, whose edges form the circle H that divides the belt from the cap, like the arctic circle; and they have this circle, lacking breadth, in place of a belt. The radiations of the second kind now obtain a belt, broader in proportion to their remoteness from I, and they advance the boundaries of that belt on the near and far sides of H. For the boundary of that belt to which the point F belongs (which point is a little lower than the boundary of the second and third class, as has just been said), closest to N, is most distant from C in the place where a line drawn from F is tangent to the sphere CM.

The last boundary of the intersections of the parallels, or the point above F, is the first that leaves no cap about N, but makes a certain sort of cap of the belt itself.

Therefore, the points of the third class, from the sign above F through the infinite line from this improperly named "belt" more and more make a hemisphere, to the extent that if there were given a point of infinite distance, its radiations would be tangent to the sphere at the opposite points M, G, and would be completing the hemisphere.

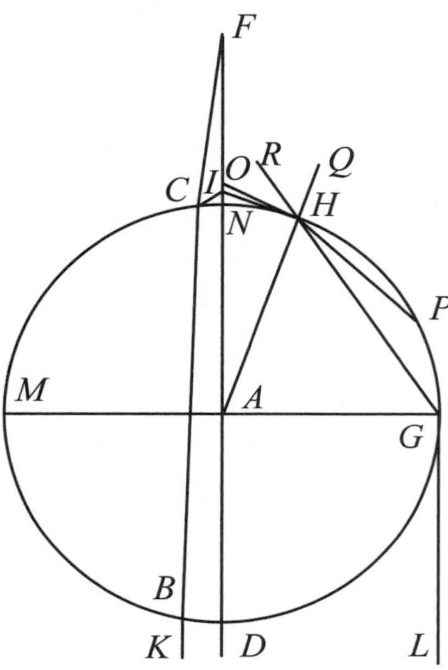

Omitted here is the demonstration that the radiations from O are still going to meet the axis when they shall have fallen upon points more distant from N than is H, in which proposition 10 cannot hold while AN remains the axis. But it is easily demonstrated, and this is done through prop. 9 and the present diagram. For because the most distant parallel LG comes through to H under refraction, all the remaining parallels, such as KB, fall upon a point closer to the axis or pole N than is H, in order that as a result at C none of the parallels will fall beyond H towards G or M. If, therefore, one of them falls at a point beyond H, it will not be parallel to AD, diverging much less from AD in the direction of D, for all those fall within NH. It therefore remains that they meet with AD on the side of D. And that was to be demonstrated. Use **442**
this same demonstration also for the full demonstration of prop. 16 following.

[end of Kepler's note]

Proposition 14. Problem

In an aqueous globe, to determine the places of intersections of any radiations parallel to the axis. In the diagram of Prop. 8, just now repeated, since LG is tangent to the sphere at G, the horizontal refraction is $36\frac{1}{2}^\circ$ by chapter 4 prop. 8, with the result that the complement AGH is $53\frac{1}{2}^\circ$. But twice this AGH, or MGH at the surface, is MAH, and as a result, its measure, the arc MCH, is 107°. And the supplement, the arc HG, is 73°, while the excess NH over the quadrant MH is 17°.[57] And because HI is tangent to the sphere, IA is the

[57] So the Latin reads. But NH is actually the excess of MH over the quadrant.

secant of the arc NH, or 104,569, and NI is 4569, about the twentieth part of the semidiameter AN.

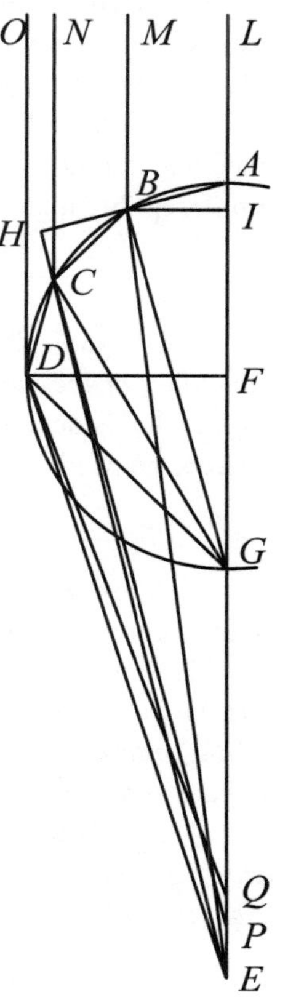

Now let KB be the incidence, or the inclination of 10 degrees, the angle of refraction from Witelo being 2° 15′. Thus since BAD is 10°, CBA will be 7° 45′.[58] And ACB is the same amount. Therefore, the arc BMC is 164° 30′, and when BAD is added, 174° 30′. The remainder, CN, is 5° 30′. And in fact if you should double the refraction of 2° 15′, CFN would be set at 4° 30′. And CNF is nearly right. Therefore, as the sine of the angle CFN is to 5° 30′, or its subtense, 9,596, so is the sine of FCN, 85° 30′, to NF. As a result, just as CFN is slightly less than the arc CN, so the semidiameter AN comes out slightly less than the distance of the cutting NF, which is shown by an inclination of 10 degrees.

Proposition 15

The last limit[59] *of the cuttings made by parallels with the axis is not greatly distant from the cutting of the radiation that maintains an obliquity of 10 degrees.* I have despaired of defining geometrically the exact point at which the last intersection occurs; I beg you, reader: help me here.[60] It certainly cannot be infinitely distant. For by 8 of this chapter, I have proved that all rays parallel to the axis meet the axis after a pair of refractions. But the reason why these points of meeting are not scattered farther, but are gathered at the end, and in fact a near

Tell at which points, and you will be the great Apollonius to me.

191 union [is made] of most of the radiations close to the axis, I shall prove in some manner from Proposition 9 preceding. For first, as was demonstrated in Chapter 4 Prop. 8, it is quite imperceptible that the refractions

---

58 Here Kepler refers to the following endnote.

To 190. For BA is perpendicular, CB is refracted, the declination from the perpendicular of the part of it that is set up in the air is 10°, the refraction 2° 15′. When this arc is subtracted, the remainder is the inclination of the refracted ray, 7° 45′. Thus in isosceles triangle CAB, the sum of the angles at the legs is 15° 30′, which, subtracted from 180°, leaves CAB, at the center A, 164° 30′. And since KB is inclined by 10° to the perpendicular AB, DA, which is parallel to it, will be inclined to the same by the same amount, with the result that BAD is also 10°. Further, CFN is generated through the doubling of the refraction, because KB is refracted twice by 2° 15′ towards the parallel DA, once at B and again at C. It is demonstrated by parallels to the axis drawn through B, C.

59 As will be seen, there are two "limits," outer and inner. The inner limit, point I, has already been found (in Prop. 13). Here, Kepler is seeking the outer or "last" one.

60 Like the marginal note to Ch. 4 prop. 6 (p. 126), this note alludes to Vergil, Eclogue 3, l. 104: "*Dic quibus in terris/Et eris mihi magnus Apollo*" (Tell in which lands, and you will be the great Apollo to me).

are proportional to the incidences near the perpendicular at a small inclination. As a result, the refraction of one more remote from the axis (which is, however, close in itself) is imperceptibly greater than the refraction of a nearer one, with respect to its own angle of incidence. But in the Lemma to prop. 9, CEB is also imperceptibly smaller than BEA. Therefore, in the case of great nearness to the axis, the refracted rays are very closely represented by the lines AE, BE, meeting at almost the same point, and CP (closest to the axis AE) will cut AE imperceptibly higher than BE, so that P, E are very nearly the same point.

Proposition 16

All the radiations of a point that is more distant from the sphere than the last boundary of the intersections meet with the axis, after making a pair of refractions. This is demonstrated in the manner of 13 above, and from the preceding; there is no need of words.

*Hence arise these corollaries**

1. *When the radiating point is infinitely distant, the boundaries of the intersections are nearest both to the globe and to each other.*

2. *When the radiating point has approached the nearly last boundary*[61] *of the intersections* (determined by parallels, through 15 preceding), *the boundaries of the intersections made by these new nonparallel radiations of the point*[62] *depart as far as possible both from the globe and from each other.*

3. *Parallels measuring out a globe of a denser medium are refracted on both sides, and meet with the axis, with boundaries of intersections at equal distances from the globe on both sides.*

*Kepler's endnote to these corollaries:

[The note begins in the middle of p. 442 of the 1604 edition.]

To 191 Prop. 16, first corollary. If you desire a demonstration of the corollary, see the diagram to Prop. 13. Let the radiating point on AD towards D be infinitely distant within the limits of sense perception. It will therefore irradiate approximately the whole hemisphere MBG, with, moreover, radiations KB, LG, approximately parallel, by the postulate of the second chapter. Therefore, by 13 of this chapter, I is the nearest possible intersection of all to the globe.

But the other boundary of the intersections, F, is also nearest of all.[63] For direct your attention to the upper part of the diagram, and suppose the radiating point to approach from an infinite distance: its rays will ever and ever all meet with the axis, until it gets close to F, by 13. Now from F certain ones no longer

---

[61] This is the boundary established in Prop. 15, where the obliquity of FC to the refracting surface of the globe is 10°.

[62] The nearest boundary is the point where the ray from F tangent to the globe intersects the axis; the farthest boundary is where the ray from F nearly parallel to the axis intersects the axis.

[63] That is, F is the nearest boundary of all the outer intersections of the rays with the axis.

meet. And since they become parallels from cutting lines, the intersection runs out to infinity, before it devolves into parallels. Therefore, when the radiating point approaches the globe all the way to the closeness of F, the last intersection runs off to infinity. But before, when the radiating point was infinitely distant, [64] the last intersection above F was distant by a little more than the semidiameter of the globe, no farther. [65] Therefore, since the causes continually increase from the infinite distance of the radiating [point] to the shortest distance, the effect, or the closeness of the last intersection, also ought to grow continually from the shortest
443 to the infinite, but without interruption. Thus, therefore, it will in the beginning be closest of all to the globe, which was to be demonstrated.

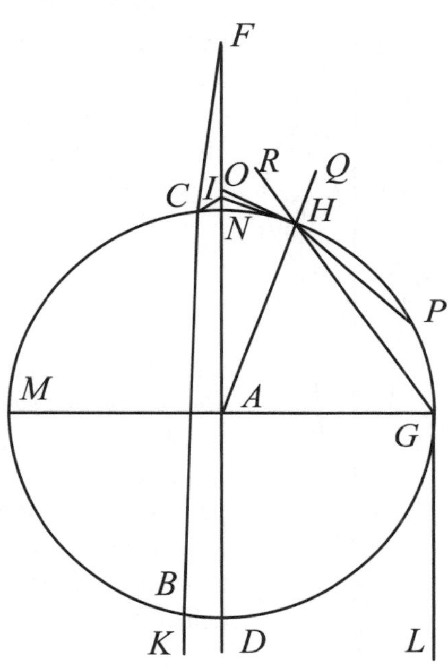

The second corollary has just partly been stated, as far as concerns the last boundary of the intersections. The rest is perceived most clearly in the diagram of p. 192. For the more α approaches, the more ν retreats. And in the diagram on p. 189, when the radiating point at the boundary is placed at F, all the radiations meet; when it descends below F, only the outer ones meet, namely, those which by their intersection mark the innermost boundary beyond the globe. Finally, when the radiating point is placed at I, the last rays also now cease to intersect, and the intersection retreats to infinity; but if the radiating point be infinitely distant on the other side of the globe, the innermost boundary is close to the globe at I. So again, when the radiating point comes forth from the shortest distance to an infinite one, the nearer boundary here shrinks from an infinite distance to the shortest.

Further, it cannot happen that as the causes continually increase, the effect increases for a while and then again decreases. Therefore, it should be least of all at the beginning, and greatest at the end. Accordingly, when the radiating point is placed near F, there is indeed some innermost boundary, and it is far away from the globe, but is farther yet, indeed infinitely distant, from the last boundary,[66] which is now not a boundary, since it does not limit because the intersections are

---

[64] We have now returned to imagining the radiating point to be below the globe, at an infinite distance.

[65] See Prop. 15.

[66] In radiations coming from a point in this region, the outermost rays all intersect the axis, while the innermost rays do not. Between these two classes is the ray (or circle of rays) which, after the two refractions, are parallel to the axis.

running off to infinity. The meaning of the corollary is not, however, that when the radiating point is at F the innermost boundary is more distant than when the radiating point is at I. For at I both the innermost and the last boundaries go off to infinity; at F, when the last boundary is now vanishing, the innermost nonetheless remains at a determined distance. The corollary therefore compares the point F here not with points F, I, but with those radiations of the third class above itself.[67]

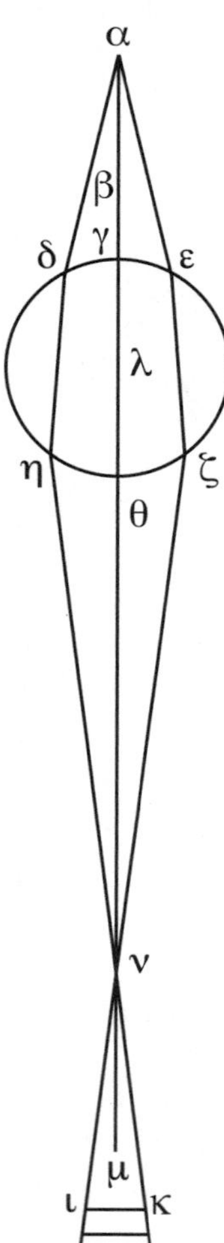

The third corollary comes from the preceding, and can be stated ocularly in the diagram of p. 192 [i.e., the diagram to prop. 17]. For if $\delta\eta$ and $\varepsilon\zeta$ are parallel, $\lambda\alpha$, $\lambda\nu$ will be equal, because the bases $\delta\varepsilon$, $\eta\zeta$, of the isosceles triangles $\delta\alpha\varepsilon$, $\eta\nu\zeta$ are equal, as are the angles α, ν, because of the same refractions at δ, ε, and η, ζ. Then, as a consequence, $\lambda\gamma$ will also be equal to $\lambda\theta$ (if γ, θ are the nearest boundaries), and $\lambda\beta$, $\lambda\mu$ will be equal (if β, μ are the last boundaries), and finally, $\gamma\beta$, $\theta\mu$ will be equal.

[End of Kepler's note]

Proposition 17

If the eye be removed from the globe beyond the boundary of the intersections of Proposition 15, whatever is placed
after the globe beyond the last boundary of the intersections, **192**
which these nonparallel radiations of the eye make, its image appears in an inverted situation on the surface of the globe. Let $\delta\varepsilon\zeta\eta$ be the globe with center λ, axis $\alpha\lambda\theta$, upon which β, γ are boundaries of the intersections of parallels, θ, μ[68] are boundaries of the intersections, not of parallels, but of those radiations that come from α, and let α be the eye, $\iota\kappa$ the visible object, the former more distant than $\beta\gamma$; the latter, than $\nu\mu$. I say that $\iota\kappa$ is going to be seen inverted on the surface $\delta\varepsilon$. For since the object is beyond the intersection, no rays will come from α to the ends $\iota\kappa$ which are not intersected between θ, μ. Let the ones that include the ends intersect each other at ν, and let them be $\alpha\delta\eta\nu\kappa$, $\alpha\varepsilon\zeta\nu\iota$. Now, by 6 above, ι will be considered to be attached to ε, and κ at δ. And thus the ends appear with their situations exchanged. Further, because the

[67] The radiations of the third class are those that originate from a point above F. All of them meet the axis, some (those tangent to the globe, or nearly so) rather near the globe, and others (those closer to the axis) farther away.

[68] Here Kepler refers to the following endnote:

To p. 191 Prop. 17. The subject matter itself leads me to distinguish between the intersections of nonparallels and the intersections of parallels, which I name from Prop. 15, or the boundaries of the two. Therefore, carefully take note of it. The boundaries of the intersections of parallels always have the same situation with respect to the globe; the boundaries of nonparallels wander this side and the other side all the way to infinity.

parts which are interposed between ικ are also removed beyond μ, they will be seen by rays intermediate between αδ, αε; and, by 9 above, will intersect each other beyond ν, but inside the boundary μ. Intersecting each other thus, they also invert all the interior parts. As a result, the whole object appears in an inverted situation. [69]

*Corollaries**

1. *Hence it is evident that if the eye is also at a greater distance than γβ, and if nevertheless ικ be between θμ, it is going to be seen partly in an inverted situation (namely, at the edges), partly (and in the midparts) in an erect situation. And the same is also going to be seen as circular, according to Prop. 43 of the tenth book of Witelo, and thus confusedly.*[70]

2. *If ικ be also within the innermost boundary of the intersection θ, the eye remaining the same, the whole object will appear erect.*

3. *But if the eye be between βγ, the intersections running off to infinity and certain radiations from the eye being refracted parallel, and if the object were then situated on the axis, and less than the distance of the parallels, it will appear erect and inverted at the same time, if, indeed, it be more distant than the nearest intersection; if, on the other hand, it be nearer, it will appear erect only.*

193 4. *Further, if it should exceed the complex of parallels, beyond the boundary the whole will appear inverted, the middle erect, and partly circular.*

5. *Finally, the eye and the object standing on the near side of the boundaries of the intersections, the former that of the parallels, the latter that of the radiations of the eye, the object will appear erect and of the greatest quantity.*

* Kepler's endnote to the corollaries

To Corollary 1, p. 192. Whatever is perceived by radiations from the cap (which we introduced in the notes to Prop. 13) will appear erect; whatever is perceived through the edge or the belt will appear inverted. Whatever is perceived through the circle which is the boundary of [i. e., between] the edge and the cap will appear circular (that is, one point—the pupil of the eye, for example—drawn out into a circular border.

To Corollary 3. Geometrically it is the same whether the eye radiates to the object, or the latter to the eye: in both cases the interposed globe does the same thing. However, when the eye is located between βγ, the radiations are divided into polar or cap-shaped parts: the circle and the belt. The radiations of the belt, such as αδην, αεζν, cut each other between θμ. An object placed here is therefore set in their way, and it flows into the eye inverted, if it be between νμ; but at the same time erect, because it also encounters the polar radiations that do not cut each other. And the whole will indeed be seen erect, because it is smaller

---

[69] Here Kepler refers to the following endnote:

To p. 192 l. 20. Supply the following corollary immediately here.

[70] *Thesaurus* II pp. 440–1. A linear segment of the axis αμ is refracted into a ring on the surface of the sphere, and thus the parts are disordered in the image. Witelo presents both a mathematical and an experimental account of the phenomenon.

than as to encounter the radiations of the circle. However, the images differ in situation. For as it is considered great when erect, so beyond the globe it will be considered at a distance, and will inhere in the actual convex surface of the globe.

Corollary 4 is when the visible object is so great that it also encounters the radiations of the circle; let it be evaluated by 1 and 3.

Proposition 18

With everything the same as in Prop. 17 preceding, the middle parts of the object will appear greater than according to their proportion, and curved. For the radiations from α nearest the axis fall with a very slight obliquity, contained within a large angle, and then suddenly fall more obliquely. Accordingly, in the middle for much space the refraction increases little; at the sides suddenly, over a narrow part of the surface. And to the extent that the refractions increase, the lines splay out and thus comprehend more of the seen object, and they set it upon a narrow part of the surface at the side. *Further*, the objects are considered *curved*, because they appear to inhere in a surface which is curved.

Definition

Since hitherto an Image has been a Being of the reason, now let the figures of objects that really exist on paper or upon another surface be called pictures.[71]

Proposition 19

It has been said what occurs to the eye in the neighborhood of an aqueous sphere. I shall now show how exactly the opposite happens on paper. For first, it is deduced from what has been said, *why, if the paper is almost touching the*

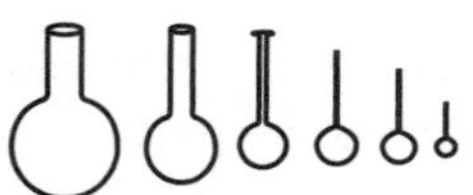

globe, the figure of the globe (if it be a urinary vessel with a long neck) is portrayed upon the paper with a very bright edge by the sun casting its rays upon the sphere. For whatever rays are within this envelope have not yet been intersected by the axis. Since the paper touches the globe, all those here are ones that have been able to pass through the globe. *The edge is then brightest* because the rays are gathered together to the greatest degree, and intersect each other circularly (but not yet on the axis), as appears in the adjacent figure at α, β, γ, δ, as the reasoning of the preceding demonstrations affirms. Now where the paper is removed from the glass by the twentieth part of the diameter (if it be the case that the illumination come from the heavens), certain rays now begin to intersect each other with the axis at the middle of the figure, at the letter ω; hence, *the middle of the figure also becomes bright. The figure diminishes, however, with the removal of the paper,* not because the cone of the same lines moves rapidly into a top-shape, for it does not decrease proportionally, but suddenly **194**
at the beginning, slowly at the end. The true cause, however, is the succession of rays. For after intersecting with the axis and dispersing, those which had

[71] On the revolutionary implications of this transformation, see Svetlana Alpers, *The Art of Describing* (Chicago: University of Chicago Press, 1983), Ch. 2, "Ut Pictura, Ita Visio: Kepler's Model of the Eye and the Nature of Picturing in the North," pp. 26–71.

previously formed a limit at the exterior points α, β, γ, δ, always move into a position below the interior ones at ε, ζ, η, θ. If that cone of refracted and mutually intersecting rays were to stand whole in the air, it would then represent a figure generated from an arc one end of which is rotated circularly while the other remains fixed on high, the arc tending inwards. For the cone would be slender in the middle, and quite acute. When the paper comes to the point of the cone ψ, *the illumination is strongest, so much so that gunpowder*[72] *in cold water is ignited when the sun is intensely hot.* For by 15 above, the intersection, not indeed of all, but nevertheless of many rays occurs at one center. The rest, τφχυ, already intersected, are scattered round about, and *give form to certain rays tending to a single center.* When the paper is removed beyond the last boundary ψ, *the figure quite vanishes,* all of *the rays having intersected.* From this it is understood how great the force of ignition will be if all the rays that can pass through the globe should come together.

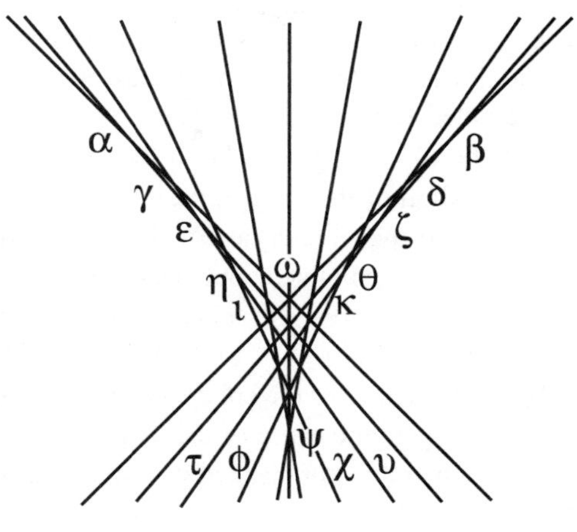

Proposition 20

Through a globe of a denser medium, any point more remote than the intersections of parallels strongly depicts itself upon paper, located at the last boundary of the intersection of its radiations, not before and not after this point; and the picture comprising all the points is seen inverted. The cause of its strongly depicting itself is that the globe gathers into one point many rays close to the per-
195 pendicular coming from the radiating point, by what was demonstrated above. Further, the cause of this not happening at the boundary of the nearest intersection ω is that the rays αω, βω falling from the sides,weakly, because they are in a great refraction, make the nearest intersection ω; while the rest of the rays (of the same point shining through the globe), which have not yet intersected, but are nonetheless gathered together and bright, also occupy the place of intersection ω on the paper, and so obliterate those gathered at the intersection ω. Finally, the cones coming from other points of the luminous object are still more to the sides, such as ι, κ here, and one occupies part of the place of another, and give birth to confusion.

In contrast, at the last intersection ψ, many rays from the same luminous point, standing close by the strongest perpendicular ψω, come together into a single point by 15 above, and because the cones are thinned out into needles,

---

72 *Pulvis pyrius.* Cf. Du Cange, *Glossarium* VI p. 568 and Partington, *A History of Greek Fire and Gunpowder* pp. 170 and 162.

they occupy no foreign region, with the result that individual points are imaged separately and distinctly. Only certain rays of other points on the visible object come down, but they are now intersected and weaker, as if another cone were to be bounded at υ. Consequently, the vertex of the new cone and the intersected line $\eta\upsilon$ of the old cone would be in the same place, and if there is any confusion in the formation of this image, these give it birth. The image, however, is inverted, not because the boundaries of the intersections have crossed before the glass, as previously explained, but because any point of the visible object, radiating perpendicularly upon the sphere, makes an image of itself beyond the globe with a very strong cone, whence it happens that the axes of the cones $\psi\omega$ intersect each other at the center of the sphere without having undergone refraction themselves. This is the genuine cause of the inversion of the picture that results from a denser sphere, when it is not masked by something solid, with an opening left in it.

Proposition 21

Suitable for mechanical and optical devices.

The place of the picture of nearby things is farther from the glass; and of distant things, nearer. This is evident through Corollary 2 of Prop. 16 and by 20, above. Some inventor will take note of this, like many other things in this book. You can also prove it in the daytime, by lighting a candle, which will appear to burn upon the paper, if the candle be properly located. Above, from Porta, I said that this was true, in this matter of the picture; not, however, in the matter of the Image, nor in air, but upon the wall.[73]

Proposition 22

Practical geometers will take note.

This picture retains the proportion of its distance. For it is traced out above all by the unrefracted perpendicular rays coming together in the center, by 20 **196**
above. This is evident from experience. When the candle is moved closer and the place of the picture recedes, the picture grows, and vice versa.

Proposition 23

We accordingly seek some light from Method. One form of refractions of the globe is that by which the sense of vision is deceived into imagining for itself likenesses that in reality do not exist (we have called them "images"), concerning which prior form, see Prop. 7 and Prop. 17–18. Again, another form of refractions, by which are formed real pictures of objects; and these are of two kinds: some are depicted by the bare

When a tablet, and an open slit, is placed opposite a globe between the boundaries of the sections of the parallels, and the slit is narrower than the globe, a picture of the greatest part of the hemisphere is projected upon a paper which is placed beyond the globe at the boundary of the last intersection of the rays of the luminous point. The picture is inverted, but is most complete and distinct in the middle. The variety of these things is so great, and constantly something new, that unless one is most attentive he is easily confused. So much so that I too was stuck for a very long time during which I persuade myself that the account of the different things was the same. About center *A* let *BC* be a globe of water, opposite which let there be an opaque tablet *DB* open in a small slit *EF*, narrower than the globe. Let the visible object be *HI*; the paper at *K*, where the last intersection of the radiations from *H* is. If the tablet were absent, by 20 above, *I* would form an image of itself with the last intersection of its radiations

[73] See Prop. 5, p. 193, above.

at *L*, and the point *H* at the point *K*. But they would project the lateral intersected rays of the cones, the one in the places of intersections of the other. Now, with the tablet interposed, no more rays flow down from *H* to the sphere than the amount that can flow through *EF*, and the amount, approximately, that flows together at *K*,[74] and that cone, as if by a kind of reckoning, is diminished, so that no intersected rays can be cast upon *L* or no confusion could appear there. In turn, the cone of radiations from *I* that is to be bounded at *L* is quite castrated by the opposition of the tablet *DE*, in its noblest part, that is, the radiation through the center *A*, and at its brightest apex *L* is too far from *K* for anything it scatters there to create confusion. Further, there is left to the radiations from *I* no more than what enters through *EF*. That portion, however, whatever radiations it consists of, successively intersects itself through the breadth *MN*, by 19 above, and now, after the intersection, passes through *AL* and falls upon the paper placed at *K*, nearer than is *L*, and not at a point, but spread out, because the intersection has already occurred at *MN*. For that reason, the picture is dark and confused at the sides. If you should move the paper nearer at the side, the sides would indeed be more correctly pictured, but never in every detail, because the intersections are spread out, not only in the depth *MN* (which hardly matters), but also in breadth.

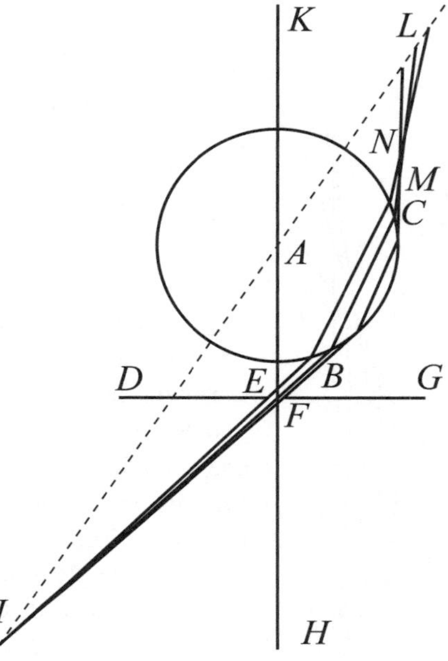

globe, on which see Prop. 19, 20, 21, 22; some by a globe covered except for a narrow slit, on which see Prop. 23. For the organization of the refractions is different. However, those things in 8 to 16 are common, concerning refractions of a globe.

197 And since the intersection of *IE* and *IF* with *HK* occurs at *ER*, but the rays are bounded[75] at *MN* before they cross *AL*, by 19 above, the right rays *I* become the left *MN*, and it is impossible for them to come out on the right through a new cutting.

The slit *EF*, moreover, must be narrow, lest it fall short of its purpose if it be made wider; and must be close to the globe, lest it radiate too little and indeed too confusedly from the hemisphere within.

Corollary

From this, the use of the opening of the uvea[76] in the eye is in part apparent;

---

[74] Here Kepler refers to the following endnote:

To 196. The amount that elsewhere too, without the obstacle of a perforated tablet, is accumulated very near the boundary *K* of the last intersection, by 15 above.

[75] Here Kepler refers to the following endnote:

To 197. They are bounded by the opposition of the paper at *KN*, the place that the rays run together.

[76] That is, the pupil.

likewise, the purpose for which the sides of the retina are moved closer to the
198 crystalline than is the fundus.

Proposition 24

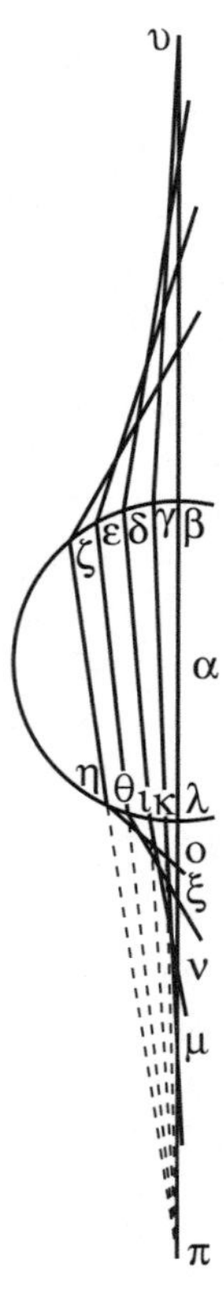

Rays converging[77] *towards some single point within a denser medium are gathered by the hyperbolic conoidal surface bounding the medium to one single point, closer than the former point.* About center α let βζ be a globe of a denser medium, in which βλ, γκ, δι, εθ, ζη converge at π. And let βαπ be the axis. Therefore, by the above, ζη will be refracted into ηο, and εθ into θξ, and δι into ιν, and γκ into κμ. I wish all rays to gather together in a single point. I will therefore need there to be a greater refraction at ικ, so that κμ, ιν come together with απ at a closer distance, and I will need there to be a lesser refraction at θ, η, so that ηο, θξ come together with απ farther away, and thus οξνμ should become a single point. Moreover, the refraction will be greater at κ if γκ falls more obliquely, and less at η if ζη falls more directly, upon the surface. But because γκ, ζη stay the same, the surface ηθικλ accordingly needs to be changed so that it not be so inclined to ζη at η, and so that it be more inclined to γκ at κ. These two things, however, are accomplished, not by one or another circle, but by conic sections. For any of them is able to cut some circle at 4 points. See Apollonius Book 4 Prop. 25. Consequently, twice in this semicircle, twice in the other one. When it thus cuts at η so that it enters the circle there, it will be presented more directly to ηζ, and when it cuts the same circle at κ, exiting again, it will be presented to γκ more obliquely. Further, of the conic sections, only the hyperbola or some line very close to it, is the measure of refractions, as was shown in Sect. 5 of chapter 4.[78] Indeed, this very thing was demonstrated there: that the surface making all the rays outside the denser medium parallel is a conoid that does not differ from hyperbolic.

Corollary

*Hence is evident nature's plan concerning the posterior surface of the crystalline humor in the eye. That is, she wished to gather all the radiations of any visible object entering the opening of the uvea into a single point of the retina, in order both that the point of the picture might be all the more evident, and that the
rest of the points of the picture might not be confused by extraneous rays, whether 199
stray or gathered together.*

It is also evident that nothing is sought in the widening of the opening of the uvea, other than that which I said above, nor that the picture be confused thereby, namely, that it only become brighter.

---

77 Here Kepler refers to the following endnote:

To 198. I use the words "converge," "diverge," and so on, for the sake of brevity, of straight lines which, when produced in all directions, all converge at the same point.

78 See above, pp. 120–123.

Proposition 25

The manner in which rays converging to the same point are led to the right point has now been explained. However, we must take pains to treat how this too happens: that *the rays which are within the denser medium converge to the same point, which nevertheless originally had flowed from the same point.* As, if in the previous diagram the rays $\lambda\beta$, $\kappa\gamma$, $\iota\delta$, $\theta\varepsilon$, $\eta\zeta$, coming from the same point π, were to be gathered at the same point υ. Moreover, since, as has often been said, there is no difference of refractions whether we say the rays go in or come out of the denser medium, the demonstration will be the same, with this sole exception, that the surface which is going to be gathering the rays on the outside that are diverging within the medium, should be more acute than the one we have investigated in Ch. 4 Sect. 5, because the latter gathers parallels, that is, neither converging nor diverging. On the contrary, therefore, that surface which is going to gather more quickly on the outside the rays that are [already] converging and inclining towards each other within the denser medium, will be more obtuse that that of Chapter 4. Besides, in such a narrow space these surfaces cannot be distinguished from each other (especially if they are near the bottom or keel of the hyperbola or ellipse), nor does it even have any perceptible effect if neither of them departs from perfect roundness. Therefore, even if there is also another conoid in the eye, it is necessary that this be in the cornea; but the surface of the Crystalline must be perfectly round, so that it might take in the incoming converging rays to the same point directly, and it is right that this roundness be somewhat depressed, so that its center shall come out to be far behind the eye.

Proposition 26

The thumb in contact with the chamber of the eye appears greater than is correct, and confused. For because vision occurs through a picture, when the picture is disturbed, vision is disturbed. But by Corollary 2 of 16 above, when the radiating point approaches (as the thumb does here), the intersection of its radiations recedes. Therefore, when the thumb approaches the eye, the intersection (which is the paintbrush of this picture, by 20 preceding) betakes itself right into the head, and the retina cuts the cone inside the vertex. Therefore,
200 the coarse paintbrush draws outlines that are not clear and distinct, and broadens the edges, drawing a surface instead of a line.

Proposition 27

When a viewable object is located beyond the point that is given to any eye by nature, from which all radiations might be gathered into a point of the retina, that object appears confused. This is evident through contrary reasoning. For by Coroll. 2 of 16 above, the point to which the radiations of a distant object and to which those of a nearby object are gathered, is not the same. However, in one person, the place of the retina with respect to the humors is always one and the same. And it was not useful to obtain for the person the most distinct vision of the most distant objects, if as a result present objects should have escaped him. But neither ought he to see distincly only the closest things, so that distant things should be excessively confused for him. Thus there was need of a balance.

Further, since it has its breadth, it varies with individuals. And it may be that the position of the crystalline with respect to the retina is the same in all, but the density of the humors is not the same. Therefore, when a point of the visible object is exceedingly remote, its radiating cone is ended before it reaches the retina, by the above corollary. And thus when the intersection occurs, it strikes upon the retina as it is now once again expanding itself. Hence it is that those who labor under this defect see a doubled or tripled objects in place of a single narrow and far distant one. Hence, in place of a single moon, ten or more are presented to me.[79] And indeed, in order that confusion also follow from this cutting of the cone, the tenderness of vision should be added, which is moved powerfully by strong radiations both at the point of the retina where the vision is distinct, and round about, through the breadth of the cone of rays (arising as a result of a defect). As a result, the cones of celestial radiation are spread out for everyone, but not everyone sees confusedly as a result.

Proposition 28

Those who see distant things distinctly, nearby things confusedly, are helped by convex lenses. On the other hand, those who see distant things confusedly, nearby things distinctly, are aided by concave lenses. It is a matter of great wonder, the use, so widespread, of so great a thing, whose cause is nonetheless hitherto unknown, so that, upon discovering the clearest demonstrations, I proclaim
it falteringly. One Baptista Porta professed to have given an account in optics, **201**
which I have hitherto sought in vain from booksellers.[80] More than once, I tormented myself in order that I might find the cause, in vain, while the means of vision lay hidden from me. And with greatest deserving I make most honest mention of Mr. Ludwig L. B. von Dietrichstein, prince of my Maecenases,[81] who has for three years now kept me busy with this question.[82] Accept at last, O man most fond of letters, arts, and professors, and also of me among the last, and on top of that most wise (let the Princes of Austria, and the most ample Provinces praise the

---

[79] This condition is now known as polyopia monophthalmica, and results from an irregular astigmatism. Kepler had nearly died of smallpox as an infant in 1575, and this visual defect was evidently a result of the disease. Cf. Caspar, *Kepler*, p. 36. Kepler was also nearsighted, as his remark in Prop. 28 shows.

[80] J. B. Porta makes this claim in *Magia naturalis*, Book 17 ch. 10 (ed. 1589 p. 269). Porta had earlier published a work *De refractione, optices parte, libri IX* (Naples, 1583), which Kepler evidently was not able to find.

[81] Patrons. Maecenas was Horace's principal patron, praised in the invocation of his first Ode (Carmen 1.1.1–2).

[82] None of the surviving letters to Kepler from Baron Ludwig von Dietrichstein, whom Kepler knew in Graz and who was his daughter Susanna's godfather, touches upon optics. The remark is thus probably to be taken as a grateful compliment, though Kepler recalls a conversation below. Von Dietrichstein remarked in his letter of thanks for the gift of the *Optics* (20 October, 1604, in *JKGW* XV, no. 295, pp. 56–7.), "That you thought of my person in your little book, now completed and in print, leads me to conceive friendly thoughts, although I can easily perceive that it has as its origin your bearing good affection for me rather than my person in reality."

rest [of your qualities])—accept, I say, the reply: if it be not sufficiently clear and indubitable, it is surely late enough! The cause is not located in this, that nearby things are rendered larger by convex mirrors, which was indeed once the only answer I had. Porta, having also followed this line, says in vain, "nearby things are made greater, but confused, by convex [glasses]."[83] In fact, they are made greater for all, but are not confused for all, just as in turn the concave glasses do represent objects smaller for all, but, for those who view distant things rightly, confused, and for those who view them confusedly, distinct. If it were the case that the greater things are, the more distinctly they are perceived, no one ought to have been be helped by concave glasses. For they diminish objects. You, Most Noble Sir, recognize your valid objection. Others locate the cause of corrected vision in the multiplicity of refractions alone. In vain: for all would be helped equally, and no one would complain that glasses blur things. Thus we should not look to entire objects, but to individual points of objects, as we always have hitherto. Therefore, those endowed with a point of distinct vision that is rather distant, when they use convex glasses, they alter the cone of radiations of the same nearby point, so as to appear to arrive as if from from a distance, and to enter the eye. The cone is thus corrected so as to be bounded at the retina, which, unless it be set right by lenses, what we have described above in Prop. 26 happens: because of the nearness of the radiating point, the cone, which is to be bounded behind the retina, is cut by the retina, and thus the cones of the radiating points, having been allotted a certain amount of breadth, each create confusion for the other.

On the other hand, those endowed by nature with a point or distance of distinct vision that is rather short and close, by using concave glasses, alter the cone of radiations coming from the same distant point, so that it may seem as if
202 it originates and enters the eye from nearby. For unless they use lenses, what is described in Prop. 27 will happen to them: that is, the cone of such a remote point will be bounded before it reaches the retina, and, proceeding farther, will again spread itself out, this falling upon the retina with breadth, and the cones will disturb and confuse each other. Experience gives me remarkable confirmation. I know two men, not of humble station, one of whom reads the tiniest fine print, but brings his eye so close that he cannot use both eyes at once. The same man cannot reach with clear vision as far as ten paces, beholding nothing but clouds. Nevertheless, lenses of deep concavity help him to see more distant things, although these same lenses completely confuse my vision, even though I also use ones that are concave but more moderate. The other man, who died some time ago, was nearly blind for nearby things, but was a Lynceus[84] for distant things, so much so that he prided himself in being able to count, on a

[83] Porta, *Magia naturalis* XVII 10, ed. 1589 p. 269. Kepler alters Porta's words somewhat: Porta wrote, *convexo propinqua* [*vides*] *maiora sed, turbida,...*,while Kepler wrote, *convexis propinqua maiora reddi, sed confusa.*

[84] Lynceus was an Argonaut, famous for his keen sight. Cf. Horace, *Sermones*, 1.2.90; Ovid, *Fasti*, 5.709 *et seq.* However, another possible meaning is "lynx-eyed," as the lynx was said to have the sharpest vision of all quadrupeds—cf. Pauly-Wissowa XIII.2 col. 2469, and Pliny, *Natural History* XXVII 122. By Kepler's time, this had become less of a literary allusion than a common noun.

house at a distance of several stadia,[85] the new roof tiles mixed in with the old. Using convex glasses, and a paper unfolded at a distance, as far as could be done with the arm, he used to read not badly. Anyone who brings less trust to my experiences should heed Aristotle.[86] In Section 31, Problems 8, 15, 16, 25, he seeks precisely this, what it is that makes the nearsighted and the elderly suffer opposite afflictions, the former bringing nearer, that latter moving away, those things that they are going to look at; and the latter, though being weak sighted, nonetheless write with the tiniest of letters.

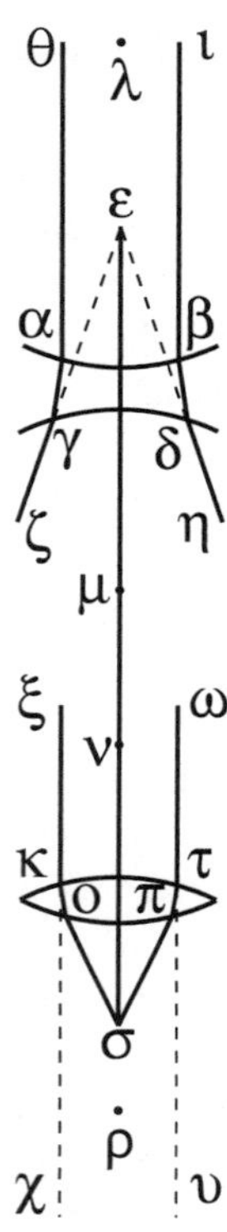

I also add diagrams, so that nothing may be lacking. Let αβγδ be a concave glass, and let the luminous point be so distant that the radiations are approximately parallel: let, I say, the radiations θα, ιβ come from the same luminous point. Therefore, because αβ is the concave surface of the denser medium, whose center λ is between θ and ι, therefore, θα will be refracted towards the perpendicular λα extended, and hence into αγ towards the outside, and ιβ will be refracted towards λβ extended, and again towards the outside into βδ, and thus θα, ιβ, previously parallel, now diverge. And because
γδ is the concave[87] surface of the denser medium, or the convex 203
surface of the rarer, with center μ, therefore αγ will be refracted from γμ into γζ, and βδ from δμ into δη, and therefore γζ and δη will diverge more, to the extent that, extended in the imagination, they would meet at ε. And so the radiations of the most remote point will flow into the eye or cornea ζη as if they were coming from the nearby point ε. What now happens within the eye itself on the retina through the approach (or virtual approach) of the radiating point, has been explained above, especially in Prop. 27 preceding. Again, let κτοπ be a convex glass, let point σ radiate, let the radiations be σο, σπ. Because these strike upon a convex surface, they come together towards the perpendiculars ον, πν, and are refracted into οκ, πτ; and these, coming out into the concave surface κτ of the rarer medium, flee from the perpendiculars ρκ, ρτ extended, are accordingly refracted into κξ, τω, and thus encounter the cornea at ξω, as if they had come in from a point of the greatest distance along χξ, υω, very nearly parallel. And so again, what is corrected by this fictitious removal of the point σ in the coming together of the radiations at the retina, see above, especially in Prop. 26 preceding.

And this completes the demonstration of those things that were said above, in Sect. 2, of the crystalline.

---

[85] The Roman *stadium* was an eighth of a mile, equivalent to the English furlong.

[86] Aristotle, *Problems*, Section 31. Problem 8 is at 958a 35–b 3, Problem 15 at 958b 34–959a 2, Problem 16 is at 959a 3–8, and Problem 25 is at 959b 37–960a 8.

[87] Reading *cava* for *causa*.

4. Consideration of those things that the optical writers and the anatomists have said concerning the means of vision

We shall not enumerate the details, but rather the most important points, nor shall we survey the opinions of everyone, lest the reader be confused. Witelo, and from him Jessenius, when they wove the spider's web, and the Crystalline in it, from the retiform tunic and the optic nerve, attribute the foremost parts of perceiving to this humor.[88] For he says that perception or vision occurs when the rays from the visible objects pass through the tunics, but particularly the web of the crystalline, and when the image is imprinted upon the crystalline to such a degree that it begins to inhere: to this end, that humor is made rather thick. But he says that in order that distinct vision occur, the rays should be perpendicular to the anterior surface of the crystalline. For the perpendiculars are ordered as the points on the object are ordered. For unless vision happen by the perpendiculars alone, the vision is going to be confused, just as perpendicular radiations, too, are confused by oblique ones. And nonetheless, most powerfully convinced by experience, he also attributes something to the oblique radiations: it seems that there is more than the amount that enters the eye perpendicularly from
204 the hemisphere. Now Witelo says that the vitreous humor is the nutriment of the crystalline, while the aqueous humor is the waste matter of the same, perhaps following the Physicians, for Fernel says the same thing from the same sources. That they are pellucid: the latter, so that the image might enter; the former, so that the image, now sensed by the crystalline, might pass through the vitreous and through the substance of spirit to reach the confluence of the nerve, and there be recognized by the common faculty or sense. And so he attributes flatness to the crystalline from its posterior part, and denies that the rays of objects should meet, lest the right parts be exchanged with the left in the replica. And above all, he thinks that the perfect image of the object seen passes all the way through to the confluence of the nerve.

This entire opinion (although it prides itself in Aristotle's being the author, who says that "the seeing power of the eye" is "made of water," in *On Sense and Sensibles*, ch. 2)[89] is knocked down when the crystalline is cut off from the nerve and from the retina, and joined with the uvea, as was shown above from Platter. What as a consequence is to be thought of this introduced form of nutrition and excretion, let the physicians consider. Surely, if none of the senses is more marvellous and subtle than vision, these things are unworthy of being declared about vision. Then, moreover, he carries the sense of light over to the tunics, on the basis of touch. If it were by the sensing of heat, and if vision were not accomplished in an instant, I would believe it. But now light and colors, which are by no means heat producing, and enter into the eye in a moment, are far too subtle to be perceived by a corporeal tunic, under the principle of touch. See what we have argued in Chapter 1 on the immateriate motion of light. And so it is not the tunic, and not even the nerve, but spirit,

---

[88] Witelo's account of vision and the eye is in *Optics* III 4, *Thesaurus* pp. 85–87.

[89] Aristotle, *On Sense and Sensibles*, 438b 19–20. Kepler quotes Aristotle in Greek.

or indeed something perhaps more divine, which receives and becomes fully conscious of light, which faculty I said above could not be investigated by Optics. For in that same place, this too is refuted, that images inhere or are impressed upon humors.

Now, as for his simulacrum which Witelo forms using perpendiculars alone, he distinguishes with tremendous subtlety between perpendiculars, and those very close to them. If light acts upon sense, and sense is affected in this action, it follows that it will itself also be affected more severely according to the strength of the action. And indeed, there is hardly any difference in illuminating between the perpendiculars and those very close to them, because of the latter being refracted very little. Therefore, the effect, i.e., the sensing of perpendiculars and of rays close to them, is nearly equal. And thus perception **205**
is confused, Witelo's labors being in vain. But no need for more words. It has been demonstrated by most certain arguments and experiments that the picture of the object or the hemisphere is situated at the concave wall of the retiform, with all confusion of rays removed. Next, it is certain that the rays of many points of the observable hemisphere come together at a single point of the crystalline. Finally, it is evident from sense that the apparition [*idolum*] or image of a contrasting object appears in the black of the eye. Thus, according to Aristotle, *On Sensibles*, Democritus of old said that vision occurs through this image or apparition.[90] Witelo said that vision occurs through the confused illumination of the crystalline.[91] I say that vision occurs through this acknowledged and irrefutably demonstrated picture. All three suppositions are certain: if it is uncertain to you by which of these vision is accomplished, follow any of them through, but pay attention to the arguments. Aristotle did indeed refute Democritus, noting that the apparition is not a "that," some object, but a mirroring (ἔμφασιν), or image, as in catoptrics, not a picture; and that it is not a reception of action by the eye in which it is thought to exist, but that by an eye placed opposite this eye, seeing itself in it.[92] Similarly, I refute Witelo by the very confusion of the rays. For his statement that oblique rays are also seen insofar as the oblique radiations flow together with the perpendiculars, implies that oblique and direct radiation is received by the same point.[93] Therefore, two objects will be considered to be in a single place. However, what confirms me is the most universal line of argument, used by Witelo himself. The effect of vision follows upon the action of illumination, in manner and proportion. But the retina is illuminated distinctly, point by point, by the individual points of the objects, and most strongly through the individual points. Therefore, it is at the retina, not

[90] Aristotle, *On Sense and Sensibles* II, 438a 5 ff.

[91] Witelo designates the crystalline (which he usually calls *glacialis*) as the sensitive part of the eye in *Optics* III Prop. 4, in *Thesaurus* II p. 87. His description of vision occurring by discrimination of the perpendiculars is in *Optics* III Prop. 17, *Thesaurus* II p. 92, where the confusion of rays is also discussed.

[92] Cf. Aristotle, *On Sense and Sensibles* II, 438a 6.

[93] This is also in II 17, *Thesaurus* II p. 92.

elsewhere, that the most distinct and most evident vision can take place. This is all the more so in that the defects of vision also follow from the perturbed arrangement [*ratio*] of the picture, just as has been demonstrated. And I do not know whether Democritus would discuss this picture under his name of "image" [*idolum*], rather than that mirrorlike "reflection" [ἔμφασιν]. It often befell the ancients that they were undeservedly refuted by Aristotle under a fabricated contriving of opinion.

As for the inversion of my picture, that might be raised against me in objection, which Witelo avoided with great care: he first attributed flatness to the
206 crystalline, contrary to obvious experience, in order to maintain that opinion. And since this opinion affirms that that surface is bulging, by the testimony of Witelo there occurs an inversion of the likeness. And I, for my part, tied myself in knots for the longest time, trying to show that left cones that are made of the right ones in entering the opening of the uvea, are again cut beyond the crystalline in the middle of the vitreous humor, and that another inversion occurs, and that the parts that were made to be left again became right before they reach the retina. Nor was there an end of this useless anxiety until I hit upon Prop. 11 and 12 among those preceding, by which this opinion was most evidently refuted. And even if I had upheld what was proposed, there was still to be a remaining complaint: the hemisphere was going to be reversed. For those who stand facing us on the outside, judging these parts to be right, those left, have images directly opposite whose right parts will be taken to be left, as is to be seen in mirrors. For the eye which to you is right becomes the left one for your image. I shall say nothing about how the picture's concavity was verging inwards towards the head, while the concavity of the object was verging in the opposite direction.

And so, if you are disturbed by the inversion of this picture, and fear that inverted vision also follows from this inversion, I beg you to ponder thus. Just as vision is not an action because of illumination's being an action, but is an effect contrary to an action, so also, in order that the places correspond, the recipients of the action must be directly opposite the acting things. Further, places are perfectly opposite when the same center forms the midpoint in all the lines of the oppositions, which was not going to occur if the picture had been erect. And so, in the inverted picture, even if from a universal perspective and with respect to some common line the right parts are transformed into the left, nevertheless the right parts of the object are perfectly opposed to the right parts of the picture, and the upper parts of the object to the upper parts of the picture (each in relation to itself), as well as the concave to the concave. Nor is there any fear that the sense of vision might err about the region. For when it perceives an elevated object it clearly turns the eyes upwards, acknowledging them to be low in position and opposite the object, with respect to place. Rather, it would have been in error had the picture been erect. For then the interior wall of the eye would in some places be directly opposite the object, in others not, as from the sides; for it would diverge from opposition. Therefore, no absurdity is committed by the inversion of the picture, which Witelo to such a degree avoided. Jessenius too
207 followed this opinion, considering that refraction at the cornea has the effect of preventing right from becoming left, which is clearly refuted in Prop. 12 of the

preceding. Finally, what is to be said of the image's inhering in the crystalline (which is itself also Aristotle's opinion, who chose, for vision, "water, because of its being more easily confined and made determinate" than air,[94] because it receives and preserves better than air), is evident from the conclusion of ch. 1. However, concerning the penetration of the same image through the extremely narrow way or concavity of the nerve to the branching of the two nerves, enough has been said just above in Sect. 3 of this chapter.

Platter is much better on the function of the crystalline, although even he did not quite hit the mark. He says that vision happens through the agency of the retiform tunic, but that the crystalline accomplishes for the retiform what convex glasses accomplish for those suffering from night blindness: it makes objects appear larger. Absolutely, Platter, the crystalline does do something like this, but it doesn't do exactly this. First, because the last boundary of the intersections of the crystalline[95] is in the proximity of the retina, it was said in proposition 18 above what is going to happen regarding an entire eye applied there. Namely, an entire eye put by itself in the place of the retina would see no more than one point, and that in great confusion and throughout nearly the whole surface of the crystalline. If objects were going to appear with greatest distinctness to an entire eye located there, the same thing would not as a result also follow in part of the eye. The cause is this, that in the entire eye there is not only a center, in which the rays of all luminous objects come together (or are so gathered as if they had come together), but also a surface, upon which the picture is distinctly unfolded. This does not, however, take place at the retina, which is a part of the eye. For if it be said that those distinct and larger objects are represented by the crystalline on the retiform, which objects project rays from all their points to the same point of the retina (to grant this point for now, in itself false), by what procedure will that point of the retina gather the distinct parts of the visible object, when the point is in accordance with what was supposed above? Further, unless you bring in the object here, so that whatever of the likenesses is represented through the whole surface of the retiform, is represented larger by the crystalline, you have now quite destroyed the analogy transferred from the entire eye and external mirrors to the retina. For many visible objects illuminate the eye in breadth, in such a way that they nevertheless both come together, as it were, in the same center, and, having made an
intersection (or something like it), are received by the same single point on the **208**
retina. Then, as a consequence, convex lenses can bring it about that a greater surface of the retina be taken up by any object you please. But between the crystalline and the retina there is no point of meeting of the rays coming from

[94] Aristotle, *On Sense and Sensibles* II, 438a 15. Kepler has ἐνυποληπτότερον, "more determinate," where the standard Greek text has εὐπιλητότερον, "more dense." Evidently there is a textual problem here: W. S. Hett, for his translation (Loeb Classical Library, 1957) p. 224, substitutes ἐπιληπτότερον, "more easily caught."

[95] That is, of the rays passing through the crystalline.

different points of the visible object, by 12 above, as a result of which the portion of the retina illuminated is made smaller, rather than larger, because of the crystalline. See 23 above, and those things that were said in Section 2 on the means of vision.[96] Finally, this is not [the same as] the use of convex mirrors to illuminate objects by enlargement, as you have seen in Proposition 28 above: indeed, the vision would be false if, in making the picture, the object were to occupy a greater quantity on the retina than is correct. For as the picture is, so is the vision.

In this opinion, Platter appears to have brought in the anatomical example which I have heard from other physicians, namely, if the crystalline humor be shelled out from all the other humors and placed upon very tiny letters, it displays them magnified. This is in fact extraneous to this enquiry. For vision occurs through the mediation of the picture on the retina. And this fallacy occurs not through the picture but through accounts of the Image. Therefore, this magnification of the letters by the crystalline (or a certain analogue to it in the eye) does not give shape to vision.

Now compare the true mode of vision proposed by me with that of Platter: you will see that the illustrious gentleman was not farther from the truth than befits one of the medical profession, who is not deliberately treating mathematics.

Cornelius Gemma, a physician and philosopher of profound hunting skill, in Book 2 p. 120 of the *Cosmocritica*, compiles and jumbles his account of the means of vision from nearly all the respected authors.[97] In this work you may see the ambiguity of words compete with the obscurity of the subject, both here and in practically the entire book, in that he prefers to rhetorize and waffle outright in a grave style concerning matters replete with majesty in the highest degree, rather than to argue, nor did he deign to instruct any but a serious and most attentive reader. First, he takes this from Witelo and from the optical writers: vision is accomplished by a cone whose base is in the visible object, and whose vertex is in the center of the crystalline. Next, this from the philosophy of the ancients, with rules that are not optical, but words that are: another cone gets in the way of that
209 exterior cone, with vertex at the center of the crystalline, and base set up within. The former is the cone of light, the latter, that of spirit or of the visual faculty; by the former, images flow in; by the latter, they are received, weak images by a large and broad cone, strong ones by a narrow one. But the riddles of this author, to the extent that they have something in common with other opinions, which I refute, I refer to those. As for the rest, since they roll away by means of a slipperiness of words, no matter how you might press, things that ought to

Does spirit clothe a shape? Can the cone exist? And for what purpose?

---

96 See p. 182, above.

97 *Ars cosmocritica* is an alternative short title for Gemma's *De naturae divinis characterismis*, Antwerp 1575. Cornelius Gemma (1534–1579) was the son of Rainer Gemma Frisius, mentioned in Ch. 2 above. He was Professor of Medicine at the University of Louvain. The passage cited by Kepler is certainly not on page 120, which is entirely concerned with the new star of 1572; Kepler may be thinking of pp. 127–8.

be done have been disregarded, primarily because we have surely found the true way.

Finally comes Joannes Baptista Porta, whom I proposed that I ought to consider in this sequence. In the *Magia naturalis* Book 17 chapter 6 he presents the first contrivance of that thing of which I provided a formal demonstration in chapter two above: by what means do all things which are illuminated outside by the sun appear in darkness with their colors?[98] Afterwards, having presented a few agreeable devices, in the course of concluding the passage, he now added these few words on the means of vision:

> Hence it is clear to philosophers and opticians in what place vision occurs, and the question of intromission that was so thoroughly ventilated in antiquity is pulled apart, nor will it ever be able to be demonstrated by any other device. The likeness is sent in through the pupil, as [through] the opening of a window, and a portion of the crystalline sphere located in the middle of the eye maintains the function of a tablet, which I know is going to be acceptable in the highest degree to those of intelligence.

Indeed, thou hast blessed us, O excellent initiate of nature! —now that that quarrel has been removed, whether vision occurs by reception or emission. For, what many have wondered at everywhere or in passing, they had either ignored when seen, or, when it was studied more carefully, they preferred to show it privately, through jealousy and to the greatest injury to divine glory and harm to the literary Republic; or, though it was considered, they were unable to give a treatment of it proportionate to its dignity (like basket weavers in Schwabia, hidden away in rather large barrels, with a little orifice as an opening). You alone have both noticed it and pursued it thoroughly, and have made it public in a worthy manner, so that, in proportion to the knowledge and love of the mysteries of nature which is in you, you might accommodate it to a most useful treatise. And so, as far as I am concerned, you have a grateful admirer and publicist of your name; as for the rest, I hope for the same things as you. I also do not think that the reception of images in the sense of vision can be more rightly confirmed in any other way, and the emission of rays refuted (for which see Macrobius's *Saturnalia*, Book 7); which Witelo examined in Book 3 Prop. 5, and Aristotle, in the book *On Sense* ch. 2.[99] Nor do I think that there is going to be anyone who is going to raise even the slightest further doubt, provided that the professors of ***210***
physical subjects shake off fatal torpor, and deign to recognize even these few things. Indeed, the Philosophers will learn how to philosophize from this discovery of yours concerning light, colors, and the transparent, more rightly than from Aristotle. To Aristotle, in the book *On Sense*, Empedocles is absurd, because he said that χρώματα are ἀποῤῥοίας.[100] For he thinks it incongruous that vision

[98] Porta, *Magia naturalis*, Naples 1589, p. 267.

[99] Macrobius, *Saturnalia*, VII 14, trs. Davies pp. 502–6; Witelo, *Optics* III 5, in *Thesaurus* II p. 87; Aristotle, *On Sense and Sensibles* II, 437a 19–439a 5.

[100] Colors are effluvia. Cf. 437b 23–438a 5.

should occur when the eye is contacted by an efflux made by colors. But let him into this camera obscura of yours: he will see the wall touched; why can the eye not therefore be touched? The motion of the medium which he introduces, caused by the sensible object, can also not be understood itself, and is far more incongruous than that emission. Not to mention that "diaphanous" is defined by him as if it were, as they say, a positive word for something necessary for vision, while it is a privative word for something obstructing vision, namely, opacity: thus, the reason why something is diaphanous is not because it is illuminated by light, but because it lacks roughness of surface, color, and high density, regardless of the light. Rather, light, illuminating the pellucid, most often impedes vision. But these are covered above in the notes to Chapter 1.[101] Meanwhile, your words, Porta, on the means of vision should be considered somewhat more accurately. You say, "Hence it is clear in what place vision takes place," and afterwards, explaining, you say, "the likeness is sent in through the pupil," through the opening of the uvea, "like the opening of a window, and a portion of the sphere of the crystalline maintains the function of a tablet." Therefore, if I understand you well, if you were asked, "In what place does vision occur?" you will respond, "in the surface of the crystalline," as if on a tablet. For you will say that vision belongs to that class of picture which you have treated in this chapter 6 of yours, and I have demonstrated in ch. 2 above. It is clear that you are on the way, for I have used the same words in Section 2 above. However, I would not have said that it has made its way through to the destined place. In fact, if you have a fixed target here, if you do not descend below the crystalline, you have erred in your opinion. You will no doubt reply, "It is stated more fully in our *Optics*," which, though I have searched for it diligently, I have not had a chance to see. Meanwhile, with the help of conjectures, I fear that even in the *Optics* you will not have attained the true object.[102] For why is there no mention of refractions here, while you assigned chapter 10 to the affects of the crystalline lens,[103] chapter 11 to mirrors, chapter 13 to the crystalline ball: why do you make no mention whatever here of the means of vision? And it is in itself easy, upon seeing that device of yours in
211 your chapter 6, to immediately be seized by that persuasion, that vision is carried out exactly through this device.[104] I myself, although I duplicated the device, i.e., that one associated with my Prop. 23 above, which was not related by you (which confirms my suspicion, since it is impossible to come to know the means of vision without it), was very much of this persuasion, that vision is nothing other

---

[101] See the Appendix to Chapter 1 for Kepler's commentary on Aristotle's theory of vision.

[102] Kepler's conjecture is correct: Porta's account of vision is described by Lindberg as being "in most respects traditional" (*Theories of Vision* p. 183).

[103] Latin *lens*, literally, "lentil."

[104] Although Porta, and Leonardo da Vinci before him, had likened the eye to a camera obscura, neither had studied the camera as fully as Kepler did, and neither managed to follow the argument through to the discovery of the inverted image on the retina. Cf. Lindberg, *Theories of Vision*, Chapter 9, "Johannes Kepler and the Theory of the Retinal Image."

than intersection, but that the aqueous globe accomplishes precisely this with its refraction, that the intersection becomes purer which is not yet perfected right at the surface of the crystalline, owing to its nearness to the opening. For I believed that with the help of refractions, it happens that in so narrow a space of the eye there is represented an intersection so exquisite as can hardly exist in some large chamber without the use of refraction. For the rest, I am immediately refuted at the eye by the aqueous globe, which had only one point beyond it at which it would distinctly portray a portion of the hemisphere: before and after, not at all. It should, however, have portrayed more distinctly at distant points, if the intersection alone generated this picture. To say nothing of what was said in Section 2 above, that intersection by itself (without the assistance of gathering through refraction) would never carry through to the most distinct vision of objects smaller than the opening of the uvea. And so, to conclude, most ingenious Porta, if you were to add to your opinion this one thing, that the picture on the crystalline is still rather confused, especially when the opening of the uvea is dilated, and that vision happens, not by the joining together of light with the crystalline, but that it descends to the retina, and by that descent the radiations of diverse points are more separated, and also the radiations of the same point are joined together, and that the place of the gathering to a point is in the retina itself, which gathering brings about the manifestation of the picture; and that by that intersection it is brought about that the image becomes inverted, and by this gathering, that it is most distinct and most manifest: if, I say, you were to add this to your opinion, you would have completely settled the account of the means of vision.

5. Those things that recoil upon Astronomy from the means of vision; or, on flawed vision

Although the perfection of vision depends upon both eyes, astronomers nevertheless do not always use vision to distinguish between objects, but most often
use it to establish the angle between a pair of stars, in which procedure there **212**
is need of a single center at which the angle is evaluated. Therefore, they sequester one eye, imitating the artisans, who, when they are going to examine
Account of sighting. the straightness of a rule, use one eye.[105] The same is done by those who aim "scorpions" and bombards at a target. For all of these, the use of one eye is sufficient, though it be for a variety of reasons. Aristotle considered it worthy for a philosopher to make an inquiry into this subject, in Section 31 Problems 2 and 20.[106] But neither does the application of a single eye have absolutely no

---

105 On the word *collimare*, translated as "sighting" in the marginal note, the *Oxford English Dictionary* has the following note (under the entry "collimate", vol. 2 p. 627).

" . . . an erroneous reading, found in some edd. of Cicero, of L. *collīneare*, f. *col-*, *com-* together + *līnea* line, *līneare* to bring into a straight line. *Collīmāre* long passed as a genuine word, and was adopted by some astronomers who wrote in Latin (e.g. Kepler *Ad Vitellionem Paralipomena*, Frankfort 1604, p. 211; Littré) and thence passed into the mod. langs. The proper word would be *collineate*."

106 957b 5–8 and 959a 38–b 4, respectively.

need of precautions against the possibility of its leading us into error, in a twofold way.[107] For in the picture by which we have said that vision occurs, two regions of the eye are chiefly considered: the center of the eye, and the surface arranged around the center, receiving the picture. For it has been said that the picture is so arranged upon the retiform tunic, that straight lines projected from individual point upon it through approximately the same center of the retina[108] are going to fall upon points of the visible object corresponding to themselves, if they be produced.

What is the center of the angles of vision?

And so, when the distances of stars are to be taken using astronomical instruments, the more careful astronomers, as has been said, do not trust the eye. For they know that although the eye be tangent to the very center of the instrument (which, however, is accomplished with difficulty), it is nevertheless not tangent, except with respect to the surface, in which the lines, in fact, drawn from the two stars through the upper sights, do not come together. Let F, G be stars, BAC the instrument, with center A, DA the surface of the eye, E the center of the eye. Since, then, it is not from A but from the center of the eye E that the straight lines falling upon F, G, are to be imagined to go forth, therefore, when the sights B, C are so applied that EBF, ECG are in a straight line, the angle BAC will give a false measure of the distance, and will be greater than is correct, because it is within BEC upon the same base. And so the arc BC is greater than is correct, because the depth of the eye EA does not permit the centers of the instrument and the eye, A, E, to be joined together.

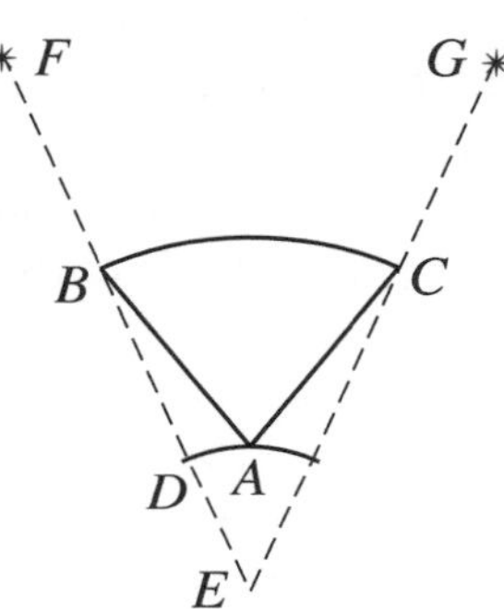

This is much more to be guarded against when very small quantities are called under the measure of an angle, such as when we seek the diameters of the luminaries; and this must be laid in the place of a foundation by all who enter into a reformation of astronomy.

Accordingly, Archimedes, in his book on the number of the sands, took this
213 precaution: First, by adopting a certain procedure, he investigated a quantity not less than the sight,[109] set a cylinder equal to it at the center of the instrument A, and drew lines from the sights B, C, tangent to this cylinder, and coming together: let them be BE, CE.[110] And so he set up angle BEC slightly smaller

---

[107] The two sources of error are, first, the breadth of the pupil and the effective position of the center of vision, discussed immediately below, and second, the apparent enlargement of bright objects, discussed afterwards on p. 232.

[108] The center within the eye, not the middle of the retina itself.

[109] That is, equal to or greater than the diameter of the pupil.

[110] Archimedes, *Sand-reckoner*, in *Archimedis opera omnia* II pp. 223 ff. Kepler was using Commandino's edition: *Archimedis opera non nulla* (Venice 1558). A summary (but not a translation) of the procedure is included in *The Works of Archimedes*, pp. 223–4.

than that angle by which the distance FG was perceived. But the angle BAC was slightly greater.

Further, the particular quantity that is not less than the sight is to be determined thus. Let AB be the surface of the cornea, and from the visible point E, let the extreme radiations which are admitted through the opening of the uvea, be EA, EB, which, refracted at A, B, meet at the point F of the retiform tunic, or close to its surface. Now let CD be a smaller quantity than that which passes through EA, EB, and closer to the eye AB than E. You see that even if CD is squarely presented to point E, nevertheless not all the radiations of E are turned aside, but they come in from the sides all about, and meet at F, the place of destination for all, by which rays, all gathered into one, vision of the point E takes place, but with an admixture of the vision of the object CD. For the right end D will image the left part N of the retina F; the left end C will image M, the right part of it. Consequently, the middle of the object CD will image F, which previously was also the image of E. Thus two objects will be seen in one place, but with this difference, that E will be thought to appear more distinctly, while CD will be like a shadow or a spider web. No wonder. For it was said above that radiations of the point C at the eye, when C is so nearby, are gathered far beyond F; therefore, they strike the retina F not yet having been gathered. And in just this way, CD will be less than the breadth of the vision. Archimedes, not much differently, sets the more distant cyliner GH equal to CD, and directs that it be white, for the sake of distinct vision. If the whole of the cylinder GH be seen, with no interference from the cylinder CD, he pronounces them to be much smaller than the vision; if, on the other hand, a certain part of GH be left out, he pronounces them a little smaller. If, on the contrary, all of GH be hidden (the two cylinders always remaining equal to each other), he states that
they are greater than the vision. There is a 214
little something lacking in this assertion. He did indeed define greater and less than the vision sufficiently, and required nothing else for his demonstration; but otherwise, the proportion cannot be established that the less CD is than the vision, the more of GH is seen. For if GH should approach, less of it will be seen than if it were to recede from CD, even though GH and CD remain equal. Not to mention that the breadth of the pupil determines the breadth of vision, which nevertheless is different at different times. This I have said only for the sake of information, so that the words of Archimedes and my opinion on vision may be compared with each other all the more rightly. As for the demonstration itself, that still remains for Archimedes.[111] For lines

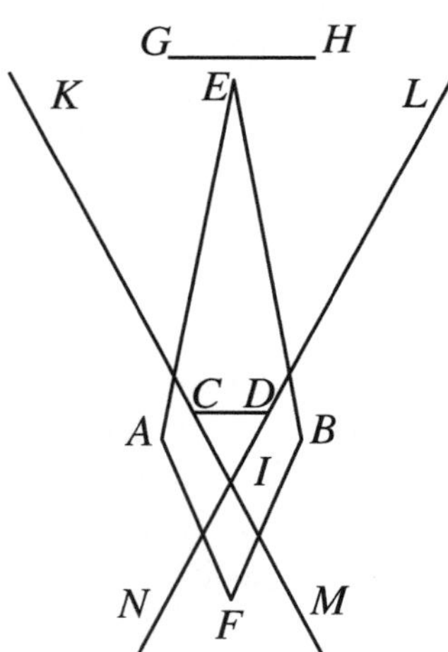

[111] Here Kepler refers to the following endnote:

drawn from the ends of the visible object, so that they are tangent to such a cylinder located at A (of the prior diagram), exactly equal to the vision, do not always coincide at the center of the vitreous humor. For it was said above that if straight lines should be drawn from the points of visible objects, they would pass through this center[112] and would be perpendicular to the retina. Accordingly, in the latter diagram, let I be the center of the [sphere of the] retina, and from C, D let CM, DN be drawn through I to the retina. M and N will be the approximate places of the picture of C, D, by what was said above, although C, D would make an image of themselves, clearly not by the lines CM, DN, but by ones that are refracted and different in place. Therefore, in order that the edges of the object seen, joined with the quantity CD by vision, might meet at the center I, it is fitting that its boundaries clearly belong at MK, NL, as if the object seen were KL, which, if it were smaller and more compact, Archimedes' lines would meet beyond I; if wider, before I. And so, as regards the diameter itself of the sun, in the measuring of which Archimedes was busy in this passage, it is too compact to be held in the span KIL or CID, with the consequence that KC, LD (if KL be the sun) meet beyond I, and the angle really is somewhat smaller than is correct.

However, that Rabbi Levi, whose words Commandino relates in comments on this passage of Archimedes, was also not negligent in investigating the depth of this point of meeting by means of the *Radius Astronomicus*: it comes out to

---

To p. 214 l. 10. Archimedes looked to this alone, that he search for an angle that was certainly less than that angle under which the sun is seen. Moreover, as regards the sun, he accomplished this with his demonstration. For the rays from opposite extremities of the sun, grazing it, have a quantity equal to the vision, and closest to it: they meet, not in the center of the vitreous humor, where the measure of the angle would be correct, but beyond it, where they really make a smaller angle, which is what Archimedes sought. Therefore, in line 13 an objection arises. For someone might say that these radiations of the extremities of the object sometimes strike upon the very center of the vitreous humor. The answer is, that this does not happen always, but only when the seen object is equal to the span of KL. And just as, when the seen object exceeds this span of KL, the meeting is before the center of the vitreous humor I (in which case the object would appear greater than is correct), so, if the object seen should be deficient from this quantity, as is the sun (the visible object, following what Archimedes proposed), then the meeting will be entirely beyond the center of the vitreous humor I. And therefore these three members are demonstrated in what follows, containing the elaboration of both the objection and the solution.

[112] Here Kepler refers to the following endnote:

To p. 214 l. 15. Read in the converse opinion on p. 212 l. 14, namely, so that straight lines may be drawn from the points of the visible to the points of the picture corresponding to them.

But you might say, "Who will measure that angle, whether it be in the center of the vitreous humor or outside it?" I answer that Archimedes taught it, as was indicated above on p. 213 l. 3. For he used another cylinder greater than one that would be equal to the vision, and placed it in such a way that (in the diagram of p. 212) it would belong between FE, GE; that is, at BC, and compared the distance of the cylinders with their quantity. See the passage itself.

be the very center of the optical globe, which he says he found himself, after "he sought it out deeply with the greatest labor."[113] In this connection, he marvellously confirms what we were saying above in Section 2. For the center of the sphere of the vitreous humor, which we marked out for the imaginary intersection (if lines were to be drawn), and about which is arranged the picture representing **215**
objects: that, I say, is simultaneously the center of the retina and the uvea and the white or scleroid tunic, and is therefore also the center of the whole ocular globe. For what that Rabbi added, that he had found that point at the center of the vision "that is at the center of the solidified moistness," he appears to have described from Alhazen, his tribesman, who "judges this to be fitting or fair," and Witelo, too, carried this over from the same source in Book 3 Prop. 7.[114] But anatomical evidence refutes this most clearly, asserting that the crystalline humor is in a position anterior to that of the center of the globe, and further, that its anterior surface has a curvature that is flattened, not protruding. In the rest, the Hebrew uses a good demonstration. For, supposing that KL and CD in the last diagram are any two quantities, the eye AB is made to approach so that when C, K are set upon one line, D, L are also in one line. For then, KC, LD being extended, a meeting occurs at I, a distance IA beyond the surface of the eye, which, compared to the globe of some eye, shows its semidiameter. For as the excess of KL over CD is to the distance between KL and CD, so is the whole KL to its distance from I, while KL is nevertheless found to be at a lesser distance from AB.

But our Tycho Brahe, at the beginning less experienced as yet in the handling of instruments, also encountered obstacles of this kind from the joint of the instrument itself. For, in the next to the last diagram, since the straight lines of the rules meet at A, and the material around A has some thickness and extends all the way to E, it was impossible not just for the center, but even for the surface of the eye to be at A; but it stands below at E, as he advises in the *Mechanical part* folium D4 and in the *Progymnasmata* fol. 341.[115] And while he supposed that, by the use of a table whose structure is found on fol. 342, he had entirely removed the observational errors that had arisen thence, he nevertheless allowed afterwards, on fol. 343, that the cheekbone kept him from placing the pupil of the eye near enough to the circumference of the head of the instrument.[116]

---

[113] *Archimedis opera non nulla a Federigo Commandino nuper in latinum conversa et commentariis illustrata.* Venice 1558. *Commentarius in librum de Arenae Numero,* fol. 60v–61r. The words quoted by Kepler both here and below, which he changed slightly from the original, are on fol. 61r. For Levi ben Gerson and his *radius*, see Bernard L. Goldstein, *The Astronomy of Levi ben Gerson* (New York: Springer Verlag, 1985), Ch. 6, pp. 51–54, and the commentary on pp.143–6.

[114] See the preceding footnote; also, Alhazen I 12, in *Thesaurus* I p. 6, and Witelo III 7, in *Thesaurus* II pp. 88–9.

[115] Tycho Brahe, *Astronomiae instauratae mechanica* (Wandsburg 1598), in *TBOO* V p. 82; *Progymnasmata* Ch. 4, in *TBOO* II pp. 334–5. Kepler evidently thought of the former book as the "mechanical part of astronomy," just as the latter part of Kepler's own *Optics* bears the title, "optical part of astronomy."

[116] *TBOO* II p. 336.

Thus far he does indeed assign a cause for the vision's still being aberrant. But while I would grant that the cheekbone also gets in the way, I nonetheless think that the chief blame lies in this, that "it is not from the pupil," which Tycho Brahe supposed, "that the visual ray chiefly proceeds," (whether you understand
216 that as the opening of the uvea or the cornea covering it, as he himself appears to mean), but from the middle point of the globe, to speak technically now of the visual ray with the ancient school of Euclid. Examining this same thing better, Tycho wrote to the Landgrave of Hesse in 1585, in this way (*Epistolae Astronomicae* Vol. 1 p. 8): "For it is not possible for the center of vision to be efficiently joined with the center of the instrument, nor does the pupil of the eye remain motionless in viewing a pair of stars, whence it turns out to be unavoidable that the center of vision also changes somewhat."[117] Up to this point he had considered the visual ray to come forth from the pupil as from a center, but when he saw that the distance of the surface of the eye from the center of the instrument did not suffice, he ascribed part of the aberration to the translatory motion of the pupil, which is plainly nothing else but this, that the center of vision lies beyond the pupil, and that when two lines are drawn from the center of vision, some portion of the surface of the eye is subtended; and the vision of the two stars cannot be direct unless the center of the pupil be carried over from one line to the other while the center of the eye stays the same.

Theory and types of dioptras

And so this quantity of the eye brings in some obstacles to accurate observations, which Hipparchus, intent on modifying the dioptras,[118] once introduced, so that, when a very narrow opening is applied to the eye, if the eye should look through it to the dioptra of either of the stars or the edge of the sun, these paired acts of vision would certainly cut each other in this opening, and the point of this cutting would not have to be sought in the breadth of the eye. And Ptolemy imitated this. Today, in place of the dioptra, they make a very subtle slit by setting up two sights nearly touching each other. But the authorities do not think this slit worthy of trust, especially at the shorter distances. Further, there is that obstacle noted by Tycho in dioptras viewing the stars, that if the openings be narrow, they occult the stars; if, on the other hand, they be wide, they do not provide the required precision in the small divisions. For that reason, some replace the more distant dioptra with a very tiny globe, supported by small handles in the middle of a perforated tablet, and line it up with the stars by covering them rather than looking at them. Others fix a pair of strings crosswise in a wider opening in the tablet, so that the intersection of the hairs might take the place of the dioptra or the little globe. However that may be, the little globes or the crosshairs disappear in the darkness, and when a candle is brought near, the viewing is impaired. Because of these impediments, in both very close and more distant dioptras, there

---

[117] *TBOO* VI p. 36. The letter is dated March 1, 1586, not 1585. Kepler changes Tycho's words somewhat, combining two sentences into one, omitting a clause, and changing the punctuation.

[118] Described by Heron, *Dioptra*, in *Herons von Alexandria Vermessungslehre und Dioptra*, ed. Hermann Schöne (Leipzig: Teubner, 1903) pp. 187 ff.; Ptolemy, *Almagest* V 14, trs. Toomer p. 252; and, more thoroughly by Proclus Diadochus, *Hypotyposes*, IV. 3 par. 72 and following. Toomer provides further references in note 52 on p. 252.

seemed to Tycho Brahe, after long experience, to be no more effective way of sighting than by two observers for two stars, sighting along a pair of straight lines directed to each star, and by the single eye, treacherous on so many counts, being ordered to be entirely unused. This procedure, first used by Tycho, was enthusiastically adopted by the Landgrave's staff, and to the great advantage of the observations: see *Epistolae astronomicae* Book 1 fol. 3.[119] Indeed, for that form of the Landgravian instrument, Tycho feared (on fol. 8) that it might be like his sextant as first outfitted, but on fol. 22 its accuracy is defended by the Landgrave. And on fol. 28 Rothmann relates to Tycho the authentic form of the other instruments, which is equivalent in effect to the Tychonic.[120] It differs in this alone, that in the Tychonic sextant each of the observers encloses the same cylinder in the center of the instrument between parallel slits in his own sight, while in the Landgravian, in the place of one cylinder, each encloses his own parallel sight somewhat removed from the center. And so Tycho, writing to the Landgrave on fol. 38, and to Rothmann on 61 and 62, greatly commends both the form and their industriousness.[121] So anyone who wants a more particular account of observing should consult these passages. See also Tycho's *Mechanica* folium H5.[122] And the center of vision apprises us of these things concerning astronomical observations. **217**

Various errors of vision regarding the light of the moon and the sun.

We need to take note much more attentively of this property of vision, why it is that, to all people without exception, all things that are luminous appear greater in proportion to things placed nearby that are less luminous. For in the first or last phase of the moon, the luminous horn appears to be enclosed by a far larger circle than the rest of the body, illuminated by the light of the earth and very clearly visible. The same in the lunar eclipe of 1603 May 15/25: certain observers were able to grasp visually the edge of the darkened part, although it remained by more than a third part. They therefore pronounced the circle of the darkened part to be narrower. On 1600 August 7/17 in the evening I saw the moon in conjunction with the heart of Scorpio[123] an hour before its setting. It was going in above the

---

119 *TBOO* VI p. 31. The letter states that the instruments had been improved under the direction of Paul Wittich; however, a letter from Tycho to Thaddeus Hagek of 14 March 1592 (*TBOO* VII p. 323) relates that Wittich had shown the Landgrave how Tycho had improved the sights. This letter is translated in Gingerich and Westman, *The Wittich Connection*, p. 58, and Edward Rosen, *Three Imperial Mathematicians* (New York: Abaris, 1986) pp. 269 and 271. Wittich's work on the sights is further described by Rothmann in his letter of 14 April 1586, *TBOO* VI pp. 55–6.

120 The passages referred to are in *TBOO* VI pp. 36, 49, and 55–6. From this last passage, it is evident that Wittich made an improvement which was not equivalent to the Tychonic arrangement, and the Landgrave's staff subsequently modified the sights to their "authentic form."

121 *TBOO* VI pp. 66 and 89–91.

122 Kepler is evidently referring to the *Supplementum de subdivisione et dioptris instrumentorum, TBOO* V pp. 153–5.

123 Antares.

heart on the northern side, so that about a third part of the division[124] projected above the heart, and while the division differed slightly from a straight line, the sun being in 25° Leo, nevertheless, the way across the luminous part appeared much wider at that place, than the distance of the star from the line of the section. This was at the common boundary of Styria and Hungary, the polar altitude being $47\frac{2}{5}^\circ$.[125]

218 Those who are weak of sight, and who are otherwise blind to distant things, imagine for themselves a rippling[126] series of ten phases in place of one phase. The same ones, seeing people at a distance with bright collars, do not recognize the people's faces, which, without this condition, are evident enough. In full moons, it is occasionally the experience, as may be seen in Tycho's observations,that when five or six people are observing the same moon, the estimation of the diameter is inclined to vary, ranging from 31 to 36 minutes, according to the vigor of each one's vision. This is, moreover, the chief controversy about the moon. On 1590 February 22 the moon was observed 22 times: twice at 31′, six times at 32′, seven times at 33′, six times at 34′, once at 36′. In the beginnings of lunar eclipses, the eclipse is noticed first of all by me, who am laboring under this defect, as well as the direction from which the darkness approaches, long before the beginning, while the others, who are of the most acute vision, are still in doubt, as happened in the month of May of this year 1603. For the rippling of the moon, mentioned above, stops for me when the moon is approaching the shadow, and is in great part removed from the sun's rays. In eclipses of the sun, the beginning is hidden for a long time, and suddenly some rather large part is seen to be missing, even to people with very sharp eyes. And the horns do not go off into a point, but are blunted or even cut off; and some people (as in December 1601) exclaim that they see precisely the shape of a horseshoe. The magnitude of the eclipse is always diminished in the eyes, as the light spreads itself out everywhere and, as is seen, enters in at the ends of the moon's edge. In fact, also evident is this proof of this dilation: if you apply the moon's body to the edge of an opaque ruler which is close to the eye, the edge, while it is an unbroken straight line, will seem to be diminished on that side when the moon is placed in between, as the moonlight effaces the image of the edge.

All these things, and whatever others there are, draw their origin from the retina tunic, but in a different respect. First, whatever of this affects those with defective vision finds its occasion from propositions 26 and 27 above. For the more distant bodies, such as the celestial bodies, gather the radiations from a single point, into a single point, before they touch upon the retiform, and, cutting

---

[124] τῆς τομῆς. Kepler uses the Greek to indicate that he means the line of division between the light and dark parts, for which the modern term is the "terminator."

[125] This was at the time when, together with the other Lutherans still in Graz, Kepler had been expelled. He spent several weeks in the area of what is now Burgenland before proceeding to Prague in September, at Tycho's invitation.

[126] Reading *crispatam* instead of *cristatam*. Later in this paragraph, Kepler refers back to the "rippling of the moon mentioned" (*dicta Lunae crispatio*), which I take to be a reference to this word.

each other at that point, they now strike spread out upon the retina. Thus, it is not a single point of the retina that is illuminated by a point of the object, but a small part of its surface that is illuminated by a point of the object, and thus it is encircled by many points: white things, however, and bright things illuminate its surface strongly. They therefore bring it about that those things 219
which are depicted less bright in the same place, where they themselves showed their own boundaries (they show them, however, either too far away, when the intersection of the rays takes place before the retina, or too near, as just now the opaque ruler did when cut by the radiant cone before the cone ends in a point; see prop. 26, 27, 28 of this chapter), become entirely invisible, and give way to the white things. And so nearly the same thing happens in the eye which, above in chapter 2, with regard to the configuration of the ray, I demonstrated to happen on a wall. However, that the image of the luminous body is not simply magnified, but is in a way multiplied, and that some one larger thing is mixed out of many distinct ones, this appears to be either because of the wrinkles in the uvea, which is dilated at night, when we are looking at the moon, and comes together into itself and into its wrinkles; or because of the gaps in the ciliary processes. For when the eyelids are shut and the brow wrinkled, many of such false images are cleaned away, but not everywhere, because by shutting the eyelids the eye is not equally covered everywhere, being left bare transversely as long as the eyelids are the least bit open.

On this account, therefore, the visual picture is corrupted, which is a necessary consequence of defective vision. If all things were equally bright, vision would be confused;[127] now, however, because the luminous things predominate, they are flawed in their quantity. And even if perchance the pictorial cone is spread out in all observers, nevertheless, not everyone uses so subtle a visual faculty as to perceive by means of all the radiations, but those alone who are affected by all radiations imagine for themselves that the luminous things are larger. Hence, those who have a conspicuous weakness of vision have double vision, not of luminous things only, but also of dark ones, if they be narrow and exceedingly distant.

Therefore, faced with the brightness of the moon or of Jupiter or Venus, the smaller heavenly bodies cannot be seen below a few degrees.

There is a kindred question in Aristotle, Section 31 Probl. 28:[128] why it is that light radiating into the eyes takes away the view of nearby objects, which returns when the light is diverted from the eyes. For, precisely as has been said up to this point, some parts of the retiform are illuminated very strongly, and thus the perception is very strong, while the radiation of the remaining parts, which correspond to nearby visibles, is in no proportion to the former, and as a result there is almost no perception. For just as the retiform's distinct picture is to distinct vision, so is strong illumination of the picture to strong perception,

[127] Here Kepler refers to the following endnote:

To p. 219. Confused vision and erroneous vision are different. Vision is confused when it sees two objects in the same place, as on p. 213 l. 21 above. It becomes erroneous place when the entire place is attributed to just one visible object alone, while the vision of another thing seen in the same place vanishes.

[128] *Problems* 960a 21–28.

in which the spirits are strongly affected. This probably looks more to another
220 defect of vision, which follows. Thus, another consideration, in which the retina enlarges the pictures of luminous things, seems to enter in beyond the laws of optics. It is indeed a commonplace among natural philosophers [physicos] that the rays of the sun are scattered by a bright color, gathered by a dark one. And in fact this does not appear to be false: if one places a white paper in the way of a globe of water or crystal at the point of burning or of intersection, it will appear to be widely illuminated, but with squinting eyes because of the brightness; if, on the other hand, the paper be black, a narrower surface will be illuminated. Who would not consequently conclude that the rays are scattered by the white, and come together by the black? Particularly if he sees that black things are for the most part ignited, not the white. But it does not follow from this that this is the nature of light and of white things. For it is impossible for any surface to have the power by its own color to make any ray strike upon it in a line that is different from that which its optical laws force it to strike with. For the cause of the apparently wider illumination of white things can be the extreme brightness, as will now be explained, and that the white areas lying around the point of burning are brightened by even a gentle ray, and are appended to the strongly illuminated part because of the squinting of the eyes. But the reason why dark things are most of all ignited (besides what we have said in ch. 1 prop. 38), seems also to be because whatever things are colored with this color smack of dryness and of burntness, and their matter is consequently more inflammable. Therefore, as to white things at the burning point appearing wider, this same thing and most of the phenomena just brought forward seem to attest that what is widened is not the ray gathered upon the white object, but the impression of the white object, and of its picture on the retina, upon the visual spirit. For, by 30 of the first chapter, there is more radiation from a white surface than from a dark one. The brightness is therefore extreme, or, if it is the sun that is under consideration, then this is self-evident. Now, as we have ascertained above that the image inheres in the spirit, because by optical laws it could not happen in a humor or tunic, so here too the image of a white thing very strongly illuminated, or the image of the sun received into the spirit, appears to pour itself out because of the kinship of its nature, no otherwise than the way a red drop that falls upon the surface of water (liquid on liquid) spreads itself out, while dark images received into the spirit withdraw into themselves, as if the drop were to fall upon dust; and this happens by laws that are not, as previously, optical.* For if you consider well, the inherence and
221 persistence [mora] of the luminous image are in exactly the same class as this spreading out. For from the persistence itself the spreading out appears to follow, as the vision flees such brightness, but nonetheless, having absorbed it, carries it around and falls upon it with others of its parts.

Whether rays are gathered by black, scattered by white.

* "Optical", i.e., having to do with refraction, in the restricted sense of the word. For if you look at the origin of the word, "optical" means anything that pertains to vision.

Whichever of these causes is to be taken into account in whatever individual case, it is certain that this spreading out of luminous things exists either in the retina, by reason of the picture, or in the spirits, by reason of the impression: this was, in fact, the thing whose cause I wanted to investigate in this whole chapter. And so, from this chapter, Astronomers will ponder this, that ocular perception or reckoning is not always to be trusted, however much they are taken into account in the quantity of the diameter of the full moon, or of the defect in an eclipse; and consequently, that other more certain procedures must not only be brought into consideration, but also one must not rashly disagree with them, on the testimony of vision, when it happens that they disagree with vision. For it has been demonstrated most clearly, from the very structure of vision, that it frequently happens that an error befalls the sense of vision, in overestimating the size of bright things.

Why the heavenly bodies appear more distinctly as dawn breaks.

It also appears appropriate to this place to state the cause why, for those with night blindness, the heavenly bodies are more confused in the depth of night, but when dawn begins to break they are seen more distinctly. For if much light interferes with them, the light added on by the light of the dawn should have interfered more.

Now the cause is this, that in the darkness of night the opening of the pupil is widened by its natural motion, while in the light of dawn it is closed more narrowly. But the cone of the radiating point enters with more power through the wider opening, and makes a strong impression, while the same cone constricted by a narrower opening moves the eye more weakly.[129] *Hitherto, however, we have always made use of this, that by an extreme sensation of one object, the weaker sensations of the others are suppressed. Moreover, in chapter 2 this served in the place of a foundation, that distinct lights upon a wall, coming in through a wider opening, are hidden, the weaker by the stronger, but when the opening is narrowed, they are not affected in proportion: although all the lights are weakened, more light is removed from the brightness of the stronger, and the weaker light emerges.*

---

[129] Here Kepler refers to the following endnote.

To p. 221. The sense is this, that although the spaces of the heavens between the large heavenly bodies have their own radiation, just as the smallest stars also have, nevertheless, the points of the retina upon which they radiate are occupied by the breadth of the stronger radiation of the large stars, with the result that those stars appear large and fringed with rays. But at first dawn, these fringes of the large stars are erased, because the radiation of the lightening heaven wins out at its place on the retina over the radiation of a bright star that occurs at the sides, and that wanders in a foreign place about the axis of the cone of vision. It can do this all the more easily as this radiation is now weakened by the constriction of the pupil.

Further, this question must be understood to concern the first rising of the dawn, when the light is doubtful, and not of the full dawn. For that now fully overcomes the heavenly bodies, just as does that brightness of the air, which comes from the moon at night, illuminating the air from on high with its full orb.

Chapter 6
On the varied light of the stars

It does not belong only to the natural philosopher to treat of the light of the stars: the astronomer too has something to say here, especially about the illumination of the moon. We shall join the two together for the sake of study.

1. On the light of the sun

222 The sun's power in the world is incredible and almost divine, for from the sun comes the life and preservation of all motion, and the embellishing of things heavenly and earthly, so much so that the more closely you study it the more miracles you find in this one power. I accordingly think it befits the philosopher to probe all the treasures of nature for the purpose of proposing theories [dogmata] befitting such a miracle. And since we are now discussing optics, come let us winnow out what is most noble in this science, and apply it to this body. We learn from the divine Moses that at first everything consisted of water, and when in the beginning there existed a rough and unorganized mass consisting of moist and dry or of water and earth, right away on that very first day light was created, which on the fourth day was distributed among the different kinds of bodies. Casting off from this harbor of most certain opinion, with conjectures that are not absurd as oars, I struggle into this most vast ocean of contemplation, and state this: that the sun's body consists of the densest material of all in the entire world, so that within its very restricted orb is enclosed as much material as is spread out in the whole aethereal air through the almost infinite amplitude of the world's solid sphere.[1]

And in fact, that the density of this body exists in the highest degree, is required by a heat-producing power, proportionately fierce and proportionately extended. For, if one may apply sublunar instances to celestial things (and that this is permissible was long ago established by Moses, who showed that all is of the same matter), among fiery things of the same quantity, a certain one burns more violently and longer according as it is denser, a coal more than a flame, burning iron more than a coal. In this regard, one of the ancient philosophers, as Diogenes Laertius has it, did not so much speak absurdly as receive an unfair hearing, in stating that the sun is a glowing stone.[2] That is, he did not fear, along with Aristotle, that it might fall to earth if it were a stone; whether he did so rightly, you will learn from the Englishman William Gilbert's magnetic philosophy, to which I fully subscribe here.[3] And when Pharnaces expressed the fear that if the moon were earthlike it would fall to earth, Plutarch answered knowledgeably, saying,

---

1 The word "solid" here obviously denotes the extension of the world in three dimensions, and not any other properties such as hardness and impenetrability.

2 Diogenes Laertius, *Lives of Eminent Philosophers* II 8, Vol. II pp. 136–7, is probably the passage Kepler has in mind, though here the sun is said to be made of metal, not stone.

3 William Gilbert, *De magnete* VI 3, trs. Mottelay pp. 325–6.

> Just as the sun turns towards itself all the parts of which it consists, so too the earth receives the stone, similar to itself. The fact that it is a body is not ascribed initially to the earth, nor is it torn away from the earth, but is something established for itself by its own special nature. What, then, prevents its subsisting separately, compacted and held in restraint by its own **223** parts . . .

and below,

> It is probable, if the world is really animate, that it has in many parts earth, in many water, fire, air, not by necessity, but by a disposing reason. For the eye is not driven into the head by its lightness, nor has the heart fallen down into the breast by its heaviness, but both are located thus because it was more expedient thus.[4]

And other most beautiful things follow.[5]

Further, my positing of a certain quantity of matter in the sun's body, making it equal to the rest of the matter by which, according to the divine Moses, an extension or insufflation is created between the waters and the waters, seems to be required by the elegance of proportion, thus: that the same thing whose force ought to have permeated that universal space has received as much body as there is in that universal space. There is also no danger that it would be impossible that either the force should be so tightly condensed or the body so widely extended without the intermingling of a vacuum. For why should this proportion not hold between extremes, if between water and air, which are intermediates and very close to each other, there exists such a proportion that a drop of water comprises the same amount of matter as is in a fairly large chamber of air, as we have proved above in ch. 4 sect. 6 prop. 10?

Besides, because it befits this matter of the solar body to be utterly simple and to the greatest degree one, it must necessarily be devoid of the two notions of the Opaque, which this word acquired in ch. 1 prop. 17 above. For it will have neither many surfaces beneath itself (in which case it would not be simple), nor color. For I defined color above in ch. 1 prop. 15 as "light in potentiality", but to the sun belongs the pure actuality of light. And things that are in themselves colored reveal an impurity of matter, while it is right that the sun is a body most pure. Consequently, notwithstanding its surpassing density, the body of the sun will nonetheless be pellucid. And so, by 11 of the first chapter, it must necessarily have been constituted of some fluid substance, and, finally, have come forth from

---

[4] Plutarch, *The Face in the Moon*, 924E ff. and 928A ff., in *Moralia* XII pp. 69 and 91. Pharnaces is an interlocutor in the dialogue. Kepler has treated the prior of the two passages with considerable liberty, changing words and omitting a phrase.

[5] Kepler has the following endnote to this paragraph, though there is no reference to it in the text.

To p. 223. I declare the sun to be a body pellucid only to its own proper light, not to extrinsic lights, and it will therefore not be transparent. For neither will the highest density give passage to the sight (or colors of things to the sight), nor will extraordinary light leave a place in the eye for seeing other objects which are beyond the sun's orb.

water that has been condensed and purified in the highest degree, which the divinely instructed Moses also implied.

Again, because by general admission the sun's function in the world is the same as the heart's in the animal (for I shall prove in the physical part of astronomy that the motion of the planets is dispensed by the sun),[6] namely, that it
224 dispenses life to this perceptible world, there must also exist in the sun's body a soul that is a handmaiden in such a task, or a vital faculty, if you prefer. Therefore, it is consistent that from the indwelling of this soul or faculty in the densest and purest body, and from its most powerful vivifying or formative faculties—that is, from the victory of the soul and the subjugation of extremely contumacious matter—light should result; by what procedure, is uncertain, but it is certain nevertheless by the examples of many sublunar things. For tell, O Natural Philosopher: where might you see a rising flame without heat, which is either from an animate faculty or was once given birth from it? Tell further: what matter might you see inflamed, which is not born through some animate faculty in its proper body, such as oil, resin, and so on. Lest you bear some suspicion about subterranean things, those are the action of an animate faculty in the globe of the earth, generatrix of metals and of rivers from sea water, warmer and protector of subterranean things from the cold of the upper world, perceiver of the harmonics of the motions of the heavens (though without discursive thought), form-giver of the marvellous figures in fossils,*[7] so that you might be able to touch it, if you are unable to make it out. In fact, light is always conjoined to the animate faculty, so much so that the foremost among the physicians compare the source of life in the heart of an animal to a flame.

Animate, or rather (in the notion of the physicians), natural, faculty in the earth's viscera.

*

This faculty of earth corresponds to the formative faculty of the maternal uterus, as it is necessary that one be the exemplar, or better, offspring, for the other. For it forms not just geometrical figures (I have seen most perfect tetrahedra and octahedra in the diamond, cubes in various materials, the anterior face of a dodecahedron in the silver bronze of Dresden, the anterior face of an icosahedron in the fossils of the Boller Bad), but also humans, wild animals,

So, this animate faculty in the sun, the producer of light, although it inflames the whole, being poured through the whole body (not as being about to consume foreign matter, but as being about to give form and protection to its own), it nonetheless will fix its seat chiefly at the center, and (by the principles of chapter 1) will spread light from the center to every body. And since the lines that are drawn from the center are perpendicular to the surface, the light will therefore everywhere rebound from the concave surface, and will be gathered again at the center, and, passing through the center to the opposite surface, will will repeat the same rebounding, by 19 of the first chapter. Therefore, whether the entire body of the sun be everywhere equally inflamed by its soul from the beginning, or more at the center, there is always more of this immeasureable conflagration at the center. This is because there is no opposition of a rarer medium occurring outside, preventing something from rebounding inward, which suspicion I removed in ch. 4 sect. 9, through the example of the mirror.

---

[6] The "physical part of astronomy" was first presented in chapters 33–38 of the *Astronomia nova* (1609), which is subtitled *physica coelestis*. See also Kepler's Preface, p. 14.

[7] *Fossibilibus*. Kepler's meaning is broader than ours, though our word, too, originally just meant "something dug up." The Boller Bad is near Göppingen in Württemberg. Its fossils are described by Kaspar Bauhin, *Historia novi fontis balneique Bollensis* (Montbéliard 1598; German translation: Stuttgart, 1602), Book IV.

and so on. Kings, popes, monks, soldiers, in that dress which they truly wore at the time; likewise, snails, tortoises, almonds, nuts, horns, and so on, exactly as if, after the manner of a pregnant woman, it transplanted those figures into the fetus by the imagination from a horror arising out of objects encountered.

Nevertheless, the greatest part breaks forth through the surface into the open 225
aether, on which note the following, point by point. First, because we attribute the highest degree of density to this body, all refractions will be right on the perpendicular, by the discussion in ch. 4 sect. 6 prop. 7. And because by Witelo X 9,[8] a form comes forth from a dense medium to a rare one in the same lines by which it enters into the dense from the rare, while the entry, as has just been said, is made only by perpendiculars meeting in the center, therefore, no radiation will occur from any point existing within the sun's body (even though it radiates in an orb) except by perpendicular lines.

In turn, because to one point of a denser medium an infinity of straight lines can be drawn from the rarer medium, following an infinity of inclinations, while in the medium of highest density all are refracted into that one perpendicular, which is drawn from the selected point to the center, therefore, by the same method of reasoning, even though the center of the solar body is hidden within the body, nevertheless, after coming out by a straight line to some particular point on the surface, it will be spread from that point through the whole hemisphere, and thus will be spread through the whole hemisphere an infinity of times by the infinity of points of its surface.

But the points outside the center, if they themselves also maintain some light from the form given them from the soul, will not be spread outside in any other line (as was said previously) except that which is drawn from themselves through the center. And thus the hemispheric radiation of any point whatever other than the center into the world coincides with some one of the infinite hemispheric radiations of the center. Nor will there be any privilege in the closeness of the center, and the single center will be equivalent to the entire body. Indeed, as the form given the center is more sublime than that given the rest of the body, so is the center's radiation also stronger at a certain point of the surface than that of all the interposed points of the body.

One might almost say that it is the center of the solar body alone that arouses such surges, and should it occur to the mind to think about what it is that you are perceiving when you look at the sun's body, do not believe that you see only the surface: a force of such magnitude does not inhere in the surfaces of bodies, but lies hidden in the depths, as in the magnet; in fact, in ch. 1 prop. 15 above, we did not even derive colors from the bare surfaces. But you, when you are looking at the sun's body, know that you are perceiving the center of the sun under refraction
everywhere on the whole perceptible surface, as it is that whose illumination is 226
strongest. If it were not so strong, you would indeed make out something of the sun's body.

The center is therefore that from which light originates; the surface, that which receives this force from the center, and dispenses it to the whole world, and that which, in itself, because of its narrowness, was going to escape the eyes, the surface spreads it out and places it before the eyes; and finally, the intermediate part is that which, for the center, mediates the passing out to the surface, and, for the surface, mediates the force of spreading light through the hemispheres.

---

[8] *Thesaurus* II pp. 413–4.

Take note of these and compare them with divine things and with what I said in the preface to the first chapter. You will see in the sun a palpable image of the world; in the world, that of God the Creator. But although from every part of the surface everywhere one center shines forth, it is nevertheless not brought about that the force of illumination is everywhere equal. For from from the middle point perpendicular light goes out, and strikes more strongly, [while] from the sides, being refracted, it is weakened. For this is an inherent property of refraction from the theory of optics. Further, a larger part of the sun also acquires a greater force of illumination; the smaller part acquires a smaller force.

The sun is an image of the Trinity.

2. On the illumination of the moon

This was the most ancient part of astronomy, to look into the causes of something obvious to everyone, namely, how it came to pass that in its monthly circuit the moon waxes and wanes, and takes on various shapes.

Berosus the Chaldean, as Diogenes Laertius reports, taught that half the moon is luminous, or, as Cleomedes reports, is half fiery, being deprived of all light from the remaining part; and that by the rotation of the globe it happens that the luminous part hidden beyond the body gradually comes out more and more.[9]

Thales of Miletus gives a better opinion of this, and was the first to say what is in fact true, that this whole hemisphere is illuminated by the sun, and the moon does not give light by itself in any part. Plutarch attributes this opinion to Anaxagoras.[10] Berosus was able to conclude what lies before the eye, or rather, to conclude from that, that it is not always the same parts of the moon that shine. For the face which we imagine for ourselves in the moon from the arrangement of the spots struggles out successively into the light, and, always motionless, is
227 turned downwards towards us, granting to the light a passage above itself. There is at the western edge of the moon, a little above the eye, towards the zenith, a very black spot, like a point, in medium and bright light, separated from the other spots the way an island is from the continents, distant from the extreme limb by barely a digit.[11] You will see this there, that is, at the western edge and somewhat towards the zenith, whether the moon is at half, or gibbous, or entirely full. Thus the circle of illumination first touches that spot, then goes across more and more, and after the full moon leaves it altogether. Consequently, it is not the same parts of the moon that are always in the light, as Berosus thought, but the same parts of its body are forever turned towards the earth, the center of its gyrating motion; however, in going around the earth, it presents one and another part to the sun, as if it were to be turned by itself as if on a spit, roasting itself every month towards

---

[9] Berosus, who was a Babylonian priest, was not mentioned by Diogenes Laertius. Since Berosus lived in the third century B. C. E., there can be no question of Thales having improved upon his opinion. The passage from Cleomedes is from *De motu circulari* II 4 pp. 180–2, trs. Goulet (French) p. 156.

[10] Plutarch, *On the Face in the Moon* XVI, 929b, trs. Cherniss p. 100–101.

[11] The spot is evidently the Mare Crisium. The digit is a measure used by Ptolemy to estimate the extent of eclipses: it is one twelfth of the apparent lunar or solar diameter. See the footnote to Ch. 2 Prop. 12.

the sun, in very nearly that manner in which Copernicus said that the earth is turned and roasted daily towards the sun, as towards a fire.

Even though this opinion of the illumination of the moon is certain and beyond any chance of doubt, nevertheless there is no lack of those who take it upon themselves to attack it using optical arguments. The opinion of these people is fully argued in Plutarch, *On the Face*, and is repeated by Cleomedes, book II (although Scaliger would attribute this same instance to a certain Arab, the son of Amram).[12] If, they say, the sun communicates its light to the moon, that will be diverted to us by reflection, so that we would not see the light of the moon, but, with eyes directed towards the moon, would take in the rays of the sun itself. This being granted, it will follow that the moon is a convex mirror. As a result, we will see the image of the sun in the moon. Besides, it will not be possible, whether in the full moon or in the halved body of the half moon, that we could see the rays of the sun all around it. For the law of reflection is this, that it is generated solely by that point of the body where the angles which the rays of both incidence and reflection make with the sphere can be equal. And so, since we neither perceive the sun's image in the moon, nor is the latter illuminated in that one particular place in which there can be an equality of angles, but it sometimes appears in its full body, sometimes halved, and sometimes horned at its extreme edges, they think it follows that the moon does not shine with rays coming from the sun.

This instance will be able to do the job for one who distinguishes absolutely **228**
all light into direct, reflected, and refracted: he does not acknowledge a fourth. It is as a result of this that Cleomedes, reduced as it were to narrow straits, concludes wrongly that the moon too has its own light which is aroused by the solar light, so far as it is touched by that light. Plutarch, on the other hand, seeks out another illegitimate way to escape. But I, by very evident reasons given above in ch. 1 prop. 22, have introduced a fourth form of light, arising from reflected and refracted light, to which I have given the name of communicated light.

As a consequence, even though it is very truly established by the reasons already presented that the moon does not make use of light simply reflected from the sun, there still remains this fourth form of communicated light, to be attributed to the moon. For the account of that light which descends from the moon to us, having arisen from the sun, is none other than the account of that which, descending from the same sun upon any wall in its way, radiates to the whole hemisphere, and, whatever darkened Camera may be set up opposite, it enters into it by the open slit, and depicts itself with its color upon the opposite white wall, as has been demonstrated above in ch. 2 prop. 7. In view of this, I bring back those words of Plutarch which he finally appends to the rest of his useless excuses, saying, "the moon has many rough places, many irregularities of shape, so that the brightness that descends from the great body may shine brightly with light reflected by the not inconsiderable heights, and be beneficially reflected, intermingled, and extend the reflected brilliance between itself, as if it were borne

---

[12] Plutarch, *On the Face in the Moon* XVII, 929F–930E, trs. Cherniss pp. 104–111; Cleomedes, *De motu circulari* II 4 pp. 180–192, trs. Goulet (French) pp. 156–61.

to us by many mirrors."[13] By these words, he has described very nearly what I am accustomed to calling communicated light. Cleomedes, however, does better, in adding this necessary condition, that the solar light is not just reflected by the moon, but is also altered at the moon, as the brilliance of fire is in iron, and thus is made its own.

And from this it has now been deduced that the body of the moon is dense, as is the earth (which Plutarch asserts with many arguments in the book frequently mentioned), and is also lavishly colored, as, with this removed, communicated light has no place (according to prop. 22 of ch. 1), and with a rough surface, so that it not be pellucid. For all the instances sought from all sources bear witness that the moon's body is opaque.

The opinion of Posidonius, as reported in Macrobius, is also proved from
229 this evidence; Reinhold recalls this on fol. 164 of his commentaries upon the theories of Peurbach.[14] While he correctly attributed to the moon the same matter which the earth also comprises (which in fact appears to have been the most ancient opinion of the Pythagoreans; nor, perhaps, did they mean by that "Antichthon" of theirs, cause of eclipses, which was given a futile refutation by Aristotle, anything other than the moon),[15] he nevertheless intended it to differ from the terrene globe in that the moon, like a mirror, sends forth again the light that is received from the sun, while the earth, overspread with the sun's rays, only becomes bright, and does not give out light. In fact, however, the account of the earth and the moon is the same. For the moon is not a mirror, as Plutarch and that Arab and Cleomedes correctly proved in refutation above, and it is false that the earth does not shine with communicated light, which is manifestly proved in ch. 2 prop. 7 by the instances of all walls, with Plutarch adding the instances of clothes.[16] If Cleomedes suffers that which he himself celebrates, the moon's proper light, to be called "color", which I have defined above as "light buried in matter", he speaks completely with me in the rest, and reaches the same conclusion with me, finally attributing, with me and against Posidonius, proper colors to both the moon and the earth, which, aroused by the sun's light, radiate on half the orb, so that in that matter of resplendance the earth and the moon are equivalent. And so the astronomers are perfectly certain about this illumination of the moon: that it comes from the sun, and that here the moon's body itself shows itself to be no different from any wall that is in the way of light.

Once this foundation is laid, astronomers now derive from it various ways this illumination occurs. They argue, first, that since the sun is of spherical form,

---

13 Plutarch, *On the Face in the Moon* XVII, 930E, trs. Cherniss pp. 110-11.

14 Georg Peurbach, *Theoricae novae planetarum* (Wittenberg 1542 and other editions). Later editions incorporated revisions; the edition Kepler used, which he cites by page number, is that of Wittenberg, 1553. There is also a Paris 1553 edition, in which the cited passage appears on fol. 103v.

15 The Antichthon, or counter-earth, was discussed by Aristotle in *On the Heavens* II 13, 293a 23 ff., and *Metaphysics* I 5, 986a 12.

16 Plutarch, *On the Face in the Moon* XVIII, 931B, trs. Cherniss pp. 114–5.

the moon likewise of spherical form, but smaller, the moon is therefore illuminated over more than its half by the sun, and the boundary of the illumination is a circle, but one that is less than that which is the greatest described upon the spherical body of the moon, by Witelo II 27.[17] Now, from this careful preparation, the distances and diameters of the two celestial bodies being known, they show how to investigate the size of the circle of illumination.

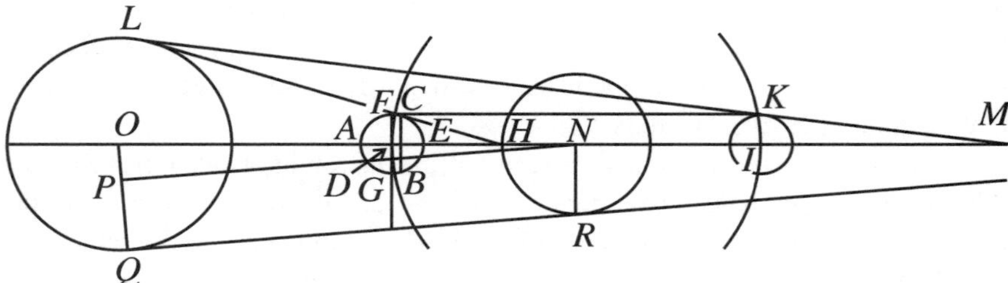

3. On the circle of illumination of the moon and the earth

Knowledge of this will be useful to us later in the problems. Reinhold, in his most erudite and worthy commentary on the Theories of Peurbach, which should be read carefully by those who study celestial subjects, gives this quantity for the circle of illumination on fol. 165.[18] Let ABC be a great circle of the moon's body, drawn through the pole of the circle of illumination. Let the pole or the middle of the illuminated part be A; let FG be another great circle from the pole A through D, the midpoint between AE; parallel to it, the circle of illumination BC cutting the globe of the moon below the center D. Now Reinhold gives the quantity of arc BAC as 181° 45′. But according to Ptolemy, the moon at apogee appears equal to the sun,[19] whence it happens that the circle bounding the vision coincides with the circle bounding the illumination, if it should happen that the moon run precisely beneath the sun. Therefore, at the same time, Reinhold gives the remainder from CAB, namely, the arc CEB, which measures the part seen through the middle, 178° 15′, so that the two together make up 360°. However, the other foundations upon which this computation of the most learned gentleman rest, are unclear. The quantity itself (I would say for the benefit of the art despite this man) is in error, which I demonstrate thus. Since Reinhold, from Ptolemy, assumed a distance of the moon from the earth at which the body of the moon is viewed in the same visual angle as the sun, let the straight line CH accordingly touch the body of the moon, at the point C of the circle of illumination CB, and let the axis of illumination AD be extended until it meets CH at H. Now, because CH is tangent to the moon at the circle of illumination, it is also tangent to the illuminating sun, whose center is by supposition high up on the axis of

230

231

---

[17] *Thesaurus* II p. 71.

[18] Georg Peurbach, *Theoricae novae planetarum*. In the Paris 1553 edition, it is on fol. 104v.

[19] *Almagest* V 14, trs. Toomer p. 253.

illumination DA; and this is by Witelo II 27.[20] And because the sun and the moon are viewed under the same angle, while the line AD through the centers is one, and the line CH to the edges of the two luminaries is one, and further, the angle CHD is also one and the same, CHD is therefore the angle under which the semidiameter of either body is viewed. But, from Ptolemy, the semidiameter of either body is 15′ 40″, as Reinhold himself assumes, on p. 209 of the book mentioned.[21] Therefore, the angle CHD is 15 minutes, 40 seconds. And because FD is perpendicular to DH by construction, while CH is tangent to the circle at C, whence DC, from the center D to the point of tangency C, is perpendicular to CH, therefore, FDH and DCH will be equal and right. Next, in right triangle DCH, the remaining angles CDH, CHD together are equal to the right angle DCH. Moreover, CDH, CDF together are equal to the right angle FDH. Therefore, the common angle CDH being removed, the remainder FDC is equal to the remainder CHD, and consequently the arc FC is 15′ 40″, while GB is equal to it, and the sum of the two is 31′ 20″. And FAG is a semicircle, or 180°. Therefore, CAB is 180° 31′ 20″, and the remainder CEB is 179° 28′ 40″. This is the greatest quantity of all of the arc CAB, that is, when the moon is both at apogee and full. For consider that H is the vertex of the moon's shadow, and likewise also the place at which the sun and the moon are viewed under an equal angle. And since a longer shadow of the same globe becomes more acute, if, that is, the globe to be illuminated by a larger illuminating globe recedes to a greater distance. Therefore, where the moon has receded farther from the sun in a straight line, the vertex of the shadow will also be more acute there than is the angle of vision at apogee of 31′ 20″. As a consequence, when the moon is new at perigee, since according to the sense of the word "perigee," it has come closer to the earth, and has correspondingly receded from the sun, it will bound the shadow with a more acute angle. Again, when the moon is full, when it is nearly 60 semidiameters of the earth farther from the sun than the earth is (while at new moon it had been the same number of semidiameters closer to the sun than the earth was), it will close the shadow with a much more acute angle, and therefore the arc FC will be much less than 15′ 40″. And this is following the opinions of Ptolemy, which, in accordance with Reinhold's views, we have followed here.

232 But in order that we may know the magnitudes of the arcs between FC or GB, when they are at absolute minimum, let DH accordingly be extended to I, and let HI, HD be made equal (ignoring for now the semidiameter of the earth), so that the moon may again be at apogee, and, in turn, full. In this place, let the straight line KL touch the moon and the sun, meeting with HC extended at L, and let LK and DI be extended to their meeting at M, and let HL, from the opinions of Ptolemy, be 1210 semidiameters of the earth, while HC and HK are 64, so that CK is about 128, and, CH being subtracted, CL is 1146.[22] For here there is no need for precision. Therefore, in triangle LCK, the sides LC, CK are

---

20 *Thesaurus* II p. 71.

21 Paris 1553, fol. 131v.

22 The moon's distance is calculated in *Almagest* V 13, trs. Toomer pp. 247–51; the apparent diameters in V 14, pp. 251–4; and the sun's distance in V 15, pp. 255–7.

given, as well as the angle LCK, for since CK and DI are parallel, DHC and HCK will be equal; consequently, LCK will be the supplement, 179° 44′ 20″ . Hence, the angle CLK comes out to be 1′ 34″. Next, in triangle HLM, angle HLM is 1′ 34″. But the exterior angle LHD or CHD is equal to the interior and opposite angles HLM, HML. Therefore, HML, half the angle of the sharpness of the full moon's shadow, is less than LHD, half the sharpness of the new moon's shadow, by the space of HLM, 1′ 34″. It will therefore be 14′ 6″, twice which is 28′ 12″. As a result, CAB will be 180° 28′ 12″ here, less than before. But the arc CEB, measuring the circle of vision CB, when the moon is again at apogee and full, remains the same as before, namely, 179° 28′ 40″, if indeed it is true, as Ptolemy taught, that the apparent diameters of the new and full moon are equal.

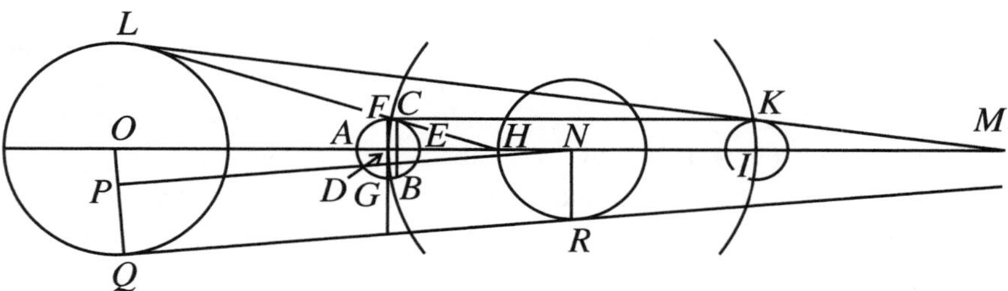

On the earth's circle of illumination

Since the full moon measures the earth's shadow by twice one and a half of its diameters,[23] the earth's shadow will be more acute and longer than the shadow of the full moon; therefore, the circle of illumination will be closer to a great circle. Witelo shows how to investigate this in X 59, in approximately this way.[24] He assumes, from the astronomical writers, that the distance of the sun from the earth is 1,210, where the semidiameter of the earth is one part, while the semidiameter of the sun is $5\frac{1}{2}$ of the same parts. In the above diagram, let N be the center of the earth HR. Let the straight line RQ be tangent to the bodies of the sun and the earth, and let perpendiculars NR, OQ be drawn to the points of tangency R, Q from the centers. Finally, let NP be drawn from N parallel to RQ cutting OQ at P. Now since OQ is $5\frac{1}{2}$ where NR is 1, OP will be $4\frac{1}{2}$ of these parts, and ON 1,210. Therefore, in the right triangle OPN, the sides OP, ON are given, as a consequence of which the angle ONP, 12′ 48″, will not be hidden. If QR be extended, it will meet ON in the same angle, forming the boundary of the point of the shadow. Therefore, the angle of the shadow is 25′ 36″. And **233**

23 *Duplo sesquiplo suae diametri.* I have rendered this literally because is so odd. In the *Almagest* V 14, trs. Toomer p. 254, Ptolemy calculates that the shadow's diameter is about $2\frac{3}{5}$ times the size of the moon's; however, according to Tycho Brahe (*Progymnasmata* I ch. 1 p. 135, in *TBOO* II p. 148), it is larger, averaging almost three times the moon's diameter. However, on p. 251 below, Kepler says that the shadow's semidiameter is $2\frac{1}{2}$ times the moon's semidiameter; perhaps that is what he means here.

24 *Thesaurus* II pp. 451–2.

further, the earth's circle of illumination will cut the great circle through the axis of illumination in an arc of 180° 25′ 36″.

4. On the phases of the moon

The manner in which it now happens that, as a result of the rays coming from the sun, the moon first appears horned or sickle-shaped, afterwards bisected, then gibbous or convex on each side [*amphikyrtos*], and finally full, and thence in the opposite order gibbous, bisected, and horned, and finally hides itself again beneath the sun's rays, Reinhold has sufficiently shown from Witelo, and the books on Spherics repeat it everywhere.[25]

The summary in Witelo is briefly this. In Book 4 Prop. 65 he proves that the moon's disk necessarily appears flat (which same thing Aristotle also [proves] in ch. 7 Section 15 of the *Problems*); in prop. 66 and 70, [he proves] that the moon appears less than a hemisphere; and that the boundary of vision is a circle, [he proves] in prop. 67. [He proves] that the nearer the moon gets to us, the greater it appears, but the less is grasped than the real thing by the vision, which Witelo borrowed from Euclid's *Optics*.[26]

Now, in prop. 74, he takes up the phases themselves, and first, concerning
234 the full moon, he argues that it appears full when the vision is between the sun and the moon, and the circle of vision is entirely contained in the illuminated part, or is tangent to the circle of illumination. In Proposition 75 he demonstrates that when the circles of illumination and vision cut each other, while the axes make an obtuse angle, the moon appears gibbous. In Prop. 76, that when the axes cut each other at right angles, the moon appears bisected. In Prop. 77, then when they cut each other at acute angles, the moon is now placed almost between the sun and the vision, or approaches nearer to the sun [than the earth], and appears sickle-shaped.[27]

These things, solidly demonstrated by Witelo, do not require many words. I would only add a few short notes. To Prop. 65, therefore, Reinhold makes the following remark.[28] When the sphere of the moon is narrow, the moon is very close to us, "there mostly appears in the body of the moon," not just a plane, but "something swelling and protruding, somewhat brighter, from the middle of which, to the extreme circumference, certain spotted parts, like cracks or

---

[25] Reinhold proves this in an exhaustively on fol. 105r–110r of the Paris 1553 edition. The books on the *Sphere* were used in introductory astronomy courses in the universities. They mainly presented the fundamentals of cosmology, together with elementary spherical geometry. The foremost of these textbooks was the thirteenth century *Sphaera* of Ioannes de Sacrobosco, which was widely used well into the 17th century. The relevant propositions from Witelo are IV 74–77, in *Thesaurus* II pp. 150–2.

[26] *Thesaurus* II pp. 147–149; Euclid, *Optics*, Prop. 3 and 56; cf. Aristotle, *Problems* XV 7, 911b 35–912a 27, trs. Hett pp. 334–7.

[27] These relations are summarized in the latter part of Prop. 77, *Thesaurus* II pp. 151–2, which is in effect a corollary to Prop. 74–77 (pp. 150–1).

[28] Peurbach, *Theoricae*, ed. Wittenberg 1553, fol. 165^{v}; Paris 1553 fol. 105^{r}. Kepler has made a few insignificant changes in quoting Reinhold's text below.

fissures, run out." That is to be credited to Reinhold's sharpness of vision, which Plutarch appears to confirm, in the book *On the Face*, in these words: "Dark spots appearing on the moon, sort of like isthmuses dividing the shaded parts from the bright, are so distinguished that, separated from themselves and their surroundings, they are circumscribed, and the penetration of the bright parts into the shaded ones makes the figure of a kind of weaving."[29]

5. On knowing the age of the moon from the quantity of the phases

Now, as regards the moon's phases themselves, Reinhold again remarks that the increase of light approximately corresponds to the moon's departure from the sun, which is borrowed from a certain obscure passage in Pliny, his true meaning being hunted out by conjectures.[30] Pliny seems to me to speak somewhat differently than he seemed to Reinhold, with the gist of his opinion nevertheless tending towards this same axiom. I shall therefore first explain Pliny's passage. He says, "the moon, shining, adds three quarters [plus] twenty-fourths of the hours" (accumulating them to its bright part) "from the second" day of its age; for before that it is hidden or dormant "to the full orb, and taking away" from the
full orb "in waning," and finally to total extinction. That is, because at opposition 235
the moon comes to mid-sky twelve hours after the sun, and further, the whole body of the moon, like the pound,[31] is itself also divided into twelve parts or digits, therefore, however many hours the moon is distant from the sun, it shines with that many twelfths or digits. But since from its first rising it is filled, not in twelve, but in fifteen days, it is therefore not a whole hour farther from the sun for each day of its age, nor does it add a whole digit to the luminous part, but three quarters plus a twenty-fourth, that is, 45 minutes plus two and a half minutes, 47 and a half minutes in all, for one digit, because it departs $47\frac{1}{2}$ minutes of an hour from the sun daily; for, when 12 hours are distributed among 15 days, 48 minutes come to each day. However, a more precise denomination than three quarters [plus] a twenty-fourth did not occur to Pliny. And so Pliny also gives the farmers and the household heads a rule for finding the age of the moon, that is, its departure from the sun, from the breadth of the shining horn, in its mean motion.[32] If anyone should wish to follow this rule exactly, Reinhold warns that it

---

[29] Plutarch, *On the Face in the Moon*, IV, 921b, trs. Cherniss pp. 42–3. Kepler's Latin departs considerably from Plutarch's Greek: Cherniss's translation reads, "The dark spots in the moon do not appear as one but as having something like isthmuses between them, the brilliance dividing and delimiting the shadow. Hence, since each part is separated and has its own boundary, the layers of light upon shadow, assuming the semblance of height and depth, have produced a very close likeness of eyes and lips." No significant textual variants are noted.

[30] Peurbach, *Theoricae*, Wittenberg 1553 fol. 173^v; Pliny, *Natural History* II vi 45–6, trs. Rackham pp. 196–7.

[31] The Latin is *as*, which can be either a unit of money (= penny) or a unit of weight.

[32] Kepler restates this more clearly in the *Epitome astronomiae Copenicanae* VI (Frankfurt 1621) VI 5 3 p. 839, in *JKGW* VII pp. 478–9, as follows. "However many days there are in the age of the moon, it will shine three-fourths plus one twenty-fourth times

is somewhat in error. We shall demonstrate both its closeness to the truth and its slight error by mean of the following diagram.

About center A let a great circle BCD of the moon's body be drawn, BAC being its diameter and EAT at right angles. And let E be the center of the sun; therefore, the circle of illumination will be parallel to BAC: let it be FG. Now let a point H be taken not on the line EA, and let this designate a point upon the earth's surface at which the observer stands, and let the line AH of the axis of the moon's circle of vision KL be drawn, and let EH be joined. Now, since in triangle AEH the side AE is about twenty times as long as AH, the angle AHE, determining the
236 departure of the moon from the sun, will be much greater than angle AEH, while the two together are equal to the exterior and opposite angle TAH. Therefore, TAH is certainly greater than AHE, the departure of the moon from the sun, but only slightly greater. But TAH is the angle by which the axis of vision AH and the axis of illumination AE are inclined, upon which angle the breadth of the shining part at the horn is consequent. Therefore, the breadth or increase of the shining part at the horn does correspond approximately, but not completely, to the departure of the moon from the sun. Reinhold shows that the difference at the quadratures, where it is greatest, is three and a half degrees, more or less.[33]

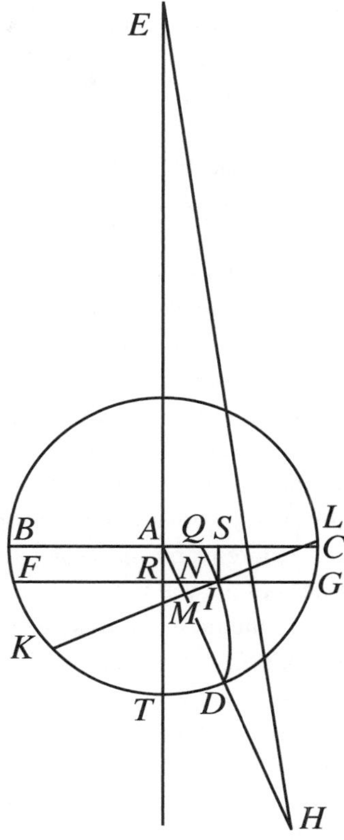

If it be desired to pursue all the subtle points, there is another thing wanting in this procedure, namely, that the circle of illumination FG is not often equal to the circle of vision KL, but is generally smaller. Therefore, as a result of this cause as well, the ratio between the increase of the shining part and the departure of the moon from the sun is somewhat perturbed.

But even if the ratio were not at all perturbed, neither by the latter nor by the former cause, the sense of vision would nevertheless not be free from error if it should wish to reckon scrupulously the days of the moon's age, or the

---

that number of hours (of which any night has twelve) until the time it sets, and this holds from the new moon to the full moon, or to the moon's age of 15 days. But from that time, however many days are added to the moon's age above 15, three-fourths plus a twenty-fourth of the same number must again be subtracted from the number of hours 12 accumulated with the age of 15 days. Three fourths of an hour is 45 minutes, a twenty-fourth is 2 $\frac{1}{2}$ minutes, the sum is 47 $\frac{1}{2}$. This taken fifteen times makes nearly 12 hours."

33 *Theoricae novae* (Paris 1553) fol. 110v.

departure from the sun, from the three quarters plus twenty-fourths of the breadth of the shining part at the horn. For even though the digits withdraw within the boundaries of the sense of vision equally, nevertheless, to the degree that they are closer to the circle of vision, they appear so much the narrower, and to the degree that they approach the center of the lunar disc more closely, appear so much the wider. The demonstration of this fact is very nearly the same as the one that shows why the chords subtending arcs of a circle do not come out proportional to the arcs, but to the degree that the arcs are smaller, the chords subtending them are so much more nearly equal to them; and since at the beginning of the quadrant 2909 small parts of the diameter add one minute, while at the end, all of 13 degrees 51 minutes correspond to the same number of small parts.[34] That is, the whole visible hemisphere of the moon appears flat, as we had said above from Witelo IV 64,[35] and further, any of its semicircles drawn through the pole of the visible hemisphere appears to be a straight line. From this it happens that the parts of the circle correspond to parts of a straight line; and for equal parts of one straight line, upon equal arcs, those are wider that are presented to the vision directly, and narrow that withdraw to the parts of the globe that slope away towards the boundary of vision.

Moreover, this is the cause for the very slow appearance of the first increments of the waxing moon, and of the last decrements of the waning moon, while 237
where the moon is divided in two, by the careful viewing of the division of the moon's face, you may perform a judgement of the true quadrature within a few hours.

In 1602, on the evening of 11/21 December, when I had not looked up the hour of the moon's quadrature, the moon appeared to me still concave, at six and one half hours after noon. And indeed, it had now come to the boundary of dichotomy.[36] For the sun's position was 29° 30′ Sagittarius. The moon was clothed in a narrow halo, such that the diameter of the moon might be placed between it and the edge of the moon, and it had passed through a certain small star* in the constellation of Pisces, so that this star was embedded in the halo.[37]

* When I look into which star this might have been, I find none in this position, certainly none of so great a magnitude that it could be perceived next to a half moon. Nevertheless, there is still no doubt whatever about the observation. Could this then be some new phenomenon?

---

34 The error in this statement evidently arose from Kepler's having used different numbers of units for the diameter at opposite ends of the table. The correct arc should have been 1° 23′, not 13° 51′.

35 *Thesaurus* II pp. 146–7.

36 τῆς διχοτομίας.

37 Here, in the marginal note, Kepler refers to the following endnote:

To the margin of p. 237. I do not propose this suspicion of a new phenomenon rashly, or without precedents. For it seems to be not so rare that, like comets, stars are also seen wandering in this way. David Fabricius sent Tycho Brahe certain observations made in Friesland, having measured the distance of Mercury from a certain bright star in Cetus, which could no longer be found either by Fabricius or by anyone else. [Translator's note: This is the first recorded observation of ο Ceti, or "Mira Ceti," a true variable star with a period of 332 days. It was rediscovered by Johannes Phocylides Holwarda in 1640, the first instance of a supposedly "new" star that had disappeared and then reappeared. Fabricius's observation is in *TBOO* XIII pp. 114–5.] Similarly, Justus Bürgi, the Landgrave's maker of mechanical devices, when he was sculpting

It is also apparent from this how much certainty there is in the drawing of the waxing and waning moon which Albategnius teaches in ch. 30 and 40, and to what extent it does not correspond to the number of visual digits (which Reinhold appears to grant all too easily).[38]

6. The paradox that never ever was there a real new moon

Now, as regards the phase itself of the full moon in particular, Witelo himself, and Reinhold after him, advises, in the explanation of Prop. 74,[39] that this phase has its own breadth of time. That is, because at full moon the sun illuminates 180° 28′ 12″parts, while we see the moon at apogee under an angle of 30′, at perigee a few minutes greater, and the arc of the great circle through the pole of vision is that much less than a semicircle, when the former excess and the latter defect are added, the total is about one degree. And so beneath the most exact instant of opposition the two circles of vision and of illuminationa are everywhere distant by about 30 minutes. Therefore, the phase of full moon lasts until the moon completes one degree; that is, for about two hours; that is, from that time when the circle of vision is tangent to the circle of illumination at the moon's eastern part, to that point where the tangency occurs at the western part. This is the reasoning of the authors just cited.

However, since they are splitting hairs[40] here, I should also be allowed to split hairs in contradicting them. Let them hear this new and marvellous voice, which I build upon these foundations of theirs: a perfect full moon has never been seen, nor can any ever be seen. So far from one full moon's lasting for
238 two hours! For because the semidiameter of the shadow is equal to two and a half semidiameters of the moon, when one semidiameter of the moon is added, so that the centers of the moon and the shadow allow the right amount of space outside the shadow, about 64 or 66 minutes will be used up. That is the amount that the moon has to be distant from opposition to the sun, whether in longitude or latitude, unless one wants it to have a taste of the shadow. But at so great a distance from opposition, the circles of illumination and of vision now have come to intersect, because they are tangent to each other at a distance between the moon

---

a celestial globe with the most notable fixed stars and was comparing it with the heavens themselves, found a certain star in Antinoous, overlooked by Ptolemy. But of the one which today is shining in the breast of Cygnus, he steadfastly asserts that it was not seen at that time when he was occupied along with the Landgrave's staff with the description of this star. In fact, Tycho's observers likewise did not note it (while simple mention of the one that was seen in antiquity in the breast of Cygnus, which was solitary, was made six hundred times, though this new one was close to it in both position and magnitude), nor did Ptolemy and Geber note it in their most accurate description of the Milky Way, while this new star is nevertheless located in a very obvious position at its edge.

38 Albategnius (i.e., Al-Battānī), *De numeris et motibus stellarum*, Nürnberg 1537. For Al-Battānī, see the footnote on p. 158.

39 This is Book IV Prop. 74, *Thesaurus* II p. 150.

40 ἀκριβολογοῦσι.

and the point opposite the sun of 30 minutes. Therefore, the moon either cannot be made full because of northern or southern latitude, or, if it can be made full, in traversing the points of opposition, it begins to be eclipsed before it becomes full, passing beneath the earth's shadow. Unless perhaps when it is off to the north, the southern parallax gives it support, or the contrary.

And so much for the full moon. Very nearly the same things might be said of the new moons. For in whatever place the moon passes by the sun untouched at new moon, the horn has a remnant, for the causes just stated. In fact, at an exactly central conjunction, if it happen to appear less than the sun (of which below), the circle of illumination will descend within the circle of vision, and (if the sun should meanwhile happen to be hidden from our vision) the edge will seem illuminated circularly. But now, as I said by way of preface, I am splitting hairs with the authorities. For the fact of the matter is that this defect in the new moon is extremely slight. You will easily understand this from what I just said, that those parts that slope away near the circle of vision are imperceptible, even though they be fairly large. For let the distance of the luminaries be 179 degrees, and let there correspond to that a whole degree upon the great circle of the moon's globe drawn through the poles of both the circle of illumination and the circle of vision, by which degree the two circles shall have overlapped each other. Now when 90 degrees is subtended by 100,000 parts, 89 degrees is subtended by 99,985. Therefore, the remaining degree will be subtended by the remaining 15 parts, which is barely the seven thousandth part of the moon's semidiameter. From this it appears that there is no danger to the observations when the moon's altitude at the moment of opposition has been investigated by the highest and lowest edge of the moon, even if, because of northern latitude, the moon were not yet completely full from the south, or the opposite. I would not deny, however, that this cause makes the beginnings and ends of eclipses **239**
obscure, just as they usually are very doubtful, because at the part where the moon usually enters the shadow, it is usually still more denuded of light because of the intersections of the circles of illumination and vision. Therefore, in the neighborhood of the shadow, where a very narrow part of the sun's body illuminates this edge of the moon, the light of the moon correspondingly also grows very faint.

7. On full and partial illumination of the illuminated bodies, and the earth's penumbra

For it is an additional consequence that on that side of the moon which is just about to have a taste of shadow, the circle of illumination is broken, and withdraws within the circle of vision farther, and for this reason is less distant from the pole of illumination, than the great circle. This I demonstrate thus. About centers A, B let great circles be described, CD of the sun, $EFGH$ of the moon. And let the straight line DH be tangent to the right sides of the circles, CG to the left sides. And again, let the straight line DE be tangent to the right side of the sun and the left side of the moon, the straight line CF be tangent to the remaining sides, cutting DE at point I. Therefore, whatever is between the points EF is illuminated by the whole visible hemisphere of the sun. But for the inquiry's sake, let it be called the "full illumination."[41] Let EF and GH, representing complete lesser circles, be joined. Therefore, whatever is between these two circles is indeed illuminated by some bit of the solar body, but no point of this edge is illuminated by the whole observable body of the sun, insofar as something of the sun is always missing, and is hidden behind the body of the moon. Let this be called "partial illumination".[42]

Should you understand $EFGH$ to be the earth, the same terms would apply, and when the lines DE, CG, CF, DH are extended somewhat to K, L, M, N, respectively, let that which is between KL, NM be called the penumbra, but let LN be called the umbra. Therefore, since the parts C of the sun illuminate the moon GE all the way to F, while the parts D illuminate EF
240 all the way to H, it is evident to the eye that when point D is blocked, point H would not receive any light, and when the parts DA towards C are gradually blocked by the interposition of the earth, the light will be gradually extinguished from H all the way to F. As a consequence, the circle of partial illumination GH, which in the preceding was the circle of illumination, will be broken from the side H and will withdraw towards the circle of full illumination F, doing so long before the eclipse begins. But to investigate the quantities, let us again suppose what Witelo and Reinhold supposed, that the sun's distance from the earth is 1210, the moon's from earth 64, and thus that of the full moon from the sun 1274, which is the line AB.[43] Further, let the semidiameter of either luminary appear to be 15′ 40″. It is evident without calculation that the point I is going to be in the earth itself.

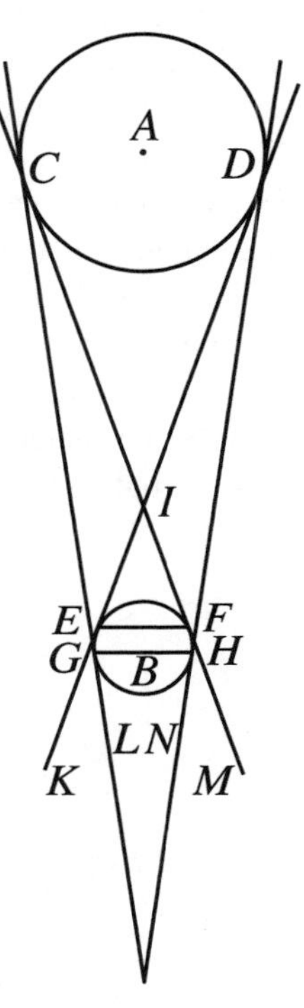

For if at true opposition EIF and CID be equal, I representing the observer, then EID will be one straight line, and so will FIC. But they should also be straight lines by construction. Therefore, I is the earth, and EF is the circle of full illumination; the same is also the circle of vision. If, that is, it were possible that the luminaries be truly opposite without darkness. Hence, as before, EG or FH, an arc of about half a degree, is given. Accordingly also, when the circle

---

41 συναυγείας. Although this word (which also appears in the title of this section) is used in *Placita philosophorum* to mean "meeting of rays" (cf. *L&S* p. 1700 and *Placita philosophorum* IV.13.11), it appears that Kepler has in mind the root meaning of the components συν "complete" + αὐγή "light of the sun."

42 λιπαύγεια (also in the section title). This as it stands is not a classical Greek word; however, there is an adjective form, λιπαυγής, which means "deserted by light, dark." Again, it appears that Kepler had in mind the root meaning of its components λιπ "remaining" + αὐγή "light of the sun."

43 *Theoricae novae* (Paris 1553) fol. 110v. The figures ultimately derive from Ptolemy, *Almagest* V 15, trs. Toomer pp. 255–7.

of vision EF is inclined so as to be tangent to GH at H, the portion FH extinguished before the beginning of the eclipse, as before, will hardly appear to be a forty thousandth part of the moon's semidiameter. I wished to add this so as not to strike fear into the less experienced by this pruning back to the quick. For at the beginning or end of an eclipse there is no danger, even if the circle of partial illumination be broken before its time, and perhaps vanishing completely and gathering entirely within the circle of full illumination; this is only useful in providing causes of that matter concerning which astronomers are now already in agreement, namely, why it is that the beginnings and ends of eclipses of the luminaries are so doubtful and untrustworthy, and why the light of the moon at the edge of the shadow appears so pallid and as if diluted by water, to such an extent that Witelo, in Book 4 prop. 77 asserts that the moon often seems to be partly eclipsed without having entered the shadow at all.[44]

Let us, however, also explore for the earth how much the breadth of full illumination and how much that of partial illumination is. As above in Section 3, let N be the center of the earth HR, and on it let RS be the circle of illumination or of partial illumination. It has been demonstrated that when the center of the sun is on the line NH so that one who is at R sees one edge of it, one at S the opposite edge, the arc RHS is 180° 25′ 36″. The arc of full illumination THV is sought. Now since the sun is supposed by Ptolemy to subtend 31′ 20″ of a great circle, it is evident that one who proceeds from R to T through 31′ 20″ of a great circle, if he previously at R saw the upmost edge of the sun graze the horizon, is now, at T, going to see the whole sun. For the sun's parallax does not confuse this at all, which in the degrees close to the horizon is nearly unchangeable. Further, the same thing will happen to him who saw the highest edge of the sun on the horizon at S, for, having gone forward towards V through 31′ 20″of a great circle (because to arcs on earth there correspond similar arcs in the heavens), he will see the whole sun, and will be standing at the edge of the region of full illumination. When T and V are joined, this will be the circle of full illumination, differing from the arc RHS by 62 minutes 40″. Consequently, THV will be 179° 23′. **241**

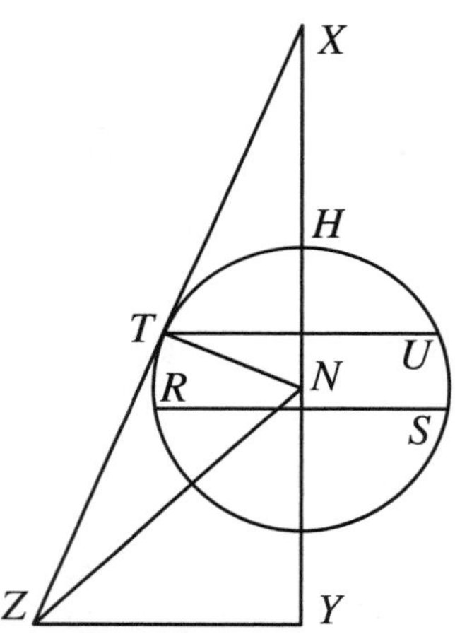

Now, from this, let the following also be investigated: how great the penumbra is in the moon's passage. Let NH be extended and let some line touch the circle at T, cutting NH at X. And let XT be extended as far as necessary, and let TN be joined. TXN will therefore be as great as half of the supplement of THV, which is 37′, by what was demonstrated in Section 3. Therefore, TXN is 18′ 30″. Therefore, where TN is 1, NX, which subtends the right angle T,

[44] *Thesaurus* II pp. 151–2. This is not quite what Witelo wrote: what he actually said is, "however, the light is somewhat darkened as the moon approaches the shadow."

is 186. To XN let there be appended the straight line NY, to denote the distance of the full moon from earth, and let it be according to Ptolemy, 64, as before. Therefore, the whole XY will be 250. Let the straight line YZ be erected cutting XT extended, and forming the boundary of the depths of the penumbra and the umbra at the same time. Therefore, in triangle XYZ, the angle X is known, and Y is right, while the side XY is 250. Therefore, YZ will become $1\frac{345}{1000}$ semidiameters of the earth. ZN being joined, a new right triangle ZYN is formed of given sides about the right angle. Therefore the angle ZNY is given, $1^\circ\ 12'\ 14''$ or $72\frac{1}{4}$ minutes, the depth of the penumbra with the umbra. But that of the umbra is about $45'$, the difference 27 more or less. For when the umbra shrinks, from the earth's approach to the sun, the penumbra grows, and where the moon is going to traverse a thicker umbra, it meets with a narrower penum-
242 bra. And so, while the moon is not yet really beginning to be eclipsed, nearly the whole body of the moon is already in the penumbra, and a slim portion of the circle of partial illumination remains, and finally also an extremely slim portion of the moon enjoys the full light of the sun. Whence it happens that as the eclipse begins the moon is very pale, and the brightness usual in a full moon is exceedingly darkened.

It is also worth while considering here whether the moon's light could appear of equal brightness everywhere. Reinhold had indeed asserted above that it appears brighter in the middle. And reason seems to require that where the light of the sun is more scattered, it is also more attenuated. But it is more scattered near the circumference of the apparent disc, for the account of both vision and illumination is the same, but we said above in Section 5 that if the angle under which the body of the moon is viewed be divided into twelve equal parts, a smaller part of the surface is seen by the middle parts of the angle, and a greater by the outer ones. So let the angle of illumination, or the light itself falling upon the moon, be divided into the same number of parts; therefore, the same amount of light will correspond to the large outer parts of the surface as to the intermediate narrower parts, with the consequence that the light will be dispersed more widely in the outer parts, and will fall more obliquely. Finally, since partial illumination surrounds the outer parts, while full illumination covers the middle, the middle will be correspondingly brighter.

I, however, am moved by these arguments not at all, or at least very little. For as regards full illumination, that covers the whole moon in such a way that the circle of partial illumination, even granting an exact opposition without darkness, vanishes in comparison with the rest of the diameter, while in oppositions that are not perfect it is to a large extent hidden, and in the place where it contacts the full illumination it hardly differs from it. But even if the sun's light is more scattered at the outer parts of the moon, it also in turn makes its way into our eyes more densely at oppositions, and indeed under the same angle by which it is distributed over that greater surface. This is confirmed by experience that is less fallible. For when the light of the full moon is let into a camera obscura/dark chamber by that arrangement which I have described above in Chapter 2, it displays on a white floor an edge brighter than the middle, because of the spots that extend over the middle. But the moon's being perceived as brighter in the middle to one looking at it, appears to be an affect of the eyesight, which when directed towards the

center is everywhere surrounded by bright parts, but if directed to the edge is now 243
deprived of brightness on one side, while the blue color of the sky rushes in.

But when the moon is already waning, or not yet full, the proposed causes are quite valid, as experience also attests. For the light which forms its inner, concave or gibbous line is very weak and diluted, because, being spread over a large surface, occupies a greater angle in vision than in illumination in this situation.

8. On the Lines of the Moon's Phases

Albategnius[45] taught how to draw the image of the waxing moon from the known motions of longitude and latitude. That that method was hardly dependable, you have seen above. If you have been taken with eagerness for a more reliable picture (although you will be asking this of yourself), and one which fits not just the horned but also the gibbous moon, you will need to know the lines by which the viewable shape of the moon is bounded. Of one of these, indeed, the one facing the sun, there it is agreed that it is an arc of the circle of vision, but those which face opposite the sun are not arcs of a circle. Witelo said in book 4 prop. 25 that the gibbous part of the moon is irregular in kind, while in prop. 77 he asserts that the horned shape is contained by two arcs of virtually equal circles.[46] Let it therefore be known, that each is an arc of the conic section that is called "ellipse", as Aristotle notes in Section 15 Problem 6:[47] I prove it thus.

Apollonius defines the conic surface to be this: whenever some line, bound immovably to some point, goes around a circle (the point not falling in the extended plane of the circle) from one part of it, creating a surface, until it returns to that point of the circle whence it began to move.[48] Now, indeed, the boundary of the moon's illumination is a circle, as was said above, but our vision has the ratio of the required point. For it does not fall upon the extended plane of the circle of illumination throughout the whole month, except about each quadrature of the moon with the sun, where the cutting of the moon is also considered to be a
straight line, hardly curved at all. Therefore, whenever you look at the moon, you 244
make a cone with the circle of illumination. For vision occurs in straight lines, coming together in the one center of the eye from all points of the visible object.

But it has just been said that the globe of the moon appears to be a plane disc, perpendicular to that line which is drawn from the eye to the center of the moon.

---

45 The Arab astronomer Al-Battānī. See Chapter 4, note 172.

46 *Thesaurus* II pp. 129–30 and 151. In Prop. 77, Witelo writes, "[the shape] seems to be contained as if by two equal arcs, because of the imperceptibility of their inequality." (p. 151).

47 Aristotle, *Problems* XV 6, 911b 6–34. There is, of course, no mention of the ellipse in Aristotle's works, as the application of this term to the geometrical figure was first made by Apollonius, about a century after Aristotle's time. See Apollonius, *Conics* I 13.

48 Apollonius, *Conics* I def. 1, trs. R. C. Taliaferro (Green Lion Press, 1998) p. 3. Kepler abbreviates Apollonius's more formal definition.

Therefore, an imaginary plane, or rather, the quasi-plane surface of the eye, cuts the cone just mentioned. Now Apollonius defines the ellipse to be this: when a cone is cut by a plane which is neither parallel to the base of the cone, nor placed subcontrariwise, so that the section meets with any straight line whatever that is drawn from the vertex of the cone to the surface; that is, such that the whole cone is truncated.[49] All these conditions are satisfied in this cutting of ours. First, because the imaginary plane is perpendicular to the straight line from the eye to the center of the moon, it will never be parallel to the circle of illumination except at the most accurate conjunction or opposition of the moon with the sun, when this cone is right, because the axis or straight line from the eye to the center of the illuminative circle or base is perpendicular to the base. And then the moon does not appear horned or gibbous. Next, when the moon departs from the sun, the circle of illumination is inclined to the straight line through its circle and the eye or axis, and then the cone becomes scalene. And it is cut by a plane perpendicular to the axis; therefore, the section cannot be subcontrary. Third, the entire cone is also severed. For that always happens when the plane is perpendicular to the axis. Therefore, all conditions being satisfied, the appearance of that boundary of light in the moon, which is seen in a place farther from the sun, bounding the concave of the horned moon or the gibbous side of the gibbous, will be elliptical: which was to be demonstrated.

Consequently, if the whole circle of illumination could be seen, it would appear with the figure of a perfect ellipse. But because half or a little more of it is hidden behind the moon's body, the amount that is seen is therefore only the arc of an ellipse. This not only must be known for the picture of the configuration, but is also useful for a particular problem below. It might be proved from Witelo IV 56, but the demonstration does not accomplish what the proposition promised, that a circle beheld obliquely appears to be a cylindrical section, be-
245 cause not everything that is close to a cylindrical is cylindrical.[50] If Witelo had demonstrated this, Serenus would already have supplied the rest, who showed the the section of a cone and a cylinder are the same.[51] Thus the defect of the horned moon differs from that of the eclipsed moon, in that the former is bounded by an ellipse, the latter by a perfect circle.

But in the same way that the section of the horned or gibbous moon is elliptical, that of the half moon necessarily had to appear straight. See Witelo Bk. IV pr. 76 and Aristotle in the passage cited.[52] However, it seems to

---

[49] If the cone in question is oblique, there are two planes that create circles as sections: one parallel to the generating circle, and the other, called the 'subcontrary,' at an angle to it. Cf. Apollonius, *Conics* I Prop. 5, trs. Taliaferro pp. 9–10. The ellipse is defined in Prop. 13, pp. 24–6; however, Kepler's definition is quite different from Apollonius's.

[50] In IV 56 (*Thesaurus* II pp. 143–4), Witelo only shows that not all diameters of the obliquely viewed circle appear equal, not that the circle appears to be an ellipse.

[51] Serenus Antinoensis, *De sectione cylindri*. Serenus points out the identity of conic and cylindrical sections in his introductory letter.

[52] *Thesaurus* II pp. 150–1.

him that something slight was lacking as to straightness. In 1602, in the evening of 11/21 December, having observed carefully, I judged the upper horn to be acute, the lower somewhat obtuse. It is open to anyone to investigate this more carefully.

This must also be considered: whether the moon is exactly bisected. For first, even if the circle of vision were a great circle, nonetheless the circle of illumination, which represents the section, goes beyond the middle circle. In Section 3 above, CAB was 180° 31′ 20″ at new moon, 180° 28′ 12″ at full. Therefore, at half it will be 180° 29′ 46″. And the circle of illumination goes beyond the middle by 14′ 53″. The sine, 433, is the two hundred fortieth part of 100,000, the whole sine.[53] Thus from one side (the bright) there would stand 241, from the disappeared and dark side 239.

Now, however, the circle of vision is also not equal to the great circle, and in this the ratio of the bright part to the dark becomes greater. Let $FAGE$ be a great circle of the moon, the circle of illumination FG, and AE at right angles to it through the center, so that the ratio of breadths of the dark and bright parts be that which AC has to CE. Now let $BIDH$ be the lesser circle of vision from the same center, so that the bright, apparent part may be $ICHD$, the dark part $ICHB$. Therefore, since from the ratio of AC to CE the equals AB, DE are removed, the ratio of the remainders BC, CD will be greater.

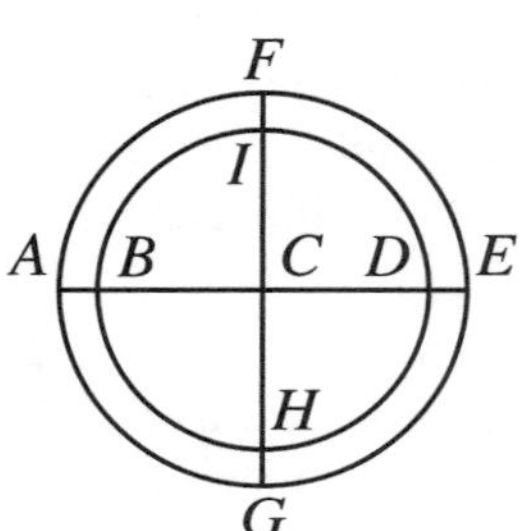

These things can indeed be said against bisection as to the appearance. But since the breadth of AB with respect to BC is completely imperceptible, as was said above, the ratio will also be imperceptibly different. But also, the ratio of 239 to 241 cannot be distinguished by sense from 246
the ratio of equality. For the one would exceed the other by about 15 seconds. Consequently, it can safely be ignored. Moreover, this will be helpful below in freeing us from this suspicion.

It is superfluous to repeat here the farmers' rules of how to distinguish, from what has been demonstrated, the waning from the waxing moon, by the orientation of the horns, which are always turned away from the sun. On this subject, the passage from Book 18 ch. 32 of Pliny is usually adduced.[54]

Nor is it very important that astronomers be informed how to imagine the zodiac rather crudely for themselves using the inclination of the horns, and how, when the horns are standing on the perpendicular, they may be made more certain of the ninetieth degree of the ecliptic.[55] Finally, how the line through the horns, extended, leads the vision to the pole of the ecliptic. For these things are encountered everywhere.

---

[53] Recalculation shows that this should be the 230th part.

[54] Pliny, *Natural History* XVIII, Ch. lxxv, no. 323, trs. Rackham Vol. V pp. 392–3.

[55] If the crescent moon is at the 90th degree or the 270th degree of the ecliptic (0° Cancer and Capricorn, respectively), the horns lie on a great circle which passes through the north pole and is perpendicular to the equator.

9. On the moon's spots

Even if it is left to the natural philosophers to argue what the spots on the moon are, nevertheless, since these chiefly shine forth in the full moon, this consideration should also be added to the enumeration. And really, the natural philosopher, even if he were fully to treat of the question, that would hardly be of benefit to the astronomer (as perhaps he might learn the same thing from the position of the spots when the moon is full as from the inclination of the horns when the moon is horned, since the circle of illumination, passing over the surface of the moon, does not cut it squarely into right and left, nor into an upper and lower part, but transversely from the right eye to the left corner of the mouth); however, the astronomer supplies the natural philosopher with many things for untangling the arguments of the question. Even though Plutarch considers this argument in an entire book, with the title *On the face of the Moon.*[56] In it, the first opinion is that of those who think the face of the moon is a product of vision. This is refuted there by arguments that are many and tightly packed. I am adding ocular evidence.

On 1602 21/31 December at 6^h in the morning, through a device described in Ch. 2 and an instrument made for this purpose, a description of which is furnished below,[57] the moon made an image of itself brightly upon the paper lying below, inverted in situation, just as it was in the heavens, gibbous. The edge all around was very luminous and bright, except from the gibbous side, for there it was
247 perceived as more washed out. The middle, however, was one continuous spot or darkness, darker in one place, lighter in another.

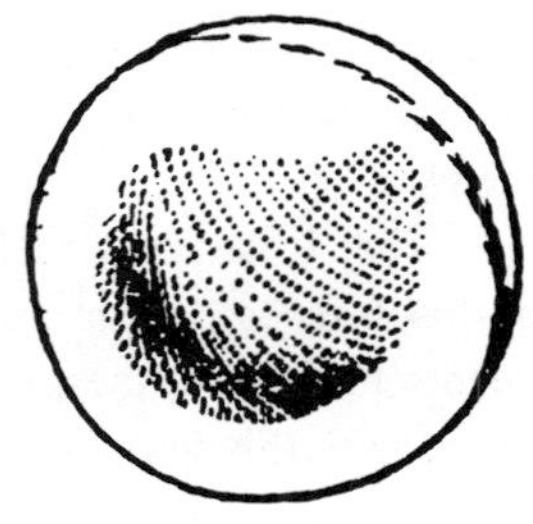

You should not think that what I would consider to be in the moon's ray was in the paper, for both the gibbous face and the spot in its middle were carried over to all parts of the paper whatever that were placed beneath it; rather, indeed, it was from moving the paper that the spot was first discovered. The shape of the spot, so far indeed as was able to be formed somewhat confusedly through a sufficiently wide opening at a distance of twelve feet, was the representation of the Hebrew samech [ס] with its belly full: that angle (for at the other sides it was nearly round) being pointed approximately towards the middle of the gibbous side. Nor was it just this one time, but, a trial being made of it more frequently, the spot always thrust itself forward along with the light, so that it could not possibly be a reflection of light.

Plutarch in that book brings forward many opinions on that face of the moon, Reinhold proposes others also, of which he most approves this one of

---

56 Plutarch's first chapter seems to begin with the "first opinion" mentioned in the next sentence (though in all surviving texts the beginning is fragmentary). The refutation of it begins in Ch. 2. Cf. Plutarch, *The Face in the Moon* pp. 35–7.

57 The instrument is described at the beginning of Chapter 11 (not Chapter 9, as in the note in *JKGW* II p. 451).

Witelo, which attributes to the moon the transmission of the sun's rays, differently in different parts.[58] Those which appear to us to be spots are the denser parts, to which the sun can pour in only a little light. Hence, he even thinks it happens that the moon is seen even when it removes the entire sun from us by the interposition of its body, for then the sun's rays pass through to some extent, and flow into our sense of vision. True as it seems, Witelo was not concerned with spots when he chose this opinion, but with this phenomenon, why it is that in a total eclipse of the sun it is possible for the moon to be seen, on the side turned away from the sun. The spots came upon him in the role of something extra. And so if we will accommodate another demonstration to this phenomenon below, we shall be freed of this opinion, which has a very great deal of difficulty, and shall bestow the cause of the spots elsewhere without these authorities objecting. For Plutarch elegantly concludes, from that surpassing brightness of the moon borrowed from the sun, that its body should be of the highest density, that sends the least light into its depth. This is easily evident through a comparison carried out with other things that shine with reflected light, among us, of which, so far as any one is more transparent, it reflects correspondingly less of the rays. Nor does the moon thus withdraw from us by many diameters of the earth, so that we may **248**
not be able to argue from terrestrial things to lunar ones, under the guardianship of vision. If our air, with a depth of a few miles, placed against the setting sun, so weakens its rays that opaque objects set out in the sun almost lack a shadow; and, as the same will be said below, when this space is doubled, it extends the shadow all the way to the moon's body; then would we not have made the moon rarer than this very air of ours if we should assert that the sun's rays can pass through five hundred German miles of the moon's body (for the moon's diameter is about that great), and, indeed, pass through it in such a way as to enter into the eyes?

That little work of Plutarch's is most elegant and lively, and worthy for a philosopher to delight himself in at some time when more weighty studies have been set aside. Furthermore, it is the cause of my at length concurring, not unwillingly, with this very author in this opinion, of which, for me at least, my teacher Maestlin too was already the author; and I would say that the moon's body is such as this our earth is, forming a single globe out of water and landmasses. Plutarch in fact persists in saying this, and fortifies it both eloquently and cleverly with many reasons against various objections, so that some Peripatetic might well wonder that so many solid things can be argued against the opinions of his sect. What especially convince me are the following. First, it was said above that the moon, when it displays a bisected face, shows an uneven cutting, and to some extent a twisted one. This is evidence that some of its parts are low, others more raised up, and these to such an extent that it can be perceived from sixty semidiameters. Next, in certain lunar eclipses there appears a great unevenness, and one not coming from the earth's shadow. For it is known that the peaks of the mountains are very rarely raised by the distance of one mile, of which there are 1600

---

[58] *Theoricae novae* (Paris 1553) fol. 103r–4v. Witelo mentions the moon's transmission of rays in the introduction, *Thesaurus* II p. 3.

in the earth's diameter. And so if there is some roughness in the earth, this must be imperceptible in the moon, which is 60 semidiameters away. For let the height of a mountain which the circle of illumination transits be a mile, that is, about the eight hundredth part of the semidiameter, and let this quantity of its shadow also remain at the moon. And since it is sixty times eight hundred, that is, almost fifty
249 thousand miles, out to that shadow of the mountain, the mountain's shadow will subtend barely 4 seconds. Consequently, an altitude of fifteen miles would in the end add one minute to the shadow, of which there are up to 90 in the diameter of the shadow.[59] Consequently, if something irregular occurs in partial eclipses of the moon, this has to draw its origin from the body of the moon itself.

In 1599, on the night between 9 and 10 February, new style, when the sky was very beautiful, I observed an eclipse of the moon. And although I lacked instruments, I at least did not neglect to take note of those things that are made out by the naked eyes. In the morning, after the third hour of the town clock,[60] the moon turned towards Cor Leonis[61] in such a way that it was reckoned that a perpendicular from Cor Leonis would fall in the middle side of the face.

The face of the moon was inclined in that direction nearly like a portrait of a person whose right ear is hidden. Compare that with what I said above in Section 2 about the face of the moon. For this takes place in every full moon, which serves as indisputable evidence that the same face of the moon always faces the earth.

Further, there appeared to be a sort of gap above the right eye, which for us was opposite our left, as if at the edge of the circle something were missing in roundness.

When quarter to four was striking, it was reckoned to be the beginning of the eclipse. There was, however, doubt whether it was a diminution of the circle or a fissure first come into shadow, just as valleys are first clothed in shadow. And this was a little below the point that stood opposite Cor Leonis.

When it was striking four, a good part was already gone. And so that the clock might not have made me err too much, (as they are usually set carelessly), I made a conjecture of the altitude of Jupiter, by a comparison of my distance from the window with the elevation of the edge of the window above my eye. And so the altitude of Jupiter was considered to be about 6 degrees. In the quarter-hour past four it had already gone behind the mountain, but could not yet have been on the horizon. It was not yet half in the shadow.

I have added this so that the little observation would not lack the circumstances of its time, and so that a reader who is attentive to horological error (if there was any such) would not be thrown into difficulties by the exact statement of the time. Other things which I have noted in this eclipse will be said below in
250 Ch. 7 Sect. 3.

---

59 That is, the earth's shadow at the place where the moon passes through it. Tycho Brahe, *TBOO* II p. 148, gives apparent semidiameters for the shadow at different distances of the moon, ranging from $43'$ to $47'$.

60 This was in Graz.

61 Regulus.

In 1598 on 11/21 February, in the morning, when a little before five on the town clock the moon stood erect to the perpendicular, eclipsed by half its diameter, and when, from that moment inclining gradually more forward, it was traversing the northern part of the terrene shadow, and when, finally, a little before the sixth hour, still appearing to decrease, it removed itself beneath the clouds with the very meagerest of light: during this whole time (but especially when dawn was breaking and the light of the remaining part was diluted), it was seen torn to pieces or cut up, as it were, by certain luminous channels working their way into the shadow; and that which ought to have been an arc, dividing the shadowed from the luminous part, portrayed the edge of a broken timber with its unevenness. I believe that dawn was indeed a contributing factor. For when the moon was affected by the same northern part of the shadow in December of 1601 and in May of this year 1603, it experienced nothing of the sort, because the eclipses fell in deep night.

Add to these that if you look carefully at the moon even exactly at full, it seems to depart perceptibly from roundness.

All these things provide me with evidence for my statement that the moon was correctly described by Plutarch as the kind of body that the earth is, uneven and mountainous, and that the mountains are even greater in proportion to its globe than are the terrestrial ones in their proportion. And to jest along with Plutarch, because among us it is usual for people and animals to follow the spirit of their land or province, there will consequently be living creatures in the moon, with much greater bodily size and ruggedness of temperaments, than ours, for the simple reason that they endure a day fifteen of our days in length, and indescribable tides, because of the sun's being so long overhead—if, indeed, there are any there. For that place is not absurdly credited by the superstition of the people with being destined for the purification of souls.

But to the point. When Plutarch said that the moon is an earth, he then stated that the moon's spots are seas, which almost absorb the sun's rays transmitted into the deep, and do not reflect them as strongly as the earthlike parts usually do. His words are:

> Just as our earth has certain large gulfs, so we judge that the moon too is opened up with great depths and rifts, containing water or foggy air, into . ***251***
> which the sun does not penetrate with its light, but abandons them, making a diffuse reflection.

Thus Plutarch, with whom I do not agree in this part. It is more fitting that the bright parts that are in the moon be considered seas; those that are spotted, lands, continents, and islands. For this which is to be demonstrated by the optical writers is most thoroughly confirmed by experience: that watery surfaces most of all become radiant with light, if they are placed next to the land, I believe because of the uniformity of the surface as a whole, but the roughness and rippled state of the tiny parts, or because they participate less in dark color than the earth. The former of these makes it suited for reflecting the sun's light in almost all directions; the latter assists the light that is communicated. For to the extent that it is less black, it is more white. But white is suited to taking light in and vibrating it back out again, by 30 of Chapter 1, and so is the moderately transparent, by 22

of the same chapter. When I returned to Styria in 1601 for the sake of business, I climbed a mountain of considerable altitude, called Schekel,[62] of the Stubenbergs' dominion, intending to measure the orb of the earth from two mountains by means of a suspended plumb bob.[63] And I in fact demonstrated that the mountain itself rises above the peaks of the other mountain, which has the Wildon stronghold placed upon it,[64] by the height of five of the towers of the Strassburgers[65] And truly, looking down on the other hills of the lower region through ten German miles and more, it provided no other opinion to the one observing, than as if I were looking down at some meadow in which hay is distributed in sheaves. Therefore, from this mountain the earth presents itself to the observer with incredible brightness, so much so that when a cloud suddenly arose and cut off the face of the sky from me, a paper spread out was in many places more brightly illuminated from below than from above: the cloud covering the mountain, but the region illuminated by the sun. And the land as a whole showed this brightness, partly verging towards black in the forests, partly exulting in greens in the fields and meadows, in some places also, with plentiful plowed ground, being red. But that which furrowed the middle region, the river Mur, at that time overflowing into pools and turbid, easily overcame the dimming brightness of the earth with its exceeding splendor. These things happened because the precipitous mountain-had raised me up somewhat on the perpendicular, where more direct rays from the lands lying below were able to fall upon me. What indeed was not about to
252 happen if I had been able to perceive the whole orb of land in lines that are nearly direct? And these things really made me believe my eyes that there is usually more brightness from water than from land, not by simple reflection, for the sun's position, the same as mine, at the left side of the downstream current, completely ruled that out, but also by communicated light.* And so I conclude that in the moon the bright parts are a watery matter, but those that are dim are land masses and islands, while the whole moon, as will be said below, is surrounded by a certain airy essence, which transmits the rays of all sides.[66]

* Moreover, among contributing causes was the brightness of the air by day, which, since it surrounded the river on all sides, had rays that were powerful in all directions, and were thus also bounced back to me on the mountain from the smooth surface of the water.

10. On the mutual illumination of the moon and the earth

It is widely enough admitted that the principal light of the moon is from the sun, the modes and varieties of which light have been the subject of discussion so far. But there are some who nevertheless ascribe to the moon some

---

[62] The Schöckel (1446m) lies about 15 km north of Graz. For more on this outing, see Kepler's letter to David Fabricius, 2 December 1602, *JKGW* XIV no. 239 pp. 330–1.

[63] This method is described fully in *Epitome astronomiae Copernicanae* I 1, in *JKGW* VII p. 38.

[64] The Wildoner Berg (550m), with the ruins of the castle Oberwildon (452m), is on the south side of Graz.

[65] The single spire of the Strassburg Cathedral, built in the 15th century, 139m tall, was one of the tallest structures in Europe at the time.

[66] After reading Galileo's *Sidereus Nuncius* (Venice 1610), Kepler reconsidered this passage in his published response, *Dissertatio cum Nuncio Sidereo* (Prague 1610), and reversed his opinion. Cf. *JKGW* IV pp. 297–8.

slight proper spark of light, by these arguments. First, because in total eclipses of the moon there nonetheless remains to it some redness that is bright enough. Next, because in total eclipses of the sun the moon is again seen brightly in full face. Third, because in new moons, and two days and more thereafter, and even up to the quadratures, the moon is seen, not just with a whole circle, but completely with the whole round face. Really, none of these causes is sound. Of the light in an eclipse of the moon, it will be said below in Ch. 7 that this is not proper to the moon. But of the light of the new moon and of the moon covering the sun, an account is to be given immediately below. Moreover, that nonetheless the moon was sometimes seen when it had snatched the sun away from mortals, it is to be believed, on the authority of Witelo and Reinhold, who were impelled from that phenomenon to provide causes, which they would not have done unless they were certain of the fact.[67] Let them pay attention, who live in the Tyrolean Alps and who live in Italy, and the southern part of France, together with all of Spain, let them (I say) pay attention to this matter and the face or color of the moon, in the coming year 1605, on 2/12 October of which year, as indicated by the calculation of Tycho Brahe, the moon, nearly at perigee, will be set centrally against the sun in the regions mentioned. For concerning Tycho's opinion that the moon's diameter in eclipses of the sun is smaller than that which could cover the whole sun, something has been said in the appendix to Volume 1 of his *Progymnasmata*, **253**
and something will be said below in its own chapter.[68] But Cleomedes, from the opinion of Posidonius, appears to deny this exact thing of the moon's light in an eclipse of the sun.[69] For he asks, since (in his opinion) it is settled that the moon's body is translucent, why the sun's rays do not therefore pass through that globe in a total eclipse of the sun, as they ordinarily pass through a cloud, and make it visible and make brightness pour forth from it; why therefore the moon is dark, escapes the sense of vision, and day is turned to night? And once the question is posed, he gathers little reasons from everywhere, by which he strives to establish that the sun's rays are not obligated to pass completely through the moon or to procure brightness for it. Thus he says exactly the opposite of what Witelo says; perhaps he is reasoning from the effect itself, the conversion of day to night, and not from the actual appearance of the moon covering the sun, as Witelo did. For although for me the truth of this matter is not established by any experiment, nonetheless the light makes this happen, that the nascent moon, with

The great eclipse of 1605.

---

[67] *Theoricae novae* (Paris 1553) fol. 104r.

[68] In the *Progymnasmata* itself, the moon's apparent diameter is given as 33′ at a distance of 60 earth radii, while the sun's is 31′ (pp. 473–4, *TBOO* II pp. 422–3). However, on p. 817 (*TBOO* III p. 320 ff.), there is an anonymous appendix stating that there had been an error regarding the diameters: a lunar diameter of only $30\frac{1}{2}'$ is consistent with observations. In a letter to Magini of 1 February, 1610 (*JKGW* XVI, no. 551, p. 279), Kepler admitted authorship of the appendix—thus he managed not only to write Tycho's posthumous opinion, but also to cite it here in his own support! Cf. Kepler's measurements of the sun and moon in Ch. XI Prob. 3–5 and the appendix to 5, below.

[69] Cleomedes, *De motu circulari* II 4 p. 190, trs. Goulet (French) pp. 159–60.

its greater part turned away from the sun, nevertheless shows in its entire body, so that I have no doubt whatever that it would strike the eyes much more obviously by the same light when it had snatched the sun, which ordinarily blinds our eyes and suppresses visible things, from our eyes.

Witelo, then, and Reinhold, argue ambiguously about the residual light of the moon, as was said above.[70] Witelo said that the body of the moon is passed through by the sun's rays, and that after the passage is made they are certainly very weak but nonetheless visible and enter into the eyes; but at no other time than when the moon is nearest the sun. For in departing, it turns these traversing rays away from us, just as it does the shadow. When Reinhold saw that the days nearest new moon, in which the moon is perceived very brightly in its entire body, are forsaken by this cause, since the diameter through the luminaries has now diverged from our eyes, he said that the moon has its own spark of light in addition, which is perceived only in the days nearest the new moon, and that in an eclipse of the sun this spark of light is mingled with the rays of the sun passing through the moon's globe.[71]

In fact we have just now shown clearly enough, in Section 9, that the moon's body cannot be pellucid. Moreover, it will be clearly shown below that that spark of light is not proper to the moon: this will be most completely undermined where the authentic causes will be opened up. When these have been looked through, no one will think it necessary to beg other causes from elsewhere.

254 There are those who think that that whole circle visible in the nascent moon is nothing but the circle of illumination, which is seen as a whole because the moon's apparent diameter is so much less than the diameter of the sun. This reason is completely false. For this light is perceived not only on the edge, but in the whole body, not for one day, but for two or three, indeed even somewhat at the quadratures, where a semicircle of illumination is completely hidden behind the moon.

In Book 2 of the *Progymnasmata*, Tycho Brahe ascribes the cause of this light to Venus, which may be able to illuminate the moon so brightly.[72] But the waxing moon is always blessed with this light, [while] Venus is not always set by its side. Besides, Venus is many times higher than the moon. Consequently, even if it might sometimes bestow its light to illuminate the edge more strongly, nevertheless the ray of Venus does not reach to the middle disc of the moon, which is equally blessed with this light, being hindered by the interposition of the moon's body. Meanwhile, I do not deny that it sometimes happens that we view the moon from the earth by illumination created by bright heavenly bodies opposite it, concerning which see ch. 7 below.[73]

---

[70] See p. 260 above.

[71] Witelo IV Prop. 77, *Thesaurus* II p. 151; Peurbach edition, Wittenberg 1553, fol. 164v-165r; Paris 1553 fol. 104r.

[72] There is no Book II of the *Progymnasmata*, and neither Ch. 2 of Part I nor Part II appear to address this matter.

[73] See p. 288, below.

On the other hand, my first teacher, Maestlin, to my knowledge discovered the true cause, and taught it to me and to all attending his lectures 12 years ago, and publicly explained it in 1596 in theses 21, 22, and 23 of his *Disputation on Eclipses*.[74] That teaching is to be presented in no words other than the author's own. Here is what he said:

> Of the light which is seen in the horned moon near the horns, diffused through the whole body, it is agreed by those who see it that it is not overshadowed by the brightness of the daylight, which either remains in the evening after the setting of the sun, before twilight [that is, while it is still bright day], or in the morning, with the dawn, precedes the rising of the sun. On the other hand, it is agreed that this same light is weakened on the rest of the days farther removed from new moon, so much so that about quadrature, when the moon is a long way above the horizon in the dead of night, nothing, or very little, of it is seen to remain [and that only by people with the sharpest vision]. Therefore, from this light's being separable, it is concluded that it is not, as certain people wish [among whom is Reinhold], born along with the moon and proper to it, but, just like that great monthly light, is similarly foreign and borrowed. For unless this were so, it would assuredly be seen much more brightly in the dead of night than in the illuminated air of daylight. Moreover, the derivation of this light upon the moon is shown by its position with respect to the earth. For in the new moon, the moon, placed **255**
> between the sun and the earth, sees that face of the earth which the sun illuminates, placed directly beneath it. But we acknowledge that the strength and brilliance of the sun's rays reflected individually on the single parts of the earth is such that in sunny places it dulls the sharpness of the eyes; and furthermore, that it fills with light the inner recesses of buildings, wherever it is allowed to enter even through a little crack. Who will deny this same thing of the whole of the light itself, gathered and reflected from the whole earth together with the water? We therefore say that the earth, by its gleaming light, sent to it from the sun, casts its rays on the opacity or night in the lunar body no less than, in turn (in exactly the same way) the full moon illuminates our nights in earth with its rays received from the sun, and turns them almost to day in proportion to their brightness. It does this with all the greater clarity in proportion to the earth's circle's being greater than the moon's circle. And the ratio of the one to the other is greater than twelve times. Consequently, just as these two bodies by turns intercept the sun's light for each other, as we have said before, so in turn each lights up the night of the other.
>
> This opinion receives support from that weakening of this light. For, when the moon afterwards moves away from the new, it begins gradually to leave this illuminated central part of the earth and to perceive its remaining part more and more obliquely. At the same time, the strength of the reflected rays is both diminished and dulled. Hence it happens that whatever of this light

---

[74] *Disputatio de eclipsibus solis et lunae*, Tübingen 1596. Bracketed words are Kepler's interpolations.

> is reflected to the moon that is halved or swelling beyond that, can be seen on earth either not at all, or with greatest dificulty, because of the excessive attenuation.

So he spoke. Let them then stop seeking other causes, since they see the true one. If anyone cannot believe that the force of light that is communicated to the earth by the sun is that great, he should bring into consideration my experience, which I set forth in Section 9, where, from so slight a portion of the earth, which was barely 10 or 12 miles in diameter, and on lines that were indeed not at all perpendicular, but were incident either obliquely or extremely obliquely from the furthermost parts, so great a force of light was vibrated upward to the mountain. And what could not be accomplished by the rays from the whole face of the earth and the sea together, and departing perpendicularly? Indeed, it daily impinges on the eyes, most greatly on summer days, how great the brightness of day is, when the sun strikes the earth at a more direct angle. For when the rays of the sun are vibrated back everywhere in the air in an orb from all the earth's parts, it creates
256 such a great brilliance that the eyes close. Therefore, the peripatetics should stop being angry at Plutarch because he dragged the earth up into the heavens, that is, gave it out that the moon's body is earthlike, when they see it established by the most reliable experience that in the sharing of light, such as in fact the moon acquires shall have come from this our earth. But finally, where Plutarch, where Maestlin, shall have been received with unprejudiced ears, then Aristarchus, with his disciple Copernicus, may well begin to hope.[75]

In Book 1 ch. 8 of the *Cosmocritice*, Cornelius Gemma adduces Plutarch, from the little book on the glory of the Athenians, in which he says that the latter wrote, at the time when the victorious Greeks were at Famagusta (I think he meant to say Salamis, not the town of Cyprus, but the island in the Attic Gulf), that the moon, being still slender, was filled with a sudden addition of light.[76] Although this is not borne out in so many words in the book mentioned (for Plutarch's words are, "the full-mooned goddess shone forth upon the conquering ones"),[77] nonetheless if perchance Gemma happened to read something of the sort in another author, this has to be accepted concerning this light of the moon, that it is borrowed from the earth, to which light a greater force from some external and adventitious cause would be reconciled.

Further, by this light, however tenuous, the moon wins among astronomers favor by its nature that is not slight, when, being visible with its whole body, it nevertheless does not extinguish the light of the fixed stars which it touches most closely. Indeed, I have more than once observed it standing very close to

---

[75] Aristarchus of Samos (Third Century B.C.E.) proposed a heliocentric arrangement of the planets. His own work on this subject is lost, but was reported by Archimedes in his book *The Sand-Reckoner* (in *The Works of Archimedes*, T. L. Heath, ed., pp. 221–2). Copernicus did not know of Aristarchus's theory.

[76] Gemma's *Ars cosmocritica* is an alternative title for *De naturae divinis characterismis*, Antwerp 1575.

[77] *On the Fame of the Athenians* 350A 1, in Plutarch, *Moralia* IV pp. 518–9. Kepler quotes Plutarch (accurately) in Greek.

the Pleiades, so that it was not a whole diameter away, while all the little stars of the Pleiades were twinkling distinctly. Not to mention that the diameter, then to be measured as correctly as possible, would not fit with a single method. And so in the discussion of the moon's diameter, below, arguments are to be taken from this place as well as from the preceding chapter on the eye. For most clearly, as was also said in the preceding chapter, the horn illuminated by the sun is judged to sketch out a circumference that is greater than is enclosed by the small circle of the remaining disc, whose light is borrowed from the earth. And so the circle of light from the sun appears to contain the circle of light from the earth, on one side, at least.

Hitherto therefore, the moon not only has shown no light that it could bear, acquired of itself, but also has been convicted of denseness and opacity. But the suspicions concerning light in its total eclipses, I will dissolve below, where I **257**
shall treat of the shadow. For I am going to show that this too comes chiefly from the sun.

11. On the first phase or rising of the moon

The reason why peoples who used the lunar year paid such careful attention to the first moon, is shown in part by Reinhold in the commentaries on Peurbach,[78] but most recently and most carefully by Joseph Scaliger in *De emendatione temporum*,[79] whose polymath erudition everyone rightly and deservedly marvels at, but very few will imitate. Therefore, astronomy must have a theory of the first appearance, which Reinhold, from Peurbach, presents adequately through the causes that advance or retard the seeing of the moon. Among these causes he places the obliquity of the sphere; long or short risings and settings of the sign in which the luminaries are moving; the moon's latitude, north or south, great or small; the swiftness or slowness in its period; to which he adds a more extended or briefer twilight. And nonetheless, when Peurbach said that the moon can be seen old and new on the same day,[80] which agrees with the ancients, who used to call the day of the new moon "old and new"[81] because of it, and with Pliny's statement that it was then seen by a certain sharp-eyed person on the same day, both in the morning before the sun, and in the evening after the sun,[82] Reinhold sweats mightily to accommodate the causes to this pronouncement. Finally he appeals to experience. I—that I might not say nothing here—am of the opinion that this thing cannot be grasped by rules. For Reinhold truly affirmed that summer nights introduce a great impediment by their brightness.[83] And the whole blame rests in the air, which, if it should be pure and clean enough, as in winter, while the observer stands on the ridge of some high mountain,[84] higher than

78 *Theoricae novae* (Paris 1553) fol. 95r–103r.

79 Joseph Justus Scaliger, *Opus de emendatione temporum* (Paris, 1583), *Prolegomena.*

80 *Theoricae novae* (Paris 1553) fol. 98v.

81 ἔνην καὶ νέαν.

82 Pliny, *Natural History*, II xv 78, Vol. II pp. 222–3.

83 *Theoricae novae* (Paris 1553) fol. 100v.

84 Here Kepler refers to the following endnote:

the great part of this thicker air, then nothing would prohibit the moon from being seen at the very moment of conjunction next to the sun in mid heaven, at least if it be passing through one of the limits.[85] For although it cannot be more than 5 degrees from the sun, with the consequence that it shows hardly the seventieth part of its body to us, and that itself, being received at the edge of the moon, appears much narrower than it is, an objection which Reinhold carefully raises against himself;[86] nonetheless, it can easily happen that the whole disc of the moon, en-
258 joying in turn the full light of the earth, in accord with the condition of the place and of the one viewing, can dispel the air's brightness and pour itself down upon the senses, since it often overcomes the surpassing brightness of the air situated around the horizon. For it is known, those things that the optical writers would declare about deep wells: that stars set close to the zenith can be seen from them, since the vision has been freed from the brightness of the surrounding air. Much more so, if one were to stand on high, beyond the bound of such thick air, which everywhere intercepts the sun's rays. And not only on account of this cause, but also because of this, that the moon's light can be perceived on the brightest day while the sun is present, I judge that the altitude of the pole and the varieties of risings are pretty well in agreement, and that it can easily happen that when the moon is swift and located at the limits, it may appear both old and new on the same day. And this I do not know: whether it happens more readily right at mid sky than about the horizon, where the more obliquely and the farther the solar rays traverse the vapor, the greater the brightness they make and the more they blind the eyes.

In 1598 21/31 July, in the morning, in Graz, I saw the moon about 16 Cancer very brightly, so that there was hope that it would be seen on the next day, but clouds took over the following day. The sun was about 7° Leo.

In 1603 25 Aug. or 4 Sept. the sun being at 11 Virgo, the moon, at 24 Leo, and nonetheless cast down to the southern limit, was nonetheless seen very brightly.

In 1603 4/14 March, in the evening, I saw the moon so brightly that it seemed absolutely necessary that it could also have been seen on the previous

---

To p. 257. Take this condition in such a way that the surface of the mountain above the observer should intercept his view of the air radiating beneath him. For all these things are referred back to the eye: carefully note this on p. 258. For aside from consideration of the direction of the eyes, the sun certainly illuminates the land and the air more brightly from mid sky, and brings about a greater brightness of day, than when it is lying upon the horizon, as was shown previously. And since the question here is, from where may the moon be more rightly perceived, standing near the sun: is it from the high mountain with the luminaries situated at mid sky, or is it from the plains, while they are setting, for then the brilliance of the air around the setting sun is in all circumstances wider than around the culminating sun, and therefore takes up the eye more widely. And although the brilliance of the air is generally brighter in mid sky, we have nonetheless placed our observer higher than that material which shines so because of the proximity of the sun.

[85] That is, if the moon be at its greatest northern or southern latitude.

[86] *Theoricae novae* (Paris 1553) fol. 100v.

day, had the sky not been cloudy from the west. At hour 6 the sun was at 23° 49′ Pisces, the moon at 14° 43′ Aries. And between the sun and the observed position of the moon was 20° 10′ on a great circle.[87]

But in 1587 Tycho Brahe saw Venus on 21 Feb. old style, when it had run out to nearly ten degrees of latitude to the north in Pisces, at 6^{h} in the evening, and 24 Feb. in the morning before sunrise, when in respect to longitude it was still after the sun. [88] And to the moon and to Venus, the nearest bodies to earth, one may not unaptly accommodate Witelo's Book Four Prop. 14, so understood that for more distant things the greater depth of air set in the way would represent those things as blue, something that painters scrupulously imitate.[89] Thus, for the moon and Venus at the perigee of the epicycle, the least amount of the aethereal substance is put in front of that which they send across by the rays; consequently, they are perceived much more evidently than the higher bodies, by rays sent down **259**
through the immense space of the aether. For even aether has its matter. But to the point. Since, therefore, Venus is so small, and is perceived so near the sun, what is not to be believed of the moon's suficiently long horn? It is also not so rare for the sun's light to be dulled occasionally through certain sublime causes. Gemma, both father and son,[90] relate that before the battle of Charles V with the Duke of Saxony, the sun for three days was seen as if flooded with blood, so that even most of the stars were seen in midday. Therefore, the cause was that the sun's light was dulled, but that of the stars not at all. The cause accordingly had to be high up, by which the sun's light was rendered dull not just at one point of the earth, but for the whole visible horizon, indeed for the whole region of land,

---

[87] Here Kepler refers to the following endnote:

To p. 258. In 1583 14 March, Tycho Brahe saw the moon at 15° Aries, while the sun had set at 4° Aries. The moon was at the northern limit. [Cf. *TBOO* X p. 241 —trans.]

[88] In the observation book for 1587, 24 February (*TBOO* XI p. 199), Brahe wrote, "On 24 Febr. in the morning, before the rising of the sun, there was seen a great star in the likeness of Venus, when the sun was just then about to rise, which was perhaps Venus itself, although it was still west of the sun. But because of the excessive northern latitude, it was seen before the sun." The observations for 21 February (on the same page) do not record the latitude; perhaps Kepler calculated it from the given declination and elongation from Procyon and Saturn. Note that Frisch, *JKOO* II p. 419, reports Tycho's note incorrectly: instead of *nimiam* (excessive), he has *minimam* (minimum).

[89] *Thesaurus* II p. 124. Witelo wrote, "The comprehension of the place of a seen object consists of the comprehension of the light and color of the object and the remoteness of the object, and of the part of the universe in which that seen object exists; and also consists of the comprehension of the quantity of the remoteness, when all these are comprehended at once through the path of cognition."

[90] Rainer Gemma Frisius and Cornelius Gemma, respectively. Rainer Gemma (1508–1555), born in Dokkum in Frisia, from which he derived his surname, was Professor of Medicine at Louvain. He edited Peter Apian's *Cosmographia* (Amsterdam 1529 and many other editions), and designed a kind of cross staff, which he described in *De radio astronomico et geometrico* (Antwerp and Louvain, 1545). His son, Cornelius Gemma (1534–1579), was, like his father, Professor of Medicine at Louvain.

from which the air might be able to radiate to some place, even to a portion of the earth that is so much greater in proportion to the greater number of places it is observed. If you were to say that the air was so widely thickened, this would also have clouded over the stars. What remains, therefore, is that the cause of this dulling was nearer the sun than is the highest air, at least at the confines of the lunar course. Perhaps it was cometary matter, more widely scattered and thinner.

But to the point: with the stars sometimes shining forth by day, the moon too will be able to shine forth when very near the sun. In general, the cause for the heavenly bodies' being hidden by day is not chiefly the sun's presence, but only that this our air, poured everywhere around our eyes, all shines brightly, illuminated very strongly from above by the direct light of the sun, from below by the communicated light of earth, all the stronger in proportion to the sun's striking the earth more directly from on high. I once trifled with this matter in the following verses. My conceit was that I was climbing the ridges of Atlas, projecting above the surface of the airy region:[91]

Now beneath my feet it begins, black clouds beneath the heavens;
While a new light was on the lands, night was risen on the pole.
I speak wonders: a vaporous humor, which is wrapped over the earth,
Took the form of a heavenly body by reflected light.
Such as where the winter aether, with dark clouds,
Has painted the roofs and fields with glowing snow.

Almost as above, on Mount Schekel of Styria.

The heavenly fires were to me then as at midnight.
Hard it was to see by the torch of Phoebus, though present.

260 And now it ought to be noted that it very often comes to pass that the star Venus is seen by day, when the sun's rays, along with the sun itself, are inclined, and that excessive brightness of midday is extinguished. And by reasons and experiences approximating these it is made plausible that the moon can be seen on the same day before and after the sun.

12. On the light of the other heavenly bodies

Albategnius thinks, in ch. 30, and Witelo, in Book 4 Prop. 77, that for all of the stars there is its own measure of light from the sun,[92] and that it even follows from this that they may be considered to be increased and decreased in light. However, the cause of their not appearing halved, like the moon, is their

---

[91] Kepler's source of information about Mount Atlas is evidently Pliny, *Natural History* V i 6–15, Vol. II pp. 222–231, where the peak is said to be "raised above the clouds and in the neighborhood of the lunar circle."

[92] Witelo IV 77, in *Thesaurus* II p. 152: "And [the phases] appear to us more evidently in the moon because of its proximity to our senses of vision; however, in all the other stars receiving their light and the actuality of their light from the sun or from other stars, the same figures must be produced, from the preceding three theorems." (The discrepancy in number is present in Kepler's text.)

measureless distance, over which a shape cannot be perceived. Others persist in saying that the light of the stars is indeed from the sun, but passes through translucent bodies, and thus at last is made to vibrate. The opinion of these is weaker than the other. For really, as we said before of the moon, if the stars shine by refraction of the solar ray, they will pass through only on a diametral line from the sun, like the tails of comets, and will not project their most lucid rays to us, especially where, by the mass of their own body, they hinder their being able to be seen in the place of the rays departing from the sun, that place being turned upwards. So, near the sun, they would be invisible, the contrary of which is attested by experience and by Ptolemy's rules in the emergence of bodies departing from apogee. Venus, however, if not any of the rest of the planets, is enough to refute Witelo. For since this star goes around the sun, now above it, now below (on which point no one has anything to doubt, after Copernicus and Tycho Brahe have divided the general undecidedness in these hypotheses today into two heads, to the extent that the opinion of one of these must necessarily be true; and both with one voice declare that the path of Venus is wrapped around the sun), it was thus fitting that the light of Venus, as that of the moon, is extinguished when it comes between the sun and us, but is seen at its broadest when Venus flees to its upper conjunction. But the situation is different from this. For it is invisible for a long time at the superior conjunction, and since it is very high, it casts a weak light, and casts a shadow from bodies poorly. Thereafter, the more it descends, the brighter it gets, and emits a light that is a rival of the lunar light; and is finally of such brightness that sometimes at the actual moment of conjunction **261**
it can be seen with the form of a great star, as I have found noted in the Brahean observations, and is not made to vanish by the enormous brightness of the nearby air, spread out all around, itself occupying a single point, by which solitude it moves the vision and turns it towards itself, thus being equivalent to the strongest light. It is therefore necessary that it have this surpassing light as its own.

Thus in my theses on the foundations of astrology, which I published in 1602,[93] in thesis 25 I first made it plausible by four arguments that the planets have their own light. One was this very one which I have just presented. For Venus would really change its face and waste away, like the moon, if it only shone with light communicated from the sun. Second, I showed by examples of sublunar things that all light doesn't have to be gathered in the sun. For here many animate things (not to speak of inanimates) have something of a spark of light placed in them. Third, from a certain combination of geometrical differences and by the function of the celestial light, I showed that the stars need a twofold light, both their own and an extrinsic light from the sun. On this line of argument, I leave judgement to the reader, for it does not belong to this place. Finally, I said that it was fitting that shining and twinkling are evidence for proper light, nebulosity and sluggishness for extrinsic light. For it is necessary that all light

---

93 *De fundamentis astrologiae certioribus* (Prague, [1601]); trs. Rossi, Philadelphia 1979). This undated work contained a prognostication for 1602, and so presumably was published at the end of 1601 or the beginning of 1602. Caspar (*Bibliographia Kepleriana* (Second Edition, Munich, 1968) dates it in 1601.

that is communicated and colored in matter be weakened, which Cleomedes also teaches, stating that light communicated from elsewhere does not radiate far.[94] And by this rule, to the moon is ascribed borrowed light, to most stars proper light, least of all to Saturn. Next, in thesis 29, I stated the same thing that I demonstrated above in chapter 1, that when the light of the planets is colored, it is necessary both that they be translucent and that the essential causes of color inhere in the matter of the globes, so that in this way the power that a particular planet shows in effect has some disposition in the body analogous to it. See Prop. 15 of Chapter 1. Afterwards, I compared the colors of planets with certain bodies, and showed that if a black surface be strongly illuminated, the color of Mars is returned; if bright red, that of Jupiter; if leaden or white, that of Saturn; if yellow, of Venus; if blue, of Mercury. It seems now that the same thing must be believed of the proper light as if iron were put in the fire, or a coal, it becomes
262 red, and if from the redness much light shines through, a shining similar to Jupiter is generated. If it be most bright and translucent, a yellow body, and there be much light from it, the bright colored extravagance of Venus becomes radiant with light; if it be of sapphire or crystal, highly subtle, that of Mercury; if on the other hand it be of a coarser matter, the rays of Saturn will shine forth.

It is, however, entirely credible that these heavenly bodies are ignited by an inherent force, no less than the sun, and it is in accord with what I said of the impurity of their matter, or the visible form itself: if you look into it a little more carefully, without question no light is completely unalloyed. From Saturn's ashen color there appears a certain small quantity of purple, from the most welcome brilliance of Jupiter a little redness, like smoke through the fire; this is brightest in Venus. Thus the stars of Canis[95] and Arcturus, most greatly Canis, put on all the colors of the rainbow in succession. And, lest my vision deceive me, I always had associates present, who, whenever they thought they saw some greenness shining out, indicated it with a very brief sign. We were unanimous: whatever was seen by the one silent person, the other signed at the same moment. This emission of flashes of light also concerns Venus. They are mistaken who think all these things come from optical illusions, from the inconstancy of air. For why is the manner of emission not the same for all stars that are equally distant? Why does Arcturus show colors, itself being chiefly reddish, but Canis more so, whose color is more crystalline: why does it stir up sharper twinklings than Arcturus, why longer ones? And why does the heart of Scorpius[96] twinkle so fast, the eye of Taurus[97] so slow, you would say you saw a coal staying alive under the ashes. Are not Capella and Lyra equally bright stars? And nonetheless you will note no colors in Lyra, while in Capella, while it does indeed shine forth colors, nonetheless they are almost purple. Finally, by ocular experience, twinkling is added to the planets either by some internal alteration of the body, perpetual and continuous, which you might say is like a paroxysm, or by an external revolution of the body,

---

94 Cleomedes, *De motu circulari* II 4 p. 184, trs. Goulet (French) p. 158.

95 Sirius.

96 Antares.

97 Aldebaran.

belonging to the parts and to the surfaces, proceeding by the unfolding of some parts after others, which Tycho in the *Progymnasmata* favored.[98]

In 1602, 19/29 December, in the evening, Venus, now declining greatly, radiated most brightly through an open window into a dark chamber.[99] The air was extremely cold. Venus was twinkling most agitatedly. When I looked back at the white wall towards which the ray of Venus flowed in conformity with the breadth of the window, it wavered, as if by smoke in the way of a flame, doing so very quickly, with unsure motions. For it was not ascending, so that I might believe it to be from ascending vapors, but, shining forth now at the middle, now at the top by a sudden alteration, it withdrew now downwards, now again to the side. And, I swear, that fluctuation, so unaccustomed, of murky darkness in the yellow ray, stirred up no inconsiderable hidden horror. The speed as well as the moments of fluctuation were coordinated with the twinkling, perceived by the eyes from the star itself. **263**

On the following day the air changed, and there appeared a morning rainbow; winds ensued. And so, whatever this is, I have referred it to the alteration in the air. But a few days afterwards, namely on 5/15 January, 1603, in the evening, the sky was again calm, there were about eight degrees left to the three day old moon to reach Venus, and both were sending their rays through the same window. At that time the rays of the two were seen in an evident enough proportion, such that the moon did indeed cast rays much more brightly than Venus, simply because the horn of the moon is also much larger than it, but nonetheless the difference was evident between the edge where the rays of the two ran together and the edges receiving the rays of Venus alone. I conjectured that the ratio of brightness of the surfaces illuminated by the moon, by Venus, and by both, was 4 : 1 : 5. The ray of Venus was wavering a great deal, but that of moon not at all. Therefore, both this and the previous wavering were from a real twinkling of Venus, not, indeed, from the air then passing in the wind, as I had believed. For it would also have applied to the moon.

Further, I have inserted this mention of the light of the stars, although it is more physical than astronomical, all the more willingly, because I think it not impossible to measure the azimuth and the altitude of Venus, Jupiter, and other stars by the lights they cast upon walls. For in 1601 in the month of December, I saw the distinct lights of Jupiter and of Spica Virginis through the same window upon the same white wall. He who is studious of celestial observations should ponder this matter, and have at hand, if necessary, plane mirrors as well, down into which he may look. For the eyes are always more inclined, somehow or other, towards looking downwards than towards looking upwards.

13. On the light of comets

What it was that ignited the lights of the comets, present day theorists have adequately shown, it being recognized that the beards of comets always are **264**

---

[98] Tycho mentions rotation as a possible cause of twinkling (among other causes) on p. 404 of the *Progymnasmata* (*TBOO* II p. 377).

[99] *Obscuram cameram.*

spread on the side opposite the sun, unless, being opposite the sun or extraordinary phenomena of immense altitude, they occulted their own tails or beards by the interposition of their heads. Nevertheless, even now the wits of philosophers are kept busy with difficulties far from ordinary. For if comets are illuminated by the sun, which the direction of the tail attests generally, and which reason pronounces probable, why, then, do the tails not look precisely to the place opposite sun? Why do they almost always bend away, why are they curved into an arc? Finally, just what is that thing that receives the sun's ray, and shines by its striking, showing the shape of a tail? For if you will say that it is material, looking to the essence of the comet, you will have fashioned the most enormous monstrosity. If it be the aethereal air, I ask why it is not thus illuminated by the sun every day, so as to shine even without a comet? Even Tycho Brahe, in the *Progymnasmata* vol. 2 ch. 7, so nearly agrees to this opinion, as to say that the comet's tail is illuminated by Venus, against which very opinion, however, he inveighed shortly thereafter, hesitating in ch. 9, and urged his readers to probe the mysteries of optics.[100] Would that I may be able to give all that is required in turn here, through his salutary admonition!

But because I am not able to do all that I want to, let me therefore be allowed to accomplish what I can. First, from observations of that memorable comet of 1577,[101] it is evident enough that the question of the tail's inclination and of its curvature is the same. For the direction in which the tail declined from the diameter of the sun is the same towards which the bending also tended. For Tycho thus explicitly stated, in more than one place, that the curvature looked towards the zenith, and the concave towards the horizon, and since the comet was in the north, the line of the tail being extended to its intersection with the ecliptic, the angle made by the line of the tail extended to its intersection with the ecliptic was from 3 to 9 degrees less than if the line had been drawn through the head and the sun, with the result that the tail was farther south than the diametral line of the sun. Therefore, the inclination or deflection of the tail from the sun's diameter is nothing other than part of the curvature. And thus, that which had begun to be produced directly away from the sun, by being curved in at every point, finally gave the images of a declining line. Say, O daughters of Pierus,[102] what may be

---

100 There is no such opinion in Ch. 7 of the *Progymnasmata* (which is in *TBOO* II pp. 415–435). Kepler meant to refer to Chapter 7 of *De mundi aetherei recentioribus phaenomenis* (Uraniburg, 1588), which was described on the title page as "Book II," although no Book I ever existed under that title (Tycho evidently intended this as a sequel to his 1573 book *De nova stella* (*TBOO* I pp. 1–142); cf. *TBOO* IV p. 5). This chapter (in *TBOO* IV pp. 135–154) consists largely of a series of observations of the 1577 comet, showing that its tail faced away from Venus rather than the sun. The subject is taken up again in Ch. 9, and there is a diagram illustrating the direction of the tail on p. 203 (*TBOO* IV p. 173). Tycho's invocation of Optics is on pp. 204–5 (*TBOO* IV p. 175).

101 For this comet, see C. Doris Hellman, *The Comet of 1577: Its Place in the History of Astronomy* (New York, 1944).

102 In one account, King Pierus of Emathia was the father of the Muses; in others, the father of nine daughters who competed with the muses. Cf. Cicero, *On the Nature of the Gods* 3.54; Ovid, *Metamorphoses* 5.302.

the cause of this curving, and the question of the deflection will be settled. It is 265
not parallax. For, as will be said below in ch. 9, this cannot represent straight lines as curved. It is not refraction, unless we contrive some sort of monstrosity, that the aethereal matter by certain degrees of nearness to this celestial body is more and more thick, and unless it is so in only one direction, that towards which the tail verges. Further, if we will really to assert this, it will also be easy for us to reply about the tail's illumination itself. For the matter to be illuminated will be ready at hand. However, it cannot be illuminated by the sun itself, because the sun's light is simple and pure, without colors. But it will be able to be illuminated by the sun's light passing through the comet's body, for the reason that the sun's rays are gathered together and redoubled, and also colored, through the comet's body, which for this reason should be purely translucent and rather dense. But if we shall not be allowed to attribute to the tail of comets its own matter, we shall be forced to state, as we already had above, that the aethereal air itself is not quite entirely lacking in matter, but is made of something suitable for being so strongly tinted by the colored ray of the sun, passing through the comet's body from so near a place, that it is able to run in to the eyes from a distance with this color or brilliance. The rest will be carried out from the diagram of Chapter 5 Prop. 19: the spreading out of the tail at the end, and on the other hand the extremely compact origin of the beginning, and that which some people relate of certain comets, that right from the head the tail comes together as if in a point, and from there, as if from a point, a new tail is attached, broader as it proceeds. For all these things necessarily follow from the laws of a solid pellucid globe.

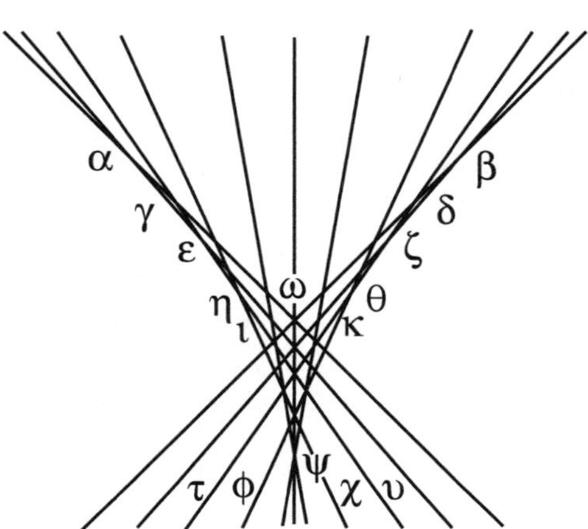

Next, this too will follow for the consideration of the natural philosophers: 266
the body of a comet consists of a certain moist substance denser than air, for that was in the definition of the pellucid in Chapter 1. But not that that globe of moisture is hard like glass, for it is gradually dissolved. Thus it is necessary that the moist substance be not solid but flowing and soft, which the comet's whitish light that Tycho Brahe attributed to it also appears to confirm.[103] This original opinion of mine is also eliminated, that comets are bubbles, that is, hollow or of a different nature on the inside. For unless they be solid and consist of the same material everywhere, such a refraction does not take place, that gathers the rays of the sun passing through and, after they intersect, disperses them again on the opposite side.

The opinion about the watery material of comets is confirmed by the new

[103] See, e.g., *De mundi aetherii recentioribus phaenomenis* Ch. 1, in *TBOO* IV p. 12.

star of 1572, which the authors, and Cornelius Gemma in particular,[104] describe in such a way that it appears to put on nearly all the colors of the rainbow in the succession of time. For he says that it began from red; next, when it was shining, it was greatest; afterwards silvery; and finally, it disappeared pallid. And this is the very order of colors in the rainbow: red, yellow, silvery or green, blue, and purple, or, what is the same thing in a star, pallid. And the colors of the rainbow are evidence for moisture, as appears from the first chapter. Therefore, it is also likely that that star consisted of moisture.

Notes

If only this be maintained, whatever cause is alleged for it, that the matter or the aethereal air is more strongly illuminated by rays of the sun passing through the comet's body than by the bare and pure rays of the sun, the remaining things will have been explained. For vision follows radiation, so that if a portion of the aethereal air beyond the comet, opposite the sun, radiates, it will as a result also be seen, and the more the former happens, the more the latter also. But the tails of comets radiate most strongly, such as the one which Cardano describes from Haly,[105] *thrice as great as Venus, whose light was so great as if a fourth part of the moon were to shine; likewise another, which they claim was seen right in the middle of day. And, lest you might think that this light is from the comet's head alone, look you at that Mithridatic comet, described by Justinus in Book 37.*[106] *In size, it had taken up a fourth part of the sky, and so shone that the whole sky seemed to be on fire: this must be understood to refer to the tail. The tail was certainly seen to rise and set without the head's being present, because rising and setting took an interval of four hours. Another one first entered into the eyes of mortals by the brightness of its tail, after sunset, before the head had risen heliacally: it at last emerged in the following days.*

Now, once the illumination of the aethereal air is granted, there will also be
267 *granted certain degrees of proximity for the location of the comet's body. From these, you might not unaptly also construe the horned shape, of which sort that comet of 1577 seems to have been,*[107] *not that the sun's rays, contrary to the nature of light, are bent into an arc, but that from one and another part of the comet's body, according to their uneven distribution, longer rays come out here, shorter ones there, all of which, arranged in relation to each other, show the shape of one curved tail.*

The same thing can also be the origin of the declination. For because we have said that by the common optical account it happens that the sun's rays, sent

---

104 Cornelius Gemma, *De naturae divinis characterismis* (Antwerp 1575) II 3, *De prodigioso Phaenomeno syderis novi* . . . The description in question is on page 139.

105 Girolamo Cardano, *In Cl. Ptolemaei . . . quadripartitae constructionis libros commentaria*, Basel 1554 and Lyons 1555, p. 353 (Book 2, text 54).

106 Marcus Junianus Justinus, *Ex Trogo Pompeio historia*, Book 37.

107 On the comet of 1577, see C. Doris Hellman, *The Comet of 1577: Its Place in the History of Astronomy*, New York: Columbia, 1944; reprinted 1971 by AMS Press, New York.

through the comet's dense body, are first driven together as to a single point, and there, upon intersecting, are again spread out: thus the lateral extremities of the tail will decline from the diameter of the sun on both sides. Grant now that one side of the head, whether because of its shape or density, does not transmit the sun's rays: that part of the tail will therefore be cut back, while the other part, declining to the other side, will by itself display the form of the whole tail, which will consequently be declining. And furthermore, when the cometary body's situation or location with respect to the sun changes, this inclination will also be able to change, as in 1596, the tail first declined to the right, next was in line with the sun, and finally went over to the left. The same also can be believed about the changeable form, and the reason will be clear why, by Pliny's account,[108] *the form of a goat was once changed to a spear at the approach of the Macedonian troops. For by an example of these same times it is known that the shape of the head tends to change: a terrible beam of light, like flaming clouds, was seen to go off on two sides and move away in different directions.*[109]

But what if we mingle the Aristotelian opinion of the tail with the more recent one, so that some luminous matter really does exhale from the head, and indeed in that direction in which it is sent forth, by the sun's rays, as it were? Then if the tail were to touch the earth, no wonder that the air be infected by a poisonous influence.

I shall not conceal this diversion from you, reader, that you might know how to form the image of a comet. Let a ray of the sun enter into the chamber described in Ch. 2 prop. 7, and place a water-filled globe in the way of half of it, so that the sun's ray falls partly on the glass, partly on the surrounding air. You will see a comet on the wall.

---

[108] Pliny, *Natural History* II xxii/90, pp. 232–3. Some sources read *hirtus*, shaggy, for *hircus*, goat. cf. *OLD* p. 798 s.v. *hirtus*. In Pliny's account, however, it was not a *hircus*, but a *iubae effigies*, or mane-shape, that changed into a spear.

[109] The source of this example is not known; it appears not to have come from Pliny.

Chapter 7
On the Shadow of the Earth

We next proceed closer to the astronomical subject of eclipses. Although the considerations of the illumination and of the darkening of the moon are so tightly interconnected, as we have now said above in ch. 6 about the penumbra, many things about illumination still had to be put off to this place, and no completely conclusive opinion was reached as to whether the moon is entirely lacking its own light.

1. On the form of the shadow

268 The shadow of the earth is a cone, or in the shape of a cone, both on account of the illuminating sun's being round and on account of the earth's being round (mountains notwithstanding) and being smaller than the sun. This is sufficiently proved by Witelo and others: see Witelo, Book 2 Prop. 26, 27, 28, where 33 following is clearly intended for eclipses. For it shows that an eclipse of the moon occurs at a central opposition of the luminaries from the earth. The ones preceding this, 31 and 32, appear to be intended by Witelo to state the reason why, when the shadow has been wrapped around it, the moon is made darker at perigee than it is at apogee, which question we shall here treat explicitly.[1]

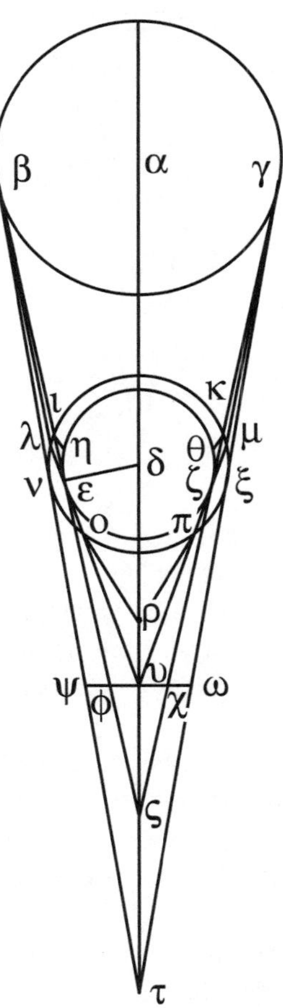

2. The paradox that the moon is not obscured by the earth's shadow

And so, these things being as they are, although ever since the birth of astronomy it has been generally admitted that eclipses of the moon result from its entering in to the earth's shadow, the computation also built upon these fundamentals did not correspond to the outcome with any great degree of exactitude. The greatest problem confronting this undertaking was presented by those things that were considered in chapter 4, on the refractions of the rays of the sun occurring in air. For if the rays of the sun tangent to the earth at sunset come to us refracted, they will therefore pass on refracted, and will shorten the shadow below the place of the moon's transit. We have to untie this knot, so that astronomers may be freed from doubt, and so that there should exist no one who dares

[1] *Thesaurus* II pp. 71–3. The theorems concern the shape and position of the shadow when one spherical body illuminates another.

quibble about eclipses from refractions, or about refractions from eclipses of the moon.

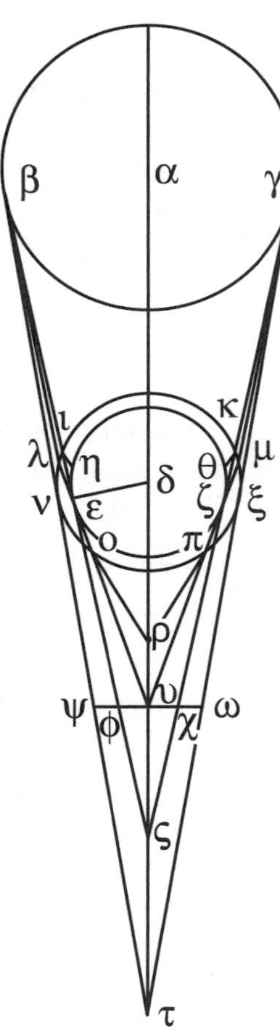

About center α let the greater circle βγ be described, representing the sun, about center δ the lesser circle εζ of the earth, and let the common tangents βε, γζ be drawn, extending until they meet at ς, with the axis αδ. It is obvious that εςζ is going to be the earth's shadow, if βε, γζ were to come through unrefracted. But now the globe of the earth εζ is surrounded by the sphere of air ικπο, whose medium is denser than the medium of the aether, with the result that βε, γς, that are going to enter the air at the points ι, κ, cannot carry through to ε, ζ, but are refracted into ιη, κθ, and strike upon the earth at η, θ, and are there hindered from proceeding any further. And since the sun illuminates as much of the entire sphere of the air as is presented to it, it sends forth other rays beyond βι, γκ, some of which, without doubt, will be tangent to the orb of the earth, and will carry on past the earth to the opposite part of the air. Let βλ, γμ be the rays, which are refracted at λο, μπ, tangent to the earth at **269**
ε, ζ. Now, from what Tycho Brahe has established by experience, the horizontal radiations λε, μζ undergo a refraction of 34 minutes. As a result, the angle βλε is 179° 26′; and γμζ is the same amount. Further, since by Witelo 10, 9 the same things happen in going out as in coming in, λορ will likewise be 179° 26′, as well as μπρ. As a result, the genuine shadow of the earth will be εορπζ, in fact still shorter, as will be clear below. But at the sides of ρ there is cast some refracted sunlight. It will be weaker by half than the light of the setting sun for us, which can indeed by no means be called darkness, even though it has come in to us through the depth of the air λε, μζ. For it makes it day for us nonetheless, and casts shadows from bodies. Therefore, even though the spaces λο, μπ are twice λε, μζ, a total extinction of light will not be able to occur.

Further, from these suppositions the altitude δρ of the earth's shadow, made by the rays λο, μπ, is easily obtained, although others present an even shorter altitude, as will be said below. In ch. 6 section 3 above, the angle εςζ was determined as 0° 25′ 36″, from the Ptolemaic suppositions.[2] But in triangle λβε, λεβ and λβε together make 34′. And because λβε is entirely imperceptible, because the extremely short portion of the air λε is imperceptible in proportion to λβ, which is equal to 1,200 semidiameters of the earth, therefore λεβ alone is 34′. Let λο be

---

[2] This is twice the angle *ONP* in the diagram in ch. 6 sect. 3, and is given at the end of the section, on p. 246.

270 extended to υ. Thus in ςευ, ςευ will be 34′, and εςυ is 12′ 48″, half of εςζ. As a result, the exterior and opposite angle ευρ is equal to the sum of the two, and is 46′ 48″.

Continuing, in triangle υορ, υορ is given as 34′, and ουρ as 46′ 48″, and ορδ is the exterior and opposite angle and thus equals the two together, and is 1° 20′ 48″. From ρ let a tangent to the earth's circle εζ be drawn, and let it be ρε. And let the point of tangency ε[3] be joined with the center δ. Thus in the right triangle δερ the angles are given, because ερδ is very small, and is imperceptibly less than ορδ, 1° 20′ 48″. Nevertheless, let it be 1° 20′. Now, as the sine of ερδ is to the semidiameter of the earth εδ, so is δερ to δρ, the altitude of the shadow. Hence, therefore, the actual point of the shadow comes out to be no higher from the center δ than 43 semidiameters, while elsewhere, from these suppositions, δς was estimated at 268 semidiameters. But when the moon comes closest to the earth, it reaches a limit of 54 semidiameters, making its transit 11 [semi]diameters higher than the earth's shadow, bounded by λο, μπ, comes to an end. And thus this demonstration is most true, except to the extent that a departure from the truth is made in the angle of maximum horizontal refraction, which is not of the same magnitude everywhere on earth.

What then is to be said in reply to the astronomers? Is not the entire theory of eclipses, and with it the measurements of the bodies of the sun, the moon, and the earth, as well as the general relative measure[4] of the celestial spheres, brought to ruin, its foundations undermined? What must be said, of course, is what is the fact—hear us with unbiased ears, worthy antiquity, for in your honor we cannot forsake the consideration of refractions, most firmly established by Tycho Brahe. This, I say, must be replied: that which comes into the computations of the astronomers, removing the sun's light from the moon, is the shadow, not of the earth (except in the middle, of which see below), but of the air surrounding the earth. Let the lines βν, γξ be tangent to the sphere of the air at ν and ξ, meeting with αδ at τ. I say, then, that ντξ, the shadow of the air, is that which deprives the moon of its light. And so, as the opaque body is, so is the shadow: the air does contain something of the opaque, especially about the edge, just as every watery globe exposed to the sun casts a shadow. But even so, just as the air transmits the rays of the sun to us, but now reddened, so its double (though in some places single, in some places half or even less) transmits the rays of the sun all the way to
271 the moon; and at the beginning of the eclipse, the moon is not in the shadow of the earth, since it is still in the rays of the sun transmitted through the air. And this is the very thing which we were trying to confirm above in ch. 4 with examples, that

3 The alert reader will note that this is the third distinct job that the point ε (or perhaps it should be "the points ε") is required to do. First, it was defined as the point of tangency of the line βε with the circle εζ, then it became the point of tangency of λο with that circle, and now finally it is the point of tangency of ρε. This redefinition allows Kepler to simplify what otherwise would have been an impossibly complicated diagram.

4 *Symmetria.* This is the often misunderstood word used by Copernicus to characterize what his system had that Ptolemy's didn't, namely, a common relative measure of the parts of the universe. The English word "symmetry" did not acquire its common meaning until after 1800.

the moon was eclipsed while the sun and the moon appeared above the horizon. For since vision occurs through the reception of rays, it will not be the case that our vision, established in the plane of the horizon, will go across in one place, while the rays of the luminaries in another, but the rays of the sun, in going to the moon when it is in the shadow of the air, will find the same path through the air that our eyes indicate.

Here someone may object that the computation of astronomers will be disrupted. For if it is not the earth εζ, but the wider sphere of the air νξ, that casts the shadow, surely, since we are standing on earth and not at the surface of the air, we measure the parallaxes and altitudes of the moon in semidiameters of the earth εδ, not of the air νδ. Therefore, having measured the shadow at the place of the moon's transit in semidiameters of the earth, not of air, even though it arises from the air, and supposing it to have the same thickness on earth as it has in our measure, arising from the earth, we have accordingly stretched it out, whence it happens that the sun comes out too high for us in our computation. Let υ be the place where the moon makes its transit; the thickness of the shadow from the earth would be φχ, but of the shadow from the air it is ψω. When we measure this by εδ, it is exactly as if, in our false imagination, we were to draw the lines εψ, ζω. If these are to subtend the sun βγ, it will be necessary to raise the sun up far higher above δ. To an astronomer who says something of this kind in objection, I have this one thing to say in reply: that it is indeed truly argued, but is beyond the limits of perceptibility. For from ch. 4 section 6 prop. 11, it appears that the altitude of the air is barely half a mile,[5] even if the horizontal refraction is just 34′. But the semidiameter has 860 of such half miles. As a consequence, when we compute one thousand seven hundred semidiameters of the earth, we are short one of these because of the altitude of the air. But we can be short one semidiameter out of 60, after we have applied all the precision of observation. For this reason, this very slight uncertainty must not be compared with others under which, for necessary causes, astronomy labors elsewhere.

3. On the redness of the eclipsed moon

Lest anyone be undecided about what was said, that the moon is still illuminated by the sun's rays, admitted and refracted through the air, even when the **272**
eclipse is total, he should carefully attend to total eclipses of the moon: he will find this illumination flowing into the eyes. This is not a case of someone's having mentioned this fact recently, for the sake of refractions, so as to fit a suitable epilogue to the tale. The thoughts of the ancients survive, and it is for just this reason that the opinion originated that that redness is the moon's own light. In the Commentaries on book 4 of Ptolemy,[6] Theon philosophizes thus:

> As regards the eclipse of the moon, the earth brings this inconvenience upon the moon. For since the moon is an eternal body, participating in the divine

---

[5] This is half a German mile, or about $2\frac{1}{2}$ statute miles (4 kilometers).

[6] Theon of Alexandria, *Commentaria in C. Ptolemaei magnae constructionis libros*, 4. 1, p. 956.

> nature, and immune to the modifications that as a rule follow upon coming to be and passing away, it accordingly will not subject itself to alterations or diminutions. But since it is endowed with its own rather weak light, which cannot easily reach our powers of vision, it appears more brilliant by another light, borrowed from the sun, so as to illuminate the earth and the air. Thus when the earth intercepts the solar radiations, the moon is then deprived of the solar light, and, darkened by the cone of shadow, it again comes to be weaker and darker. It does not itself undergo any alteration from this illumination, nor from the privation of of the solar light, but since it is free from corruption, it always preserves the same color and appearance which it has as its own, and the same magnitude. Further, that the moon, though indeed always illuminated, is in certain full moons not illuminated with its usual brightness, but is darkened, they have called an 'affect' and a 'labor', because they believed that it undergoes that alteration in its own activities.

And below:

> The moon indeed does also have a light of itself, but weaker, as we have said.

Cleomedes, in book 2 of the *Cyclica theoria*, follows Theon: he blends the light of the moon from the sun's light and its own. Reinhold, most recently, on fol. 64 of the *Commentaries on Peurbach*, says

> There is in the moon a certain special or rather dim light, and its quality is openly shown by the total eclipses of that body, in which the entire orb is perceived with a foul and horrible color, which, however, is redder in some places, when the moon is higher and lifted farther outside the ecliptic, darker in other places, that is, to the extent that it is lower and for that reason more deeply immersed in the shadows of the earth.[7]

By these final words, he transparently refutes what both he and his predecessors had said, that that light is the moon's own. For if it were the moon's own, it would appear more brightly out of the darkness. For thus also, the light of the moon that is borrowed from the sun is brightest at night, while by day it hardly
273 moves the eyes. Now, by the testimony of Reinhold, the opposite happens. For upon entering deep shadow, it appears darker, higher and nearer to apogee it is redder, as well as when it verges to the side of the shadow to the north. Whence it is understood that at the edges of the shadow the moon finds a confused light, which I say is this very light that is transmitted refracted through the body of the air.

For this reason, Plutarch, more ancient than all of these, speaks rightly in the book *On the Face of the Moon*, when he introduces Pharnaces, arguing that, like the other stars, the moon has its own light. "It does not go straight into hiding in

---

7 Peurbach, *Theoricae novae*, ed. Reinhold: Wittenberg 1542 fol. T6r; Paris 1553 and 1556 p. 104. The relevant passage is under the heading, *De Illuminatione Lunae.*

eclipses, but shines with a certain color suggesting a coal, a terrible color, which is its own." He himself objects in reply as follows:[8]

> We see the eclipsed moon take on different colors in different places. Mathematicians distinguish these on the basis of time, thus: if the moon is eclipsed in the evening, it appears horribly black, right up to the third hour and half an hour beyond. If at midnight, then it gives off that scarlet and fiery color. From half past the seventh hour, redness is shown, near dawn it now takes on a blue and harsh countenance.

This rule in fact is false, and does not do what it proposes to. The Arabs have prescribed a different one for us, on the color of eclipses, which Cardano, in the supplement to the Almanac,[9] and in the commentary on the *Tetrabiblos* from Alfonso and Linerius,[10] accommodates to the nearness to the node and to the apogee, which are more in harmony with Reinhold's tradition, although the varying constitution of the air seems not to permit a rule. But Plutarch carries on, and accomodates this objection of his: "One might say instead," (and excellently, in my opinion), "that this coal-like [light] is foreign to the moon, and is instead a mixture of the eclipsed light and that shining through the shadow, while its own [light] is black and earthy."[11] But since it is a question of the causes that spread this light within the shadow, he brings in the other stars standing about the sun. The cause of this particular redness is very weak, although below his argument will also be kept for the rest of the colors. Our Witelo, when he was about to completely deny the moon its own light in book 4 prop. 77, appears to have aimed book 2 proposition 31 at this redness of the moon in eclipses.[12] He says there, "The shadow is less shadowy at more distant parts." Sound words, if the cause be sound. For the rest, Witelo ascribes the cause to the vision, which compares the shadow and the more luminous or weaker rays, placed next to each other. For the rest,[13] the moon in this way would undergo nothing else in its body, traversing **274**
the shadow farther out at its darkest part, and so this phantasm of vision contributes nothing to this redness that it has. The cause, however, is entirely in the refractions, so that that redness is nothing else but the illumination of the moon

---

8 Plutarch, *On the Face in the Moon*, XXI, 933F and 934C–D, pp. 132–3 and 136–7.

9 I have not been able to identify this work.

10 Cardano, *In C. Ptolemaei Quadripartitae Constructionis libros Commentaria.* Basel 1555, II text 52, p. 338–9, citing the "tables of Alphonsus and Linerius." Cardano's table relates the color of the eclipsed moon to the sun's distance from the node, ranging from black (close to the node) to light yellow (at 12°). Cf. Ptolemy, *Tetrabiblos*, II 9. "Alfonso" refers to the astronomical tables prepared in the thirteenth century under the patronage of Alfonso X of Castile and Leon, which circulated in a number of versions and were the most commonly used astronomical tables even in the sixteenth century. "Linerius" was John of Lignères, a fourteenth century Parisian mathematician largely responsible for the diffusion of the Alfonsine Tables, which he supplemented with tables of his own. Cf. *DSB* VII 132–8.

11 Plutarch, *On the Face in the Moon* XXI, 934D, pp. 136–7.

12 *Thesaurus* II pp. 151 and 72, respectively.

13 Kepler has begun both this sentence and the preceding one with *caeterum.*

by the sun's rays, transmitted through the density of the air, and refracted inwards towards the axis of the shadow, as becomes clear from the following examples.

Frodoardus in A. D. 926: "The moon was eclipsed on the Calends of April, and was changed into a pallor by the small part of the light that remained, as if it were a copy, and thus, as dawn was breaking, it was all changed to a bloody color."[14] Therefore, it grazed the shadow from the side, since it carried this redness before it.

Cornelius Gemma relates in the *Cosmocritica* Book 2 fol. 64:[15] In 1569, on March 3, "at the third hour of the morning, Phoebe suffered a terrifying eclipse, marked with frightful colors. First," he says, "dusky, it next shone out bloody, soon after scarlet, and green, and slate colored, and finally unsightly in unbelievable variety." Here he practically said that in the moon were seen the colors of the rainbow. Go now, and place a spherical glass filled with water in the way of a ray of the sun entering through a small crack into a dark chamber: you will see all the colors of the rainbow on the opposite wall. Further, these colors are always the offspring of moisture, which the ray of the sun traverses. But what moisture is there that the sun's ray could traverse, other than the humid sphere of the air? Indeed, the moon was towards the perigee, but was traversing the southern part of the shadow. And so in both instances it was in the part of the shadow that, to use Witelo's words, "is less shadow-forming," that is, diluted by many rays of the sun that have undergone refraction in the thick air of the north. For when Saturn had a configuration with Jupiter, and Jupiter with Mars, and the sun with Mercury, they aroused a vaporous air from the warming recesses of the earth, as completely unvarying experience shows us. Nor did the seasonal weather happen to give moderate assistance, like that in another year (1597), which so greatly thickened the air in those places that the phenomenal refraction of Zembla occurred.[16] But what Gemma adds, "the third hour of the night," he seems to write from Stadius's Ephemerides rather than from a sure observation. For the middle of this eclipse could not have appeared far from an hour after midnight, by the evidence of Tycho's computation from reliable experience.

275 I remember when I was still a boy in Württemberg being called outdoors by my parent to the contemplation of an eclipse, at 10 p. m. at night; the moon appeared entirely reddened all over. The circumstances attest that this could not have been anything but the one that was seen on 1580 January 31, whose northern

---

[14] There is some uncertainty as to which Frodoardus is meant. According to Frisch, *JKOO* II p. 420, Frodoardus was an Italian monk who lived around the end of the tenth century, and wrote a *Chronicum a Caesare Octaviano ad annum 996*. However, Hammer, *JKGW* II p. 452, cites Flodoardus, *Annales Remenses*, *Mon. Germ. Hist. Script.* III 376. This Flodoardus is presumably the Flodoard who was born in 893 or 894, probably at Épernay, France, and died in 966, probably at Reims. He wrote *Annales*, a chronicle covering the period 919–966. His works are in the *Patrologia Latina* Vol. 135; there is also a separate edition of *Annales* by P. Lauer, Paris 1905.

[15] *Ars cosmocritica* is an alternative short title for Gemma's *De naturae divinis characterismis* (Antwerp, 1575).

[16] Cf. IV 9, p. 151

latitude Maestlin, in the *Isagoge*,[17] said was a little less than was the sum of the semidiameters, for which reason it was entirely hidden in the shadow for a very brief time. Compare it with Frodoardus's.

On 1588 March 3, in the morning, in the Tychonic observation of the total eclipse of the moon, I find, at 2^h 58^m the annotation, "The moon was now seen brighter towards the east," and because of this these words were added: "Therefore, it had passed the middle."[18] But the middle followed by a few minutes, as is clear from the beginning, end and other phases. Here, in one and the same eclipse, the moon at one part gradually puts aside, in the other puts on, that redness which the ancients considered the moon's own light. And since it had not yet come to the middle of the shadow, it shows, through the redness of the eastern hump, that the rays of the sun refracted in the terrestrial air are wider near the western than near the eastern part of the shadow.

I saw a similar eclipse of the moon at Graz between 30 and 31 January, or 9 and 10 February, 1599, concerning which see also above, ch. 6 sect. 9. For when the moon was just immersed in the shadow, with a very small part of the light from the sun remaining, it was nevertheless perceived almost totally by its redness, so obvious was the redness. Next, opposite the light from the sun (a little above), it appeared to be eclipsed again; that is, the redness itself gradually gave way, or rather, the moon on its eastern edge entered in to the darker part of the shadow. The last light from the sun ceased at the high point[19] of the redness, so that you would not know how to distinguish it, except that over a large area at the horn it had a red color, and it was equal, but not brighter, in the middle. A little before it moved itself beyond the mountains, even though it was in the bright twilight and the moon was deeply immersed in the shadow, the redness could still be seen; and further, it could even be distinguished that the western part was still the brighter. It was, nevertheless, about the middle of the shadow when it set, for no light appeared through the air, and no trace of it shortly afterwards, the air nearest the mountains being quite humid and thick.

For something similar, see also below in ch. 11 probl. 31, in the eclipse of the moon of 1603 8/18 November.[20]

But no clearer example can be supplied than what I saw in 1598 6/16 August **276**
near Graz. The moon rose in darkness, the sky being overcast near the horizon, so that the rising moon could not be seen. When it now had several degrees of altitude, the clouds opened up and it shone forth, with half of its body very bright with redness, so that it would not be considered eclipsed in that part, half being barely visible, even though the whole was in the shadow. I, being unused to this redness, was amazed, for I was unable to reassure myself in any opinion

---

[17] Maestlin, *Epitome*: Heidelberg 1582, p. 446; Tübingen 1588, p. 458; Tübingen 1610 p. 493. Citations hereafter will be from the 1610 edition.

[18] *TBOO* XI p. 258. Kepler's words are slightly different from Tycho's as transcribed by Dreyer, which are as follows: "N. B. It had now passed the middle, because the moon was brighter towards the east."

[19] ἀκμῆν.

[20] See p. 414, below.

of whether it was entering the shadow or leaving it. It also powerfully held me in doubt, why the shining part was not clearly distinguished from the dark part, for the redness did not have a boundary, but was continuously diminished and finally faded out from the greater part of the moon into darkness. While I was thus contemplating, the clouds again interposed themselves. A little later, when three quarters past seven was ringing in the town, the departure of the clouds dissipated all doubt. For the bright horn had risen below, and I saw the redness vanish and become like the darkness, so great was the radiance of the narrow horn illuminated by the sun. The redness did not lie uniformly around this horn, for the redness was spread out most widely on the upper left side, and it was narrow at the lower right, between the bright horn and the part that was competely dark. At that moment, I captured the phase of the moon like this. At the ninth hour on the town clock, it appeared completely restored, more than 5 quarter-hours being taken up in its exit, which agrees excellently with the Brahean observation taken at Wittenberg. What I am seeking here can be said to be different from the idea that this redness arose from the rays of the sun refracted in the earth's air, the measure of which refraction was different in different places on earth at that moment.

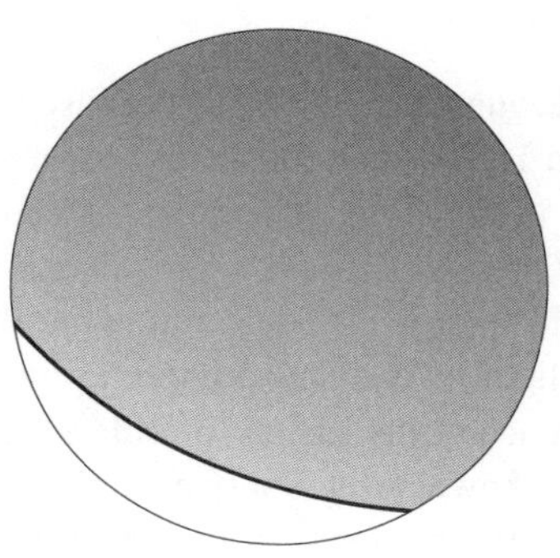

For if the heavens[21] standing around the sun were to color the moon so, they would color its whole disc, equally set out before them, uniformly.

The entirely dark part, here and in the preceding eclipses, would have been in the full shadow of the earth, the ruddy part in the shadow of the air, which was transmitting the refracted rays of the sun, and finally, the lucid part, free from
277 both shadows, would have been in the grasp of the pure sun, or of some part of it.

4. On the pallor of the eclipsed moon.

It is now appropriate to take note, though rather inadequately, of what concerning the moon's body lies hidden in the earth's shadow. Thus, in 1588 March 3 at Maulbronn in Württemberg,[22] when the moon was in the middle of the shadow, I was barely able to perceive it with the eyes because of the ashen color, and I was amazed when I remembered the one I had seen in 1580. In Denmark, when they measured its distance from the fixed stars, they had such difficulty picking out its

---

21 *Sidera.* This usually refers to heavenly bodies, but the context here and below suggests that Kepler meant the glowing "aethereal air" surrounding the sun.

22 Where Kepler was a student at the Stiftsschule from 1586 to 1589.

edges that the observed distances sometimes appeared to show it to be retrograde, near the middle of the eclipse.

So is it this pallor or ashen color that belongs to the moon's own light? We still do not have any need to say even this. In fact, we may rather reply here, with Plutarch in the *Face of the Moon*,[23] what we were unable to reply above on the redness of the moon, because of its brightness, in this way.

> Now since here on earth, on ponds and also rivers, receiving the rays of the sun from purple and crimson awnings, the shady places nearby imitate that color, and because of reflections of whatever origin, are illuminated with multiple splendors, what wondrous thing is it if an ample flow of shadow falling celestially as if upon the sea of light, which is not calm or at rest, but is stirred up by numberless stars, and receiving various mixtures and alterations into itself, carries back from the moon hither different colors at different times?

I have recounted these words because they seem to smack of an experience like that from 1569 which Cornelius Gemma had set forth above, on the colors of the rainbow in the moon. Therefore, as concerns the opinion of Plutarch, if various colors of this kind appear in the moon, I would say, as above, that they arise from the humid sphere of the air, refracting the sun's rays, but not at all from the fact that the light of the heavenly bodies in those bodies themselves is so obviously marked out by this variety of colors, and that it imbues the moon with similar colors. If, however, it is a question only of the very faint pallor of the moon, or the ashen color, whence it comes to the moon, and whether it be of the moon's own light, I shall reply with Plutarch that this spark of light is acquired for the moon from the heavens standing about the sun, or whatever else is luminous in the heavens. This I shall do, though Tycho was impartial, who believed, above,[24] (though less thoughtfully) that Venus alone is sufficient as to acquire light for the **278**
moon so strongly, in order that, when it had come out from the sun, it may thereby struggle its way out through the exceptional brightness of the twilight.

No night is so black, even if the densest clouds hold the sky, that if you were outdoors you could not distinguish between snow and coal, between sky and mountains. Accordingly, in the densest part of the earthly shadow (which brings about the night) it never happens that absolutely all the rays of the stars are excluded, or that nothing of these terrestrial objects is illuminated. Why then would it not be even more true, that in that pure aethereal air, in which the moon has its passage, existing sixty semidiameters of the earth above this misty air of ours, and with the shadow now drastically weakened, the heavens might be able to do the same to the moon, lying in the earth's shadow, that they can do here in the mountains, in clouds, in snow, in coal? Not to mention that some of the rays of the sun refracted in air reach even to the middle of the shadow. As a result, with the support of such lucid causes of this spark of light, there is no need to ascribe to the moon its own light.

---

23 Plutarch, *On the Face in the Moon* XXI, 934D–E, pp. 138–9.

24 See p. 275, above.

5. Problem: To measure the refractions in the regions at the greatest distances by observation of eclipses of the moon.

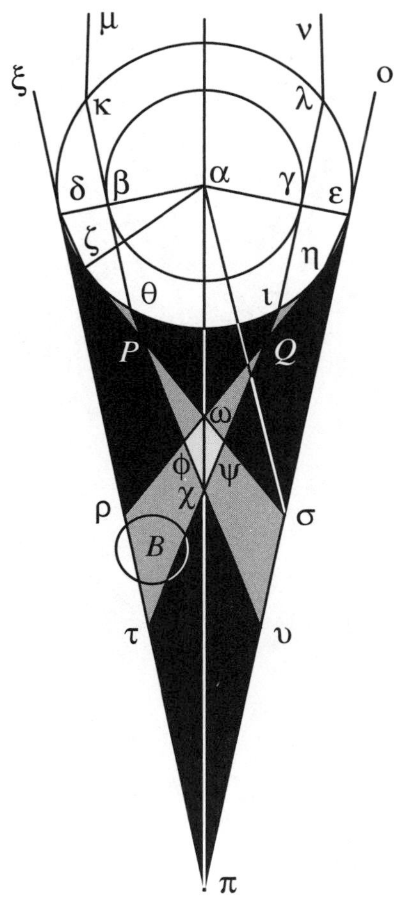

In section 2 above, from a horizontal refraction of 34′, we marked the limits of the shadow of the earth from the outermost refracted rays of the sun, tangent to the earth, it being determined as 43 semidiameters of the earth long, and it was said that when the moon is lowest it crosses 11 semidiameters higher. In this way, this shadow of the earth never would touch the moon. But in section 3, when an account was given of the examples that were presented, we fully granted to the authentic shadow the darkening of part of the lunar body. In order that no one think something contradictory was proposed there, I ask that the following be considered: that refractions are not the same in all places, as was sufficiently stated in ch. 4 sect. 8 above. For in Kassel, in an inland region, they are found to be less. Next, it must be carefully considered which particular region of the aerial shadow the rays refracted by the humid substance of the air can in any way cross. About center α let the greatest circle of the earth's circle βγ be drawn, and about
279 this the wider circle of the air δε; and let δε be the circle of the illumination of the air; and let the rays ξδ, οε from the edges of the sun come together at π, forming the boundary of the shadow of the air, which causes the eclipse of the moon. Now let the rays tangent to the sphere of the air at δ, ε be refracted into δζ, εη; and, exiting there, let them again be refracted into ζω, ηω. Thus αω will be the nearest boundary of the shadow. For as regards the parts of the sun between the rays δξ, εο, they do not reach δ, ε, but, tangent to the sphere of the air above, towards κ, λ, they make the same angles of refraction, and, since they entered at a higher place than δ, ε, the also exit at a higher place than ζ, η, and thus become farther out than ζω, ηω. Again, let β, γ be places on opposite semicircles of a meridian of the earth, at which places, when the center of the sun is placed upon the line αω, the entire sun is seen from both, refracted, grazing the horizon with its lowest edge, so that μκ,νλ are refracted into κθ, λι, and thereafter into θχ, ιχ, cutting ζω, ηω at *P*, *Q*. By what was demonstrated in ch. 5 sect. 3 prop. 9, it is necessary that if ξδ, μκ had been perfectly parallel, and likewise οε, νλ, still ξδζω and οεηω would meet at ω nearer the earth than μκθχ, νλιχ. This is now much more so when these lines are inclined, because ξδ, οε by supposition come

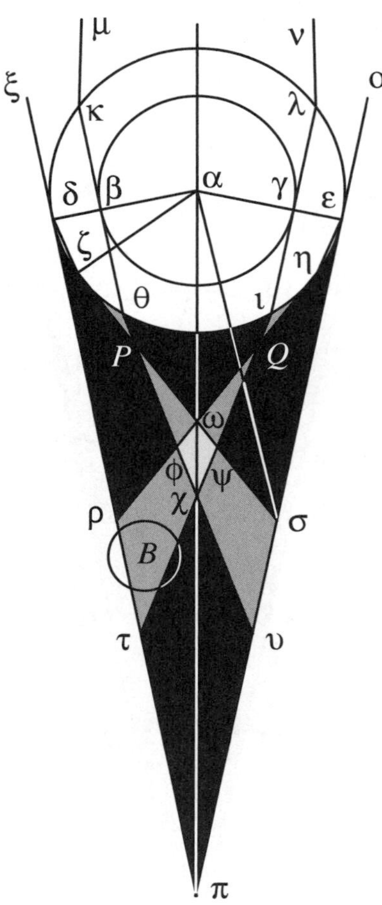

from the sun's upper edge, μκ, νλ from the lower. Thus χ will be more distant than ω, and θχ, ιχ are the other boundaries of the refracted radiations. Now whatever radiations besides μκ, νλ should fall from the upper parts of the solar body upon κ, λ, will by that intersection en- **280**
counter the earth, since μκ, νλ, are the last of the ones that do not encounter the earth, but, because of the refraction that occurs, are the last of the ones that are tangent to it, at β, γ. With these things supposed thus, let ζω, ηω, θχ, ιχ, be extended until they cut each other at φ, ψ, and farther, until they cut δπ at ρ, τ, and επ at σ, υ, conjointly. There will thus be a conical region with a double vertex φχψω in the middle of the shadow, in which the refracted rays falling down circularly are gathered, around an altitude from the earth of 43 semidiameters. Again, in the earth's shadow there will be a certain oblique garland, ending in a point, which is represented in two places by the spaces ζθ*P*, ηι*Q*, in which the refracted light of the sun is spread circularly, but more concentrated at the circle *P*, *Q*, where again the hollow conical region begins, represented by *P*ωφ, *Q*ωψ, in which again the refracted light of the sun is spread, but now more weakly, because it is over a broader region. But whether either of these serves to illuminate
Chasmata. chasmata, which nearly always face north, the natural philosophers[25] may judge. Finally, there remains the conical region of the shadow, hollowed out on both sides, represented by ρφχτ and σψχυ, through which the refracted light of the sun is much more broadly spread out, and which is much higher than 43 semidiameters of the earth, and is hollowed out in the middle by the pure shadow τχυ. This, then, is the place that the moon traverses.

In order that we may measure this region, note first that, in ch. 4 sect. 6 prop. 9 above, the refraction of a ray from the aether tangent to the air (on the supposition that the ray that is tangent to the earth after refraction receives a refraction of 34′) was 1° 1′ 30″. Therefore, angle ξδζ is 178° 58′ 30″. And hence the arc δζ is 2° 3′, by Euclid 3, 2. And since δπα (as above, and ch. 6)[26] is 12′ 48″, and αδπ is right, δαπ will be 89° 47′ 12″. When δαζ, 2° 3′, is subtracted, ζαπ

---

25 *Physici.* See note on p. 141.

26 See above, sect. 2, and ch. 6 sect. 3, p. 246.

will be 87° 44′ 12″, which is ζαω. And because ξδ is tangent to the circle, ζσ will also be a tangent, by Witelo X 9.[27] Therefore, αζω is right. But before, ζαω was 87° 44′ 12″. Therefore, ζωα is 2° 15′ 48″. Hence, where the semidiameter of the air αζ is 1, αω becomes no longer than $25\frac{1}{3}$ semidiameters. And because ζωα is 2° 15′ 48″, σωπ will also be that much. But the exterior angle ωσε is equal to the two interior and opposite angles ω and π, and ωπσ is 12′ 48″. Therefore, ωσε will be 2° 28′ 36″. But since the circumference δθε was 179° 34′ 24″, as was said in ch. 6 sect. 4,[28] and δζ is 2° 3′, ζθε will be 177° 31′ 24″, half of which is 88° 45′ 42″, which is εασ or ζασ, because ζσ and εσ are tangents. As a result,
281 ασε is 1° 14′ 18″. Hence, where the semidiameter of the air αζ is 1, ασ becomes $46\frac{1}{3}$ semidiameters of the earth. Here, then, the region illuminated by the sun's refracted rays begins.

Again, because the upper arc βγ, by those things that have been supposed (and by ch. 6 sect. 8), had to be 179° 23′, if the rays of the sun had come through unrefracted, while the two observers at β and γ see the sun 34′ higher than it should be, that arc is therefore increased by 1° 8′, and becomes 180° 31′, with the result that the lower arc βγ is 179° 29′. Half of this, 89° 44′ 30″, is βαχ, and consequently βθ extended, together with αχ, encloses an angle of 15′ 30″, and the exterior angle θχα is equal to the sum of that angle and the supplement of βθχ, or 34′. Thus θχα is 49′ 30″. And since χθ extended cuts αδ between β, δ near β, it matters little whether we take αβ or something slightly greater, yet something much less than αδ, in the place of a measure. Now, since βαχ is 89° 44′ 30″, and αχθ (let it now be αχβ) 49′ 30″, αβχ will be 89° 26′. Therefore, where a certain line that is nearer to αβ than to αδ is 1, αχ comes out to be 70. Thus χ would be distant from the earth by 70 semidiameters of the earth; that is, much farther than ρ, σ. However, the rays from all the particles of the sun are not refracted all the way to 70 semidiameters, but only those from the edge, while the point where the rays of all the particles of the sun begin to be refracted was proved above to be raised up 43 semidiameters. And since ρ, σ are 46 semidiameters distant, while τ, υ are much more than 70 semidiameters, and the region ρτ, συ is illuminated by the refracted rays of the sun, and the moon makes its transit between the stated boundaries, between the altitudes of 54 and 60 semidiameters, as astronomy bears witness, it is therefore obvious that while it is in the shadow of the air, it is nonetheless somewhat illuminated by the sun.

Again, when the simple shadow τχυ is bounded at the apex χ at a distance of 70 semidiameters, the moon therefore would not make its transit through the simple shadow, because it transits lower, and nonetheless when τχυ (if now it be understood as the region lacking any particle of the sun, even shining refractedly) is bounded by the apex χ at a distance of 43 semidiameters,[29] the moon, which transits higher, will accordingly pass through the point τχυ (taken in its second

---

27 *Thesaurus* II pp. 413–4. This proposition shows that the direction of a ray is reversible.

28 This is not from section 4, but the very end of section 3.

29 *JKGW* II, like the 1604 edition, has a period here; following Frisch I am replacing it with a comma, which the sense requires.

sense), so that when it enters into the beginning of the air's shadow, it nevertheless enjoys the refracted view of the entire sun, but where it encounters the middle of the air's shadow, it no longer will enjoy the full view of the sun, but only a certain small part. Hence what was noted from Gemma is entirely consistent, that the colors of the rainbow are sometimes seen in the moon. For in a watery globe it is perfectly certain that where a particle of the sun shining under refraction is hidden by the actual occultation of the refracted rays, colors arise one after the other, according to whether much or little of the solar body will have been shining. You thus see a remarkable consistency of nearly everything that has been presented. If anyone should wish to refute it by false assertions, he will have to be artful about it. **282**

There remains a single reservation put forward in just this context; and though it is now almost entirely removed, it befits a lover of truth not to hide anything. In the above examples, the moon was divided, as it were, into three parts, the bright, the reddened, and the hidden, and the bright was divided from the reddened by an evident line, while the reddened was not divided from the obscured by an evident line or limit, but the reddened faded out successively into the obscured. It therefore appeared completely consistent that what was perfectly reddened was illuminated by the whole body of the sun, refracted, while what was perfectly obscured (to the extent that it happens to be obscure) plainly lacked all exposure to the sun, and the uncertain part in between was exposed to a part of the sun. As a result, the simple shadow $\tau\chi\upsilon$ and its apex χ ought not finally to arise from the seventieth [semi]diameter of the earth, but far below, so that in the place of the moon's transit, some small part may be in the simple shadow. The question arises, how this is achieved? The answer is easy, if the horizontal refraction is assumed to be greater, for then $\theta\chi$, $\iota\chi$ will meet at a lower point. Thus in August 1598, if *AG* is the ecliptic, and *ACG* the earth's shadow, *CD* the southern part, then the end *F* of the refracted rays will come from a lesser refraction, *E* from a greater. Thus in the north, the refraction was greater than in the west (because *A* is the eastern part of the shadow).[30]

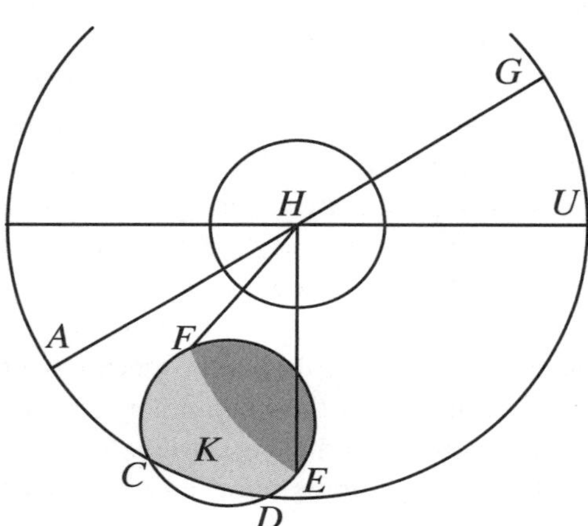

[30] Here Kepler refers to the following endnote:

To 282. The reasoning is as follows. *DE* are the southern parts with respect to the ecliptic *AG*; consequently, as you see in the diagram on p. 279, they will be illuminated by the refracted rays of the opposite region on earth, i.e., the north. And because a small surface of the moon, namely, *DE*, is illuminated there, this proves that the refraction of the north is large, by which it happens that (p. 279) the region $\tau\chi\upsilon$ becomes closer to the earth, and thus fills up the moon's transit broadly. In turn, *CF* are eastern parts

But so that it may be known, or rather, gathered from the observation of an eclipse of the moon, how much greater a refraction is required, let K be the center of the moon, 54 semidiameter distant from earth, because the moon was at perigee; H the center of the shadow; and let us suppose that it is at its nearest to the observation, and the beginning of the
283 obscured part would have been quite far from the center of the shadow, by 20 minutes at F, 36′ at E. In the following figure, let M be the vertex[31] of the cone of full illumination [συναυγείας], from chapter 6 section 7;[32] N be center of the earth; F or E be the boundary of the refracted rays, so that ONF is 20′, ONE 36′. Therefore (by ch. 6 sect. 7) IMN will be about 18′ 30″, whether MI be tangent to the earth or to the air, because the ratio of the air to the earth is imperceptible. And because IN is 1, NO 54, ION will be 1° 4′. If OI be tangent to the air, not to the earth, there will be added to IN less than a thousandth part, with the result that the angle will come out imperceptibly greater. And because ONM is one straight line, the sum of ION and IMN, which is 1° 22′ 30″, subtracted from 2 right angles, leaves OIM, 178° 37′ 30″. Subtract OIF or OIE, for it [i. e., each respectively] is approximately as great as ONF, ONE, there remains, for the former, 178° 17′ 30″; for the latter, 178° 1′ 30″. Let TR be applied, tangent to the earth, cutting the air at R, T, so that RIT is isosceles: the refraction, IRT or ITR, will become 51′ 15″ for the former, 59′ 15″ for the latter. But from Tycho Brahe we had accepted 34′, the amount it is in Denmark. However, in ch. 4 sect. 8–9 we showed that at different places and times the refractions are either greater or less, so that there is nothing absurd in this, that in a single eclipse in certain places the greatest refraction becomes 51′ or 59′, when the greatest and completely total refraction can be $61\frac{1}{2}$′, as was said above.

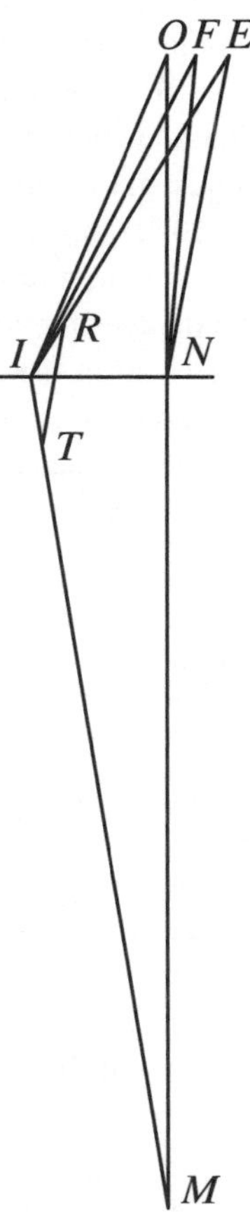

It is, however, not necessary even to state that this, however slightly it differs[33] from what would be entirely satisfactory, that what is cut off by EF, the boundary of the dark part, was in the simple shadow, nor would anyone wish

---

of the moon, since A is the eastern point of the ecliptic. As a result (in the diagram on p. 279), these parts EF will be illuminated by the refracted rays of the opposite region on earth, i. e., the west. And because CF is a wide part of the moon, this proves that the refraction is small, by which it is brought about that the region τχυ becomes farther from earth, and it barely touches the moon with a narrow point.

31 Here Kepler refers to an endnote which, because of its length, is included in the appendix at the end of this chapter.

32 See the footnote on p. 253.

33 Here Kepler refers to the following endnote:

To 283. In fact, there is another valid exception, namely, that everything is set up and measured from the estimation of sense perception, particularly p. 282 l. 34

to be content with the statement that that part lacked the greatest part of the sun, shining under refraction.

In order that it may also be shown at which regions of the earth the refractions now investigated arise, let us proceed thus. The observations of Tycho adjusted to Uraniburg give evidence that the moon began to emerge a little before $7^h\ 50^m$. I, however, when I observed it now first shining forth, saw it increased by **284**
a digit[34] or a little less: therefore that would have been at $7^h\ 52\frac{1}{2}^m$ at Uraniburg. Its degrees are about 117°.[35] The sun was on the meridian this many degrees to the west: in the region of Yucatan, in America.[36] The sun was at about 23° 20′ Leo, with a declination of 13° 47′ north. Therefore, with one foot of the compass on the intersection of the said meridian and the parallel,[37] and the other extended to measure a little more than 90 degrees, the circle of the illumination of the earth will be drawn, from whence the shadow arose.

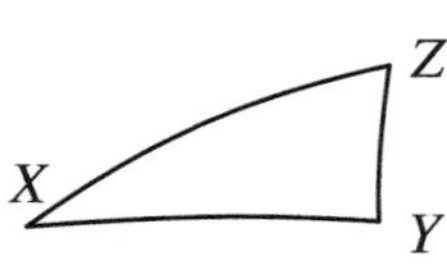

This circle cut the earth's equator to the west, between New Guinea and the Solomon Islands in the desolate ocean; to the east opposite the African Guinea beneath the meridian of the west coast of Spain. Towards the south, however, it touched upon the Magellanic landmass, which looks towards Peru; towards the north, the pole itself being held beneath the sun's gaze, the circle passed across the Batavians' winter place, beyond Muscovy.[38] And since at that time at a polar altitude of 47°, the rising point was about 18° Pisces, from Copernicus's table [of the angles of the ecliptic with the horizon] on fol. 42, the intersection of the horizon and the ecliptic, GHU in the next to the last diagram, was about 20°.[39] Therefore, in the right [spherical] triangle ZXY,[40] the angle X is given with the base ZX 25°, from 23° Aquarius to 18° Pisces.[41] Hence

---

and 284 l. 20, and here, from the reliance upon the sense of vision, we easily make quantitative errors.

34 One twelfth of its diameter, or about $2\frac{1}{2}'$. See the note on p. 71, above.

35 This is how far the moon has apparently moved in the 7 hours $52\frac{1}{2}$ minutes since noon.

36 According to the maps available at Kepler's time, most notably that of Abraham Ortelius (*Theatrum orbis terrarum*, Antwerp 1570 and many other eds.), the difference in longitude between Denmark and the Yucatan Peninsula was about 120°; in fact, however, it is only about 100°.

37 That is, at 13° 47′ north latitude.

38 This is a reference to the Barents expedition to the arctic, for which see Ch. 4 Sect. 9, above.

39 *On the Revolutions* II 10, trs. Rosen, p. 76.

40 For the sake of economy, Kepler combined this diagram with the preceding, placing the vertex X at the top in order to fit the triangle in. It has been rotated 90 degrees counterclockwise here to orient it correctly.

At this place in the text, Kepler refers to the following endnote:

To 284. Let X be the rising degree, Z the moon, Y the azimuth on the horizon, and XZY the intersection of the ecliptic with the vertical.

41 In the diagram, XY is the horizon and XZ the ecliptic. As Kepler says, the moon is at Z. Y is the point on the horizon directly below the moon; hence, angle ZYX is right. Angle X was just found to be about 20°, and the angle on the ecliptic is 25°.

is obtained XZY, about 71° 45′. However, the luminous horn CD, which is as close as possible to the observation, was also declined by about 20° from the vertical through the center of the shadow or of the moon. Therefore, the circle which passes through the centers of the moon and of the shadow cut the ecliptic at an angle of about 51° 45′. And the parts F of the shadow were closest to the ecliptic, while E was closest to the vertical. The former were cast forth by the Solomon Islands and New Guinea, while the refracted rays that were spread in between underwent refraction in the inmost deserts of Africa. The latter, on the other hand, arising from the Magellanic landmass, were spread in between by rays of the sun refracted in Finland and the places around it. Thus the refraction was greater, at this moment, in our north than in Africa.

As much as I was able to accomplish in this, not following the beaten path of any antecedents, I have presented. There is nothing to prevent this theory, having arisen from these contemptible seeds, from growing to some degree of usefulness. Thus may you, whoever you are, a lover of the truth and ardent for the knowledge of things, not trample upon these feeble first sprouts with mockery; instead, may you try to support them and draw them out with applause.

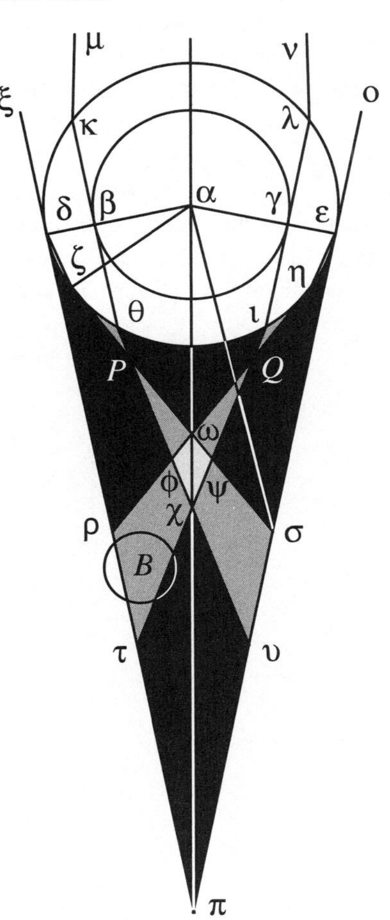

Kepler's note on the "cone of full illumination" (p. 293)

[The note begins on p. 447 of the 1604 edition.]

To page 283. The reason for my using the cone of full illumination,[42] and its vertex M, which is a point between the earth and the sun, here, is to be sought on p. 279. For when the sun is so placed that $\pi\delta$ and $\pi\varepsilon$ extended are tangent to the edges of the sun, $\tau\iota$ and $\upsilon\theta$ are drawn in such a way that (when the true refractions happen at ι, θ and at λ, κ) they would both be tangent to the earth at γ, β, and would contain the opposite edges of the sun with the lines $\lambda\nu$, $\kappa\mu$. Now if $\lambda\nu$ on the right is tangent to the left part of the sun, and $\kappa\mu$ on the left is tangent to the right, they therefore previously meet
448 (as at I, on p. 239). On this account, even if $\lambda\nu$, $\kappa\mu$ extended are not actually

By Napier's rules for right spherical triangles (with some rearranging), $\cot XZY = \cos XZ \tan YXZ$.

42 συναυγείας.

tangent to the earth βγ (for λγ, κβ are different lines), they are nevertheless nearly equidistant from the tangents because of the enormous distance of the sun. They therefore make about the same angle with each other that the lines marking out the circle of full illumination make. Further, from the same diagram on p. 279, it is apparent that those lines τι, υθ fall upon the parts of the moon which are indicated by the letter *F*, *E* on p. 282 [that is, the diagram that goes with the August 1598 lunar eclipse], for this is presupposed here. Therefore, in the same way as on p. 279, where from the known angle which λν makes with ωα extended upwards (that is, 18′ 30″, on p. 241 l. 22), and from the known quantity of refractions which the tangent to the earth λι undergoes at λ and ι, 34′ at both places, both the angle ιχα, 49′ 30″, and αχ, 70 semidiameters, and the divergence of χτ from χπ at any given altitude, are easily known; so here, in turn, the angle 18′ 30″ remaining the same while supposing a divergence of χτ from χπ (on p. 283, of *RF* or *RE* from *NO*) in the opposite direction, and that at a certain altitude from the earth, the quantity of the refraction at ι, λ (on p. 283, at *T*, *R*), is easily known. And this very thing takes place on p. 283, except that, for the sake of convenience, things that are very close to each other are taken as equal. You should not be disturbed that the angles *I* at the eye are not really on earth, or in the air, but are outside, for they do not differ perceptibly from those that ought to have been set up at *R*, or at some other nearby point.

Chapter 8
On the Shadow of the Moon and Daytime Darkness

285 For the sake of greater ease of understanding, this enquiry was pertinent to those things which we are going to say below on the quantity of the moon's diameter. But the elegance of method does not allow separation of the shadow of the moon from the shadow of the earth.

1. The occasion of this enquiry

When Tycho Brahe, the Hipparchus of our age, noticed that eclipses of the sun, whether the ray be allowed in through a notch or received by the eyes, always show the moon's diameter to be much less than it appeared at oppositions,[1] he first conceived the suspicion, which he at length defended in the place of a just opinion, and openly professed in volume I of the *Progymnasmata*: that the moon in conjunctions does not maintain the same visible diameter that it had at oppositions, but by the force of the solar light its edges are made thinner, a certain optical argument lending support.[2] As a result, as if this were a universal thing, he displayed a special table of semidiameters for new moons. And when in 1600 he wrote to Clavius,[3] he said he wondered at Clavius's having seen a total eclipse in Portugal in 1560, and night-like darkness during the day. For his own observations do not allow the whole sun to be covered by the moon, no matter how low it is. Besides, that this is not to be established by this small table of semidiameters, the eclipse of the sun has taught, which, in December of 1601, followed directly upon the death of Tycho, of which an announcement is made in an appendix to the *Progymnasmata*.[4] As regards the opinion itself, this must first in any event be

---

[1] Reporting his observations of the solar eclipse of February 25/March 7, 1598, Tycho Brahe wrote to Maestlin that the moon appeared "constricted" during solar eclipses (*TBOO* VIII p. 55, partially quoted by Rosen in *Kepler's Somnium*, p. 138). Maestlin quoted this passage in a subsequent letter to Kepler (*JKGW* no. 110, XIII p. 280).

[2] Near the end of Chapter 1 of the *Progymnasmata*, Tycho wrote, "It has been observed that in the ecliptic conjunction of the luminaries, the moon does not retain the same visible diameter that it has in other places, but its edges are contracted by the power of the solar light, optics providing this with a certain explanation." He then provides a table of diameters with a special column for conjunctions (*TBOO* II pp. 147–8).

[3] Christoph Clavius (1538–1612), Professor of Mathematics at the Collegio Romano, the preeminent Jesuit college, and author of a widely read commentary on the *Sphere* of Joannes de Sacrobosco (13th Century). Many editions of this work were published both during and after Clavius's lifetime, and he continued to revise it. In the edition published in Rome in 1581 (p. 425), he described this eclipse as having occurred in 1559; however, in the Rome 1607 edition (p. 594), he gives the year as 1560. No day or time is given in either edition. Cf. Lattis, *Between Copernicus and Galileo*, pp. 44–5. Brahe's letter to Clavius seems not to have survived.

[4] *TBOO* III pp. 321–2.

granted, that the full moon appears larger than the truth, but differently to different people. For this was the chief aim of Chapter 5 to show the causes, from the actual structure of the sense of sight, why the edges of luminous objects are enlarged, particularly in darkness. Thus, here, the edges of the moon are not made thin by the force of the solar light, but are enlarged by the force of the lunar light, communicated from the sun. And thus the astronomer should carefully take note of this from that passage, that unless he be endowed with the sharpest and most **286**
powerful sense of sight, he is not equal to measuring the moon's diameter at the full with the eyes without error, so much so that this belongs to hardly anyone.

Next, this too must be granted, that in the eclipsed sun, the boundaries, both about the sun and on the side of the moon that has entered beneath, are spread out in the vision, which was demonstrated above at the end of Chapter 5, again from the structure of the eye and the principles of sight. In addition, one has to distinguish carefully here between those things that happen to the sense of sight and those that happen when the consideration of the sense of sight is removed. For those things that happen to the sense of sight vary by individual cases, but those things that really happen are uniform within a single horizon. And since the question here is whether day or night is going to occur at the time of an eclipse of the sun, and particularly, how great a part of the light of day is going to be lost, one looks in vain to defects in the sense of sight, which, though it be very deficient in one particular person, does not thereby extinguish the light of day for the rest of the people. And so, in the second place, astronomers will now take note of this: that one must not trust the sense of sight either for the number of digits,[5] nor for the narrowness of the interior and concave circle bounding the sun, which is the circle of the lunar body. For here in fact, in our sense of sight, the edges of the moon, as Tycho says, are made thinner by the force of the solar light, or rather, the edges of the luminous solar particle are enlarged.

It cannot therefore be argued from this accident of the sense of sight to what happens outside of consideration of the sense of sight, nor can tables be established for the sake of the sense of sight, which represent neither the object itself nor the defects of all senses of sight. For the astronomer should not present anything other than those things that in actual fact occur. The sense of vision, however, we leave to the physicians to remedy.

But that, in addition, the ray of the eclipsed sun received through a notch (since Tycho used this also) and through a narrow opening spreads out the luminous boundaries, and is evidence for a smaller moon, this is so far from showing the moon's true quantity, that instead, in Ch. 2 above, the occasions became on this basis apparent, on which some deceived persons might be led into this opinion, as if the moon really were that much smaller. And below, by this device (this distinguishing of the enlargement), we shall teach how to enter in to a most certain procedure for measuring the quantities of eclipses. In this place it will also be made plain by examples, that if the device be correctly applied, the diameter **287**
of the moon appears decidedly greater than the amount that Tycho's table shows.

Thus, these things being presupposed, it is to be demonstrated by us in the

[5] See the note on p. 71.

beginning, by many examples, that the diameter of the moon is fully large enough that it has at some times covered the whole sun, and has enfolded certain regions in deep darkness. Next, I shall ascribe optical causes, by which it can happen that sometimes, when the moon intercepts the whole sun, nonetheless the sun appears to protrude all around.

2. Examples from historical accounts, that the moon's shadow brought night into the day

That the moon's shadow is conical, both because the sun is round, and because the moon is round and also is smaller than the sun, is evident from the same principles by which we have pronounced the earth's shadow to be conical, in Ch. 7 Sect. 1 above.

As to the case in which the diameter of the moon is perceptibly not greater than the diameter of the sun, but equal: the point of the moon's shadow ends right at our sense of sight, and enfolds only the most narrow little region of the air. And so in this case full darkness can by no means take place. For the sun, being turned away from only a single point of the earth by the intervention of the moon, illuminates most brightly all the surrounding parts of the earth with some part of itself. These parts emit the light communicated from the sun everywhere upwards into the air, even insofar as it is in the moon's shadow, and thus supply it with brightness nonetheless.

It is therefore necessary that as often as the darkness of night makes an onslaught during the day, at the time of an eclipse of the sun, the moon be seen with a greater angle than the sun: such that the sun may lie well hidden behind the moon, and a sufficiently great part of the horizon may be covered by the shadow that the observer's air cannot be illuminated by the neighboring air. In a word, it is necessary also to darken that matter which ordinarily lights the lamp of twilight for us.

On the matter of twilight.

As for the case where the altitude of that matter reaches 12 miles (I would note in passing), in order that this really be deprived of the sun's light by the moon, the moon's shadow in our neighborhood would have to have a thickness of 300 miles.[6]

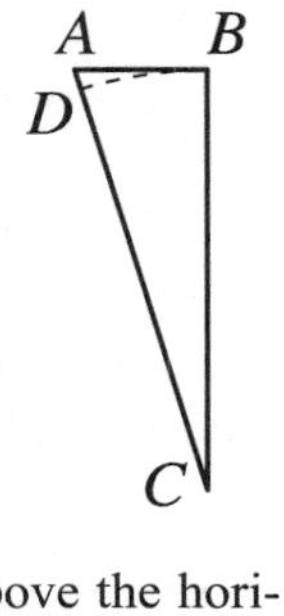

For let AD be 12 miles, of which DC or BC is 860 miles. Therefore, as BC, 860, is to the whole sine, so is AC, 872, to the
288 secant of the angle ACB, $9^\circ\ 31'$. As the whole is to the tangent of this, so is BC, 860, to AB, 144 miles, from one side, and the same amount from the other. But in order that the moon's shadow, standing at the zenith, cover 300 miles all around, the moon ought to be seen as twenty minutes greater than the sun, as much greater, that is, as the moon's parallax, when standing at the zenith, varies at minimum for those travelling through $9\frac{1}{2}$ degrees on both sides on a great circle of the earth. But even if you give it an altitude above the horizon of no more than 13 degrees, from here to 23 degrees the Moon's parallax still

[6] This case was brought up at the beginning of Chapter 4. The mile used here is the German mile, or five Roman miles.

varies by 3 minutes on this side, and the same amount on the other, as is to be seen in Tycho's parallactic table.[7] Therefore, it must surpass the sun's diameter by six minutes in order to darken the matter producing the phenomena of twilight for 300 miles lying beneath the same meridian. And yet with this quantity of diameter it will not be able to cover that much in both directions, towards the east and the west, but the light of twilight will remain at the sides. Consequently, authors who have written about twilight should see about reconciling their teaching about the altitude of the matter of twilight with the total eclipses of the sun and the complete darkness by day, without a prodigious magnitude of the lunar diameter.

For, that we might pursue more closely what was proposed, the occasional existence of deep darkness of this kind is limpidly proved by authority and historical accounts. For this reason, Plutarch,[8] in order to prove by the strongest possible argument that the moon is not pellucid, and does not transmit the sun's rays, refers to common experience: "It is," he says, "so far from being bright that not only it is itself dark [invisible] in conjunction, but also on many occasions hides the sun." And he brings forth Empedocles, who also expressed this opinion more evidently in verses:

> And robs light from a space of earth so wide,
> As wide as the orb of the yellow moon encompasses.

Understand it as a little less space, as geometrical reasoning relates. Plutarch follows up these words of Empedocles thus: "Exactly," he says, "as if the light of the sun had fallen on night and darkness, not on another celestial body." A little later, using the same argument, he made bold to compare ordinary night with this darkness at the time of an eclipse of the sun, and thence to attribute the same substance to both the earth and the moon, for the reason that each blocks the sun in exactly the same way, and brings darkness upon things. Here he brings in **289**
Theon, the mathematician of his age, defending exactly the theme that I am now treating, and by the same arguments derived from the authority of the ancients. "This Theon of ours, if you do not concede" that such darkness occurs, "will bring in Mimnermus and Cydias, and Archilochus, and Stesichorus and Pindar besides, who lament, in eclipses, that the most brilliant sun has been abducted from them, and that they are as if in the middle of the night, and that the sun's rays are carried on a dark course; and above all," (the most ancient of all, or before all), "Homer" (A Christian will append Isaiah Ch. 13), "who said that the faces of the people were invaded by night and darkness, and that the sun and moon disappeared from the heaven." [9] Indeed, this, in Plutarch, is readily admitted. Of Stesichorus and Pindar, Pliny writes in Book 2 Ch. 12 in this way: " . . . the human mind, in eclipses of stars, fearing crimes or some sort of death from the

---

7 *TBOO* II pp. 132–4.

8 Plutarch, *On the Face in the Moon*, XVI, 929C–D, pp. 102–3. Brackets mark Kepler's interpolation. The quotation from Empedocles is from Fragment 31B 42.

9 *On the Face in the Moon* XIX, 931E–F, pp. 116–119. References to the other authors cited are given in the notes to this text.

heavens—it is general knowledge that the sublime voices of the seers Stesichorus and Pindar were in this fear, because of an eclipse of the sun."[10]

Cleomedes, being of exactly the same persuasion as Plutarch, professes to give the causes of the same most bewitching phenomenon, that is, why the moon does not transmit the sun's rays, which was [sic] gathered from such deep darkness, and again, "why, when the moon is smaller, it may hide the sun, standing in the way of all its parts, and subtending its entire diameter." [11] And when Sosigenes (I think it was) reviewed the opinion of certain of the preceding authors, stating "that in perfect conjunctions, when the centers of the luminaries are perceived on the same straight line, the sun's orb embraces the moon with a circle, exceeding it on all sides," Cleomedes made bold to interpose a comment, as if to say "this is not perceived in observations. For," he says, "that limb would have been seen by us to be protruding, since it was most brilliant." Thus he clearly asserts that total eclipses were seen by him as well, and makes this theme general. Martianus Capella does the same: "Frequently," he says, "an eclipse of the sun, occurring in the clime through Meroe, overshadowed the whole of the same on every side."[12] And Albategnius takes it as generally acknowledged that the whole sun is covered.[13] Witelo also proposes a cause for why the moon might be seen in a total eclipse of the sun. And, in sum, there has never been an astronomer who doubted that this happens: however many there are who relate that the moon's diameter is perceptibly greater than the sun' s diameter, all likewise
290 have taught this. No wonder: the histories are full of examples, some of which, from Maestlin and Mercator, and the ancient historians, I will disclose.[14] For this recounting is useful not only here, but also for another work I am contemplating on the magnitude of the year, clearly necessary, if God grant life and ability.[15]

Dionysius of Halicarnassus, in Book 2, says, "They say that at the conception of Romulus the whole sun was eclipsed, and complete darkness, just like

---

[10] Pliny, *Natural History* II ix 54, Vol. I pp. 202–3.

[11] Cleomedes *De motu circulari* II 4 p. 190, trs. Goulet (French) p. 160.

[12] Martianus Capella, *Marriage of Philology and Mercury* VIII 859, trs. Stahl and Johnson p. 334. Kepler considers this passage from Martianus Capella at greater length in the papers relating to his unpublished *Hipparchus*, in *JKGW* XX.1 p. 246. His words there make it clear that *Dia* in this sentence is actually the Greek preposition meaning "through," not the Goddess or the island of Naxos, despite the initial capital. The "clime through Meroe" is one of the seven climes, or terrestrial latitudes, listed in the *Almagest* (cf. II 13, Toomer's translation pp. 122–3 and the note on p. 19).

[13] For Albategnius (i.e., Al-Battānī), see Chapter 4, p. 158. In Nallino's modern edition (Vo. I p. 56), however, Al-Battānī explicitly states that the moon at apogee does not cover the sun.

[14] Cf. Maestlin, Michael, *Disputatio de eclipsibus*, Tübingen 1596; Mercator, Gerhard, *Chronologia*, Cologne 1569. Many of the following accounts (as Kepler's words show) came from Heinrich Bünting (1545–1606), *Chronologia, hoc est, omnium temporum et annorum series, calculo astronomico exacto demonstrata, cui inserta sunt ipsae picturae ac imagines eclipsium Solis et Lunae* (Zerbst, 1590). I have not seen a copy of this book, and have been unable to document all of Kepler's citations. —translator.

[15] No such work by Kepler ever appeared.

night, took the earth; and that the same thing happened at his death."[16] This might seem a myth. However, the Prutenic computation[17] gives it confirmation, although it is not fully certain: in this uncertainty, it nevertheless points to exactly the same year and exactly the same interval of years as Romulus's age of 55 years, in which two great eclipses of the sun happened. For although the prior eclipse is given as $7\frac{1}{4}$ digits, the latter as $9\frac{1}{2}$, it can easily happen that an emended computation (for it needs emendation) would give both as total.

Herodotus, in Book 2 on the war between the Lydians and the Medes: "In the sixth year it happened that during a conflict, when the battle was closely joined, night was suddenly brought about out of day. Thales of Miletus foretold that this alteration of the day would take place, adding the time, this very year, in which that alteration was also brought about."[18] However, since Herodotus does not make mention of the moon, he was indeed inexpert in astronomy. So that no one might think this to be some other wonder, Pliny gives us confirmation. In Book 2 ch. 2, he ascribes this teaching to Thales of Miletus, first among the Greeks "because of a predicted eclipse of the sun, which took place in the reign of Astyages, in the fourth year of the 48th Olympiad."[19] Bünting accordingly found a new moon on the ecliptic in the year indicated, and the digits, as indicated by the Prutenic computation, not yet perfectly certain, were eleven and a half, as the moon was approaching perigee, in 1° Gemini, in that summer from whose middle the fourth year of the 48th Olympiad began.

The same Herodotus, in Book 7: "When Xerxes set forth from Sardis to Abydos with the army, the sun, departing its place in the heaven, was not visible, though there were no clouds, but particularly calm weather; but in place of day, night came."[20] Since Xerxes' passage to Europe fell in that year in whose summer the 75th Olympiad was celebrated (for some had related that, though Xerxes had already entered Greece, the Greeks viewed the Olympics in safety), and nevertheless astronomical computation shows no eclipse of the sun in that year, you **291**
might with good reason believe that a monstrous event was brought about in the sun while the whole army watched (for Bünting wastes his time in producing an eclipse of barely 3 digits, and that in the previous year). But two years after this crossing of Xerxes, the Prutenic computation showed a maximum eclipse of the sun, at the first hour after noon in Asia, and this on the 17th day of February, in 24° Aquarius, and thus truly the first that agrees with Herodotus's account. I

---

[16] Dionysius of Halicarnassus, *The Roman Antiquities*, trs. Earnest Cary, Vol. I (Loeb Classical Library, 1960), II 56.6, pp. 474–5.

[17] Erasmus Reinhold, *Prutenicae tabulae coelestium motuum* (Tübingen 1551 and other editions). These tables were based on Copernicus's models parameters, though they were geocentric in orientation.

[18] This is from Book 1 ch. 74, not Book 2. According to David Grene's note on this passage, the eclipse took place on May 28, 585 BC. (Herodotus, *The History*, translated by David Grene, Chicago 1987, p. 67).

[19] Pliny, *Natural History* II ix 53 (not ch. 2), pp. 202–3.

[20] Book 7 Ch. 37. According to Grene (Herodotus, *The History*, translated by David Grene, Chicago 1987, p. 483), this probably happened in April of 480 B. C. E.

therefore suspect a lapse in the report, from Xerxes' having perhaps gone forth twice from Sardis. For although Xerxes was everywhere defeated in the first and second year of the 75th Olympiad, nevertheless, in the third year, he was pestering Pausanias about a betrayal, and perhaps for the sake of that he marched to the shore with the army at the time when this eclipse happened. For in the following year, Pausanias was convicted of this betrayal, and was killed by starvation.[21] Compatible with this is that interpretation of the Magi, taken from the present matter, "For the Greeks, it signifies a revolt of the citizens."[22]

I have often thought about whether this may be the eclipse that appeared to Cleombrotus sacrificing at the Isthmus, really because of this, that it followed closely after the defeat of Mardonius, but the computation does not agree.[23] And you can't tell whether by Herodotus's words, ὁ ἥλιος ἠμαυρώθη, an eclipse is understood, or some kind of darkening or obscuring. This, then, is a little more uncertain.

Thucydides, in Book 2, describes a memorable eclipse of the sun in the first year of the Peloponnesian War, which occurred in the second year of the 87th Olympiad. "The sun was eclipsed after noon, and was again made full after it had been made similar to the crescent moon, and a number of stars had shone forth."[24] It was on the third of August, Julian style, 431 B. C. Since often under total eclipses the stars do not show themselves, the author does not seem to be speaking of a partial eclipse, in which the sun was narrowed to a lunar horn, but clearly of a total eclipse, where, after a start has been made, the sun, going next through the shapes of the crescent moon, becomes full again, as in the description which will follow below from the life of St. Louis. For since after seven years an eclipse again occurred, not a slight one, but of seven digits, the magnitude that Maestlin found from the *Prutenics*;[25] nevertheless, the author speaks distinctly of it. "Some part of the sun was eclipsed," ἐλλιπές τι ἐγένετο, indicating that it was not totally eclipsed. This is simply ὁ ἥλιος ἐξέλιπε. Further, the reason for his having added γενόμενος μηνοειδήσ appears to be the same as that which led him to add this, "It appears [to him that an eclipse of the sun] can happen only at
292 new moon." For, being about to confirm this opinion of his that the moon stands in the way of the sun's light in an eclipse (astronomy not yet being cultivated among the Greeks), he adds what he saw, that the sun, gradually having become full again (as the moon usually does on other occasions), by that very eclipse, and by its concavity, projects the roundness of the moon that stands against it.

Livy, Book 7 dec. 4: "In the consulate of Lucius Cor. Scipio and C. Laelius,"

---

21 Thucydides, *Peloponnesian War*, Book I no. 134.

22 This appears to be from Herodotus VII 37, though Kepler's Latin rendering of Herodotus's Greek differs from the text as we have it.

23 This is the Isthmus of Corinth, and the time was some time before the summer of 479 B. C. Cleombrotus was Pausanias's father. See Herodotus, *The History*, translated by David Grene, Chicago 1987, 9.1 and 9.10, pp. 612 and 615.

24 Thucydides, *Peloponnesian War*, Book 2 no. 29.

25 This is presumably from Maestlin's *Disputatio de eclipsibus*.

"During those days in which the Consul had set out for war, during the Apollonian Games, on July 9th, the heaven being clear, the light was darkened during the day, since the moon had gone beneath the orb of the sun."[26] The year was the second of the 147th Olympiad. The moon was rising from perigee. In the Julian form, March 3, 190 B. C. E.[27]

Two years later, the same Livy (Book 8 Dec. 4) says, "by the light between the third and fourth hour, darkness having suddenly arisen,"[28] at which times computation shows no eclipse. It was therefore a prodigy. Bünting brings forward an eclipse of three digits, ill suited to producing darkness, not to mention that it happened at another hour, namely, the ninth hour of the day (Roman style).

Especially memorable is that eclipse which Hipparchus had introduced in the book on the magnitudes and distances; it is not certain whether it was observed by him or by Timocharis, since it was described as having been observed at Alexandria.[29] For, as Cleomedes says about it, "The sun, totally eclipsed at the Hellespont, was observed at Alexandria with one fifth of the diameter still extant, the rest eclipsed."

Ptolemy recalled it in Book 5 Ch. 11 of the Great Work,[30] as did Theon in his commentary on this passage, in noteworthy words: "He makes use," he says, "of an eclipse in the places that are around the Hellespont, accurately made in the entire sun, so that nothing of it appeared."[31] He not only asserts that in the Hellespont the entire sun was hidden, but also attributes breadth to this phenomenon, saying, "in the places lying round about." Hence it is that Cleomedes reports, beyond doubt from this same passage of Hipparchus, "This shadow takes up more than four thousand stadia; for every place in which the sun is not seen, when the moon runs beneath it, is the shadow of the moon."[32] p. 153. For Theon reports that from this eclipse Hipparchus concluded that the moon's nearest distance is 71 semidiameters; the farthest, 83; although afterwards, for other reasons, he comes down to between 62 and $72\frac{1}{2}$. Supposing that the moon's diameter be taken to be 30′ at apogee, which Hipparchus supposed, it will become 35′ at perigee by the Hipparchan eccentricity, exceeding the diameter of the sun, in the quantity
293 used by Hipparchus, by 5 minutes. Further, in order that, at this elevation of 71

26 Livy, *History* XXXVII iv 4, Vol. X pp. 300–1.

27 In modern terms, this eclipse was on March 14, 190 B. C. E. (–189, in astronomical reckoning). The Roman calendar was then several months out of adjustment; hence, the change from July 9 to March 3.

28 Livy, *History* XXXVIII xxvi 4, Vol. XI pp. 118–9. According to the translator's note, this darkness was the result of an eclipse on July 17, 188 B. C. E.

29 Hipparchus's book has not survived, but is known from citations in other authors. Kepler's information comes from Cleomedes, *De motu circulari* II 3 p. 172, trs. Goulet (French) p. 153.

30 *Almagest*, trs. Toomer pp. 243–4. Ptolemy does not mention any particular eclipse, but describes Hipparchus's procedure generally.

31 This, as well as the distances given below, appears to be a quotation from Pappus's *Commentary*, p. 68.

32 Cleomedes, *De motu circulari* II 3 p. 172, trs. Goulet (French)

semidiameters of earth, the body may change its place by 5 minutes, there is need for it to go through 6° of the terrene circle from the place where the Moon is vertical, and about the Hellespont, at an altitude of the sun (in Cancer) of 71°, by $6\frac{1}{2}$ °. But in that time Eratosthenes had fixed the circumference of the earth as 250,000 stadia, of which at least 4,000 make up 6 degrees. Therefore, whether by a simple consideration of the extent of lands through which the whole sun was hidden in this eclipse, or, from this line of reasoning just presented, Hipparchus had taken away those 4,000 stadia, of which Cleomedes writes, under either head it is strongly proven that this eclipse was total, the sun being hidden beyond the moon with some lapse of time.

In the consulate of C. Marius and C. Flaccus, "in the third hour of the day, an eclipse of the sun darkened the light," Jul. obs.[33]

In the consulate of M. Vipsanius and Fonteius, in 59 A. D., Cornelius Tacitus says (in Book 14 of the *Annals*),[34] "The sun was suddenly darkened." And the not insignificant author counts it among the great prodigies that, to his wonder, were without consequence. That the eclipse was indeed noteworthy is concluded from Pliny, who said (Book 2 Ch. 70) that it was also noted by Corbulo in Armenia.[35] The day was April 30.

Plutarch, in *On the Face of the Moon*, reminds his interlocutors "Of that coupling of the sun and moon that recently had occurred," (about 100 AD), "upon the beginning of which, immediately at noon, stars shone forth everywhere in many places of the heaven, and the air was of that condition such as there is under twilight in unsure light."[36] That it was total is concluded from the preceding as well as the following. For giving causes, "why darkness is not so deep in eclipses, and the air is not taken over by it as it is by night,"[37] he is not summoning us towards the view that there is something of the sun remaining, not covered by the moon, but is seeking other detours, of which later. By which it is proved that this eclipse seen by Plutarch was total, not otherwise: what he saw in one case, he believed to occur in all. But the histories give us sufficient confirmation of plain darkness. At least, this eclipse makes for this, that we may believe that the whole sun can be covered.

At the time when Giordanus the Younger began to rule, in 237 AD, on April 12, "there was such a great eclipse of the sun that it was thought to be night, nor could anything be done without lamps being lit." —Julius Capitolinus.[38] This had to have been total; the Prutenic calculation shows less than 11 digits.

Ammianus Marcellinus, in Book 20 on the consulate of Constantine X [and] Julian III: "Over the Eastern lands, the heaven was perceived to be dimmed with

---

33 Julius Obsequens, *Prodigiorum liber*, pp. 276–7. The second consul's name was Flavius, not Flaccus.

34 Tacitus, *Annals*, 14.12. The consuls' names were Caius Vipstanus and Caius Fonteius. This eclipse occurred on April 30.

35 Pliny, *Natural History* II lxxii 180, Vol. I pp. 312–3.

36 Plutarch, *On the Face in the Moon* 931E, pp. 116–7.

37 932A, pp. 120–1.

38 This is presumably in Julius Capitolinus (fl. after 320 C. E.), *Vita Giordani*.

darkness, and from the first rising of dawn until midday," (I understand this to mean the place in the heaven, not the time), "stars twinkled forth continuously. To these terrors there was added that when the celestial light was concealed, the light having been utterly snatched away from the view of the world, the terror-struck minds of the people considered that the sun was eclipsed for a long time, at first narrowed into the likeness of a horned moon, then into the enlarged form of half a month, and aftwerward was restored to its entirety."[39] Consequently, here too the entire sun was eclipsed; the Prutenic computation groundlessly leaves a circle about the moon, the moon being at apogee. For after the explanation of causes, the author also adds these words: occasionally, "The heaven is overspread with dense darkness, the air being thickened, so that we are unable to see even the closest things, and those placed in front of us."[40] Which very thing beyond doubt this eclipse itself teaches, as Plutarch's previously taught him. The year was 360 AD, 28 August. **294**

Since the histories make extremely rare mention of eclipses for a number of centuries, it is believable that only those were noted that struck the eye by turning day into night, and were total. It is credible that Liechtenberger in the time of St. Martin;[41] Jornandus and Marcellinus on 19 July 418; the *Annales Constantinopolitani*[42] on 19 March 592; and Bede on 1 May 664; were reporting such eclipses. This is not much in disagreement with the Prutenic computation, which gives the former as $10\frac{1}{2}$ and the latter as almost 12 digits. In 787, "A maximum eclipse of the sun occurred."—Cedrenus[.] Computation shows $11\frac{1}{2}$ digits on 16 September. But that it was total is deduced from the words.[43]

On 5 May 840, an anonymous author said there was an eclipse so great "that even the stars were seen because of the darkness of the sun, and for objects on earth the color was changed." Another author says it was "total"; Annonius, that it was "maximum." The moon was at perigee. The following notable words concerning this eclipse are included in the life of Louis the Pious:

> An eclipse of the sun happened on the third day of the greater letania in an unusual way. For in the withdrawal of light, darkness had prevailed to such an extent that it seemed to differ not at all from the reality of night. For the established order of stars was so perceived that [note:] no star suffered

---

[39] Ammianus Marcellinus, *History* XX 3 1, trs. John C. Rolfe, 3 Vols. (Loeb Classical Library, 1963), Vol. 2 pp. 6–9.

[40] *History* XX 3 5, pp. 10–11.

[41] In a letter to Herwart von Hohenburg of 10 February 1605 (*JKGW* XV no. 325, pp. 145–7) Kepler writes, "The words of Liechtemberg were taken from a small German book, in quarto, without mention of the place, publisher, or year of printing. It is a book of prophecies, where, in Chapter 23, he speaks of the eclipse of 1485, which happened upon him as he was writing."

[42] Frisch, *JKOO* II p. 422, writes that Kepler's main source for these eclipse reports was Gerhard Mercator, *Chronologia*, Cologne 1569

[43] *Georgii Cedreni Annales, sive historiae ab exordio mundi ad Isacium Comnenum usque compendium*, Basel 1566. This book lists many eclipses, though they do not seem to tally completely with those mentioned by Kepler.

> weakness from the solar light; rather, the moon, which had set itself forth against the sun, restored it to its usual self in procuring for it a gradually rising light [note:] from the western side, first in a hornlike manner, when it was first or second perceived, and thus by increments the whole wheel of the sun received its whole delightful appearance.

The author is worthy of credence, because he had experience of astronomy, knew
295 all the stars exactly, and was accustomed to show them to Louis, as is apparent from the conversation which he had with Louis on the comet of the preceding year and its signification: he seems to emulate Marcellinus.

On 29 October 878, the author just mentioned said, "The sun, after the 9th hour, was so darkened, for over half an hour, that the stars appeared in the heaven, and everyone thought that night was impending upon them." Gemma asserts that it was total, which is in agreement with the words. The moon was beyond perigee.

In the third year of Lothair,[44] about 957 AD, December 18, "An eclipse of the sun was brought about, such that the stars appeared from the first hour to the third hour."

Compare with that of Ammianus Marcellinus. For morning eclipses have a special reason why they are more horrible: because of the moon's shadow being above the horizon, as if striking with a certain impact, and dousing the dawn. By which I believe my grandfather the more readily, who used to relate to me how on one occasion (namely, 29 March 1530), the heaven being extremely cold, when it had barely become light, the light of day was suddenly extinguished, and turned into night. The Tychonic computation in fact does not allow this eclipse to be total. Cyprianus[45] also calls it "horrible"; I don't know whether from personal inspection. The moon was near perigee.

On 2 August 1133, the Historians report an eclipse of the sun that was "maximum"; computation allows it to be total.

On 4 September 1187, "An eclipse of the sun so great that the stars were seen by day, as if by night." It was therefore total, as comparison with the preceding shows. The moon was at perigee.

In 1191, Albericus: "A total eclipse of the sun was brought about." And yet the moon was at apogee.[46]

On 6 October 1241, a "great" eclipse of the sun. Gemma said it was "total." This too was in the morning.

On 7 June 1415, "there was a horrible eclipse of the sun," (Leovitius). That the "whole" sun was covered is deduced from the Polish Historian. The moon was about perigee.

On 16 March 1485, the sun was eclipsed, by Walther's reckoning, to 11 digits, which doubtless the sense of sight got somewhat wrong, for the reasons set

---

44 Lothair I, King of France from 954 to 986.

45 This is Cyprianus Leovitius of Bohemia (d. 1574 in Lauingen), mathematician to Otto Heinrich, Elector Palatinate. Kepler's source, according to Frisch, is Leovitius's *De conjunctionibus magnis insignioribus superiorum planetarum, solis defectionibus et cometis* (Lauingen 1564).

46 The 13th century chronicler Alberich the Monk is cited from Mercator.

forth above. For elsewhere, by the testimony of Lycosthenes, "It was simply dark, to such an extent that candles had to be lit; the chickens in the towns, and other 296
kinds of animals in the country, betook themselves to the nighttime places of their accustomed rest." The moon was at perigee.[47]

On 14 January 1544, Gemma Frisius observed an eclipse of 10 digits (through an opening),[48] and thus it is false, as was shown in Ch. 2. And the eclipse was greater. Funk reckoned 11 digits, but here too the eyes were in error, as was said in Ch. 5. In all events it was a little more, and in some places the entire sun was hidden. Hence, Funk said, "Day began to become night again, as if at evening twilight, and the flying creatures of the heaven, which at first light were lively, began to become silent when so great a darkness occurred suddenly." The moon was ascending from perigee.[49]

On 21 August 1560, Clavius attests, "At Coimbra the sun was hidden, being covered up for a not inconsiderable time about noon, and the darkness was in a way greater than that of night. For one was not able to see the place where he would set his foot, and the stars appeared in the heaven with great brightness. Even the birds, wondrous to tell, fell from the air to earth, in horror of such a terrible darkening."[50] The moon was at perigee. Thus, to give credence to this one, which Tycho denied it, one required the accumulation of so many antecedents. In Vienna, Austria, as Mercator reports, Tilmann Stella and Paul Fabricius observed $5\frac{1}{2}$ digits at $1^h\ 40^m$. The Tychonic computation shows the true conjunction at Uraniborg at $1^h\ 20^m$ p.m.[51]

Cornelius Gemma at Louvain noted the beginning at just after 11, the end at $1^h\ 23^m$, $7\frac{1}{2}$ digits, nearly.[52]

---

47 Walther's solar observations are included in Regiomontanus's *Scripta* (Nürnberg 1544); however, this eclipse is not included among them.

48 Rainer Gemma Frisius, *De radio astronomico et geometrico liber*, Antwerp and Louvain, 1545, Cap. XVII, *De eclipsium magnitudine*, fol. 29v–31r.

49 I have inspected Funk's *Chronologia, hoc est, omnium temporum et annorum ab initio mundi usque ad hunc praesentem anno Christo M.D.LXI. computatio,* Wittenberg 1570. This is the second edition of a work originally published in 1552. It list many eclipses, and one would suppose that Kepler is referring to this book, yet no eclipse is listed for 1544 (fol. 162). Frisch, *JKOO* II p. 423, states that this eclipse was also not included in the 1578 and 1601 editions of this work.

50 Clavius, Christoph, *In Sphaeram Ioannis de Sacro Bosco Commentarius*, Rome 1581, p. 425; Rome 1606, p. 594. Kepler's quotation differs in many details from the original, though the sense is the same. In the 1581 edition (the second), the year is given as 1559; in the 1606 edition (the fifth), he corrected the year to 1660. No date or time is given in either place.

51 Paul Fabricius (ca. 1500–1588) was a Viennese physician and astronomer. Tilmann (or Tielmann) Stella is said by Frisch to be the author of a small work titled *Begriff vom Nutzen und Gebrauch der neuen Landtafeln*. Tycho makes a critical remark in the *Progymnasmata* p. 770 (*TBOO* III p. 284) apparently directed at this book.

52 On the eclipses of 1544 and 1560, there are numerous notes by Kepler in Vol. XV of the Kepler Manuscripts at St. Petersburg, some of which are transcribed by Frisch (*JKOO* II pp. 423–6).

Unequal brightness of nights.

At all events, from these examples, it is as evident as can be that when the moon is closest to earth, night-like darkness follows, from optical or astronomical causes. But that these are not the only causes, but rather that they are very greatly helped or hindered by the conditions of the air, the unequal brightness of ordinary nights shows. For if in pure night the air, because of its whitish color, a sign of thickness (as Aristotle asserts of Pontus),[53] sometimes so imbibes the light of the stars that for the whole night a certain illumination of twilight is emulated, what can even the slenderest particle of sun not accomplish by means of this assistance upon thick air, and illuminating lands of this kind, beyond a few miles from us, and communicating light to them, which is reflected again to us? Here are arguments of this unequal brightness of night deriving from the air. In 1599
297 in the eclipse of the month of January or February, mention of which was made above, the diameter being not yet half covered by darkness, I was able to see the whole edge of the moon. In December of 1601, with the thinnest horn remaining, I nevertheless did not see the dark part. It was nevertheless the same season of the year. In the month of May, 1603, when the third part of the diameter remained, some people still saw the dark edge. In the following November, when not a fourth part was in the shadow, the edge was nevertheless seen.

3. Whether it can happen that in a central conjunction of the luminaries, the sun is still not entirely hidden?

The middle of this eclipse at Uraniborg, by the Tychonic computation, was 12h 10m at midday, the quantity of digits 6 29 . Maestlin at Tübingen observed almost 10 digits from the south. Gerhard Mercator, at Duisberg in Clivia observed the beginning at 10h 25m, the end at 1h. Therefore, the middle was 11h 43m. Cornelius Gemma, at Louvain, observed the beginning at 10h 12m, the strength at 11h 40m, the end a little past 12 1/2h, the points nearly 9. The light was very

And so hitherto we have proved from the histories that the entire sun was hidden more than once, even by the moon at its highest.[54] What is more of a wonder, there exists in all history but a single contrary example, which Clavius reports in his *Commentary on Sacrobosco*: on 9 April 1567, "the sun at Rome was not all eclipsed, but there remained a certain narrow luminous circle all around."[55] The moon was nevertheless between its highest and lowest. It is a wonder, I say, for here the visible diameter of the moon should definitely be smaller, and you could not ascribe this circumstance to the sense of sight, as was done at the beginning of this chapter. For those things that are in the eye, or in a dark chamber, are enlarged because they are luminous: it is necessary that they first project a ray to that place; but those things that do not radiate, prevented by the interposition of opaque bodies, as the sun is by the placing of the moon in the way, will also not be able to spread out. And so this phenomenon is not to be ascribed to the sense of sight.

Therefore, the same thing that Tycho Brahe attacked in the one of Clavius's eclipses, I have attacked in the other; that is, in calling into question this very thing, whether the remaining circle was completely unbroken, and not instead an

---

53 *Problems* XXV 6, 938a 7–b 4.

54 (To the marginal note:) Maestlin, *Disputatio de eclipsibus*, p. 3.

55 Clavius, *In Sphaeram . . . commentarius*, Rome 1581, p. 425. Again, Kepler's words are a close paraphrase of Clavius's. Despite Kepler's implication, below, that Clavius was reporting someone else's observations, Clavius wrote, *conspexi*, "I have seen."

pale, but there were no stars; the time was nevertheless like evening. It began from the west, from the lower part of the sun. Tycho Brahe said in a letter to Clavius that as an adolescent he had observed this at Rostock on the shore of the Baltic Sea, pretty much right at noon, of not quite 7 digits. And in the *Progymnasmata* it is 12h 0m, 6 digits 29 min., the amount also shown by the computation. On another slip of paper, however, I have found 11h 0m, 9 digits 0′.

extremely thin horned edge at one side, the centers still being not fully aligned. For it can happen that in the beginning these things might really have been recorded in one way, but transcribed differently by Clavius, whether by lapse of memory or by a wrongheaded understanding of the attested document, especially if Clavius were recalling things seen by others. Indeed, even he added, doubtfully, "it perhaps had never before happened."

And in fact, that Clavius committed this phenomenon to memory with suf- **298**
ficient reflection, having considered the circumstances, these very words of the author show. Consequently, let us reason as follows, as about something completely certain.

And first, I do indeed not deny that the hypotheses of certain astronomers are so constructed that the sun at perigee can be covered by the moon at apogee in this manner, with the edge left protruding. For him whom we suspected above to have been censured in the plural by Cleomedes, i.e., Sosigenes,[56] —to him this form of hypothesis is openly attributed by Proclus Diadochus, after first recalling Ptolemy's opinion.[57] Ptolemy had said that it was obtained with certainty from Hipparchus's dioptra that the sun's diameter does not vary perceptibly from apogee to perigee, but is seen under the same angle, but the moon's diameter is only seen to be equal to the sun's diameter when the full moon is situated at the apogee of its epicycle, while at perigee the moon appears to be greater. Proclus then adds to this, "If this is true, then what Sosigenes the Peripatetic narrates is not true, in what he wrote on the Revolutions, that in eclipses brought about at perigee the sun is not seen wholly running forth to anterior places," (I understand this to mean betaking itself beyond the moon), "but at the edges of its own circumference it goes outside of the moon's circle, and gives light with minimum hindrance." It appears that Hipparchus said the same as what Ptolemy said of the diameters of the luminaries; Sosigenes, however, added this interpretation: if at apogee the luminaries are seen under the same angle, surely the sun will be seen under a greater at perigee, even if this is not captured by the dioptra because of the subtlety of the difference. Thus it can happen that the moon at apogee (and thus under a smaller angle) might be set against the sun at perigee, and thus the sun might not be covered entirely at all extremities. For that reason, Proclus adds, "For if anyone will admit this, the sun will do a variation of apparent diameters." Really, however, Sosigenes will reply to Proclus, because it is absolutely necessary that the sun do a variation of apparent diameters, because it approaches and recedes, I myself have therefore considered this as something to be admitted. Thus Proclus teaches us that Sosigenes declared that eclipses such as the single one Clavius presented to us definitely are seen. Indeed, in this passage Proclus proposes for himself an interpretation of that which he had said in the fifth "question of marvels" at the beginning of the *Hypotyposes*. "Also since," he
says, "certain differences of the moon are perceived in perfect solar eclipses; for **299**
sometimes the whole sun seems to our sight to be covered, sometimes at the very

---

[56] See p. 301.

[57] Proclus Diadochus, *Hypotyposis astronomicarum positionum*, Latin trs. by Georgio Valla, Basel 1541 and 1551, Cap. 3, "De Luna," p. 400 (ed. 1541); p. 352 (ed. 1551).

moment at which the two centers and the eye fall upon the same line, the moon is observed within a ring of sun."[58] He shows, I say, in the passage first quoted, that he is speaking here, not from experience, but from the traditions of Sosigenes. For if he had had personal, or at least certain, experience of those things he described here, he never would have appended those things which we have adduced in the first passage: he never would have called those things into doubt which Sosigenes had handed down, since, naturally, they would be confirmed by experience itself.

Could therefore this eclipse related by Clavius be justified by the Ptolemaic hypotheses taken over from Hipparchus and Sosigenes? Absolutely not. First, the moon was right at its mean longitude: the moon's diameter was greater than the sun's, even at perigee. Next, the sun itself was going towards apogee, and was seen under nearly its least angle. Therefore, another cause of the phenomenon is to be sought. And I do not know whether Plutarch, in the book on the face of the moon that has by now so often been beat upon, might have fully uncovered it when, having begun from the eclipse which he had seen himself, he added a general account, saying, "Although sometimes the moon occults the entire sun, this eclipse is nevertheless lacking in breadth and time,"[59] (he makes a rule of his example; you have seen otherwise in the examples cited), "but a certain brightness shines out around the orb, not allowing a deep and excessive shadow to happen." It is thus clearly the case that although the sun be covered in its entirety, the air standing around the sun is nevertheless more brilliant the nearer it is to the sun. Consequently, if the air shall be thicker, and some slight particle of it, of a magnitude intercepted by the observed sun's cone, shall be darkened, the the air in the vicinity will be seen to shine, in a circular form. Other experiences are at hand. Mr. Jessenius, whom I mentioned in Ch. 5, fully confirms for me the eclipse of the sun seen by him through the clouds at the Torgau Town Hall on 25 Feb. or 7 March, 1598, clearly ringing the moon with brightness, and says that he followed up that form of the laboring sun with an epigram and an allegory. However, that eclipse in the Brahean computation could not have been total, not even through the legitimate diameter of the moon departing from perigee. Therefore, Jessenius was unable to see the sun everywhere, but what he saw everywhere standing around the moon was the brilliance of the air.

300 On 8/18 January 1603, at one hour after sunset, when a pillar of a tower was blocking the whole moon, nonetheless a sufficiently bright white orb was apparent, so that you would think the moon to shine through a watery cloud, or that the moon's brightness was diluted as if in water. However, it had no apparent end, but vanished into a cloud. And this brightness was easily distinguished from the actual body by moving the head. Something similar happened to me on 6/26 January following, in the evening, the moon being near perigee, with the greatest wonder. I set up a bronze wheel, precisely circular, clasped with a spike to one end of a pole twelve feet long, applied the eye at the other end, and before the eye a very narrow opening in another bronze sheet, so that the eye would have a perpendicular view to the wheel through the opening, and nothing might happen

[58] Ed. 1551 p. 334. The words are altered slightly.

[59] *On the Face in the Moon* XIX, 932B, pp. 120–1.

because of the width of the eye. And since, where the distance between the eye and the wheel was 10,368, the width of the wheel would be 104, covering an arc of $34\frac{1}{3}$ minutes, I was hopeful that the moon was going to be completely covered by this wheel, because of other ways of observing, which I was applying at the same time. And in fact, the moon was seen to protrude all around. Here I was troubled by anxiety that perhaps other ways of observing, in which I had the greatest trust, might be false. But that there was a fallacy became immediately plain when the eye was brought nearer the wheel. For all that brightness still did not betake itself behind the wheel even when the eye came to a distance of 10 feet. In this way the moon would have had to represent more than 41 minutes, which everyone knows is false. I was also unable to recognize a determinate distance from which the moon would be covered, for I always saw something bright on the circumference, even from a distance of seven feet. Something not much different was set in my way in 1600 at Graz regarding the observation of the sun's diameter admitted through an opening. For when two holes were opened, one the size of millet, the other the size of a pea, and a pair of circles were painted on the opposite wall, one of which exceeded the other by an interval that was as great as the difference between the larger and smaller holes, the ray of the sun admitted through the larger hole was indeed equal to the greater circle, but when the larger hole was blocked, the ray that came through the smaller did not maintain an evident boundary, and had the edge gradually passing over into a dusky color, and finally, far exceeded the smaller circle. For the ray of the sun, greatly weakened **301**
through such a small hole, was unable to illuminate the paper much more brightly than the rays from the air standing about the sun, whose continuation with the solar rays portrayed a breadth greater than the truth, and a brilliant color. Further, the air on that day was more than usually bright, far from the sun.

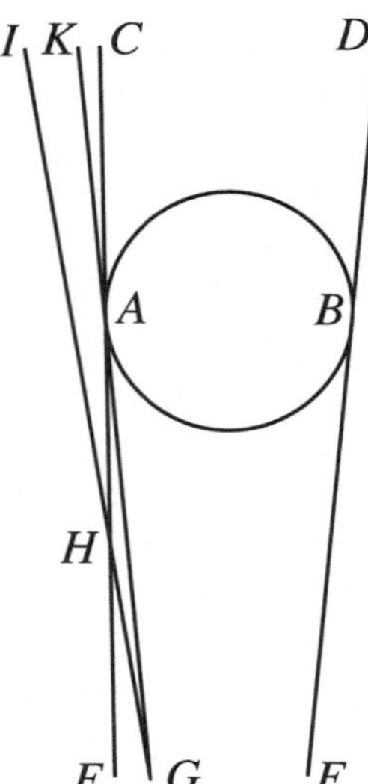

It is believable that something like this can also happen around the sun, when the moon is interposed upon it: that either the air or even the aethereal substance, which is not just nothing, but also has its own proportion of density, when illuminated by the sun, takes on a brightness that, since the sun is hidden, represents the solar light. Just so, above, the moon's redness, in the absence of the solar illumination, seemed to me to be the authentic light of the moon received from the sun. For this can be deduced from Prop. 23 of Ch. 1 in this manner. Let AB be the moon's body, CA, DB the outermost rays of the solar body bounding the moon's shadow AE, BF. And let G be the position on the earth's surface in the middle of the moon's shadow. Let GK be projected touching the moon's orb, and GHI farther out. Now, since anything of the aethereal substance beyond KA towards IH can be seen at G, but anything beyond CE is illuminated by the sun, therefore, something illuminated by the sun will appear from G. Consequently, by Prop. 23 of the first chapter, there will be communicated to that substance (which is by supposition a partaker of color to some degree) something of the light of the sun, and it will radiate that into the region away from the sun. Therefore, the particles which are in the space KA will radiate at G more strongly than those which are in the space IH, because the former are of a more direct radiation and are nearer to the actual solar radiation CE, while IHG is more oblique to CE.

302 If those things that Clavius saw in the eclipse of 1567 belong to this kind of effect, it is necessary that to Clavius too there appeared its bright edge, gradually less intense towards the outside, and not notable for a meticulous edge on the outside.

But what if Clavius should also deny this, and what if he clearly saw a cut off and bounded image of light? Are we really ready for this case? In fact, if Clavius could say this most truly, then I both would finally concede that the sun itself was seen around the moon by him, and will assign causes that are probably more geometrical than hitherto. Further, because in Chapter 6 Section 9 above we went along with Plutarch so far as to have dared to ascribe to the moon continents, seas, mountains, and valleys, such as this our earth has, how much more is it also to pour air around the moon, such as is poured around this our earth? For then, and even if it is not poured densely, that which we have demonstrated above in Chapter 7 Section 5 concerning the terrestrial air will happen: that the rays, approaching from the edges of the sun, go around the lunar body in a kind of bending through the refractions in the lunar air, and thus are bounded at our vision by a shorter cone. As in the first diagram in Ch. 7 Sect. 5, if the eye were set between the point χ and ω. So, in this way, it is not the moon's diameter that would appear less than it should, but the sun's diameter that would appear greater than it should.

In fact, this very thing was suggested to me, in a word, by a certain great proponent of these arts: "the outer parts of the moon appear to be pellucid."[60] But this could not have been as easily defended as it was easily said. For what will help them to be pellucid, if they are of a denser substance than the aethereal air, and the moon is round? Now if this should happen, the path for the solar rays is not straight, and nothing certain can be demonstrated without that diagram of Chapter 7. Besides, it might be asked why, if the outer parts of the moon are always pellucid, a circle of the sun does not always remain. I accordingly prefer to give the moon air, not of a great depth, so that sometimes, but extremely rarely, it could proceed as far as to this effect. But I much prefer that that cause proposed in the penultimate place be valid, while this last one not be put to use, but to remain in storage.

4. A number of corollaries on eclipses of the sun

303 Now, in order not to end this chapter without profit,[61] among the rules mentioned previously we shall take note of the following.

---

[60] This was Herwart von Hohenburg, as his letter to Kepler of 16 May 1599 (no. 121, in *JKGW* XIII p. 332), containing this remark, shows. Later, in a letter of 11 November 1602 (*JKGW* XIV, no. 235, p. 312), Herwart retracted this opinion.

[61] *Caput*, "chapter," can also refer to the "principal" of an investment.

1. As many times as any eclipse of the sun brings darkness similar to that of night upon the lands during the day, it is certain that the whole sun was concealed from view by the moon.

2. However, the astronomer should not as a result predict that there will be darkness even if he sees that the whole sun is going to be covered, unless the moon is also near perigee and the air shall have been pure. The cause of this you have in Ch. 6 Sect. 11, above.

3. It also does not follow that if the stars have appeared the eclipse was total. For we read of stars having been revealed by many partial eclipses.

4. However, this is certain enough: if a total eclipse is going to happen in a calm sky, the stars will be revealed.

5. Morning eclipses are always more inclined towards darkness before the middle, evening ones after the middle, the former particularly at the descending node and on the descending semicircle of the zodiac, the latter at the ascending node and semicircle. For they extinguish the light of that portion of the air illuminated by the sun, which part, being above our heads, brings most of the light to us.

6. Color, however, is changed for things, not so much by eclipses near the horizon as by those near the zenith, in summer, in a calm sky. For the clearer the air, the more strongly the sudden darkening of the light enters into the eyes, which of course were strongly pervaded by it before, and retain strong images of the light. On the other hand, if rain or clouds occupy the heavens, and bring about a darkening beforehand, nothing unusual will happen to the eyes, even though the darkness is increased by the eclipse. Thus, on 21/31 July 1590, to the reapers in Schwabia, the eclipse, which was not particularly great, showed a yellow color everywhere, and a reddish one in Styria. But one much greater than this, on Feb. 25 or March 7 1598, in a turbid heaven with rain, in winter, when the sun was lower, gave rise to no disagreement about colors, even though it made the light of day abate to the dimness of evening.

7. There are many that are not completely total that bring upon the lands a darkness greater than total eclipses when the moon is at apogee or when the air is bright. This is because the sun is not deeply hidden all around beyond the moon at apogee. But if the sun be almost fully covered near the horizon by a moon that is slightly higher, the sun nonetheless protruding somewhat below, the air above **304**
our heads is the more deprived of a view of the sun the higher it is, so that sheer night, or the moon's shadow, hangs over our heads.

5. On the mutual occultations of the other heavenly bodies

Though it is tenuous, there is nonetheless some connection joining this consideration as well with eclipses of the sun and moon.

First, in order that any star be eclipsed in the same way as the moon, it must lack its own light, which is denied above in Ch. 6 Sect. 12. If, however, this be granted, the star Mars will seem not to be entirely free of the suspicion that it may run into the earth's shadow. For, concerning the length and breadth of the shadow belonging to earth, and the sun's parallax, if you look at the quantity, there is even now grounds for cautious consideration. And since the point of the lunar shadow falls exactly on the earth, it seems fitting that the point of Venus's shadow come to an end at the moon when nearest the sun, the point of the Mercurial shadow at Venus; likewise, again, the earth's shadow at Mars, Mars's at Jupiter, and Jupiter's at Saturn.

See also whether that darkness which we said in Ch. 6 Section 11 came over the earth in 1547[62] might have been the more diffuse tail of some comet, a suspicion which I also raised above.

But in the same way as with the sun, we daily see all stars except the moon become eclipsed, with the lower ones always covering the higher, if they intercept them in their path.

In the year 45 according to Dionysius, which was the year 241 before Christ, on the 10th day of Virgo, or September 4, at dawn, the star Jupiter occulted the Southern Ass. Ptolemy Book 11 ch. 3.[63]

Aristotle saw the same star come together twice with the other of the two in the feet of Gemini, and occulting—completely hiding[64] —it.

In Dionysius's year 13, 25 Capricorn, which is 18 January 272 B.C., in the morning, the star Mars was reckoned to be placed beneath, "imposed upon", the northern star of the brow of Scorpius. Ptolemy Book 10 ch. 9.[65]

In the year 13 of Philadelphos, between 17 and 18 Mesore, which is the same as 272 B.C. 12 October, the star Venus was seen in the morning to have caught up with, or, as Theon expounds it, to have occulted, the star opposite Vindemiatrix, which is located at the end of the southern wing. Ptolemy Book 10 ch. 4.[66]

305 On 16 September 1574 at 4^h in the morning, Maestlin saw Cor Leonis covered by Venus. I saw the same at Graz on 15/25 Sept. 1598 at hour 3 in the morning, when Venus had barely risen. At the fourth hour, more than one [diameter of] Venus could have fit in between; nevertheless, the line from Venus to Cor fell a little below Jupiter.

That Jupiter should cover Saturn can hardly happen once in many centuries; nevertheless, it was seen to have been done in 1563. And events followed in proportion to the greatness of the sign, which even today we perceive before our eyes.[67]

---

62 See p. 270

63 The Calendar of Dionysius is known only through citations in the *Almagest*; its months were named after the signs of the zodiac. Cf. Ptolemy, *Almagest*, trs. Toomer, pp. 13–14. The Southern Ass is the fifth star in Cancer in Tycho Brahe's catalog (δ Cancri in modern tables). See *TBOO* III p. 347.

64 ἀφανίσαντα. The cited passage is in the *Meteorologica* I 6, 343b 30, pp. 46–7.

65 Trs. Toomer, p. 502. The star is the first star in Scorpius in Tycho Brahe's catalog (β Scorpii in modern tables). See *TBOO* III p. 350.

66 Trs. Toomer, p. 477. Vindemiatrix is the thirteenth star in Virgo in Brahe's catalog (ε Virginis), and, as Ptolemy explains, the "star opposite" is the sixth in the catalog, η Virginis in modern tables.

67 Kepler believed that the efficacy of a planetary conjunction increased in proportion to the nearness of the apparent mutual approach of the planets. He discusses in a general way the effects of the so-called "great conjunctions," i.e., those of Saturn and Jupiter, in *De stella nova* (Prague, 1606) Ch. 8, *JKGW* II pp. 184–9.

That some comet was covered by the star Jupiter, certain people conclude from Proclus; see Tycho's *Progymnasmata* Book 1 fol. 619.[68]

Maestlin, at Tübingen, and I with him, saw Jupiter entirely eclipsed by Mars, on 9 January 1591. The fiery ruddy color of Mars argued that Mars was the lower.[69]

Proclus said that Venus was observed to run beneath Mars, just as Mercury was observed running beneath Venus.[70]

Of Venus and Mars, the same Maestlin relates an example on 3 October 1590 at hour 5 in the morning: Mars entirely occulted by Venus, the brilliant color of Venus again indicating that Venus was the lower.

Mutual occultations of Venus and Mercury are possible. Now Venus is higher, now Mercury.

When, on 21/31 May, 1599, Mercury was about one degree beyond Venus, it was nearly the same amount farther north, but on the following days it descended, both towards the ecliptic, with decreasing latitude, and towards the earth on its epicycle, with the visible diameter increasing. It nonetheless happened on 29 May or 8 June (for at Graz the whole intervening eight days were rainy and cloudy, as were the following days also, up to [June] 3/13) that, looking with greatest care at Venus, I nevertheless saw no Mercury, while I saw the Twins and Capella. I was in fact persuaded that I saw certain rather long and thin rays from the eastern part of Venus; Venus, however, did not change color. The analogy of the diurnal motion and the preceding observations definitely put Mercury very close to Venus.

In this century, Venus cannot cover the sun. It was, however, able to about two hundred years ago, and will be able to at some future date.[71]

The nodes of Mercury are at the beginning of Taurus and Scorpio or the end **306**
of Aries and Libra, and today, and at almost all times, they can carry this planet beneath the sun.[72] It is consequently the less to be marvelled at what we read in the life of Charlemagne, noted at the year 807, in these words, "On the 16th day before the Calends of April, the star of Mercury was seen in the sun like a small black spot. It was, however, a little higher than the middle center of the same heavenly body, which was thus seen by us eight times [as I read barbarously, not

---

68 *TBOO* III p. 134. In this passage, Tycho is quoting Peucer, who is citing some unknown persons who related something Proclus remarked. Dreyer, commenting on the passage (*TBOO* III p. 396), suspects a fabrication.

69 Maestlin, *Disputatio de eclipsibus* p. 18. In this place, according to Frisch (*JKOO* II p. 431), Maestlin also described occultations of Regulus by Venus and of Mars by Venus (mentioned below).

70 *Hypotyposes* 7.22–23 (Citation from *TLG*).

71 Kepler later changed his opinion on this point, announcing a transit in *De raris mirisque anni 1631 phaenomenis Admonitio* (Leipzig 1629), in *JKGW* XI.

72 In *Phaenomenon singulare seu Mercurius in Sole visus* (Leipzig 1609), Kepler remarks that he was in error about the location of the nodes: "For today the node of Mercury is not at the beginning of Taurus, but of Gemini." Cf. KGW 4 p. 85.

'eight days']. But when it first entered or departed, was impossible to note, clouds being in the way."[73]

The author was experienced in astronomy, which is evident from placing on record so many eclipses, and that he grasped the connection of the sun and Mercury with a computation, from which he knew that that spot was Mercury. The year, however, was written erroneously, by some chance. For it was on March 17 808 (probably because the year begins with Easter). For at the noon hour of that day in Königsberg, the Prutenic computation shows the sun's position as 0° 45′ Aries; Mercury's, 0° 31′ Aries, with a latitude of 2° 9′. It is certain that this latitude was so arranged by Ptolemy, without guidance of observations, in order not to lead these two planets beneath the sun. But it was already said that the node today is at the end of Aries, so that it is not beyond what is reasonable that at that time it was at the beginning of this sign.

Therefore, we no longer put faith in Averroes alone about this phenomenon, after a professedly Christian man has also added his own computation.[74]

The encroachments of the moon over the fixed stars occur almost daily, and are usually carefully noted by the astronomers. Thus Agrippa of Bithynia and Timocharis saw the Pleiades occulted by the moon; Menelaus the Roman, the Brow of Scorpius; and the same authors, Spica Virginis more than once;[75] Tycho Brahe and Copernicus, Palilicium;[76] I, the Heart of Scorpius[77] (as was said above in Ch. 5 Sect. 5);[78] Maestlin, very closely spaced tinier stars.

Likewise, Walther noted Saturn covered by the moon (see below, Ch. 11 Prob. 30); a monk historian saw Jupiter covered by the moon in January of the year of Christ 807; Aristotle and Maestlin saw Mars; Copernicus (Book 5 Ch. 23) saw Venus.[79]

307 Aristotle's words in *On the Heavens* Book 2 Ch. 12: "For when the moon was so divided into two parts, that on one part it was darkened, on the other, it shone, we saw it apparently come together with the star that is called Mars, and indeed, when it was covered by the moon's dark part, we saw it emerge from the

---

73 As a source for this, Franz Hammer cites "Frankische Reichsannalen" (*Annales Laurissenses maiores*), *Mon. Germ. Hist. Script*. I. 194. Kepler also mentions this instance in *Mercurius in sole*. Later (e.g. *Ephemerides* for 1617) he thinks what was seen was really a sunspot.

74 Frisch (*JKOO* II p. 431) reports that, according to Mercator, this computation was by a certain Benedictine monk named Anonius (Aimoin), who wrote a "Historia Francorum" at the end of the 10th Century. Kepler's source for Averroes's observation is Copernicus, *On the Revolutions* I 10, trs. Rosen p. 19.

75 These observations are reported by Proclus, *Hypotyposes*, 5.7–9 (Citation from *TLG*).

76 Aldebaran. Copernicus relates this occultation in *On the Revolutions* IV 27, trs. Rosen p. 218. Brahe's discussion of this passage, and his report of his own observations, is in *Progymnasmata* p. 245, *TBOO* III p. 245.

77 Antares.

78 See p. 232.

79 Copernicus, *On the Revolutions*, trs. Rosen, p. 276.

moon's luminous part."[80] This therefore could not have been at any other time than in the third year of the hundred and fifth Olympiad, on the night of 4 April 357 B.C., the sun being in 10 Taurus, the moon with Mars in 3 Leo, with the same latitude; when Aristotle, a youth of 21 years, would have been studying with Eudoxus, as is known from Laertius.[81]

In that place, Aristotle commends the industry of the Chaldeans, from whom he says many such observations came down to the Greeks.[82]

[80] Aristotle, *On the Heavens*, 292a 4–7. This occultation is also discussed by Kepler in *Astronomia nova* (1609) Ch. 69 p. 323 (trs. W. H. Donahue (1992) p. 641). A modern account puts the occultation on 4 May: see Aristotle, *On the Heavens*, trs. Guthrie, p. 205.

[81] According to Diogenes Laertius, *Lives* V 9–10, Vol. I pp. 452–3, Aristotle was born in the first year of the 99th Olympiad, which would make him about 26 in the third year of the 105th Olympiad.

[82] Aristotle actually cites the "Egyptians and Babylonians".

Chapter 9
On Parallaxes

And so the bodies whose appearance and measurements astronomy considers are, as we have said, the sun, the moon, and the stars, to which are added the shadows of the earth and the moon. Aside from that, the foremost thing we seek in these bodies are their motions, so much to be wondered at. But in order that an astronomer be able to fish these out with geometrical demonstrations, one first needs to measure their position with instruments. Now in geometry, when we don't know how to describe a spiral or a conic section in a single process, we imagine some points which the line passes through, from which the whole trace of the line may be distinguished. So likewise, in astronomy (I speak to those inexperienced in it), we by no means perceive the motions of the heavenly bodies with the eyes, but we compare various positions with each other, and we thence seek the form of the motion, by which all those places previously noted, and similar ones that are going to be, may be set in order. Once this has been put forward, we astronomers have fully carried out our task. And thus the various positions of the stars at certain times are, as it were, like the elementary components, or rather the created objects,[1] of the motions.

And position is in fact in the category of relation, and, insofar as we are now considering it, has reference to the places of the stars. For we are enclosed along with all the stars by the same outermost orb of the world. And since, from the nature of the spherical figure which the world has received, there are three regions of this worldly edifice, the center, the surface, and the in-between; while
308 we occupy the center, whether in reality or in our perception; therefore, there are left to the stars the two remaining regions, according to which their positions are considered in two respects, namely, either in respect to the in-between or the diameter of the world, or in respect to the outermost surface.

1. On the observable position or place of the heavenly bodies, or the reckoning of it beneath the fixed stars

And indeed, the position of a heavenly body with respect to the diameter of the world comes closer to the thing itself, while the position with respect to the surface is almost entirely visual.

For since everything that is seen is comprehended and seen with a certain angle of luminous rays flowing together from the edges of the object seen to the center of the eye, we then finally consider ourselves to have seen, i.e. observed, a heavenly body rightly, when we shall have measured the angle of vision accurately. But the angles are measured by the circle described from the point of the angle. And that enormous circle, or the spherical surface of the outermost world, is described from this point of our visual angles, that is, from the earth, because

---

[1] Kepler uses *apotelesmata* here, which is a transliterated Greek word meaning "things that are brought to completion," but which can refer to the astrological effects of the motions. Cf. *L&S* p. 222.

the little space by which the surface, entrusted to our care, is distant from the exact center of the world completely vanishes in such vastness. It is established, therefore, that the spherical surface of the outermost heaven is employed by us to measure the visual angles formed at our eye; and conversely, that our visual angles adequately give evidence of the position of the stars with respect to the surface of the world. Now this consideration is also twofold. First, it will be said in what follows that the stars along with their distance from us cannot be perceived with the eyes. Therefore, as regards place, we do not distinguish the stars that are very near us and those that are fixed in the most distant orb, unless one be covered by another. And so, on this consideration, we note with our eyes the conjunctions of the planets with the fixed stars. Accordingly, the angle by which two fixed stars, in conjunction with a pair of planets, are perceived, is the same angle, and the same arc of a great circle, the measure of the visual angle, by which we say the planets themselves are also distant from each other with respect to the surface. Likewise, the angle by which the outer edges of the sun or the moon are distant from each other is said to be the magnitude of the sun or moon. So much so that this leads to the sun and moon being considered equal, even though the sun exceeds the moon by many thousands of times, just because they are per- **309**
ceived by about the same angle. See Witelo 4.19.[2] These distances of the fixed stars, as well as of the planets, and even of the edges of the sun and the moon, from one another, the astronomers measure with instruments, such as the astronomical radius or backstaff, the sextant, and others, setting up a comparison of the arc interposed between a pair of stars with the whole circle, or with the fourth part of it. Although this subject is related to the optical faculty, it belongs particularly to the mechanical part of astronomy, which that most noble Tycho cultivated with greatest liberality, and on the other hand treated with greatest diligence.

Next, since the nature of things does not allow, nor does practice ever permit, that we take note of the approach of the heavenly bodies to the fixed stars by eye (for by day the heavenly bodies are hidden), astronomers had proposed for themselves other marks from which they could number the arc distances of the stars, which do not go out of sight as do the fixed stars, subject to risings and settings, but are absolutely inseparable from their places on the ground. These are the horizon of any place, which, extended as if to infinity, forms the plane surface of the earth, dividing the whole sphere of the world into two hemispheres, equal to the sense, and the pole of that horizon, or the point that stands above the vertices of whatever place at any moment of time, and which is shown by the line of the perpendicular, following which all heavy things are carried downward, and we stand upright. In this way, sea captains take the sun's altitude with respect to the surface, whether the arc of vision, or the angles which the rays coming both from the sun and from the horizon lying beneath make at the eye. Thus the astronomers take note of the angle which a ray of the sun or a star forms with a line perpendicular to the surface of the horizon, using rules and quadrants constructed for this purpose, with a suspension of weights. But I have begun to babble about

[2] Witelo IV 19, *Thesaurus* II p. 126: "All things seen under the same angle, whose distance from each other is not taken into consideration, are seen as equal."

On the mathematical disastronomer Tychonomast [Tycho-whipper].

industry with ignorant people, as the occasion was given me to exclaim upon a certain Censurer of Tycho,[3] who, while he would have been able to be numbered among the astronomers on account of his happiness of intelligence, preferred to babble what are more irrelevancies than puerilities along with the crowd of tinkers, out of lust for quibbling, such as for example the inexperienced may have, under which master they may dare to pluck the good arts. He denied that the Tychonic observations are of the certainty and accuracy that were attributed to them
310 by the author (thinking of these numberings of arcs or visual angles, through the most precise instruments), for the heaven (which likewise before instruments we had made the measure of this visual angle) can be divided into many more parts, and these of fully noticeable magnitude, about which the astronomer must still have some doubt after all the precision of the instruments. And this cunning detractor of the art pretends that this position of the stars, with respect to the surface, and these distances, these altitudes, and by whatever other name they come, are not set up primarily on account of the objects or the stars themselves, between the centers of which there are not arcs of this kind, but on account of our vision, and that this whole enterprise in astronomy rests upon optical reasons; and besides, that it is stupid to wish to affect a precision that is different from what can be accomplished by the sense of vision, but proud and barbarous to reject this visual precision, which is for us the first approach to truth. Therefore, neither Tycho nor any sane person professes that by such an easy undertaking he lays open the genuine and most true distances of the stars, and the positions on the world's diameter, through these visual distances. But he does profess this: that he uses the arithmetic numbering and the geometrical division of the visual arc to imitate the precision of the sense of vision, and by these visual distances constructs a road to finding out the truest places of the heavenly bodies on the diameter of the world. And let it be sufficient to repeat and to point out these things in a timely manner concerning the position of the stars, as regards the surface or the visual angle, for the sake of the more inexperienced.

2. On the altitude of the heavenly bodies from the center of the earth and the parallax resulting from the distance of the eyes

Next, I should speak of the position of the stars in regard to the diameter of the world, proceeding on the same track, so that in a place where I cannot propose anything new for the learned, I may assist the learned in language that is popular, in proportion to their abilities, as much as the established brevity allows. I shall begin from what is best known, starting with those things previously established in Chapters 3 and 5. Nature gave two eyes to animals, not just, as is commonly
311 believed, to assist with a loss, should the animal be destined to have incurred such a loss of one eye, but for the grasping of the distance of visible things by the eyes. For since the centers of the two eyes are distant from each other in a certain proportion to the body, that is, by a breadth of about one palm, while no certain vision occurs except when the diameter of the two eyes, which passes

[3] This was the Heidelberg professor Jacob Christmann, alluded to above in Ch. 4; see the note on p. 139.

through the centers of the humors and the pupils, is directed towards the object set before the vision, it thus happens that these diameters between the object to be seen are not completely equidistant from themselves, but nod towards each other in proportion as the object seen stands nearer the sense of vision. This nodding together, natural to the eyes, Aristotle describes in Probl. 7 Section 31, when he says that each eye is put in position from the same beginning, implying that each eye sees the same thing, in one vision and image, because the cognitive faculty communicates the same things to the two eyes.[4] And in Probl. 17 he asks by what means it happens, and why, we should see the same image with each eye even when looking at an object from an oblique angle? Let the centers of the eye be A, B, distance AB, object presented for viewing at D. The diameters of the eyes EF, GH will be directed towards it, so that if extended they would come together at the visible D. Now let the object be seen nearer the eyes at C. Again, the diameters of the eyes will be directed towards it, and will be positioned at IK, LM. And because the two isosceles triangles ADB and ACB are set up upon the same base AB, the angle at C will be greater than the angle at D. Therefore, CAB will be less than DAB, and CBA less than DBA. Consequently, the ends of the ocular diameters have passed over from F, H to K, M, approaching each other, and the posterior ends E, G have diverged to I, L. And thus it is necessary to employ a motion and a turning together of the two eyes, when the vision is transferred from the more remote D to the nearer C. And by the use and perception of this motion or animate action, the animal becomes accustomed to distinguishing the longer and shorter distances of visible objects from itself; and this only when there is some perceptible proportion of the distance of the eyes AB to the removal of the seen 312
object from the eyes AD or BD. These things do not in fact have any effect in astronomy. For the distance of the heavenly bodies is so great, and further, so great is the narrowness of the distance of the ocular centers AB, that the eyes, when looking at some heavenly body, and even at a rather distant mountain, extend diameters AF and GH that are parallel within the limits of perception. For the rest, ordinary conceptions are to be elevated to those exceptional and subtle matters by examples of small things. And these are genuine examples of them. I therefore proceed. In the same figure let the visible again be C, the other things as before. And opposite, beyond the visible C let there be a wall, cutting the straight lines AC, BC at N, O.

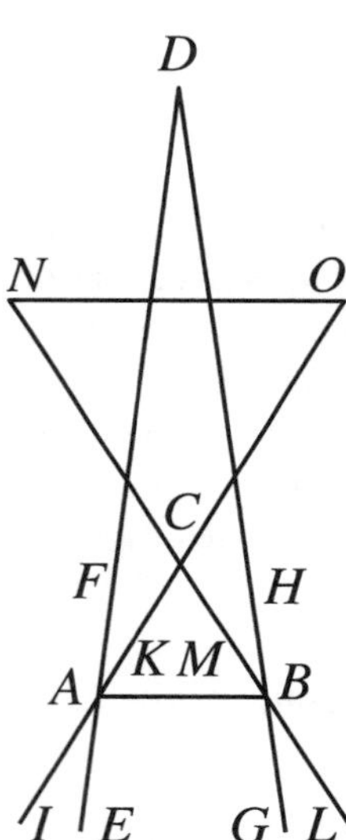

There still remains for a single long accustomed eye a slight ability to distinguish distances

Now when one eye is closed, once the medium is removed, the function stops. That is, there is no longer any distinguishing between distance, the company and spacing of the two eyes being hindered. Consequently, the eye A will consider the visible C and the point of the wall O to be in conjunction, because

4 *Problems*, XXXI 7, 957b 7–958a 34, trs. Hett vol. II pp. 184–9.

they are situated upon the same line *ACO*. So when the eye *A* is closed, the visible *C* will be seen by the eye *B* joined to the point of the wall *N*, because there is no distinguishing between the distance *BC* and the distance *BN*. In this manner, when the eyes *A*, *B* are alternately opened and closed, the visible *C* will change its seen place from *O* to *N* and *vice versa*, as if by a leap. And the optical writers, imitating the words of the astronomers, will call this "commutation of vision," or, using the Greek word, parallax. The demonstration is almost the same as the one I brought in above in Ch. 5 Prop. 7 on Aristotle's problem on the double image. And so I would say that something similar happens to us in the contemplation of the heavenly bodies, so that it may become completely clear what it is that astronomers call parallax.

that are very close, because of the width of the eye and the motion of the opening in the *tunica uvea*, most greatly because of the translation of the head, but this does not hinder us here. See above, ch. 3.

3. On the diurnal parallax resulting from the distance of the earth's surface from the center

And it was said shortly before that the diameters of the two eyes proceed parallel within the limits of perception, if the sense of vision be directed to some heavenly body. Thus, with the inclination of the diameters removed, the discovery of the distance of the heavenly body by the eye, or its position with respect to the diameter of the world, is taken away. And if two heavenly bodies shall have been by our vision in the same line, our vision, although making use of both eyes, will be unable to distinguish the different distance of the two, and will consider them in conjunction.

313 In this way it happens that we reckon the moon and all the planets to be located under the sphere of the fixed stars, since the distance of the eyes [from each other] does not help us. And in this way, our vision may be completely in error in judging the positions of the planets with respect to the diameter of the world.

This defect in the sense of vision Nature removes through a wonderful device. For it was by all means the will of God the Creator that the human being, His image, should lift up his eye from these earthly things to those heavenly ones, and should contemplate such great monuments of His wisdom. Hence, the entire arrangement of the fabric of the world tends to bear witness to us of this will of the Creator, as if by a voice sent forth. For that reason, the ratio of the earth's globe to the orb of the moon has been made perceptible, so that what has deserted the eyes of individual humans, the attentiveness of all of them living on the whole surface of the earth, assisted by its magnitude, might supplement, and might in this way teach the position of the planets on the diameter of the world by those prior positions with respect to the surface; i.e., the distances of the angles. In the above diagram, let *A*, *B* now not be the two eyes of the same person, but two places on the earth's surface, *A* in Europe, *B* at the extreme promontory of Africa; let the moon be at *C*, and let it be seen at the same moment by people at the two places. And so, since the ratio of the distance of the places to the distance of the moon from earth with respect to the diameter of the world, is perceptible, which is now our supposition, the inclination of the lines *AC* and *BC* will also be perceptible. Further, let *NO* be the sphere of the fixed stars at night, or the body of the sun by day at the moment of the new moon. Therefore, since those who are at *A* cannot

distinguish between the distance of the moon and the distance of the sun or of the fixed stars with respect to the diameter of the world, the moon C will seem to be in conjuction with the fixed star O, or the edge of the sun. But to those who are at B at the same time, the moon C will seem in conjunction with the fixed star N, or the edge of the sun, so that the former observers have the portion NO at the right side of the moon, the latter, at the left side. Therefore, the position of the moon imagined by the people at the two places will be switched from O to N. The description is exactly as before, concerning the conjoined vision of the two eyes. And in this way in general the parallax of the heavenly bodies is taken, that the arc NO of the outermost sphere, or the angle which that arc measures, is the parallax of the moon. But because it is not essential on every occasion, that
people have been set out upon the earth's surface who may look at the moon at 314
one and the same moment of time, astronomers proceed by a somewhat different road. For by another method, about which this is not the place to speak, they compute the fixed stars among which the moon ought to appear, or the distance from the zenith under which it ought to appear, at any given moment to an eye set at the center of the earth. At the same moment that distance from the zenith is measured, which makes its way into an eye set at the surface of the earth. The difference of the two angles they call by the special name of Parallax or the commutation of vision, so that the parallax is in fact sought in this way in order to learn the position of the heavenly body on the diameter of the world, although in itself it is comprehended under the genus of visual angles, or place with regard to the surface.

About center A let there be described a great circle BC on the earth's surface, and another DEF on the surface of the outermost sphere of the fixed stars, and let A, B, D be in the same straight line, passing through the zenith of the place. And in an intermediate place let there be a heavenly body at G not on the vertical line AD. Next, from center A and the place B of the earth's surface let straight lines be drawn out to G and extended to E, F. Finally, about center A with radius AG let the arc GH be drawn in the orb of the heavenly body. Astronomers, then, as I have said, first find the angle DAG or the arc DE, the measure of that angle, that is, the arc of the body's apparent distance from the zenith, which is DE, by what was said at the beginning of this chapter. Next, they find the angle DBF or the arc DF, the measure of that angle (since B is imperceptibly distant from center A, when BA is compared with BD), that is, the arc of the heavenly body's apparent distance from the zenith from the place B of the surface. The reason for this is that if the body were to fall right on the vertical line DA, such as if it were at point H, beneath the zenith D of the place, it would clearly not change its visual place at all. For if the body H, the sense of vision B, and the center of the earth A, were in the same straight line, BH and

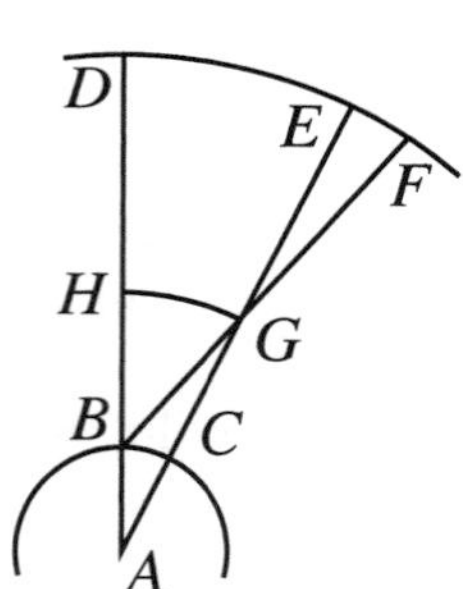

AH extended would fall upon the one point D, and AH is the visual line from 315
the center, while BH is the visual line from the surface. And so from either place of vision, the same place D beneath the fixed stars would be indicated.

And so astronomers choose this terminus of the arcs on which the parallaxes are numbered, because from this common terminus the parallaxes first begin to exist in all directions. For example, the heavenly body now at G will appear from A at E, from B at F, so that DE or DAE is less than DF or DAF, or the place F seen from the surface B is elongated from the zenith D more than the place E from the visible center A, which otherwise they call the true place. For it is an optical axiom that those things that are more distant appear smaller; that is, are perceived by a smaller angle of vision.[5] If HG be the arc of the heavenly body's distance from the zenith in its own orb, which, since it is perceptibly closer seen from B than from A, because the semidameter AB is given as having a perceptible distance [in relation] to the distance of the body AG, therefore, angle HAG will be perceptibly less than angle HBG. Consequently, its measure DE will also be perceptibly smaller than DF, the measure of the latter angle. And therefore, their difference EF is called by astronomers the parallax of this heavenly body placed at G.

Further, EF is the measure of the angle EGF or BGA, because the ratio of BA to AE is imperceptible, while the ratio of the same BA to AG is perceptible by supposition. Therefore, the ratio of GA to AE is also imperceptible, and further, the angle at G is sensibly equivalent to the angle which can be set up above EF at the center A. The same is also evident thus: since DF is the measure of angle DBF within the limits of sense perception, but angle DBG is equal to the opposite internal angles BAG, AGB taken together, therefore, DF is the measure of the angles BAG, AGB together, within the limits of sense perception. But DE is really the measure of angle DAG by itself. Therefore, the remainder EF is the measure of the remainder AGB, or EGF, equal to it to the sense.

Consequently, it follows from these things that the angle BGA is likewise said by astronomers to be the parallax of the heavenly body at G.

Now this parallax is of godlike use in astronomy. For whenever the arc DE of the heavenly body's true distance from the zenith at some certain moment can be obtained from astronomical theory (which is done in various ways), and the arc DF of the visible distance from the zenith is taken with suitable instruments
316 and with the required accuracy at that moment, in order that the parallax EF be thus obtained, the ratio of AG, the most true position or removal of the heavenly body with respect to the diameter of the world, to the semidiameter of the earth AB, has then become known. And indeed, this theory of the parallaxes of the sun and the moon gives support, both to all the rest of astronomy, and particularly to the theory of eclipses of the sun. And so that parallax which is considered on the vertical circle is distributed in different ways: either in longitude and latitude through the ecliptic and the circles of latitude, by which name all of this kind of parallax is called latitudinal/longitudinal,[6] and the rest are called by some other name; or it is distributed through the equator and the circles of declination, into

---

5 See e.g. Witelo IV 7, in *Thesaurus* II p. 120.

6 μηκοπλατής.

the parallax of right ascension and of declination, according to what serves the uses of astronomers.

4. A most easy and succinct derivation of the diurnal parallaxes in longitude and latitude, using a new parallactic table

The theory of parallaxes makes its main trouble for astronomers in the eclipses of the sun. So much so, that just on account of this labor alone, it is no surprise if astronomy is neglected. And thus theorists have always led themselves to be praised for this: to assist everyone with a succinct method. Ptolemy produced an outstanding piece of work, with a lucidly described delineation[7] of the varieties that exist in parallaxes.[8] Reinhold, with tables very laboriously worked out at many elevations of the pole, considered that he had come to the support of the less expert.[9] In fact, he oppressed the zeal of the intelligent with the tedium of seeking the proportional part. And so Tycho Brahe called astronomers back to the triangles, by showing a number of ready ways of solving them.[10] However, even this labor is enormous. And so I think that if, having imitated the examples of these authors, I at last shall roll this Sisyphean stone on across the ridge, so that it may not again fall back, I may hope to deserve a certain amount of gratitude from astronomers.

I say, first, *that all parallaxes of latitude in whatever degree of the ecliptic the sun or moon happens to be, are equal, provided only that the same point of the ecliptic is at rising in the same altitude of the pole, and the moon is equally distant from the vision*. About center A, representing the center of the earth, let **317**
the great circle of the earth BF be drawn, on which let B be the place of the observer; but let F be the point lying beneath the pole of the ecliptic, so that both this circle and all the lines cutting it are thus in the plane of the circle of latitude,[11] which is the same as the vertical circle, because it passes through the observer's place B. Further, let the line AB be drawn stretching to the zenith point at L; again, line AF stretching to the pole of the ecliptic, and in the same plane let AG be erected at right angles to it, which will stretch towards the nonagesimal degree,[12] because GAF is right and AF stretches toward the pole.

---

[7] ὑποτυπώσις. In its general sense, this word can mean "sketch" or "outline," but also had a technical meaning in astronomy very much like our word 'model.' Cf. *L&S* p. 1900.

[8] Cf. *Almagest* V.11, 12, 17, 18, and 19.

[9] Cf. Reinhold, *Prutenicae tabulae coelestium motuum* (Tübingen 1551 and other editions), pp. 99–121

[10] Cf. *Triangulorum planorum et sphericorum praxis arithmetica*, *TBOO* I pp. 281–305; and *Progymnasmata*, esp. pp. 118–124, *TBOO* II pp. 130–136, and pp. 587–90, *TBOO* III pp. 102–105.

[11] The circle of latitude is the circle along which celestial latitude is measured, perpendicular to the ecliptic.

[12] The nonagesimal is the ninetieth degree on the ecliptic measured upward from the rising point, i.e., the point where the ecliptic and the eastern horizon intersect. A great

But because the proportion of AB to the sphere of the fixed stars is imperceptible, let BI be drawn through the place B parallel to AG, and let GA be extended to H, IB to K. Therefore, IK will now be visibly on the ecliptic. Further, let the moon be right under the ecliptic at point G, which is closest to the zenith.

It will be seen from B on the line BG, and for that reason GI will be its parallax on the vertical. For AB to BG is perceptible, and is the cause of the parallax. I say that however many lines are drawn from B in the plane of the ecliptic, equal to BG, represented by the line GH, make equal angles with the plane of the ecliptic. For from B let a perpendicular be drawn to GH: let it be BD. Then, because BD is in the plane of the circle of latitude, and this is at right angles to the ecliptic, BD is therefore in a plane perpendicular to the plane of the ecliptic. And because GH is the common intersection of the planes, and BD is perpendicular to it, BD is therefore universally perpendicular to all the lines in the plane of the ecliptic. But it is supposed that all lines BG all around are equal, and BD is common to all BG, and D is everywhere right. Therefore, it is
318 necessary that all the remaining angles DGB be equal. But DGB are the angles of parallax of latitude, because B is in a plane tending towards the pole of the ecliptic, and DGB, GBI, which formerly measured the vertical parallax of the highest point of the ecliptic, are equal. Therefore, what was proposed is evident: the demonstration of this could in fact have taken their occasions from Copernicus and Alfraganus on parallaxes.[13]

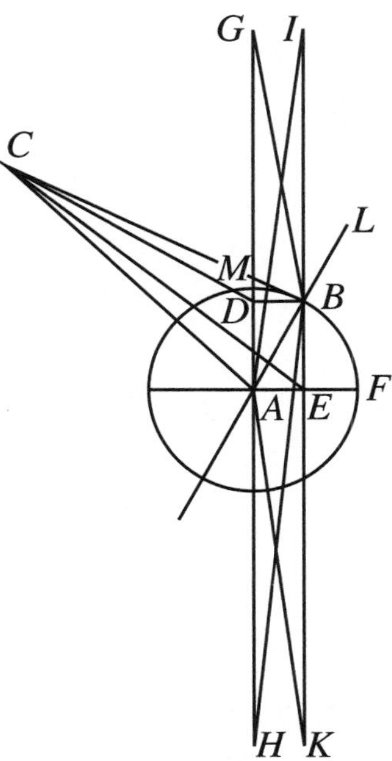

---

circle drawn from the nonagesimal to the zenith will make right angles at the ecliptic, and therefore, extended as necessary, will pass through the poles of the ecliptic. Although this may at first seem implausible, it is readily shown as follows.

The nonagesimal and the zenith are both 90° from the rising point, the former by definition, and the latter because the rising point is on the horizon. Therefore, a great circle drawn through the nonagesimal and the zenith will have the rising point as its pole, and will consequently be perpendicular to all great circles drawn through the rising point. The ecliptic is one such circle, and the vertical circle through the rising point is another.

[13] Copernicus, *De revolutionibus* IV 26, trs. Rosen pp. 216–218; Alfraganus, *Alfragani astronomorum peritissimi compendium, id omne quod ad Astronomica rudimenta spectat complectens*, Latin translation by John of Seville (in 1135) (Paris, 1546), Cap. 7 pp. 98–101. This book is often referred to as the *Rudimenta*, or as *Elementa astronomica*. For Alfraganus, or (correctly) Al-Farghānī, see A. I. Sabra's article in the *DSB*. The *Rudimenta* was also translated by the "Tycho-whipper," Jacob Christmann (Frankfurt, 1590).

I say again: *Even if the moon is not equally distant from the vision, provided that it is equally distant from the earth's center, again the parallaxes of latitude are approximately equal*. For if the place B be at M right under the ecliptic, all parallax of latitude is absorbed; and if at the pole of the ecliptic F, then if the moon G be equally distant from the center of the earth A all around, it will also be equally distant from the place F, and this case reverts to the preceding one. For all GA are by supposition equal, and AF remains everywhere the same, and GAF is everywhere right, because AF is the axis, therefore all GF are also equal. And thus if there is some inequality, it ought to be at a maximum where B is the middle place between M and F at degree 45. It will be useful to look into this. So, where AF is 1, let AG or AH be 54, as much as is the moon's least distance, when it has the greatest parallax. And because MB is 45°, DB or BE, that is, DA, will be 70711 where AF is 100,000, and AG 5,400,000. And thus GD is 5,329,289, and DH, 5,470,711. Hence BGD comes out to 45′ 38″. But BHD is 44′ 26″. The difference is not greater than 1′ 12″. But we have not used BHD, the deepest parallax beneath the earth, for we do not go below the horizon in seeking into parallaxes. Therefore, from point D let DC be erected at right angles to the plane DBA, and let AC be extended equal to AG, determining the length of DC, and let C, B be joined. Therefore, because AD, DB are equal, and DC is the same, and ADC, BDC are right, ACD, BCD will also be equal, representing the horizontal parallax of latitude.[14] For because CDG is right, the moon C will be at the horizon, or a little below. And because CB or CA is 5,400,000 where BD is 70711, angle BCD will be 45′ 1″. And thus the difference from BGD is 37″, which is the greatest of all possible.

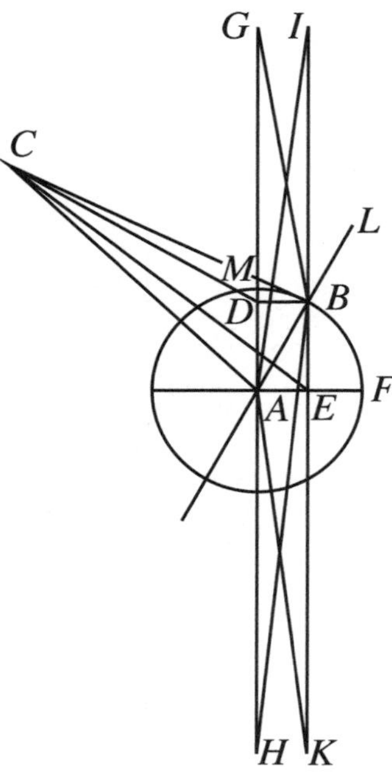

I say third, at the same altitude of the pole, *when the same point of the ecliptic is rising, and the moon is at the same distance from the center of the earth, when it is seen on the ecliptic, all parallaxes of latitude are really equal at any degree of the ecliptic*.

For, other things staying the same, and GH representing the true ecliptic **319**
from the earth's center A, and IK the visible ecliptic from the place B on the surface, let the moon be at I or K so it may be seen from B on the lines BI, BK on the ecliptic, visibly, at any rate. Now since AI, AK are supposed equal, and IBK is one straight line, AIK, AKI, and all these angles set up with the upright

[14] The horizontal parallax is the parallax at the horizon; the horizontal parallax of latitude is the latitudinal component of the horizontal parallax.

plane through IK will be equal to each other. But these represent the parallax of latitude, because the place B is in the plane of the circle of latitude.

Fourth, as regards *the parallax of longitude*, let C, E be joined. And let CBL now be right: CBI will also be right, because CB is in the plane perpendicular to the plane ABE. Therefore, ACB will be the parallax where it is the greatest of all, when the moon is visually setting, and it will be the parallax on the vertical circle, or the circle of latitude/longitude. And because CEA is right (for C is now in the plane ICK perpendicular to the plane ABF), and ACE is the parallax of latitude, ECB will be the parallax of longitude on the horizon, because BEA is right. But as the whole sine is to to the sine of the distance of the zenith from the pole of the ecliptic, so is the sine AB of the total parallax at the horizon to the sine BE of the parallax of longitude at the horizon. But also, as CE is to the sine of angle CBE or CBI, the visible distance of the moon from the nonagesimal, however much it may be, so is BE to the sine of angle BCE or the angle corresponding to the parallax of longitude, and that from trigonometry.

Next, as for the procedure, from the hypothesis of the lunar motion, the ratio of AC to AB is given. But from the altitude of the nonagesimal, by the theory of the *primum mobile* or the table on fol. 42 of Copernicus,[15] AE is given, and AIE, by operating either by the sine or by Tycho's parallactic table on fol. 120 of the *Progymnasmata*.[16] And because AEC is right, as is AEB also, and ACE, AIE are equal, therefore EC and EB are given. And because CBE is right, from this and through CBI the parallax of longitude BCE is given at any altitude. But because AC is slightly longer than EC, it is worthwhile to see again how much error is incurred if AC be taken instead of EC. And again, when BF is 45°, the error is greatest. For if B be at M, then EC and AC coincide. If, on the other hand, it be at F, then there is no parallax of longitude. Let AI be 54 semidiameters: AIE will be 45′ 1″. Consequently, EI or EC will be 5,399,532.
320 This line, with EB, lends not half a second to angle ECB compared with the case where we would have used AC and angle ACD instead of ECB.

Parallactic table

And because Tycho's table has few columns, and does not operate down to the genuine first minutes, but has appendices of the seconds, even at the top; then too, because the entry into it is properly made through the true altitudes, while we shall need the observed ones; and finally, because in the margin it has, not the distances from the zenith, but the altitudes, which fact made confusion for us; for these reasons, I have added here a more universal table of parallaxes.[17] Hence, the following instructions.

A facsimile copy of Kepler's two-part Parallactic Table may be found in the pocket inside the back cover. —Translator.

---

15 Copernicus, *De revolutionibus* (Nürnberg 1543), II 10, fol. 42^a, trs. Rosen p. 76. The altitude of the nonagesimal is the same as the angle the ecliptic makes with the horizon, which is the title of Copernicus's table.

16 *TBOO* II pp. 132–4.

17 The parallax table computes the products of sines, allowing entry and extraction through the angles themselves. The number in the body of the table is the angle whose sine is the product of the sines of the angles at the top and the margin. As Kepler remarks below, the table can be used for all kinds of sine products, not just parallaxes, and in fact he uses it for other computations in the *Astronomia nova*.

* * *

By the theory of the primum mobile, *compute the distance of the nonagesimal from the zenith, and its complement, that is, the altitude of the nonagesimal or the angle between the ecliptic and the horizon*. In fact, Copernicus's table on fol. 42 shows the latter in a rough way, while Reinhold's parallactic tables on fol. 99 and following of the *Prutenics* shows the former at the beginnings of the signs and a certain number of altitudes of the poles.[18] *Next, enter in to our parallactic table from the margin through the distance of the nonagesimal from the zenith, but from the top through the greatest parallax of the heavenly body, which applies at the horizon* (and this is located at the top immediately below the distance of the body from the earth's center), *first, by the whole number of minutes, then by the seconds, if there are any, and the body of the table will show the parallax of latitude, in the former case in minutes and seconds, and in the latter, in seconds and third parts, as is usual in these tables*. And these are obtained most correctly when the heavenly body is visibly on the ecliptic, as the moon is in an eclipse of the sun. The parallax *of longitude* is extracted by a double marginal entry in this way. *First, enter from the margin by the altitude of the nonagesimal, from the top by the greatest horizontal parallax: the body of the table will show the greatest parallax of longitude. Then enter by this, again from the top, but from the margin by the visible distance of the heavenly body from the nonagesimal: the body of the table will show the corresponding parallax of longitude at your moment*. You thus see that this table of mine is chiefly useful when the visual place of the moon is known from observation, while the Tychonic is chiefly useful when the moon's true position is had from computation. Nevertheless, each can be obtained approximately from the other, by adding or subtracting the parallax, first taken roughly to a precision of half a minute, next, by the transformed place, extract the perfectly accurate parallax, from whichever table you please. How little this procedure may be in error, even if the moon has latitude, and how it is to be remedied, and how the same table is to be accommodated to the equa- **321**
tor and the circles of declination, and thus to the first motion, and the problems that arise thence of finding the altitudes of celestial bodies from the earth, would be too lengthy to treat here, the reader himself may imagine. I think, however, that to prepare for future use, that the column which has at the top the horizontal parallax of 60 minutes, should be marked in red.

5. On the parallax resulting from the distance between the sun and the earth, or the annual parallax

But in fact the most wise Architect of the world does not cease to edify humankind with this. For just as, when the distance of the eyes was not sufficient

[18] *Prutenicae tabulae*, Wittenberg 1585. In this edition, which is the one Kepler was using, there are three series of page numbers. First, there are eight unnumbered front pages together with folia 1–68, comprising the *Logistike scrupulorum astronomicorum*. Then there is a new title page, inscribed, *Initium canonum Prutenicorum*, followed by folia 1–14. Finally, there is yet a third title page, *Sequuntur igitur nunc canones . . . .* It is fol. 99 of this third series that Kepler means.

to a person for knowing the true distance of the moon (which is the lowest heavenly body) from earth, it was fitting that the breadth of the orb of the earth come to the aid of the narrowness of the sense of sight, likewise, when even this remoteness of the surface of the earth from its center has vanished compared to the incredible altitude of the higher planets, in order that those heavenly bodies might not be spread out in vain and unobserved through the circumference of the heavens, but that instead the human mind might even carry on through to them, God has constructed another, much broader class of parallaxes, in case perchance there might be destined to exist among humans one who might want to follow this line of reasoning without offending piety, and disregarding false criticisms of his works. Copernicus, and Reinhold in the *Prutenics*, call this the parallax of the annual orb. However, Tycho Brahe carried that line of reasoning over from the earth's mobility to the sun's mobility so that the optical theorist has nothing by which to choose the former or the latter reasoning. I shall enunciate them both in the diagram in Sect. 3 above, arising from the Copernican reasoning. Let A be the sun's body, the common center of the annual orb BC, which carries the earth, and of the sphere of the fixed stars DF, and let the ratio of BA to AD be imperceptible. Now let the heavenly body be at G and the ratio of BA to AG be perceptible. Therefore, when the earth is placed at C, at the midpoint between the sun A and the body G, the line AG and CG will coincide, and, both being extended, will arrive at one point E among the fixed stars. Therefore, whether
322 the eye be placed at the sun A or at the earth C, the heavenly body will be seen in the same place under the fixed stars. Here, then, namely at opposition of the sun and the heavenly body, there is no parallax of the body from the annual orb. With the other things remaining the same, let the earth be at B outside the line GA. Therefore, the heavenly body's place will appear from the earth to be at F beneath the fixed stars, but from the sun A at E, because BA to AG is perceptible, and consequently the inclination of the lines BG, AG is also perceptible. Therefore, since the center of DF is equally B, to the sense, and A in truth, the arc EF will again be the measure of the angle EGF or BGA, and the former as well as the latter will be the parallax of the annual orb, when the earth is located at B. And so, when the heavenly body's place E, which a line projected from the sun determines beneath the fixed stars, is known by other astronomical principles, while that place F of the heavenly body, which the vision coming from earth determines beneath the fixed stars, is known, and so also the parallax EF or BGA, it will be impossible for the ratio of AG, the distance of the heavenly body from the sun with respect to the diameter of the world, to AB, the distance of the sun from earth, to be unknown, however much it may be exceedingly great. And so it was evidently not fitting that the human being, destined to be the inhabitant and watchman of this world, should reside in its middle, as if in a closed cubicle, under which circumstance he would never have made his way through to the contemplation of heavenly bodies that are so remote, but rather, by the annual translatory motion of the earth, his domicile, he circumambulates and strolls around in this most ample building, so as to be

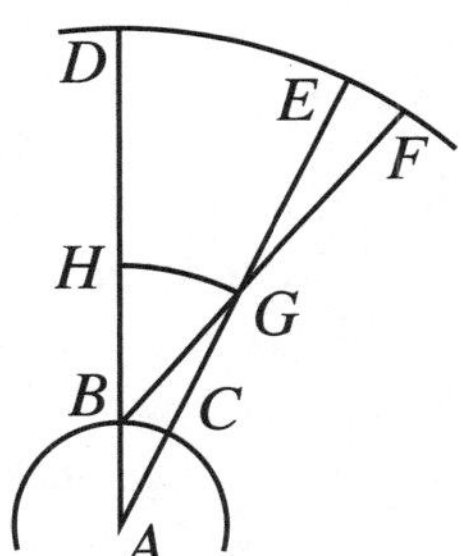

able more rightly to perceive and measure the individual members of the house. The geometrical art imitates something similar in measuring inaccessible spaces. For unless the measurer passes from one station to another, and turns his eyes sideways at each, he is unable to arrive at the measure sought.

In the Tychonic reasoning, let A be the earth, the center of the sun's orb BC and of the sphere of the fixed stars DF. Let the heavenly body be at G, and let the ratio of the three lines BA, AG, AE be perceptible. First, let the earth A, the sun C, and the heavenly body G be in the same straight line: there will be no parallax, because lines CG and AG drawn from the sun and earth to the body coincide. Now let the sun be, not at C, but at B, and let the line AB show at D the place of the sun beneath the fixed stars. And because the sun at B is that point to which Tycho refers the eccentricities and apogee, and the simple motions of the planets' orbs, it may therefore come to be known through astronomy how great is the angle DBG between the line from the sun through the earth, and the line from the sun through the heavenly body. And at that very moment, the place E of the body G beneath 323

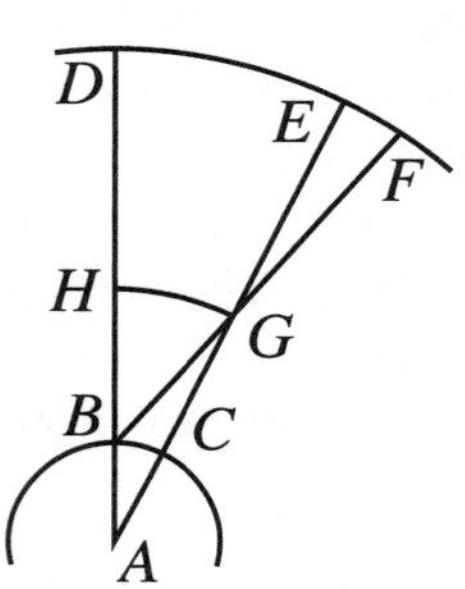

the fixed stars that appears from the earth A may become known through instruments. Therefore, the arc between the sun's place D and the observed place of the heavenly body E will be obtained, which is the measure of angle DAE. and because the ratio of BA to AG is perceptible, while B is outside the line AG, the lines BG and AG will therefore be inclined, and the angle DBG will be equal to angles BAG, AGB together. Thus the parallax BGA will be known, and again as before, the ratio of BG, the distance of the sun from the heavenly body, or AG, the distance of the earth from the heavenly body, to BA, the distance of the sun from the earth, will come to be known. The only difference is this, that here EF is not parallax, because it is not the measure of the angle BGA or EGF. For because the ratio of the lines BA, AD is perceptible, and A is the center of DF, B will therefore be perceptibly distant from the center of DF, and consequently DF will not measure the angle DBF within the limits of sense perception, nor are BAG, AGB together equal to it. But DE measures the angle DAE or BAG by itself; therefore, the remainder EF does not measure the remainder AGB, but there is a perceptible difference between them, which needed to be said in place of a warning, lest this ratio escape someone who is changing over from one form to the other.

The optical principles upon which rests the theory of parallaxes, that is, the distance of the stars, have I think been explained sufficiently in accord with the brevity that has been established. It remains for us to speak as well of the motions of bodies, so that in this book nothing of those things in astronomy may be overlooked that are to some extent to be decided from optics.

6. Short appendix on the curved tail of comets

Those who argue about the curvature of the tail in the comet of 1577 from the theory of parallaxes as if this illusion of curvature were based on the different

parallaxes of different parts of the tail, and were not really in the tail itself, do not sufficiently consider the subject of parallaxes. Nor do they rightly bring in the optical theorists as witnesses, for in those passages they are dealing, not with the aberration of vision about curvature, but with the true and legitimate vision of obliquity. If it really were true that parallax could show a curve from straight lines, it would not now be true that every straight line, no matter how extended, viewed from the perceptible center of the world, corresponds in all its part with
324 some great circle. As a result, the ways of observing the places of the heavenly bodies with a thread and with rulers would be false. Therefore, as was said above in chapter 6, another occasion for this curvature must be sought, or if one cannot be found, this phenomenon must be left among nature's mysteries.[19]

---

[19] See Ch. 6 Sect. 13, p. 274

Chapter 10
Optical foundations of the motions of the heavenly bodies

Since in astronomy, our aim is the contemplation of the motions of the heavens, while everything that we know previously came into our sense,[1] therefore, it is worth our while to consider whether the motions of the heavens come immediately into the perception of the eyes, and what kinds of optical illusions occur in the celestial motions. Let us begin thus.

Everything that is moved is moved in place, for motion is change of place.[2] And place is attributed to the surface with contains the movable body.[3] But the container is greater than the contained, and makes room for places because it occupies places. For the whole is greater than the part, while that which contains and encloses together with that which it encloses is a kind of whole.

It therefore follows, conversely, that, of two objects which are separated from each other by motion, that which appears to be greater performs the visual role of the place, while the remaining one performs the role of that which is placed. For as motion is in a place, so is the motion of a visible in a visible place. For this reason, since rest belongs to place, what is viewed to be greater will be thought to be at rest, while what is viewed to be less will be thought to move, even if it might happen to be the contrary in actual fact. For there exists no way to grasp motion visually except by comparison to some things at rest.

But the cause of this fact is obtained more evidently from the form of vision. For since the eye is spherical, and in addition makes use of multiple refractions, it happens that in one and the same glance the image of more than a hemisphere flows into the eye at once, and nonetheless, from this entire hemisphere only the slightest little part is perceived directly and distinctly, namely, that which is in the middle of the hemisphere, while the parts lying all around are all perceived
325 more and more obliquely and confusedly. On these points, see chapter 5. Hence it happens that what is perceived to be greater takes up more of the eye, while that which is perceived to be less takes up a lesser part of the ocular surface. Therefore, when a separation occurs, such as, for example, that of some cloud from a star shining in between, then the object that is smaller to the sight, the star, more conspicuous through the effect of the separation, attracts the eye's vision to itself. Thus the star, since it is perceived under a small angle, is observed directly by the eye, while the cloud, which is seen under a greater angle, and which takes up nearly the whole eye, is observed obliquely by the same eye. It therefore ascribes the act of separation to that object which is perceived directly, namely, the star. In this way, the sense of vision is in error about the movable.

---

[1] Cf. Aristotle, *Posterior Analytics* II 19, 99b 26–100b 4.

[2] Here Kepler departs from Aristotle, who held that there are four kinds of motion, corresponding to the four kinds of being (thinghood, amount, quality, and place). See *Physics* III 1, 200b 35, trs. Sachs p. 73.

[3] Cf. Aristotle, *Physics*, IV 4, 212a 2–9, trs. Sachs p. 102.

For if a cloud tends from east to west with a swift motion, the star, even though it tends to the west, albeit slowly, will appear to be carried to the east, in the path of the cloud.

But the cause of this aberration that is the most obvious of all lies in this, that the eye, being attached to the posterior part of the head, sees nothing under an angle greater than the projecting parts of the face. These, however, maintain the same place with respect to the eye, with the result that all the rest will appear to move, if by some device one might turn it about without its knowledge, so that it might not be the assessor of the motion which it creates itself. And just as the eyes are attached to the head, so, through the head, they are attached to the body; through the body, to the ship or the house, or to the entire region and its perceptible horizon. Since these are nearby, and appear large, and are seen under a large angle, and hold the same place with the eye, it is necessary that all the remaining things, whose place changes with respect to that which contains the viewer (be it a ship or the surface of the earth), appear through themselves to be in motion. For they play the role of the located, while those things that lie nearby about the eye are likened to the place and the container.

From these things it follows that even if someone were to carry us across to the moon or to another of the wandering stars, and the moon's motion is most highly perceptible because of its swiftness, of which more later, nonetheless the moon is going to appear to be at rest along with us, while the sun and whatever heavenly bodies are at the right distance are all going to be thought to be moved with those motions which were proper to the moon itself alone, in addition to their own motions.[4] Therefore, the optical writers do not have anything to produce from their arsenal against Copernicus when he maintained that our home, the earth, moves.

And, to add this too in passing, the absurdity of the triple motion is slander- **326**
ously exaggerated. For the motion which Copernicus attributes to the earth has no other form than that of a wheel on a cart moving in a straight line. First, the wheel rotates, then by the rotation itself the axis is carried along, and third, the axis points in the same direction, and remains parallel to itself. Who is it that conceives the opinion of a third motion, from the fact that it is really rest?

For the fact that in the passing of the ages the earth's axis finally also inclines, does not come into this reckoning, and would instead be a fourth motion, if a completion of the libel were desired. Astronomers know it to be of that kind of motion by which the apogees and nodes of all the planets are moved along, not by a true motion, but by a kind of difference, so to speak, of two other acknowledged motions.

And now the optical writers will have much less against Copernicus if I should remind them that the speed of the stars is not even perceptible. Aristotle, Sect. 15 q. [i.e., problem] 12, takes that as acknowledged, saying, "the motion of

---

[4] Kepler later elaborated upon this theme in his science fiction novel *Somnium* (Dream), which describes a journey to the moon and gives an account of the astronomy of its inhabitants. See *Kepler's Somnium*, trs. Rosen (Madison, 1967).

the sun is not evident."[5] And Cleomedes, Book 2: "since it appears to be standing still."[6] For it is in conformity with reason that, between two motions, the one that is slower is thought to be more like rest. Therefore, since those animate motions and rotations of the body, of the neck, of the eyes, are perceptually many times faster than the celestial motions, it is evident that the motions of the heavens are to be thought instead to be like rest by our sense of sight. This all the more so from the fact that any visible object, just as it has a quantity, ought also to have a certain speed, and a perceptible proportion to the eye, and so ought its motion.

Whatever there is besides this cannot by grasped by the sense of vision, as Optics attests. See Witelo IV, 3 and 110.[7] Now the speed of the stars has no proportion to sense perception, as will be apparent if you consider that your sense of vision is carried around by the sun about the center, either of the eyes, or of the head, in the space of twelve hours, by not more than 180 degrees, through which space we are in other circumstances accustomed to turning the eyes faster than in one second part of hourly time. And in one hour there are 3600 seconds, in 12 hours the sum is 43,200 seconds. And what proportion is there between one and
327 forty thousand? Quite imperceptible. The same is derived from Witelo IV 112, when the object seen stays in the same perceptible place over a perceptible time, it is thought to be at rest.[8] But this is true of the heavenly bodies. For a moment, or one second part of an hour (which is usually set about equal to the arterial pulse) is a perceptible time, but the place or arc of the eye with a size of one degree cannot be sensed without the use of an instrument. And a star stays for four first minutes, i.e. 240 second minutes,[9] in any one degree, that is, in some space that hardly has the ratio of a point. And so whatever is in our senses concerning the motion of the heavens, we have absorbed thanks to the intervention of reasoning. The sun was there before, now it is here. It therefore moved from there to here.

Moreover, just as the vision generally attributes motion erroneously to things at rest, so it also contrives for itself the forms of motion. Hence, the vision attributes rising and setting, that is, ascent and descent, to the stars, and Ovid,[10] imitating vision, attributes it to Phaethon, in the most delightful fable of book II of the *Metamorphoses*, because the vision finds these differences of places within a person, and in that person's upright position with respect to the visible horizon, since there is nothing of the sort in the heavens themselves. Hence, when Witelo showed in Book IV Prop. 10 that bodies that are set in order in a continuous

---

5 ἄδηλον τὴν μετάβασιν ἡλίου. This is actually from *Problems*, XV 13, 913A 7–8. Kepler has changed the Greek slightly, and omitted a few words.

6 ἐπεὶ ἑστὼς φαίνεται, *De motu circulari* II 1 p. 130, trs. Goulet (French) p. 137.

7 *Thesaurus* II pp. 119 and 167. The former establishes the limits of the angle under which an object can be perceived; that latter describes how visual perception of motion takes place.

8 *Thesaurus* II p. 168.

9 The hour is divided into 60 "minute parts," which are the "first minutes," each of which is divided into 60 parts, the "second minute parts," or seconds.

10 Ovid, *Metamorphoses*, II 63 et seq.

line with the vision are evidence for a distance of the last that is greater than if they had not been set in order in a continuous line, he showed in prop. 13 why the earth's horizon appears to be in contact with the heavens; and why that part of the heavens appears more distant from us, and the distances of the heavenly bodies greater, than those that are at the head's zenith.[11] These are accordingly the necessary terms of vision, which we would not be able to do without even if we were really carried about in the globe of the moon.[12] We ought all the less to be amazed that Copernicus dared draw a distinction between those things said in Holy Scripture, rightly indeed, to give an account of vision, and those things which, examined astronomically, are understood to hold otherwise. For scripture does not speak falsely, but affirms with perfect truth that the sense of vision says this, or better, that it accommodates to its own purpose this thing suggested by the sense of vision. The astronomer, on the other hand, or rather, the Optician, convicts the sense of vision of error without any affront. Indeed, when we read in a thousand places the mention of the ends of heaven, to which the people Judea are scattered and thence called back, there is no one who doesn't see that these are the things explained by Witelo IV 13.

But since by Witelo IV 111, motion itself is judged by the space over which **328**
the visible is moved,[13] while all motion takes place in a line, and this is either straight or circular, it is therefore evident that if some error befalls the vision in the lines of motion, the same happens in the motions themselves. Errors of this kind, or rather, phantasms,[14] are of two kinds among astronomers, following their attribution of two main circles to one planet, one the eccentric and the other the epicycle; but Ptolemy does it in one way, Copernicus, whom I follow, in another, and Tycho Brahe, most recently, in another. For the planet itself traverses the eccentric circle with its own body for Copernicus and Tycho, while for Ptolemy it is not the body of the planet, but the center of the epicycle; on the other hand, for Ptolemy the planet itself traverses the epicycle with its body, each on its own epicycle; for Copernicus, the single circuit of the earth removes all the epicycles; for Tycho, both the sun and the whole planetary system have one and the same circuit, common in all parts, which likewise removes all the epicycles. For that reason, for Ptolemy and Tycho, the planet's motion is not simply about some immobile point, epicyclic for Ptolemy and eccentric for Tycho, but it is

---

[11] *Thesaurus* II pp. 121–4. Kepler's sense here is that in proving the latter proposition, Witelo cites the former. His restatement of Witelo IV 10 is somewhat different from what Witelo wrote. He said, for example, "Therefore, if we are lacking the comprehension of the bodies set in continuous order, or if we are lacking an intermediate remoteness, we will never comprehend the remoteness of those bodies by a true comprehension, by only following guesswork." (p. 121).

[12] That is, terms such as "rising," "setting," "moving ahead," "falling behind," are indispensable to the way we see things, applying only to our perception and not to how things really are.

[13] *Thesaurus* II p. 167. Witelo actually says that the *quality* of motion is comprehended by the space.

[14] φαντασίαι.

truly spiral and compounded of eccentric and epicycle: for Copernicus alone, that eccentric which he hypothesizes is truly the simple motion of the planet itself about a motionless point or the sun (except insofar as I am going to be introducing a small correction to this opinion, applicable to all the authors, in the *Commentaries on the motions of Mars*),[15] nor is it varied by an epicyclic motion, since according to this author, that is nothing but an illusion of motion, in that the earth's motion is not noticed on the earth and is ascribed instead to the individual planets.[16]

However this difference of hypotheses be compared, there exists, as I said, a twofold perception of motion, vulnerable to error. One, which is from the eccentric, is common to all the authors; the other, arising from the epicycle, in Copernicus is a mere hallucination of vision about the planet, in the other authors has something joined together from the true motion of the planet.

Because of the eccentric, the planets appear either slow or fast. The cause is partly physical, partly optical. The physical part of the cause does not give the
329 sense of vision a reason for error, but also represents to the vision that which in fact occurs, [an account] of which is in the *Commentaries on the motions of Mars*. The ancients used to represent it through the equant circle. Part of the optical cause is located in this, that since the planet's motion is eccentric with respect to the vision, some parts of this circle are accordingly farther away from the vision than the rest. Therefore, through those things that were said in ch. 9, and through Witelo IV 7 and 131, equal arcs of the planet's circuit will appear unequal, small about the apogee, large about the perigee. Therefore, even if the planet itself were equally fast in all arcs, it will nonetheless appear slow where the arcs appear small, fast where they appear large.

---

[15] The *Commentaries on the motions of Mars* appeared in 1609 as the *Astronomia nova* (translated into English by W. H. Donahue under the title *New Astronomy*, Cambridge 1992). At the time of writing the *Optics*, Kepler was in the midst of his work on Mars. Although he had not yet decided that the orbit must be elliptical, he had known since April of 1602 that it must be some kind of oval, not a circle. This is the *correctiunculam* to which he refers. See W. H. Donahue, "Kepler's First Thoughts on Oval Orbits: Text, Translation, and Commentary," *Journal for the History of Astronomy* **75** (1993) pp. 71–100, and "Kepler's Approach to the Oval of 1602, from the Mars Notebook," *Journal for the History of Astronomy* **89** (1996) pp. 281–295.

[16] Kepler's account here contains another simplification. As Copernicus stated in *On the Revolutions* V 4, fol. 142v, the planet's actual path in Copernicus's theory is not a perfect circle, but "differs imperceptibly" from a circle. The noncircularity is not based upon observations, but is created by an extra epicycle which Copernicus used to replace the Ptolemaic equant. In chapter 4 of the *New Astronomy*. Kepler shows that the resulting path bulges out slightly at the sides, a change from circularity contrary to that which Kepler's theory required.

Kepler evidently did not count this departure from circularity as a true epicyclic motion, even though Copernicus used an epicycle to produce it, because it did not result in the loops and reversals characteristic of the motions produced by the much larger Ptolemaic epicycles.

This optical magnification or diminution of arcs is best evaluated in semicircles, in the following way. Let A be the center of vision, as well as the center of the world, B the center of the eccentric ECD; AB being extended to C and F, C will be the apogee, F the perigee. At the points A and B let perpendiculars GH and DE be set up. And let the planet now be constant in its strength of motion, throughout all the arcs of the eccentric. It will be on GCH and HFG for an equal time, and will consequently be on DCE for a greater time than on DFE. But at D and E it will appear from center A to be at opposite places on the sphere of the fixed stars, whose center is A; therefore, it delays more on DCE than on DFE.[17] Further, because the vision does not know that DCE is greater than DFE, because it does not distinguish the remoteness of the parts of the two circles, but considers them equally distant, it therefore considers the planet to be slower above the line DAE than below it.

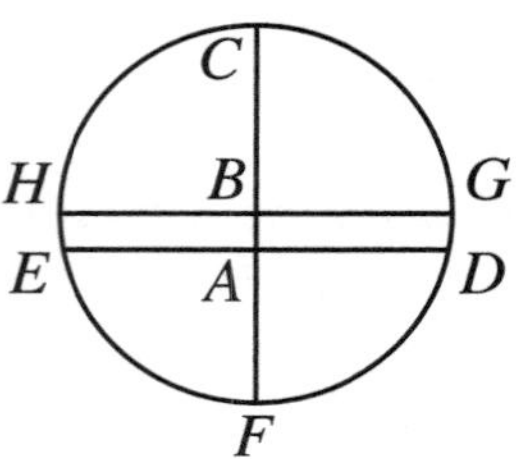

But you ask, by what argument is it known that the planet is seen at opposite parts of the circle? I answer, first, from the center A, we imagine for ourselves a great circle, which is called the equinoctial[18] Next, we know from observations that the same center of our vision A also lies in that plane in which any of the planets you please carries out its eccentric path. We also know that this plane is even with itself, not twisted, and is inclined to the former circle, that is, that it cuts it. And when two circles cut each other, the line of common intersection is straight, by Euclid XI 3. Since this goes through the center of the equinoctial, as through A, which by supposition is in both planes, it will therefore cut the equinoctial at D and E, places opposite when seen from A. Therefore, whatever **330** contrivance is used to ascertain that the planet is on the equinoctial, is also used to ascertain that it falls at opposite places. And that is taught in astronomy and in the theory of the *primum mobile*.

Further, from the occasion of turning their attention to this phantasm of optical speed and slowness, the astronomers run into a different and contrary error, not so much of vision as of reasoning. For when astronomy teaches us this axiom (in Aristotle's words), "the more distant moving bodies appear to be moved more slowly,"[19] the astronomers convert it with great plausibility to say "the slower moving bodies seem to be carried along farther out."[20] To the extent that a planet delays for a longer time in some arc or semicircle than in

[17] The verb *moratur* could perhaps more naturally be translated "spends time." However, the notion of *mora*, "delay," plays a crucial role in the *New Astronomy*, so the present translation has been chosen accordingly. For Kepler's use of this term, see Peter Barker and Bernard R. Goldstein, "Distance and Velocity in Kepler's Astronomy," *Annals of Science* **51** (1994) pp. 59–73, and W. H. Donahue, "Kepler's invention of the second planetary law," *British Journal for the History of Science* **92** (1994), 89–102.

[18] This is another name for the equator.

[19] τὰ ποῤῥώτερον φερόμενα βραδύτερον κινεῖσθαι φαίνεται. This appears to be a paraphrase of *On the Heavens* II 10, 291a 32–34.

[20] τὰ βραδύτερον κινούμενα ποῤῥώτερον φέρεσθαι δοκεῖ.

the rest, they argue that that arc has receded that much farther from the vision. But the conversion is not necessary, and is in part false. For there are also other causes of retardation mixed in with these optical ones. Thus, in the beginning, Ptolemy was deceived when he raised the planetary epicycles too high on the one side and pushed them down on the other, because the slowness of the one place, and the speed of the other, seemed to require as much. But the error was immediately obvious from the apparent magnitude, for the epicycles increased less at perigee than accords with such a close approach, for which reason another cause for the slowing down was seized upon, which, as I have just said, Ptolemy ascribed to the circle of the equant.[21] In the sun, no epicycle was needed, and as a consequence this error has remained to this day. It was, however, first discovered by me, through an exact observation of the visible diameter, as I shall say below, and then by Tycho's most precise observations taken of the star Mars, as I shall make plain at the proper time and place.[22] By both arguments it is established that the sun recedes from us by only half of the eccentricity which was attributed to it by Albategnius[23] and Tycho, and thus that an equant circle governs the motion in the sun, too.

Because of the epicycle, or, in Copernicus, because of the circuit of the earth, and of the vision along with it, the planets do not always go forward for us, but sometimes are seen to stop, sometimes even to go back. They stop when they are perceived to stay in the neighborhood of the same fixed stars; go back, when they are at first perceived in the neighborhood of more easterly fixed stars, then after a number of days they are perceived in the neighborhood of more westerly ones, just as, on the contrary, they are seen to go forward, when the
331 contrary occurs. But, as I have said, in Ptolemy the planets really go back on their own epicycles when they traverse the epicycles' lower semicircle; since, as Aristotle teaches in the *Mechanics*,[24] a circle moves with one and the same motion in opposite parts, but in different directions, and this retrograde motion of parts surpasses in speed the slower forward motion of the center. The same happens in Tycho, when the planet does in fact go forward somewhat on the eccentric, but, being driven back by the motion of the sun itself along with its entire eccentric, it is carried far faster in the opposite direction. But since, by Witelo IV 4, a straight line appears to be a point when it is extended directly away

---

[21] The procedure sketched here appears in the *Almagest* X 7–8. For Ptolemy's determinations of the eccentricities and sizes of the epicycles of the outer planets, see Otto Neugebauer, *HAMA* pp. 172–80, and Olaf Pedersen, *A Survey of the Almagest* pp. 273–86.

[22] Kepler's observations of the apparent diameter of the sun appear below, ch. 11 problems 2 and 3. The demonstrations to which he alludes here establish that the earth (for Copernicus) or the sun (for Ptolemy and Tycho) does not traverse an eccentric orbit with uniform speed, but instead moves on an orbit with a smaller eccentricity, and, like the other planets, moves faster when closer to the "center of vision." The demonstration from Tycho's observations was later (1609) to appear in Part 3 of the *New Astronomy*, chapters 22–30.

[23] See Ch. 4, footnote 172.

[24] *Mechanica* 847b 16–848a 37.

from the vision,[25] the judgment will also be very nearly the same concerning the apsides or points of tangency of the circle of the epicycle, which arcs, directed nearly straight upwards from our vision into the depth of heaven, appear, if not as points, at least with the image of a very small quantity, as a consequence of which they are perceived as slowest about those parts. Thus it can happen, despite the epicycle's being faster than is the eccentric, that nonetheless the forward motion of the eccentric and the backward motion of the epicycle become equal to the senses, so that when one is subtracted from the other, the planet appears to stand still, and stays in the same longitudinal position in the heaven itself, with respect to the sphere of the fixed stars, even though meanwhile it either labors at an ascent into the aether away from the vision, or sends itself downwards from the deep aether to the earth, in a nearly straight line.

As regards Copernicus, however, this whole illusion[26] of standing still and retracing of steps is demonstrated most beautifully from optics. And although these things are more appropriately learned from the author himself, nevertheless, so that nothing might here be said which would affect the reader negatively, I shall repeat the fundamentals in three words from Euclid himself. It is indeed my judgement that if we had not had other arguments by which antiquity had tested this Copernican opinion, this passage alone would have been sufficient to vindicate Copernicus from the truth of Pythagoras. First, it is evident, not only in itself but also from the commentary of Proclus,[27] that all of Euclid's geometry is Pythagorean and aims at the knowledge of the five regular figures which are called "cosmic":[28] Euclid was therefore a Pythagorean. Next, consider for me the bundle of Euclidean propositions in his *Optics*, namely, 53, 54, 55, 56, 57, 58, which Witelo carried over into his Book IV Propositions 134, 135, 136, 128, **332**
132, 133, 129.[29] In these propositions, Euclid propounded pure, unadulterated Copernican astronomy.

And in fact, Proposition 53 appears to seek an exemplar of celestial objects in nearby objects, and calls that into consideration. For it states,

> Of those things which are carried with equal speed and are in the same straight line, that which is nearest the eye appears to follow; that which is farthest, to go ahead. But where the line of the movable objects gets more distant from the right of the vision to the left, that which had gone ahead before now appears to follow; the one which had followed, to go ahead.

This appears to have in view a cart going across in front of the eyes, in order to show that it is not absurd for many things to happen in the heavens which, though not completely the same, are similar or of the same kind. One might, however, use this for his own ends, to demonstrate that even if for Saturn, Jupiter, and

[25] *Thesaurus* II pp. 119–20.

[26] φαντασία.

[27] Proclus, *A Commentary on the First Book of Euclid's Elements*, trs. Glenn R. Morrow (Princeton: Princeton University Press, 1992) pp. 52–3.

[28] *Mundana*. I have, however, reverted to Proclus's original word.

[29] *Thesaurus* II pp. 176–9.

Mars, the epicycles were so adjusted as to be exactly the same size (which one would do if he were to correct the Ptolemaic form from the Tychonic observations), nonetheless, the epicycle of Mars would turn out to appear greater, and that of Saturn less, than Jupiter.

And now, in Proposition 54, he smacks of nothing but Copernicus. "If," he says, "some things be carried with unequal speed, and the eye also be among them, those which are carried with the same speed as the eye will be thought to stand still; those which are carried more slowly than the eye, to be carried in the opposite direction; those which are carried more swiftly, to go ahead." I shall change nothing but the words. If the planets and the earth, the lookout post of our vision (which, moreover, takes place on the semicircle of the terrene orb that faces the planets), are carried forward, and it should happen than the earth and a planet are moved forward equally (with respect to some identical straight line), the planet will seem to stand still;[30] but if the planet be slower, it will seem to be carried backwards; and if it be faster, it will seem to be carried forwards. If there be anyone so nitpicking, so particular, as not to be able to hear this, let him substitute the moon in place of the earth, and locate upon it some viewer of celestial objects, then the same things would follow in the moon: this earth of ours, even if it really be at rest, will appear to move, but the moon will appear to be at rest, although it moves, and those things will not be able to be overturned by any solution.

Proposition 55 appears to make noises about the diurnal motion. "If," he says, "some number of things be carried along together while some one thing is at rest, it will seem to move in the opposite direction." The eye may indeed be supposed to be in the middle of the world, because of the earth's evanescent
333 ratio to the world, and the earth may be rotated from west to east, with the diurnal motion: therefore, the mountains, which seem large and contiguous, being thus carried along, and the stars, which appear small and scattered, being at rest, the stars will seem to move in the opposite direction, that is, from east to west.

Again, Prop. 56 distinctly expresses the spirit of Copernicus. It says, "When the eye approaches the thing observed, the thing will be thought to grow larger." I accordingly bring in under this, that when the earth carries our vision towards the bodies of the planets, both the lines of motion and the bodies of the planets themselves will be seen as large. Thus this phenomenon is explained, not only by the approach of the star to the eye, which is imagined to occur through an epicycle, but also by the eye's coming closer to the object. And that is powerfully obvious. Melanchthon states that in July and August of 1529, Mars was seen with such a prodigious image that it was believed to be a new star. The same ought to have occurred in August, 1561, and nearly the same in August, 1593; and it will happen in July, 1608. And in the month of February and March of this year 1603, we saw the star of Venus, setting through clouds, unusually large, and many were

---

[30] Here Kepler refers to the following endnote.

To p. 332. This is therefore the definition of a station in Copernicus, because it is made to happen when a line through the earth and the planet, being moved along, does not incline, but remains parallel to itself.

asserting that they had seen a new star. For at these points of time, the perigees of the eccentric and the epicycle came together, which in these two planets is of the greatest importance; it has a lesser effect in the others.

Proposition 57 is greatly accommodated to that illusion of motion that is ascribed to the eccentric. For it states that "Of things that are carried with equal speed, those which are farther away appear to be carried more slowly."

Finally, Proposition 58 states, "When the eye is carried along, those things seen from farther away appear to be left behind," where he openly uses the astronomical term, ὑπολείπεσθαι. Further, in astronomy, ὑπολείπεσθαι is the same as to move forwards, "in consequence": I believe the term is relative to the other contrary, προηγεῖσθαι, to go ahead.[31] For if some things go forwards, it is necessary that the others be left behind. And thus he is very obviously speaking of astronomical matters.[32] The experience drawn from mountains and fences is properly subsumed here: to one walking by a fence, the nearby fence appears to go in the opposite direction, and the distant mountains appear to accompany him. You might be able to accommodate an example drawn from this to the sun's motion among the fixed stars, but with terms bent slightly from astronomy to everyday usage. For let B, A be fixed stars, C the sun, the earth at D, at which place let the earth be moved in the direction of B, just as when it is opposite it is moved in the direction of A. Therefore, when
the earth D is going towards B, the sun C, being at rest, 334
will seem to move towards A, but the fixed stars in conjunction with the sun, as E, will seem to be left behind by the sun and to accompany the earth in this position in the same direction, which in this place comes to be indicated by the word ὑπολείπεσθαι in common usage, although

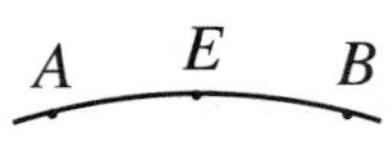

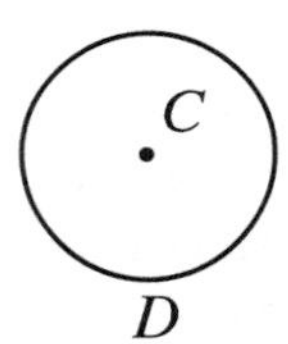

[31] "[T]he Pythagoreans, when they shared out musical tones among the stars, gave the lowest (the *hypate* among the strings of the lyre) to the moon, because the motions of both are slowest. Hence have originated the words *proegoumenos* and *hypoleiptikos*. The former of these terms originally corresponded to a star which, on the next day, comes to its setting before another (the sun . . . is said to be *proegoumenos* with respect to the moon . . .). The latter term corresponded to a star that is slower in the first motion (such as the moon here), which is, as it were, abandoned and left behind . . . by the swifter ones . . . For more on this subject see our *Optics*, ch. X." —Kepler, *Astronomia Nova* (2015), Chapter 1, trs. Donahue. p. 76.

[32] Here Kepler refers to the following endnote.

To p. 333. The argument is this: Euclid uses an astronomical term; therefore, he is presenting astronomy. Further, he speaks of eye being carried along. Therefore, he is presenting the kind of astronomy in which the vision is so moved that the heavenly bodies appear retrograde: that is, of the Copernican. I nevertheless add to the present account the following: by the term προηγεῖσθαι, Euclid appears to want to present a reason why the mountains do not just appear to accompany one running in the plains (which is the only thing I am considering), but even seem to run on ahead. Thus the origin of the term προηγεῖσθαι will become more popular.

subsequent people used it more strictly in astronomy, of motion BA "in consequence," without respect to the earth's position, by which procedure it does not accord with the fixed stars. They would be able to offer this justification of the astronomical terms: that in the combination of the first and second motions,[33] those things that are moved with a retrograde motion outrun their position of the previous day among the fixed stars by an earlier approach to the meridian, and thus they go ahead.[34] But those which move with a direct motion, their places beneath the fixed stars which they held the previous day come to the meridian or the horizon first. They are accordingly thought to be left behind,[35] just as if two runners run for the very same goal, but one, being slower, complains that he is being left behind. And so an optical cause has given birth to these terms. For even though, as was just said, not even the first and daily motion of the stars comes under the bare senses, nonetheless, by an easy process of reasoning, the eyes, being raised up, now easily take notice of this first and daily motion from east to west; but of the second motions by no other means than the difference of the diurnal motions.

The fallacy giving birth to these terms is nothing other than if one should see, by the shore, a boat going downstream, following the river, and in it two people, one of whom stands still in the bow, and the other walks against the current from the bow to the stern; the spectator, however, ignorant of the facts, would consider those two people to be carried in two different boats, one slower. For he will be in error, and the motion which is in the person on the same boat, he would ascribe instead, erroneously, to the boat itself, in the opposite direction. And those who first introduced these terms into astronomy erred similarly, thinking that the same first motion is weaker, for example, in the moon than in the sun, not knowing that the moon itself, by its own motion, strives against the common
335 first motion (to present for now the commonly accepted hypotheses).

And these are about all the things that the sense of vision, and incautious reason following it, attaches to the stars, contrary to the truth, and that must be freed from obscurity optically.

Appendix on the Motion of Comets

Those who demonstrated the motions of the comet of 1577 with circles took on an extremely difficult task, nor did it entirely succeed, because they did not think they had to investigate more carefully. They will encounter much greater difficulties if they take up the demonstration of the same thing in other comets. I met with success more easily in many for which I acquired observational records, in this way: if, as the nature of things urges, I should attribute to them straight lines, which they mostly traverse uniformly in equal times, only a little slower at the beginning and end, and closer to rest, as is usual in other trajectories. For the earth's motion, insinuating itself, easily obtains circularity for them. For example, the one in 1577: if it had ascended in a straight line originating from

33 That is, the diurnal rotation (parallel to the equator) and the various planetary motions (along the zodiac), respectively.

34 προηγοῦνται.

35 ὑπολείπεσθαι.

the plane of the Tropic of Capricorn towards the north pole, or slightly more inclined, but still in a straight line, then, just as in going around the motionless sun, the earth indues it with an image of circular motion, so, by the same process, in going around the comet which is practically at rest (for by this supposition it tends almost entirely crosswise), the earth will obtain for it an image of circular motion. Thus that comet of Regiomontanus, carried from the deepest aether close to the earth in a straight line, and passing by it rather close, will find a most beautiful occasion for completing 40° of a great circle in one day in the middle of its appearance, a very small amount before and after. Here too, the cause will be evident why, at that section of time when the comet was so fast, the tail equalled 50 degrees in length. By Witelo IV 22.[36]

Using this support, some have asserted of the star of 1572 that it was taken up into the depth of the aether with a rectilinear motion, using the evidence of its decreasing magnitude, for whom Witelo IV 4 and 132 was of service.[37] *In fact, they are optically correct, if you grant their assumptions; as for the rest, those things which Tycho Brahe argued to the contrary in Book 1 of the* Progymnasmata *from other sciences, soundly and with great judgment, see in that author.*[38]

---

36 *Thesaurus* II pp. 128–9.

37 *Thesaurus* II pp. 119 and 177.

38 Brahe did not give much considerations to comets in the *Progymnasmata*, which was largely concerned with the 1572 nova. Kepler may have had in mind *De mundi aetherei recentioribus phaenomenis* (Uraniburg, 1588), in *TBOO* IV.

Chapter 11
On the observation of the diameters of the sun and moon and eclipses of the two, following the principles of the art

Problem 1

To construct an ecliptic instrument

Let a curtain[1] be set up in the open, with so many layers of black cloths that no light can break in. If this convenience is lacking, let a chamber be chosen facing the region from which the eclipse of the sun will be viewed. Let this chamber have a wall that is not thick, that offers a window, and let it be possible
336 to block off this window as well as all cracks against the entry of light. Next, let a rule [*A*][2] be made of as great a length as can be, all lines of which are straight, whose thickness is as much as the planed lumber provides, and whose breadth is half a foot. And let it be so adjusted that, because the wood is flexible, it may lie on its back, and at a place intermediate between the ends, so that it may be less bent. Moreover, let it not have any hole through the middle of its breadth, lest, being weakened, it break under its weight. Instead, to the line of its back, upon which it is to rest, let a board [*B*] be joined, so that a hole or socket[3] can be set on the line of the joint. Let the socket have its own axle [*C*]. Next, let there be a post [*E*] capable of being turned above an axis [*D*], divided at the top, so as to receive the thickness of the rule in the slot, drilled so that it can be transfixed by the same axle as that of the rule. From a board [*F*] on which is a socket [*D*] receiving the post, let them set up the post in its (the board's) perpendicular with equal pieces of wood [*G*], arising crosswise on both sides, surrounding the rounded post in a raised position with their hollow embrace. Let three other boards [*H*] be joined with this one, so that a right angled parallelogram is made of all of them, in the place of the azimuthal circle. Now, upon the head of the column,

---

1 *Scena.* This can also mean "platform"; however, in the "first example" of Problem 32 below, Kepler refers to the pole from which the *scena* is hung. I therefore believe he is describing a sort of tent, much like the one whose use by Kepler is related by Sir Henry Wotton in a letter to Sir Francis Bacon in the autumn of 1620 (no. 892, in JKGW XVIII p. 42).

2 The bracketed letters refer to the diagram on p. 349, below. I have inserted these into both the diagram and the text, to help clarify the relation between the two. The diagram is otherwise Kepler's. —trans.

3 *Matricula.* This certainly does not have its late Latin meaning of "a roll or list." There appear to be no similar uses in Du Cange's *Glossarium.* I am conjecturally reading it as the diminutive of *matrix*, which is to be understood in its late Latin sense of "womb, receptacle." The context, in any case, indicates that Kepler means something like "socket."

from which the axis goes out into the board, let there be attached and joined a crosspiece [*K*], in one of their planes,[4] of the post which has a slot on top, and let this crosspiece likewise be bonded to a transverse post [*L*], so that the post may attach to the crosspiece at a right angle, and so that if the crosspiece, resting upon the parallelogram, be moved, it rotate the post. Let the crosspiece be of a suitable length, routed out in the middle, so that it may hold the thickness of the rule in this slot, and so that the rule may be carried around together with the crosspiece and the post, and, at the same time, the rule can be raised towards the zenith, or lowered towards the horizon, as much as the sun's altitude at the beginning and end of an eclipse requires. For this purpose, it is appropriate both that the post be tall enough that the parallelogram never get in the way of the rule, and that the crosspiece, and the rule itself, be long enough that they do not come apart as the sun is setting, and that the slots be configured in both directions to the same end, and that the square [i. e., the parallelogram *H*] be high, so that when the [back end of the] rule is lowered it may not hit the floor, and in the plane of the horizon, which a perpendicular suspended on the post will easily indicate. Now this form cannot indeed be universal, unless either a geometrical square be made of the crosspiece [*K*] and the post [*E*], or a full quadrant be available. However, this construction was adequate for me for impromptu use.[5]

Now to the rule, whose use is the main thing here. And so, measure on it a **337**
certain space of length, from the point of attachment downwards, no more than is the height of the post; and there you should make grooves on both surfaces of the breadth, perpendicular to the length; and similar ones at the head of the rule, which will be above, beyond the axis, the upper parts being about 12 feet away from the lower. Next, you should prepare two wooden panels, a palm or a little more in breadth, of a length that is its own width and that of the rule combined, of a thickness that fits the grooves of the rule. They should be routed

---

4 The Latin is puzzling here. "Their," I take to refer to the boards H forming the base. I believe he says, "one of their planes," because they have two common plane surfaces, one on the top and one on the bottom. I am uncertain, however, about how the next phrase fits into the sentence.

5 As Kepler's description may be hard to understand, the following supplementary description is included in the hope that it might be helpful. Bear in mind that it is based only the translator's understanding of the text, not on inspection of any existing instrument.

In the base board *F* is a socket that is the same size as the bottom end of the post *E*, which has a round peg or axis *D* that can turn in the socket. The crosspiece *K* is solidly attached to *E*, so that as *K* is swung around on the other base boards *H*, *E* rotates on the socket. There is a cross brace *L* going from *K* diagonally upwards to *E*, to which it is fixed so as to hold *E* perpendicular to *K*. Above the joint where *L* meets *E*, there is a kind of bearing consisting of concave surfaces hollowed out of the two braces *G*. These braces are attached to *F* and to each other at the top, while *E* is free to rotate in their "hollow embrace." The top of post *E* has a slot cut in it that receives the rule *A*, and an axle peg *C* goes through the two prongs of *E* and through the socket made at the bottom edge of *A*. The small piece of wood *B*, which is hard to see in the diagram because it is mostly inside the slot, attaches to the bottom of the rule *A* and serves to hold the axle *C* in place.

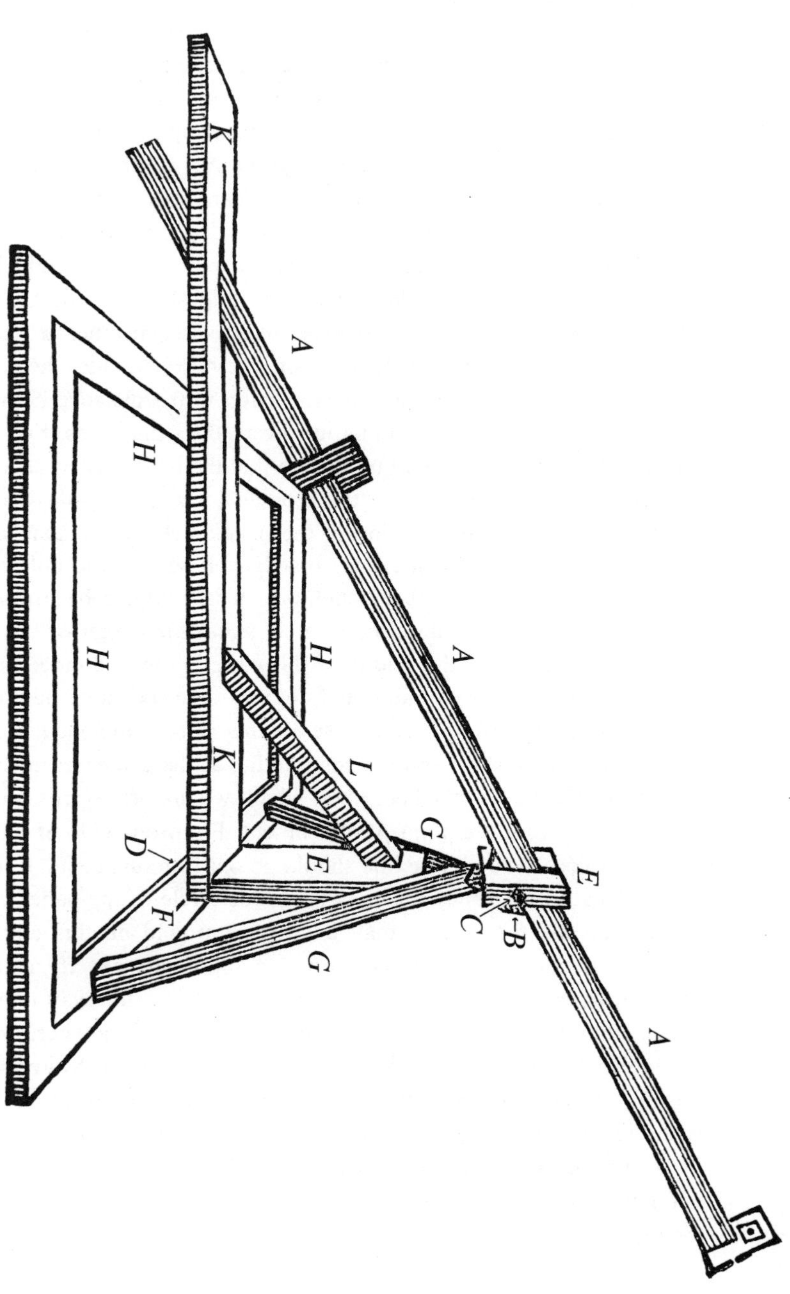
K
A
H
H
A
H
K
L
G
E
E
D
F
C
B
G
A

out in the middle of the breadth from one part of the length, so that the height of the slots may equal the breadth of the rule, and the opening remaining may pinch the thickness of the rule which it has after the grooves that were made. And so, when the panels have been set in the grooves, they will be both parallel to each other and perpendicular to the rule in length and breadth. On each panel, let a line be drawn lengthwise, beginning at the middle of the slot and from the thickness of the rule, and let another line be set up perpendicular to it, bisected at the point of intersection, drawn through the full width of the panel. Afterwards, you should cut out the middle parts of the panel that is going to be upmost, with a rectangular opening, about two fingers wide, in which the intersection of the straight lines occurred; but let the residual part of the straight lines remain at the edge of the opening. Now, in a very thin sheet of bronze, well smoothed and not unyielding, let a pair of lines intersect each other at right angles, and, with the intersection as center, let a small circle be made, the size of a pea, so that this diameter may be less in proportion to the distance between the panels than the diameters of the luminaries are to their distances, by 6 of the second chapter; and let it be drilled through, so that the hole may be accurately circular, and be at the middle between the crossed lines, the width of the sheet being a little greater than the opening in the panel. Let this be attached to the panel with the cut-out, so that the lines match, and so that the hole is in the center of the opening. On the other panel, which is going to be lower, and upon its upward facing surface, describe a circle about the center of the intersection of the lines, as great as the width of the panel allows, divided from the top to the middle parts into 90 individual degrees, and the same number from below on both sides. Or, if you prefer, use another numbering sequence. Next, let the panel hold a very short style in the middle, at the center or intersection. About this let a volvelle be attached, upon which we are afterwards going to describe small circles from the point of attachment. From one side of the volvelle let a pointer project, whose edge line should go out from the center of the volvelle, of **339**
such a length that it may reach as far as the circle of the panel, and by rotation of the volvelle the pointer may be carried around on this circle. Once these things have thus been built, measure everything very exactly, as I have done on my instrument. When it is going to be used, cover the way lying between the panels with a tube that is black on the inside, so that nowhere is entry allowed for light other than by the opening in the upper panel. Moreover, you should position the instrument so that the part of the rule that is above the post, along with the tube and the panel, may be outside, beyond the curtain in the open air, the rest in darkness inside, and let it be able to be turned freely. And let the window, which lets the rule out into the air, be well sealed all around against the light.

Problem 2

To measure the sun's diameter with the instrument

Really, this might in general be done, by 8 and 9 of the second chapter, on any ray let in through any opening, even a square one. Besides, the senses here do not follow mathematical precision, but by 29 of the first chapter, they do not grasp the edge of the image, to which few particles from the sun radiate, in the face of the brightness of the parts of the image in between, to which all the particles of the sun radiate. This happened to the Tychonic observers, when they did make use of the ray, but with an opening that was quadrangular, and too wide for a complete intersection to occur, by 6 of the second chapter.[6]

So let the instrument be set in order, and the rule be brought to bear on the sun's body, which is done by a double rotation, one of the rule upon the post, the other of the post and the crosspiece upon the azimuthal quadrangle. And thus the sun, striking the upper panel directly, will send rays down through the hole in the sheet and the tube, onto the curtain and onto the lower panel, and will create a round image of the illuminated surface, by the eighth proposition of the second chapter. Where you shall see this image (we shall henceforth call it the ray), first use a compass to take its diameter, as carefully as you can. But because unsteadiness will hinder you, with half of what you took with the compass, from the center of the panel (which takes the place of the wall), draw a circle, and one slightly smaller than that, and again one slightly larger, as many times as
340 you think necessary. Then investigate again which of the described circles the ray equals.

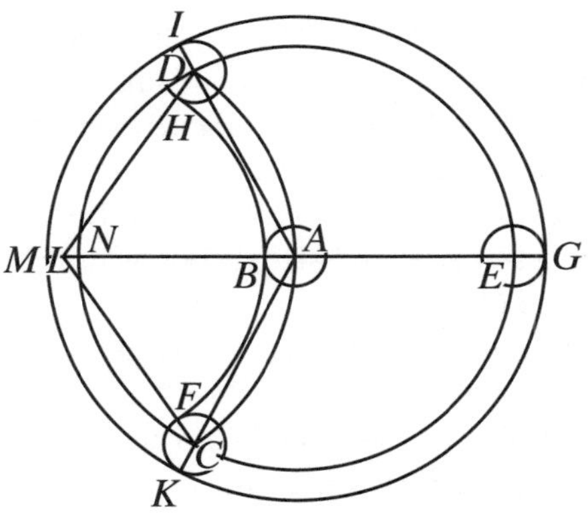

So, let AB be the radius of the opening, AK the radius of the ray, and let KC, BA be equal. Therefore (by 6 of the second chapter), AC is the size of the image that will come down through a single point of the opening. But since the edges of the ray and of the sun's body are touched by the same straight lines (for by 4 of the first chapter, the lines of light are straight), therefore, the angles of the imagined point of the opening are at the same vertex, and are equal. Therefore, by an eye located at the place of the opening, both AC below, and the radius of the sun above, will be seen under the same angle.

And so, when the radius AB or CK is subtracted from the radius of the ray AK, the remainder will be AC, which, together with the distance of the panels, shows the angle of vision. For as the distance of the panels is to AC, so is the whole sine to the tangent of the angle under which the body of the luminary is viewed.

---

[6] The term, *plenaria intersectio*, did not appear in Ch. 2 Prop. 6. The sense of it appears to be that, under the ideal conditions described in that proposition, the rays coming from the midpoint of the luminous surface would spread out exactly as much as would the rays coming from the entire luminous surface and passing through only the midpoint of the opening. These rays intersect at the wall, at the points K and L in the diagram for that proposition.

On 1601 13/23 December, the diameter of the ray was 72 digits or units, and 38 units in addition, that is, 110 in total, the half being 55, the radius of the opening $8\frac{1}{4}$. Therefore, AC is $46\frac{3}{4}$. Consequently, as the distance of the panels, 10,368, is to $46\frac{3}{4}$, so is 100,000 to 451, the tangent of the arc 15′ 30″.[7] Twice this is 31′. This is the diameter of the sun at perigee, with which a repeated examination in December 1602 agrees.

In 1602, in the month of June, using the same opening and rule in an equally dark place, the sun's radius on the panel very obviously fell short of the quantity in winter. And when the quantity in winter was divided into its 12 digits, the ray in summer fell short by about 2/5 of one digit, as far as could be judged in this small breadth. Since, therefore, the whole diameter amounts to 31 minutes, that is, 31/60 of one degree, therefore, 1/12 of 31/60 is 31/720. 2/5 of this make 62/3600 of one degree, or 62/60 of one minute, that is, about one minute. And the diameter in summer is 30 minutes. I was, of course, able to proceed as before in **341** the wintertime, but this way is the safest of all, because it connects the summer radius in a ratio with the winter one.[8]

Indeed, in 1600, in the month of June, at Graz in Styria, through the same opening and the same distance, the units of the ray appeared to me to be $105\frac{1}{2}$. Half of this is $52\frac{3}{4}$. The radius of the opening, $8\frac{1}{4}$, subtracted from this leaves $44\frac{1}{2}$. And as 10,368 is to $44\frac{1}{2}$, so is 100,000 to the tangent 429, whose arc is 14′ 45″; its double, 29′ 30″. This quantity falls short of the previous examination by half a minute (or $\frac{3}{4}$ of one small part of which there are 72 in a digit). On the preceding days, with a brighter heaven, the ray, coming in through an opening 40 small parts in diameter, exceeded a circle of $129\frac{1}{2}$ small parts, by about 1 or 2 parts. When 40 is subtracted from $129\frac{1}{2}$, the remainder is $89\frac{1}{2}$, its half $44\frac{3}{4}$, which indicates an arc of 14′ 51″. But by the addition of two parts, 15′ 10″. And thus the mean of these is 15′, its double 30′. But there is no reason for me to doubt the later year 1602, and my observers were expert in astronomy, while of the earlier year 1600, I wonder whether I was so precise in dividing the digit that I did not commit a greater error. For in addition, the curtain that I had set up did not show as much darkness as I wished, so I was unable to draw the line accurately enough at the edges of the ray.[9]

Tycho found nearly the same quantity in 1591.[10] The tube was AB, the opening had 10 units and was square. Therefore, AE was 5, AB 1000, CG $18\frac{1}{2}$. Consequently, BG was $9\frac{1}{4}$. And IG (after BI or AE was subtracted) was

---

[7] Kepler is using a foot divided into 12 inches, each of which is divided into 72 digits. The distance between the panels is 12 feet, or 12 · 12 · 72, or 10,368. Note that below, in dealing with the diameter of the image, the digit changes meaning, becoming the Ptolemaic twelfth part of the diameter.

[8] Franz Hammer notes here that, while Kepler's value for the difference is correct, his value for the perihelial diameter is too small: it should have been 32′ 32″, the corresponding aphelial diameter being 31′ 28″.

[9] In a note to this passage, Frisch (*JKOO* II pp. 435–8) includes an extensive series of Kepler's observations of this eclipse, from Vol. XV of the Kepler Manuscripts in St. Petersburg.

[10] These observations are in *TBOO* XII pp. 108 and 118.

$4\frac{1}{4}$. Hence, the radius was 14′ 37″. But it came out variously from 14′ 20″ to 15′ 40″. But on December 5, and with different tubes, he thrice found 15′ 30″ (he did this with me), as I in fact computed from his observations. For he had not drawn anything from this, and it is known that, by persuasion of the hypothesis of eccentricity, he makes a perigeal diameter of 32 minutes and more.

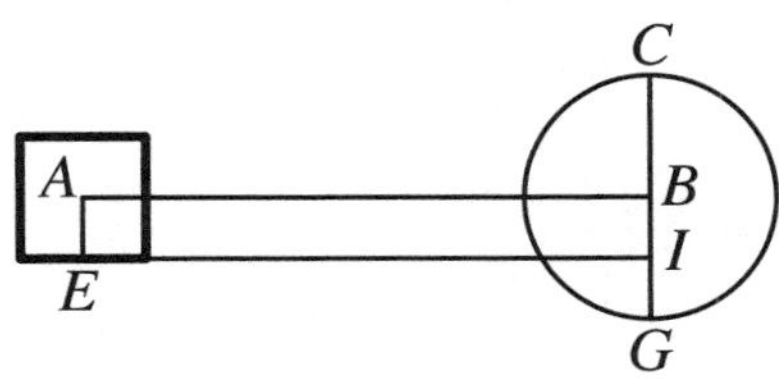

Now as for there being no more than one minute's difference between the summer and winter diameter, and it is wonderful how well this fits with the true and geometrical eccentricity of the sun (if you separate out a fallacy that arises from a physical cause, on which
342 see ch. 10). For Tycho Brahe and the Landgrave's observers, by unanimous computation, show the eccentricity to be 3,600 out of 100,000. But through a physical cause, the half pushes in to replace this, as will be shown geometrically in the *Commentary on Mars*,[11] therefore, the real eccentricity is 1,800. And when the sun is at apogee, in the month of June, its distance is 101,800; but at perigee, in the month of December, its distance is 98,200 of the same units, of which the mean distance is 100,000. And as 101,800 is to 98,200, so conversely is 31′ to 30′, approximately. For as regards Theorem 8 of Euclid's *Optics*, it does not apply to such small arcs.[12]

Problem 3

To observe the sun's diameter through a slit

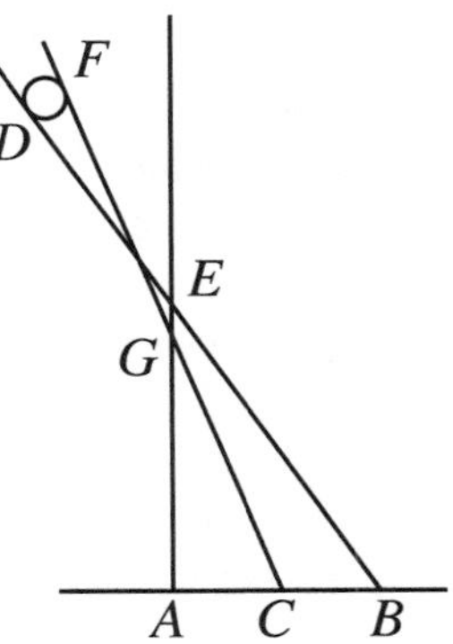

I found this in Tycho's observations, with an appended Encomium of the rectangle as the master of the universe of mathematics.[13] Let AB be a plane equidistant from the horizon, AE a wall perpendicular to AB, in which GE is the slit, DF the luminary body. Then, from its highest part F a ray descends through the lowest edge G of the slit, and is extended to the point C, closest to the perpendicular A. On the other hand, from the luminary's bottom edge D through the top E of the slit, the ray DE is extended to the most distant point B. Therefore, as CA is to the whole sine, so is AG to the

[11] This is the subject of Part III of the *Astronomia nova* (chapters 22–40), and especially of chapters 22–29.

[12] Euclid, *Optics*, 8: Equal magnitudes at unequal distances from the eye do not have the same ratio of angles as of distances.

[13] *TBOO* X 56–7. If the "Encomium" of the rectangle was anything other than the words, *per rectanguli instrumentum instrumentorum*, it was not included among these observations. Dreyer's numbers are in part different from Kepler's. For 15 March, he has: (1) 30′ 48″, (2) 30′ 6″, (3) 30′ 44″. For 14 June, he has: (1) 30′ 4″, (2) 30′ 8″, (3) 29′ 30″.

tangent of the angle GCA, which measures the height of the highest edge F above the horizon. Again, as BA is to the whole sine, so is AE to the tangent of angle EBA, which measures the height of the lowest edge D above the horizon. Then by subtraction of the smaller arc from the larger, the remainder is the angle that the body of the luminary fills here on earth. This procedure is good, too, if one can make sure that in some building EA is exactly perpendicular to AB, and is sufficiently high. On 1578 March 15, Tycho observed the diameter in this way, and the results were 30′ 40″|30′ 6″|30′ 44″|30′ 50″|, and on 14 July,[14] 30′ 4″ twice|29′ 30″ once.

There exists among Tycho's astronomical papers a letter of my teacher Maestlin to a councillor, of Augsburg, if I am not mistaken, in which he makes the sun's diameter about the same, except that, being an adherent of the hypothesis of simple eccentricity, he enlarges it, and in fact says that at apogee he caught **343**
it at 29′ 36″, at the middle longitude 30′ 11″, at perigee 31′ 45″.[15]

Gemma, too, does not differ much from this method in his *Radius*, if it is possible to find out anything certain using so coarse an instrument.[16]

The ancients, however, fully go along with me. Archimedes reports of Aristarchus that he said that the sun's diameter is 1/720 part of four right angles, that is, 30 minutes.[17] Hipparchus, however (Albategnius complains about Ptolemy's following him in his computation), denied that the sun perceptibly changes its magnitude from apogee to perigee.[18] This is undoubtedly because it does not vary by more than a minute, and one gathers that he himself also supposed it to be 30 minutes (chapter 8, above),[19] so far as is clear from Ptolemy and his expositor Theon. Proclus seems to relate the same thing about Sosigenes.[20]

Ptolemy was the first to revolt simultaneously from the ancients' ways of observing and from the perfectly true quantity itself of the diameter of the luminaries, with a procedure that is absurd in measuring, and as well as impossible in practice, which the authorites up to the present, and even Copernicus himself, followed. For this reason, I have used that measure in Chapter 7 above, though certainly with no danger. But it is a certain thing, and evident to anyone to investigate, that the sun's diameter at apogee is 30′, at perigee 31′.

---

14 These were made in June, at the solstice, as Tycho noted.

15 The letter, which (according to Franz Hammer, in *JKGW* II p. 455) was addressed to Hieronymus Wolf of Augsburg, seems not to have survived.

16 *Gemmae Frisii medici et mathematici, de radio astronomico et geometrico liber* (Antwerp and Louvain, 1545). The *radius* is a cross staff, whose construction is described in chapters 1–4, folia 6v–12v.

17 Archimedes, *The Sand-Reckoner*, in *Works*, trs. Heath, p. 223.

18 Ptolemy, *Almagest* V 14, trs. Toomer p. 252; Albategnius, *Opus astronomicum*, ed. Nallino, Vol. I p. 56. For Albategnius (i.e., Al-Battānī), see ch. 4, note 172.

19 See p. 304.

20 Proclus, *Hypotyposis* IV 3, Latin trs. by G. Valla, ed. 1541 p. 400, ed. 1551 p. 352. What Proclus says here is that Sosigenes believed it possible for the moon not to cover the sun completely in a central eclipse of the sun (an annular eclipse).

Problem 4

To observe the moon's diameter through the instrument

A difficult task. For the moon is not so bright as to give our eyes adequate sureness of vision. And if it shines in through the instrument into the chamber and the darkness, the difference between the the ray and the nearby edges of the paper is made out with great difficulty. Nevertheless, you shall make the attempt as follows. You shall draw on the paper some number of circles rather close to each other, but increasing somewhat in sequence, each by itself, and you shall fill the enclosed area with black, especially near the edges, so that the blackness may extend over the whole surface, or at least for some width from the edges towards the center. You shall then attach them in order to the panel on the instrument, considering which particular one of them the ray of the moon encircles in such a way that the illuminated whiteness of the paper around the black circles enters into the eyes to some extent. For whichever one may be the first to be perceived to
344 be smaller than the ray, it soon establishes itself as the greater measurer of the ray. Nevertheless, after all your carefulness, there will still remain some uncertainty in this method. What you get from it is this alone, that you will be able to make judgements about huge errors.

On 15/25 July 1600, at Graz, I found the ray to be more than $105\frac{1}{2}$ small units, less than 110. Therefore, the apparent diameter was greater than $29' \, 30''$, less than $31' \, 12''$. It is also always presumed to be less than it should be, because the edges of the ray, having very weak light, are not well discerned by the eyes. The moon was full, at its mean distance.

On 16/26 January 1603, in the evening at Prague, the ray held a quantity intermediate between two black circles, one of which, the enclosed one, had 113 small units, the other, enclosing the ray, had 120 small parts. Therefore, the apparent diameter also appeared definitely greater than $32'$, but less than $34' \, 18''$. The moon was closest to earth; every precaution was taken, especially the darkness of the chamber.

But in the previous month, when it was gibbous[21] and revolving through the same perigee, with light diffused through a turbulent sky, and an insufficiently sealed chamber, it was considered by this method barely to surpass 30 minutes, even over the whole illuminated area, to such an extent that it easily eluded the yawning observer. And so other ways of observing the diameter of this luminary must be tried.

Problem 5

To pass judgement on the ratio of the apparent diameter of the moon to the diameter of the sun from the horn of the moon in the first phase

In chapter 6 section 11 above, I recalled the moonrise that I saw on 4/14 March 1603, at Prague, at the 6th hour,[22] when the sun was in 23° 49′ Pisces,

[21] ἀμφίκυρτος.

[22] In the evening. See p. 269. Perhaps by *exortus* Kepler means the moon's first appearance in the evening.

the moon in 14° 41′ Aries, by its motion on the ecliptic, however. The arc through the moon's apparent position and through the sun cut the horizon at an angle of about 78°. For the moon's horns clearly faced closely to the left of the zenith. And thus the apparent distance of the center of the moon from the center of the sun was 20° 10′.

So, then, the moon in this position, when, availing itself of the earth's light (as was said above in Chapter 6 Section 10), it was seen most clearly in its entire body, it was surrounded by the shining horn on less than half its circumference. For this was plainly deficient from a semicircle to the senses. I say that from this it is shown that the moon's apparent diameter was noticeably greater than the sun's apparent diameter, although the sun was moving towards its mean distance, **345**
nearer to perigee, while the moon had descended, not two signs from apogee. This works towards the end that we may here as well fight for Chapter 8, and may believe with all the more certainty that the moon's diameter is decidedly greater than the sun's diameter, and consequently that the entire sun can be covered by the moon.

Demonstration

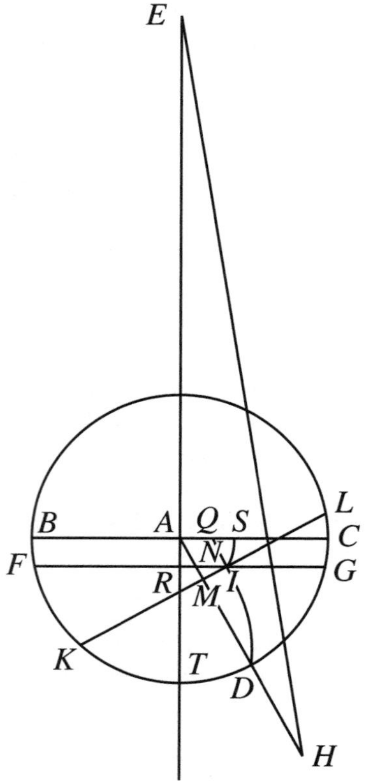

About center A let the great circle of the moon's body BCD be drawn, in it the diameter BAC, and EA at right angles to it. And let E be the center of the sun; therefore, the circle of illumination will be parallel to BAC: let it be FG. Now let a point H be taken not on the line EA: let this be a place on the earth's surface, Prague, say. And since HA is the axis of the circle of vision, this will be at right angles to HA: let it be KML, cutting off from the moon the small part illuminated by the sun, or the horn, whose true breadth is LG. And since this horn has fallen short of a semicircle, half the horn will fall short of a quadrant, and it therefore will not reach all the way to M. Therefore, let the circle of illumination FG cut the circle of vision KL at a point beyond M towards L: let this be I. Let the axis of vision HA, on the other hand, cut FG at N, and the axis of illumination cut the same at R. Therefore, if the horn had come all the way to N, the circle that passes through N, equidistant from KL, and representing the circle of vision, will come out to be less than the circle of illumination FG. For AN, subtending the right angle ARN, is longer than AR. Therefore, a circle through N would be more distant from the center A than is the circle FG, as a result of which it would be smaller, since that which is through the center is alone the greatest of all.

346 Now, however, the half horn did not go as far as N, so as to be perceived with the length of a quadrant, but fell short at I. As a result, KML stood at a greater distance from A than N did. It was therefore much smaller than the circle through N. But the one through N was also smaller than the illuminatory circle FG, so the visual circle KL is much smaller than the illuminatory circle. But by Theorem 24 of Euclid's *Optics*, the less of a globe is seen, the greater its diameter appears to be. And above, in Ch. 6 Sect. 3, it was shown that if the visual circle KL and the illuminatory circle FG had coincided, the two luminaries would have been seen under the same angle. Therefore, because KTL is now smaller than FTG, the moon's apparent diameter is therefore greater than the sun's, and quite perceptibly so, even though it is near apogee.

Now, so that the theorem may be perfect, let it be that the ratio of the horn to the remaining circumference of the body be known without error from estimation by eye, which is indeed quite difficult: I too was unable to estimate anything more exact than that the horn is between a third and a half of the visual circle, that is, between 120 and 180 degrees. Let it be proposed for us to find by computation from this the moon's apparent diameter. Therefore, the horn LI will be half of the estimate. Let D be the pole of vision, and let the arc DI descend, and let it cut BC at Q. Thus the angular length QDC of the half horn is given on DQC. But DCQ is right. And the side DC is given. For because the sun is at its middle distance, and the moon is 56 degrees past apogee, the ratio of HE to HA will thus be given from the hypotheses of the astronomical authorities. But the angle AHE comes to be known from the visible positions of the two luminaries, so when A is produced to T, TAH will be given, whose measure is TD, and its remainder in the quadrant is DC, which is what is required. So, in DQC let there be found from the givens, first DQ, then DQC. Next, from I let an arc fall perpendicular upon BC, and let it be IS. Thus in triangle IQS the angle S is right, Q is given, and side IS is 15 minutes, that is, as much as is the distance of the circle of illumination from the greatest circle, Ch. 6 Sect. 3 above. Therefore, QI will be given, which is to be subtracted from QD, so as to obtain the arc between the pole and the circle of vision, whose complement shows the apparent semidiameter, by what was shown in Ch. 6 Sect. 3. One should not pretend not to notice that the moon's apparent diameter is made enormous here, if you say that even the least per-
347 ceptible amount is wanting from the semicircle of the waxing horn. And so another cause has to contribute, namely, the points of the horns disappear in the sense of sight because of their narrowness against the brightness of the parts in between.

Appendix

Other ways of measuring the moon's diameter are compared with the preceding

Even though the most certain and reliable measure of all will finally be introduced later through observations of solar eclipses, nevertheless, the remaining methods should meanwhile be set down for examination here. And indeed, the dioptra of Hipparchus is known from Ptolemy, which Ptolemy himself also used, though he despaired of eliciting a reliable magnitude from it.[23] He found only this, that the moon at apogee shows a diameter equal to the solar, and increases it at perigee, which Hipparchus also had said.

On 29 March or 8 April 1598, at hour 8 of the evening in Graz, I saw the moon in conjunction with the western stars in the quadrilateral of the Pleiades, in such a way that the moon's edge was removed from the nearest by no more than a sixth part of the lunar diameter. At the farther edge its distance from the shining star of the third magnitude was as great as was the width of its body.[24] Thus the two stars there on the western side of the quadrilateral were too far apart to allow them both simultaneously to cover the moon, had it happened to pass over them. The moon was three days old, its mean distance surpassed, tending towards apogee; and close to its greatest northern latitude: it was seen most clearly, in its entire body.

On 17/27 July following, between hours 2 and 3 in the morning, the moon, ascending again towards apogee, and closer to it, and near the northern limit, was standing at the Pleiades with the horns reversed, so that a line drawn from the lower horn to the perpendicular of the intersection would graze the Pleiades from above. It had passed through the Pleiades, and was distant from the brightest by more than Palilicium is from the neighboring star of the Hyades towards the nostrils of Taurus, less than the latter is from the lowest in the nostrils.[25] Thus the distance of the two bright transverse stars of the Pleiades appeared to be equal to the diameter. Comparisons of this kind to nearby fixed stars with each other ought to be more vigorously sought out, because this measure is universal to all people and is permanent.

The same thing can be tried, either by the fixed stars, or by the radius, or by the dioptra, when the moon is passing through the earth's shadow, while still bright with its ruddy glow. But when immersed in the middle of the shadow, as in March of 1588, it is seen poorly, and the observation is untrustworthy.

Let us see, however, also *when it is observed simply by full light with in-* **348**
struments, whether that be done with the radius, or by comparing the edges with stars on opposite sides, or by the altitude of the two edges found with quadrants.

On 14 June 1592, when an eclipse was impending, the Brahean observers, having measured the diameter of the moon with the astronomical radius, following the teaching of Gemma, recorded 32 or $31\frac{1}{2}$ minutes.[26] The moon was at apogee. The method is both unreliable and presumably gives results greater than the truth.

---

[23] See note on p. 231.

[24] The "shining star" is η Tauri, described by Tycho Brahe (*TBOO* III 345) as *media et lucida pleiadum.*

[25] Palilicium is another name for Aldebaran (α Tauri).

[26] This was Gemma Frisius. See the note on p. 354, above. The observation is recorded in *TBOO* XII p. 196.

In the same year, on February 12, when the moon was not at its lowest, they recorded 35′.[27]

On 6 January 1587, the elevation of the moon's top and bottom edges above the horizon, observed by day when the viewing of the moon is more reliable (for it was bisected, at quadrature with the sun), showed a diameter of 30′.[28] The moon was at apogee.

On 2 March 1588, before an eclipse in the evening, the difference of declination of the edges, by armillary spheres, repeated many times, was 33′, plus or minus one half.[29] On the meridian, the difference of altitude of the edges was 31′, $32\frac{1}{3}'$, $30\frac{3}{4}'$.[30] The same on the preceding day was 33′.[31] Evidently, this method is somewhat more uncertain. The moon, ascending from perigee, was approaching its mean distance.

Observed in this way on 22 February 1591, at it mean distance, it appeared to be 31′ twice, 32′ six times, 33′ seven times, 34′ six times.[32] This variety came about partly because of different eyes, partly because of the sights, partly because of the abundance of light seen at night. For all these means make the procedure of this observation a little less reliable. And although by my instrument the moon's diameter may perhaps appear a little less than the truth, there is nontheless greater consistency in it, than in this procedure of observing through sights. I would nonetheless not deny that *if anyone should apply the dioptra of Hipparchus skilfully*, he will sight more surely.[33] All the more so that since Hipparchus's dioptra, Tycho's daytime observation, and my instrument agree closely in this quantity, which I will present below through the moon's shadow, I therefore conclude more easily, at least now, that the moon's diameter at apogee is $30\frac{1}{2}$ minutes. But how much it is at perigee is not as easily deduced from this as
349 it was in the sun. For since the moon has a twofold maximum equation, one of 5 degrees, the other, at quadratures, of $7\frac{1}{2}$ degrees, the quantity of these, considered physically, is 4336 for the one, 6520 the other.[34] It postulates an eccentricity:

---

27 *TBOO* XII p. 190. The record indicates that this diameter was supposed rather than measured. On the following page, it was recorded as 32′ once and 36′ twice.

28 *TBOO* XI p. 142. However, there are three such measurements on that page, the resulting diameters being 30′, $29\frac{5}{6}'$, and 31′.

29 *TBOO* XI p. 257. The actual recorded differences on the armillary were $33\frac{1}{2}'$, 27′, 27′, $27\frac{1}{2}'$, and 27′.

30 Kepler omitted the first measurement, taken with the rotating quadrant, recorded as 26′. The others are reported correctly.

31 On the previous day, three such measurements were recorded, one of 33′, a second of $30\frac{2}{3}'$, and a third of $25\frac{1}{2}'$. *TBOO* XI pp. 256–7.

32 *TBOO* XII pp. 120–1. Although the number of observations related by Kepler tally with the number in the observation book, the values are differ considerably. However, Kepler's point that this method produces inconsistent results remains valid.

33 For the verb *collimare*, translated here as "sight," see the note on p. 226.

34 Kepler divided the whole equation, or correction to be applied to a celestial body's mean motion, into two parts: the physical part and the optical part. The latter represents

you would not know whether you ought to follow the one or the other, or the mean, 5428, so that the two physical causes thus harmonize. And thus the moon's diameter at perigee is either 33′ 20″, or 34′ 0″, or 34′ 40″. And the mean* is either 31′ 55″, or 32′ 15″, or 32′ 35″. You see that the observations made with sights in the years 1587, 88, 91, 92, being reviewed, for the most part play around these; but about the selection of one of these three, we are left in uncertainty. However, if you trust my instrument, from the observation of 16/26 January 1603, the last perigeal quantity will be set aside. And surely, physical arguments also do not particularly favor the greatest eccentricity of these three.

* Albategnius agrees, stating that the moon's mean diameter is 32 min. 25 sec. However, he distinguishes the apogeal and perigeal diameters by the whole eccentricity to which he also gives credence, so that the former would be 29 min. 30 sec.; the latter, 35 min. 20 sec.

Problem 6

To estimate the quantity of the defect in an eclipse of the moon, or also of the sun

This is commonly done without an instrument, by an imaginary division of the diameter into 12 parts.

We make use of this method most safely when the quantities in the defect and in the light [i.e., the bright part] are about equal. Thus, on 25 June 1572, Maestlin estimated the maximum defect to be exactly half the diameter. (However, Gemma Frisius, in Book II fol. 233 of the *Cosmocritica*, wrote that the defect at Louvain was 8 digits).[35]

Further, it is more helpful for that eclipse for which the shadowed circumference is in view at the same time. Outside these cases, the procedure is variable and risky.

On 29 November or 9 December 1601, although there was a good part of the moon's body remaining, it was not possible for anyone to discern its quantity reliably. Ambrosius Rhodius, from a Tychonic computation for a considerable time, estimated the defect to be 10 digits at Wittenberg.[36]

On 8/18 November 1603, some argued that it was more than one fourth deficient, while I judged that somewhat less than this was missing. And nevertheless, the bright part of the circumference was darkened.

---

the apparent increase and decrease of speed resulting from the observer's eccentric position alone, and the former expresses the real variations in speed along the orbit. Each part is responsible for about half of the whole equation, the exact proportion being determined by the form of hypothesis adopted and the parameters used. Kepler does not tell us what geometrical model he is using, but the numbers are approximately the tangents of $2\frac{1}{2}°$ and $3\frac{3}{4}°$, respectively, the radius of the moon's deferent being taken as 100,000 units, as was usual at the time. These numbers represent the eccentricities (or radii of the epicycles) that would be required to produce the two equations. The exact numbers are not important in the present instance: Kepler is arguing that the equations imply real eccentricities of about the sizes given, and that these in turn imply observable variations in the moon's apparent diameter.

35 Michael Maestlin, *Epitome Astronomiae* (Heidelberg 1588), p. 460; (Tübingen 1610) p. 494. The *Cosmocritica* was written by Cornelius Gemma, not his father, Rainer Gemma Frisius.

36 The computation is in *TBOO* II 141–3. Tycho predicted an eclipse of 11 digits.

350 Since, however, it is of great interest to Astronomy to take note of partial eclipses correctly, Tycho Brahe, from the precepts of Cornelius Gemma, *was accustomed to measure with the radius, both the moon's diameter before the eclipse, and the remaining part at the greatest defect.*[37]

By the way, the difficulties by which this method is encumbered are recounted at great length in the preceding.

On 14 June 1592, Tycho Brahe observed the diameter of the moon by the radius, at the beginning and still whole, to be 32 minutes.[38] Then, around the middle of the eclipse, he found 14 minutes remaining, so that the defect was 18 of these. But this quantity diverged greatly from his own computation, for he adopts $26\frac{1}{4}$ minutes for the shadow of the lunar body. And so I attribute this to the difficulty of observing. For although someone might call into doubt the Tychonic computation or the diameter of the shadow (on which more elsewhere), he will nonetheless never make this eclipse, with regard to this quantity, square with the others in one norm.

And so, in order that here too I might rest on more sound defenses, *I make a practice of estimating the missing arc of the lunar circumference.*

For once this is given, and the ratio of both the moon's diameter and the shadow are known approximately, the quantity of the defect is itself also given. About center D let the circle of the shadow FEG be described, and about center B, the circumference of the lunar disk FAG, cutting the shadow at F, G. Let the centers be connected with each other and with F, and likewise also the points of intersection, with the lines FB, FD, BD, and FG, which will cut each other orthogonally at C. Let FAG be, for example, one sixth of the circumference: FA will be half, that is, 30°, namely, the angle FBC. Consequently, BFC will be 60°and the secant BF will be 200,000 where FC is 100,000. But let the ratio of BF to FD be given, which is 1 to 3; FD will be 600,000, the secant of angle 80° 24′. But also given are the tangents of the same angles: BC, 173,205; CD, 591,236. Which, subtracted from BA, 200,000, and DE, 600,000, leave CA, 26,795, and CE, 8764. And so, where BF is 200,000, the deficient part EA is

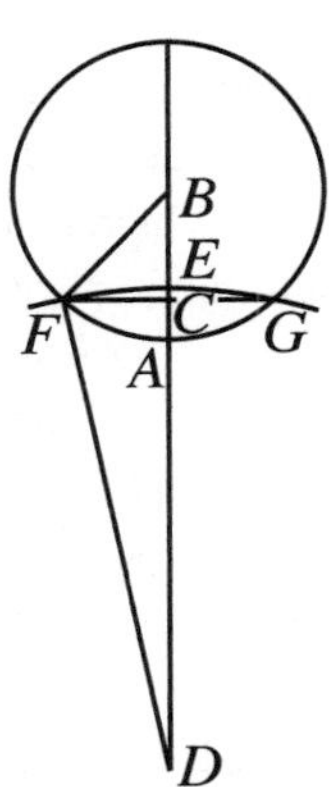

351 35,559, made up of CE and CA. Once this ratio is given, the digits or minutes of deficiency are easily had afterwards. For if the semidiameter of the moon is 16 minutes, there will be 2′ 51″ in the defect. You will find instances below, in the eclipses of the year 1603.

Here, although both the sense of vision and the estimation of the sense of

---

[37] This time, it is Gemma Frisius's precepts in the *De radio astronomico* that Kepler means here. It is not clear why this clause is in italics; possibly it is a quotation, though I have been unable to trace its source.

[38] *TBOO* XII pp. 195–7. In the midst of these observations, Tycho remarks, "one must not trust the radius too much." I have not succeeded in locating the computation referred to later in the paragraph.

vision can be in error to some extent, only a slight portion of that error figures into the estimating of the moon's diameter.

I want astronomers to work at setting up some more reliable ways of making this observation. For dependent upon this one thing is what is commonly made the chief point in astronomy, the measurement of the solar altitude and the solar body. That is, if the sun's maximum parallax is 3 minutes, and in the estimation of the lunar defect an error of one third of a digit is made, this is a matter of 600 semidiameters of the earth; but if the sun's maximum parallax is 2′, we shall be in error by 1700 semidiameters of the earth, by omission of one minute in the moon's defect, as is to be seen at the head of our Parallactic Table.[39]

Problem 7

The true image of the eclipsed sun being set forth, to find the true ratio of the diameters of the sun and moon, and the true [number of] digits of the eclipse

This problem is from my teacher Maestlin.[40] We went up under the roof of the church, and, the doors being shut against the light, someone climbed up into the highest beams, in order to remove a roof tile in a suitable place so that a very slight crack could make an allowance for light, now one tile, now another, according to the way the beams were intercepting one or another ray. Thus the opportunity of the decking provided us a much larger ray than my instrument or rule, which is no longer than 12 feet. He received this ray, eclipsed along with the sun, on paper (by Chapter 2 Sect. 9). And because the whole radiation is a right cone, whose vertex is about at the opening, it is obvious that the ray formed on the paper does not come out circular, unless the paper be perpendicularly opposed to the radiation, by Apollonius I 9. Therefore, he drew upon the paper a number of circles of different sizes, about as many as he saw that the ray was going to take **352**
up, and, after drawing the diameters, divided them into 12 equal parts, or digits. Next, he received the ray with the marked circles set opposite, in such a way that the edge of the ray should coincide everywhere with the circumference of one of the circles, by changing the circles or by putting them nearer or farther from the opening until this should happen. This was evidence that the cone of radiation was cut perpendicularly by the paper. Now he directed the divided diameter by rotating the paper so that it should bisect the horns of the sun. The interior arc of the deficient ray, where it cut the divided diameter, thus then showed the digits of the eclipse. This teaching, presented by Reinhold, he refined with greater care,

---

[39] One third of a digit is about one minute. At the head of the Parallactic Table (in the back pocket), Kepler has the distance in earth radii corresponding to that amount of horizontal parallax. To a parallax of 3′ there corresponds a distance of 1145, to 2′ corresponds 1718, and to 1′ corresponds 3437. The effect of an error of one minute could therefore have an effect as great as that described by Kepler.

[40] See Problem 28 below. This observation took place in Tübingen, while Kepler was a student.

following the advice of the author.[41] For he extracted the ratio of the diameters at the same time, in this way. When the edges of the ray precisely coincided with the circle, he marked the interior circumference of the ray with three or four points. From there, the circle continued through these points easily showed what ratio it had to the former one, which represented the sun.

Let $ABCD$ be the ray of the eclipsed sun, and the true image of it; let that square with the outer surface ABC, in a circle described about E. Now let any three points be marked on the interior circumference ADC, and let them be A, D, C. Therefore, by Euclid III 24, let a circle with center F be continued through ADC. And because the cause of the defect in the sun is nothing but the interposition of the moon between the sun and the sense of vision, therefore, the interior arc of the horned sun ADC is a small part of the circle by which the moon is viewed. Consequently, the circle ADC with center F represents the lunar body, and the ratio between FD and EB, which is found mechanically, is the ratio of the visible diameter of the moon to the visible diameter of the sun.

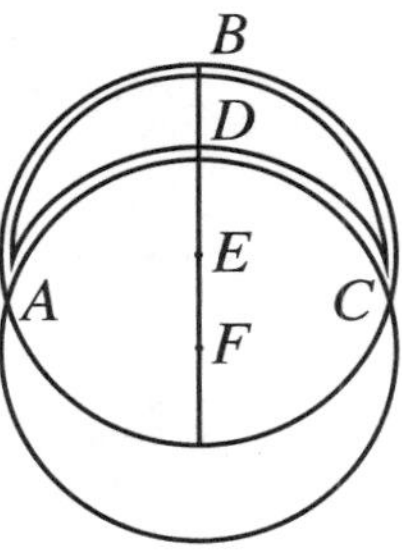

Incidentally, the moon's diameter, observed in this way, appeared much smaller than the solar diameter. But by Chapter 2 Sect. 11 it is clear that unless the opening is very tiny indeed, almost like a point, the ratio of the diameters is always faulty and the moon appears smaller than the truth, and (by Sect. 12) the digits appear to be fewer. Consequently, this method is untrustworthy, applied
353 simply in this way. In addition, the shaking of the hands, and the rapid motion of the sun, greatly disturb the marking of the points. Therefore, to my instrument.

Problem 8

To capture in points the image or ray of the eclipsed sun, as it is formed in the instrument

It is done as in the preceding, and with somewhat greater certitude. For the instrument made there be no need for me to change the circles, nor for moving in and out. For on the day on which the eclipse is to happen, the quantity of the sun's ray is noted accurately on the instrument, and a circle of such a quantity is described, which is carried all around attached to the panel, and for that whole day certainly equals the ray of the sun. In the marking of the points, however, there is the same objection as before. For at every moment the sun's ray is being carried across, and so both the crosspiece and the rule on the crosspiece have to be carried across between the marking. Nevertheless, take this precaution: mark one point when the outside circumference of the ray falls exactly on the circle. Then you should get ahead of the ray by moving the instrument; and while the ray is catching up, you shall note the place with your eyes for marking the other point. And it will be advantageous that a spectator be at hand, who, while you

[41] *Theoricae novae*, Paris 1553, *Methodus doctrinae Eclipsium,* fol. 142r–143v, esp. no. 2, fol. 142r–v. Hammer reports that this is on fol. 198 of the Wittenberg 1553 edition.

are intent upon the other edge, can indicate when the edge on the other side again falls on the point already marked. You shall again apply the procedure as a test, and do it quickly, before the defect grows or shrinks perceptibly. Besides, this image is adulterated. There follows therefore:

Problem 9

From the image on the instrument or the ray, however much adulterated, to extract the true image of the eclipsed sun

For let the circle in which the three points have been marked (namely, *H*, *B*, *F*) be extended, its center *L* being established. This will be the center of the lunar body. Next, to *LB* let there be applied the radius of the opening *BA*, and about center *L* with radius *LA* let the circle *DAC* be drawn, representing the moon's 354 true body. In the same way, from *AG* let there be subtracted the radius *GE* of the opening, and about center *A* with radius *AE* let the circle *DEC* be drawn, representing the sun. Therefore, *DECA* will be the true image of the eclipsed sun, by Chapter 2 Sect. 10, 11, and 12. See the computational procedure below, in the examples.

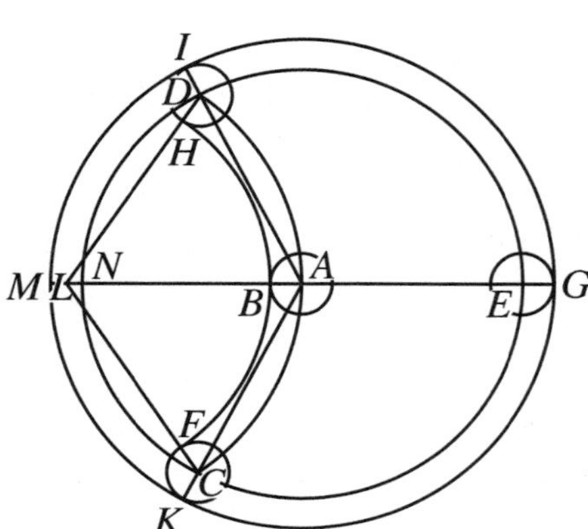

Problem 10

To extract the image of the eclipse more skilfully using the instrument's transverse rule

Because there is difficulty in marking points, as was said, and because many things have to be taken care of at once, especially if the observer is alone, I thought up another, easier procedure. Above the volvelle mentioned in the description of the instrument, I placed a pointer of solid parts, by which it might project slightly above the plane of the paper wheel, in such a way, however, that it should not cover the diameter divided into digits, but should enclose it with a pair of pieces of wood equidistant from the diameter. Above this solid pointer let a transverse rule rotate, perpendicular to the pointer and to the marked diameter, with the length of the diameter. Specifically, let the groove in this rule be equal in its depth to the thickness of the wood of the pointer, and let it attach to the wooden pieces with a clamp, so that one may push it forward. And let the rule have a sharp edge. Now, with the wall and the opening correctly oriented to the 355 sun, let the pointer with the volvelle be rotated on its axle, as well as the transverse rule up and down above the pointer, until the sharp edge comes into contact with both obtuse horns. And then let the point be noted at which the sharp edge of the transverse cuts the marked diameter; and let the point also be noted at which

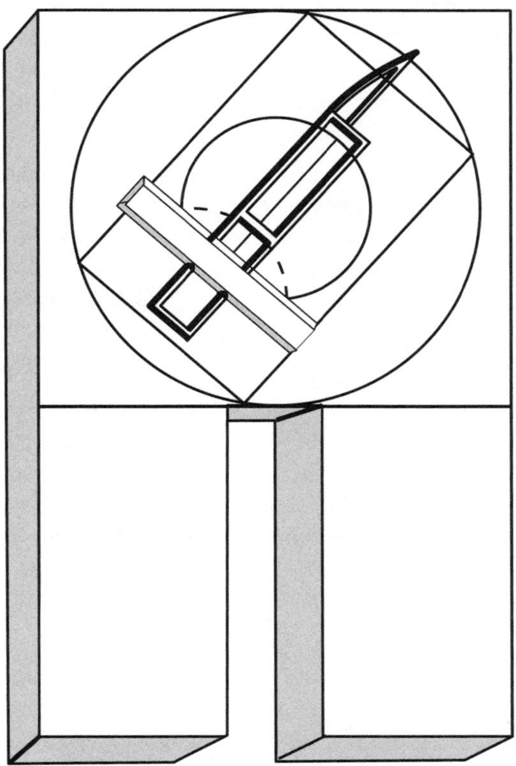

the interior circumference of the ray, or the moon's shadow, cuts the same diameter. If this will be managed correctly, there will be no need to mark the points. For there follows:

Problem 11

From the transverse rule, to show the true image of the eclipse

Let $CDEF$ be the image of the eclipsed ray, whose center is G, the marked diameter DF, and with it, the line of the pointer IDH. And let there be the line CHE touching the obtuse horns C, E, and cutting the marked diameter at H; and let the segment CFE cut the same diameter at F. Next, let the radius of the opening be extended on the diameter DH, and let this be DL, from D to L, as well as from F to M and from H to P, and through P let the straight line NPO be perpendicular to the diameter and equidistant from the rule. Then, with center G and radius GL let the circle NLO be drawn, which will represent

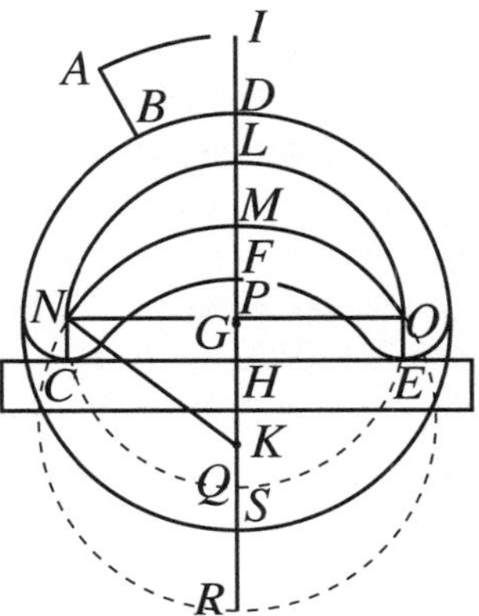

the true image of the sun, as in 9 of this chapter. This will cut the line NPO: let it cut at N, O. Next, through the three points N, M, O, let a circle be described, and let its center be K. I say that $NLOM$ is the true image of the eclipsed sun. For let the perpendiculars NC, OE be drawn. So, since NC, OE are perpendicular to the same CE, they will be equidistant. And because DK, that is, PH, is also perpendicular to the same CE, from the structure of the instrument, therefore, NC, PH, OE will be equal. But PH is the radius of the opening. Therefore, NC and OE are also. But since, by 10 of Chapter 2, the small circle of the opening is also drawn around the points of the ends of the horns, and since CHE is tangent to **356** those small circles (for they are the obtuse horns of the ray), it is necessary that NO be those points of the ends of the horns. For from any point of the circumference NLO above N or O, a longer line would be dropped, and from others below N, O, a shorter line, than NC, OE, the radius of the hole in the opening. Therefore, if N, O, are the points of the horns, the circle of the moon cuts the solar circle there. But M is also on the moon's circle, as in 9. Therefore, NMO is the moon's circle. And NLO is the sun's. Therefore, the true image is $NLOM$.

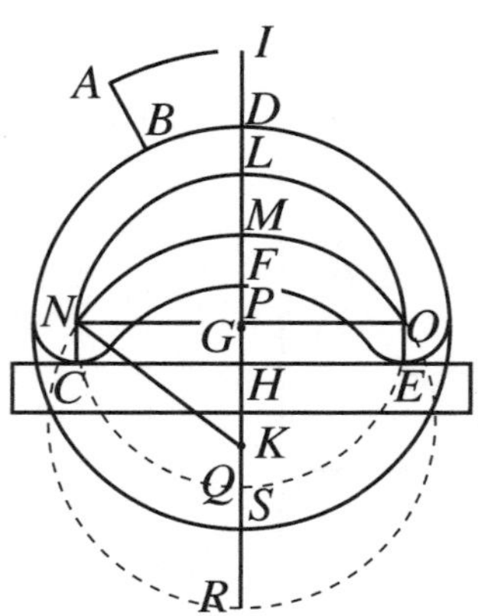

Problem 12

From the ray or image of the eclipsed sun marked by the instrument, to find the ratio of the diameters, the visible distance of the centers, and the quantity of the eclipse

First, from the adulterated ray, let the true be extracted, by 11 or 9 of this chapter. Next, let it be done as in problem 7; that is, let the true circumferences of the image be extended, and the diameters be compared mechanically, as well as the distance of centers and the digits. But if you have proceeded by 11, all these things *are different, in the observation itself and in the computation.* In the preceding diagram, the points F and H are given. Therefore, FH is given from the observation itself. I say that this is equal to the line MP. For MF and PH are equal, and FP is common. Further, ML is obtained at the same time by subtracting FM, LD from FD. Consequently, PL is found. But QL is found from 2 of this chapter. And this is the diameter of the circle QNL; therefore, both QP and their [i.e., PL and QP's] mean proportional PN are found. Moreover, PN is also the mean proportional between MP and PR, and PM was given from observation, i.e., FH. Therefore, as FH or MP is to PN, so is PN to PR, which, with PM, makes up the visible diameter of the moon. *However, the quantity of the eclipse is also given more simply.* For because SQ and FM make up the radius of the opening, SF and QM are equal. Further, as QL is to 12 digits, so is SF or QM to the digits of the defect, or if you prefer, you shall say that as LQ is to 12 digits, so is the shining part LM to the digits not covered, which, subtracted from 12, leave the digits of the eclipse. Examples of the computation are in Prob. 32 below.

Further, those things presented in Problem 10 are a little more uncertain to apply if one should examine them using this [procedure of] 11 and 12. For it is
357 with difficulty that one simultaneously attends to the two points of intersection, that of the rule and that of the shadow, with the marked diameter. And the little circles of the points on the horns are very diluted, since they are spread out from a point, and in estimating the intersections we err in the small things. As on 30 June or 10 July 1600, when 3 digits were eclipsed on the ray, the transverse rule cut off $1\frac{2}{3}$ digits. Therefore HF is $1\frac{1}{3}$ where FD is 9, and in the dimensions of the opening, where the whole radius had $105\frac{1}{2}$, but 89 with the shell removed (to allow this quantity for now to be true, which, as was noted in Probl. 2, is a little less than the truth), three digits are $26\frac{3}{8}$ of these units. And HF or PM are $11\frac{13}{18}$. But FD, 9 digits, is $79\frac{1}{8}$ units. Therefore, the whole HD is about $90\frac{61}{72}$ units, and with HP, LD, $16\frac{1}{2}$ subtracted, PL will be $74\frac{25}{72}$, and PQ $14\frac{47}{72}$. These multiplied by each other make $1089\frac{1}{4}$, NP squared. But PM was $11\frac{13}{18}$. Therefore, division of NP squared by PM gives the remaining part PR, about 93, to which the segment PM being added makes the moon's whole diameter $104\frac{2}{3}$, where the sun's is 89. Therefore, the moon's diameter would have been $34\frac{3}{4}$ minutes, even though it was a little below its mean distance. Consequently, so that you may see the fallacy, let the moon's diameter have been $32'$, and 3 digits obscured on the ray: it is asked, what ought the transversal to have cut off, that we reckoned the digits to be $1\frac{2}{3}$? If, then, $29\frac{1}{2}$ minutes give 89 units, 32 minutes will give a little more than $96\frac{1}{2}$ small units of the moon. And because 3 digits on the ray are $26\frac{3}{8}$, it is also deficient by so many small units from the ray (89). So, since as PL is to PN so is PN to PQ, and at the same time, as PR is to PN so is PN to PM, therefore, as PL is to PR, so is PM to PQ. And, *separando*, as ML is to QR, so is PM to PQ, and *componendo*, as ML, QR together are to ML, so are PM, PQ together, i.e., MQ, to PM. Therefore, since QL is 89 and QM, that is, SF, is $26\frac{3}{8}$, ML will be $62\frac{5}{8}$. And since RM is more than $96\frac{1}{2}$, while QM is $26\frac{3}{8}$, RQ will be $70\frac{1}{8}$. The sum is $132\frac{3}{4}$. Therefore, as $132\frac{3}{4}$ is to $62\frac{5}{8}$, so is $26\frac{3}{8}$ to PM, $12\frac{4}{9}$, approximately. But ML is $62\frac{5}{8}$, and PH, LD are two semidiameters of the opening, that is, $16\frac{1}{2}$. The sum, HD, is about $91\frac{5}{8}$.

Since, therefore, $105\frac{1}{2}$ makes 12 digits, the remainder, $13\frac{7}{8}$, makes $1\frac{3}{5}$, approximately; but how does this differ from $1\frac{2}{3}$ in reckoning by eye?

358 So, since here too the senses have left me nearly bereft, I thought of a third procedure for measuring the moon's diameter in eclipses of the sun, which now follows.

Problem 13

In an eclipse of the sun, to estimate the moon's diameter easily and surely using prefabricated small moons

The procedure is clearly the most reliable of all that can be thought up, which even in itself is a reason for my bringing this publication to completion: I want mathematicians, in whose care astronomy is, to observe the eclipse that will occur on 2/12 October 1605, near perigee, in this way.[42] For in 1601, right at apogee, a most beautiful calm shone upon me, and it can happen that some people, no matter how much their desire, are hindered by clouds or by health. It is therefore expedient that everyone be ready. For by comparison of the diameter at apogee and perigee, it will be possible to conclude something certain about the true and geometrical eccentricity of the moon at conjunction, which occupies a fundamental place in the discussion of physical causes.

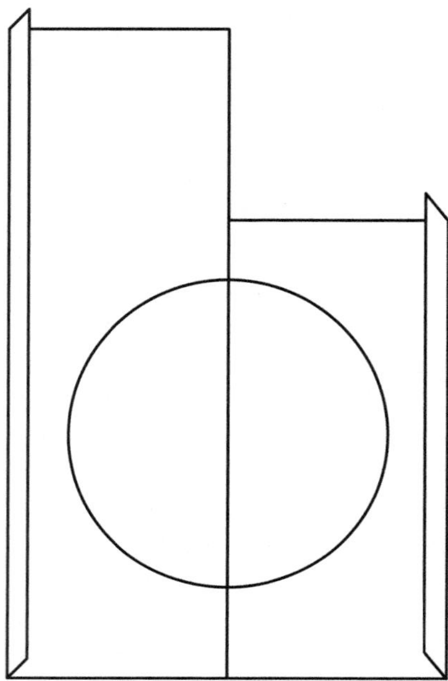

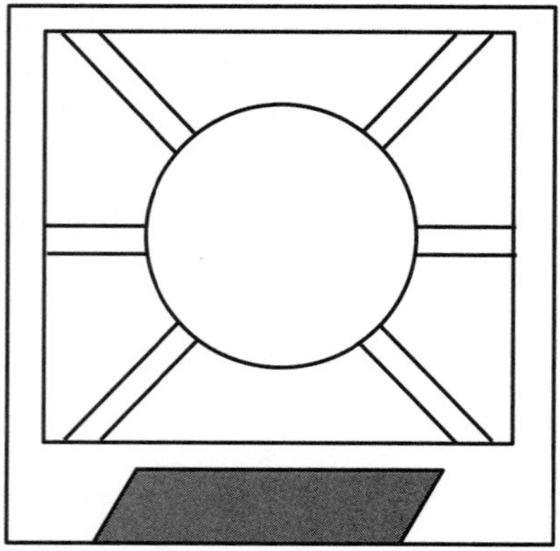

To the point. And so let our paper volvelle be taken in place of the pieces of wood and the somewhat solid pointer, and in place of the transverse rule, two folded back handles, likewise parallel to the marked diameter, but outside the circumference of the circle, which measures the ray on that day. Between these handles, in place of the transverse, let there be fastened small squared off sheets, in the middle of which are little circles, as great as we suspect the diameter of the moon will be, diminished all around by the radius of our opening. Also, let the volvelle nonetheless bear its own pointer from its center, made of paper, in place of the previous wooden one. For I shall show the use of this later. But let the small sheets be enclosed in a grid, with parts of the paper cut back from the circumference of the small moons, except at the sides, where little arms are to be left for keeping the small moons in the center of the squared off sheet. It will thus happen that you may nonetheless see the diameter marked on the wheel, unhindered by the squared-off sheet. With these things thus prepared, when the eclipse of the sun is beginning and the rule of the instrument along with the panels pointed at the sun, let the volvelle be turned around its axle, while the sheet is moved up and down between the handles of the

[42] To this end, Kepler published a small book, *Ad rerum coelestium amatores universos, . . . de solis deliquio, quod hoc anno 1605 mense Octobri contigit, epistola* (Prague 1605), in *JKGW* IV pp. 37–53. The contents of this book are closely related to the present chapter.

359 volvelle in a straight line, until the top edge of the small moon sits upon the deepest apex of the shadow. Thus, in a rather large eclipse, if your small moon is larger than it should be, it will cover the horns of the ray from the sides, before it can touch the inner circumference of the eclipsed ray from the front. If, on the other hand, it is less than it should be, when it has been applied to the edge of the moon's shadow from the front, it does not reach at the sides to the circumference which it touches in the middle. And so, change the small moons, until some one of them fully matches the interior circumference of the ray. That one will give you the moon's apparent diameter.

On 14/24 December 1601, at Prague in Bohemia, by my usual instrument, without changing the opening at all, I studied an eclipse of the sun in the way just described. The quantities of the circles were exactly as those you see here.[43] For I had taken for the small moon a radius of 37 small units, which with the $8\frac{1}{4}$ of the opening make $45\frac{1}{4}$, that is, so as to make 15 minutes, or 30 minutes for the whole diameter. For the sun also at apogee has 30′ in its diameter, but the moon is said by Ptolemy and Hipparchus to be equal to the sun, and Tycho himself assigns a minimum diameter of 30′ to the moon at oppositions. However, those things he had said about the diminution of this diameter in eclipses of the sun, I have held
360 suspect for many reasons. And so, with the small moon set at that size, to the Brahean observers, when the eclipse had now grown to its middle, it appeared most evidently to all of them that my small moon not only was not greater than the shadow, but still fell short of the sides of the shadow, if it was tangent to it at the front. And so the apparent diameter of the moon is greater than 30′. It would also be much greater if it were to appear at this place on the epicycle in Cancer and at mid-heaven, because of the greater nearness of the observer, for which cause it can appear almost half a minute larger at the zenith than at the horizon.

For the rest, I only had a single small moon, and so I was unable to proceed exactly in the way prescribed. I did, however, manage to pull the moon somewhat away from contact with the interior circle of the ray, so that the circumferences were at a uniform distance. And thus there were then estimated to be about 2 small units between the circumferences of the small moon and the ray. And so the radius of the moon would be $47\frac{1}{4}$, in minutes, $15\frac{2}{3}'$. Therefore, the apparent diameter was $31\frac{1}{3}'$. And this was at apogee. But to avoid my attributing too much to the sense of vision in determining the edges of the shadow, let there have been just a single small unit in between, and let the diameter at apogee be $30\frac{1}{2}'$, as great as is the sun's mean diameter, so that thus, following the teaching of Sosigenes, the moon at apogee cannot cover the whole sun at perigee.[44] And so let there be this most certain axiom: the diameter of the moon at its most remote in the eclipse of 1601 did not appear to be greater than $30\frac{1}{2}$ minutes. Now supply the rest, O seers of matters celestial, and seize, all of you, the occasion which the year 1605 will offer, with the moon at perigee.

---

43 The adjacent diagrams are copied from the woodcuts in the 1604 edition, and are the same size as those woodcuts (as are the figures in *JKGW* II p. 306).

44 In chapter 8 section 3, above Kepler, arguing against Sosigenes, denied the possibility of an annular eclipse of the sun.

Problem 14

To extract the inclinations of solar eclipses: the method of Maestlin, and a caution

Of the farm ewe, it is proverbial that there is no part that isn't useful. The same thing is true in general about the phases of eclipses, especially of the inclinations, which Ptolemy, in the final chapters of Book VI, calls προσνεύσεις, as if they were librations. If these are observed with greatest certainty, they provide us evidence of the most important things in the moon's motion, and serve as a short cut, as will be said in the other part on use. But also in this category there exist phenomena worthy of admiration 361
and display, not adequately explained by our predecessors, and which had kept me tied in knots for a long time.

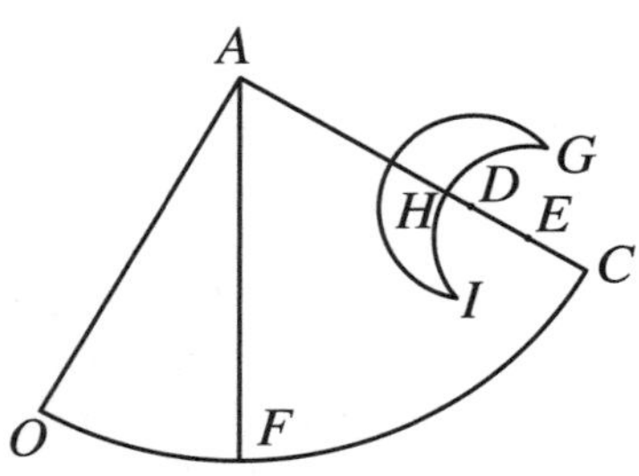

Now my teacher Maestlin uses a small quadrant in this way. Let the eclipsed ray fall perpendicularly on the paper in a chamber, with an assistant supporting the paper. Next, let the side of a small quadrant be set perpendicularly in the way of the ray, so that it cuts the horn of the ray through the middle, for which the estimation of the eyes is to be believed. From that side, thus directed, let the quadrant be hung, and a plumb line on the quadrant; and the plumb line on the edge will show the degrees or angle by which the vertical circle (represented by the cord) cuts the circle through the center of the luminaries, represented by the side bisecting the horns. But immediately, and practically at the same moment, the sun's altitude must be taken, without which the inclination is almost useless. Let GHI be the eclipsed ray, let the side AC of the quadrant AOC bisect this through its shadow, and let AF be the plumb line: FAC or FC will be approximately the angle sought. For it is beyond controversy that AF is on the vertical. But AC passes through the centers D, E. For it is posited that it bisects the horns GI, that is, the straight line GI described in the two circles, since it bisects both arcs. It is therefore necessary that AC pass through the centers. For the rest, in the example of the eclipse of 21 July 1590, in the observation of which eclipse Maestlin used this method, the sun was low, and the angle of inclination was either fully a right angle or almost none. And so that method was certainly of value then, and was satisfactory for the accuracy of sense perception, but if the sun had been higher and the inclinations in between, there is no doubt that I would have been advised by Maestlin to be cautious. For although AF represents the vertical, AC the circle through the centers, FAC nonetheless does not measure the angle of those circles, but is smaller. This is readily evident, if the point A be right at the intersection of these circles on the sphere, and from the 362
same point there go out two lines, one tangent at the same point to the vertical and the other tangent to the circle through the centers. For the former will enclose an angle the same as that of the circles, while the latter will enclose one that is clearly smaller. The demonstration and diagram and the computation of the necessary precaution you will find below in Prob. 29.

Problem 15

To extract the inclinations using the eclipse instrument

The method is easy, and not troubled by the trembling of the free hands as the former one is, and is demonstrative. For with the rule pointed at the sun's center, the wheel is turned with the pointer, until the transverse rule either touches the horns, or is parallel to the tangent, and the pointer will show on the outside circle the degrees of inclination with the vertical, and also immediately given is the sun's altitude by the intersection of the rule with the crosspiece, by the same distinct mark made on the rule and the crosspiece, so that you may measure after the observation, if the crosspiece and rule are not divided beforehand according to the sun's degree of altitude. If you like to use the small moons instead of the transverse, as in Problem 13, the volvelle is again turned until the eclipsed ray embraces the small moon on all sides equally with its horns, which both the divided diameter and the small arms holding the small moon, which can take the place of the transversal, easily indicate.

The demonstration is in this, that the two slots, both of the crosspiece and the post, locate the rule in the plane of the vertical circle, and the line of length on the panel by the same means also follows the rule, from which line there begins the division of the outer circle on the panel.

If the whole instrument is not always at hand, as, if there be nothing present but the rule with the panels, you shall point the rule in the plane of the vertical thus: you attach to the rule at its base a transversal two feet in length, or longer, and at right angles to the rule, and afterwards you allow the rule with the transversal to lie out on a floor that is even and parallel to the horizon. For the transversal will set up the rule on its back, so that it fits in the vertical plane, exactly as it was
363 kept before by the slots in the crosspiece and in the post.

Problem 16

To mark out the inclinations even on the floor

It was said in Chapter 2 that a ray of this kind in a *camera obscura* is a right cone, if the opening be circular and perpendicular, whose vertex is above the opening, and the base on the surface illuminated. If this surface be directly presented to the opening, the base of the cone, or portion illuminated, will be a circle, by 8 of Chapter 2. But if the surface be obliquely presented to the ray, the base of the cone, or the portion of the oblique surface illuminated, will be a conic section, and in fact an ellipse, if the surface intercepts the whole ray, by Apollonius I 13. Since in our latitude[45] the plane of the horizon or the floors always receive the oblique ray of the sun, the figure of the ray will therefore always be elliptical, and this even in nearly all shapes of openings, provided only that there be a full intersection. For by 6 of Chapter 2, the opening then communicates the least of its shape to the cone, the sun the most. Therefore, since the sun is circular in form, it nevertheless will bring about a cone that is

[45] The Latin word, *climatis*, used here, refers to the division of the globe into bands of latitude each of which is called a "climate."

approximately right, even if the shape of the opening be something very different from circular.

Therefore, if you are without instrument, as happened to me on 25 February or 7 March 1598, mark out ellipses upon a paper laid out on the floor, in the extent to which the sun may shape them; and since they leave their place quickly, let three or four points be made, at the top apex A and the bottom apex B of the ellipse, as well as at the points of the two horns. Now the length of the ellipse AB extends itself into the vertical circle. Consequently, by the mediation of the knowledge of the ellipse, the inclination of the eclipse CD to the vertical AB will be given in position. This is most of all useful at the beginning and end of the eclipse, where the points C, D become one.

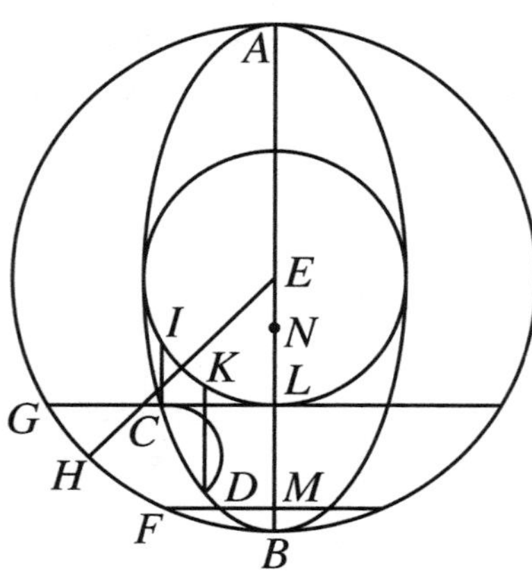

Likewise, for determining whether the shadow stands exactly at A or at B, or at the positions at right angles. Outside these cases, it is a little more risky. **364**

Problem 17

From the marking out of the ellipse, to learn the inclination of the eclipse

First, mechanically, as follows. In the preceding diagram, draw the longer diameter AB, and, it being bisected at E, about center E with radius EA or EB let a circle be drawn, while from the horns C and D let perpendiculars be dropped to AB, extended to that part of the circumference of the circle, and let them be CG, DF. And let the circumference GF be bisected at H, and let EH be drawn. Therefore, HEB is the angle of inclination, to be learned from the arc HB, if AHB be divided into 180. For since that cone is extremely acute, its angle being no more than half a degree, the middle of the ellipse is distant from the axis of the cone by a completely imperceptible amount, and for that reason, no matter from what point of the ellipse's circumference the perpendicular to the longer diameter is dropped, it cuts off the versed sine of that arc on the circle whose diameter is BA, beginning from the vertex.

If you have also marked a pair of points for the shorter diameter, or if you have used the sun's altitude to derive the ratio of the shorter diameter to the longer, as a test you shall describe about E a circle with the size of the shorter diameter, and from C, D, you shall draw lines parallel to AB, so that they may cut the small circle at I, K: if the middle of this arc falls on the line EH, the marking out is certain. For again, the shorter diameter of the ellipse is imperceptibly longer than the line perpendicular to the axis at the point of the axis, which is in the plane or the base; consequently, no matter from what point of the elliptical circumference the perpendicular is drawn to the shorter diameter, it cuts off the sine, between itself and the center, of that arc which, on the circle having the size of the shorter diameter, lies between the vertical point (or the one on the line AB) and the given point.

Indeed, so that those who are experts in the elements of conics may find nothing lacking in me, I also add a *raffinement*[46] from Apollonius. Let N be the point at which the axis of the cone falls. Therefore, NA and NB will be obtained in this way. Increase and decrease the sun's distance from the zenith
365 by 15 minutes, or the amount of the sun's radius, and take the tangents of all three arcs. The differences of the tangents will show the required portions in proportion, where the whole sine is 100,000.[47] The shorter radius, however, going out from the point N (applied ordinatewise, in Apollonius) is obtained thus. As the whole sine is to the secant of the sun's distance from the zenith, so is the tangent of fifteen minutes to the quantity of the shorter radius, in the measure in which AN, NB is given from the tangents. This, then, applied ordinatewise, is the more exact size of the smaller circle. And because there is a danger that you might not have marked the points A and B exactly in the middle of the circular vertices, especially as, owing to the intervening delay of time, the vertices of the ray depart from A and B before you have punched the points C and D, thus vitiating the quantities CL and DM: therefore, you will note, by *Conics* I 21, that as the longer diameter AB (Apollonius calls it the axis of the figure) is to the shorter (in general, as the transverse side of the figure is to the upright side [*latus rectum*]), so is the rectangle AL, LB to the square on LC, and the rectangle AM, MB to the square on MD. And in this way, CL and DM are restored, if they have an error in the marking: provided only that AB be correct, and the place of the points C, D between A, B be correct with respect to longitude, where there is less danger in the marking. There is an example below in Prob. 23.

Problem 18

To take lunar inclinations with the instrument

On the perpendicular of the lunar rays, set up a small circle such as is on the outer panel, of solid parts, and cut back on all sides, divided from the top by any procedure into 360 parts, so that the beginning of the division is exactly at the highest place, at the moment at which you wish to take note of the inclination of a lunar eclipse. Let a rule rotate on the circle, longer than the circle's diameter, and on the rule let there be a transversal, perpendicular to the rule. And so, with the small circle positioned as you were instructed, transfer the eye to the rule, and turn it and move they eye until the transversal subtends the moon's horns equally. The rule will show on the circle the degrees of inclination required. The
366 craftsmen who make instruments of metal are used to attaching a quadrant, fitted to the plane of the vertical circle, to a tripod, and to set the quadrant into a circle,

---

46 ἀκρίβειαν.

47 Let the center of the opening be O, and from O drop a perpendicular OP to the plane of the floor, intersecting AB extended at P. The angles AOP, NOP, and BOP, are the distances from the zenith, and the lines AP, NP, and BP, are proportional to the tangents of those angles.

with an open trough, but to attach the circle to the same tripod from two opposite points on a line parallel to the horizon. This instrument is extremely useful.

It should be used in this way in an eclipse of the moon, where our ecliptic instrument does not clearly enough present the ray of the eclipsed moon in the darkness, and the ray itself does not allow the presence of a lighted taper for getting the numbers of the degrees of inclination on the panel, or if, in the often repeated switching of the tapers with the darkness, leads to an impediment for the sense of vision in perceiving the ray. In this way I observed the eclipse of the moon which happened on 14/24 May 1603, the series of observations of which I will append to Problem 21.

Problem 19

To take note of certain inclinations of both kinds of eclipses as is ordinarily done

Look carefully at those moments when the shadow is either exactly at the zenith or exactly at the lowest part of the luminary, or on either side; that is, when the horns are turned exactly supine upwards, or prone downwards, or stand erect on the perpendicular. For the estimation does not easily err here. Take note of these moments carefully in those methods that follow. Many examples of this precept are supplied from the observations of Tycho, and also from those which, observed by me, I shall now directly add.

To take note of the inclinations of a lunar eclipse without an instrument in another way

Almost as before, note the times when the moon's horns coincide in the same straight line with one of the fixed stars or of the planets, or are turned perpendicular to it. But then in the extraction you should remember the moon's parallax, especially if it was compared to a nearby star. For the line was going to be applied differently to the horns if it had been viewed from the center of the earth. If you withhold this one thing, you have an easy summary of the rest of the method here. For up to now the inclinations of the circle through the centers were noted with respect to the vertical, not because of the vertical itself, but so **367**
that through the vertical the inclinations of this circle through the centers might be known with respect to the ecliptic itself. Here, however, the inclinations are referred directly to the sphere of the fixed stars itself. The foundation of these problems, as regards the horns falling on a straight line with a fixed star, is this, that the line joining the ends of the horns, that is, the intersections of the pair of circles, is perpendicular to the line passing through the centers. Examples are in Prob. 21 and in a few places in the preceding problems.

Problem 20

To take note of the phases as is usually done

Regiomontanus once taught, from Albategnius,[48] how to take note of the

[48] The Arab astronomer Al-Battānī. See Chapter 4, note 172.

sun's altitude. The altitude of some bright fixed star is taken, most correctly, at night. If the sky be cloudy, and the fixed stars are hidden, *the altitude of the moon itself* is noted. The notations of the *azimuths* of all of these is also helpful. But how the time may be obtained from these, is taught thoroughly in the theory of the *primum mobile*. I hear that the illustrious gentleman Melchior Iostelius has in hand an excellent work, of forty problems of the *primum mobile* through the bare addition and subtraction of arcs and chords, a kind of computation familiar to Tycho for many years, and developed to a considerable extent by Clavius, now finally perfected by Iostelius.[49] I therefore refer the reader to it. For the present problems are observational, while those which will follow in the other part will be used for the second movables, specifically, the moon's motion, not the *primum mobile*. Tycho Brahe teaches how to use *clocks* more efficiently. First, a test of the clock is made over a number of days preceding. Next, once the beginning of the course has been made, from any mark, there are noted on it the indicators, not just of the phases, but also of the conjunction of the sun or moon or of the fixed stars with the meridian, or with another circle that is the same and certain. Thus the phases, with the moments of the day through determined time intervals indicated by the clock are connected. There are examples in Prob. 21 and the following.

I shall, moreover, state a number of ways by examples. On 13/23 April 1595, at Graz in Styria, when the beginning of an eclipse was noticed, the altitude of Arcturus was 44° approximately, with a latitude of $\frac{3}{4}$ by a paper astrolabe
368 and a pendulum, while the moon's altitude was $15\frac{1}{2}°$. When now no more light appeared through the thick air, the moon evidently being deeply immersed in the shadow, the altitude of Arcturus was 34°, of the moon, 6°, approximately. Therefore, the beginning would have been seen by me at $14^h\ 59^m$, from the altitude of Arcturus. From the observations of Tycho at Hven, the beginning was concluded to be at $14^h\ 51^m$. And so my position was 2 degrees farther to the east. But it is certain on other grounds that the difference is a little more. And so my observation was not perfectly certain.

Difference of the meridians of Graz and Uraniburg.

On 11/21 February 1598, at the time when the town bell was striking three, Spica was $56\frac{2}{3}°$ from the zenith, with a wooden quadrant of $\frac{3}{4}$ of a foot. Then, at the town time of $4\frac{1}{2}^h$, I was not yet able to notice the defect, but it could

[49] Although this work by the Wittenberg astronomer Melchior Jöstel was completed in manuscript, it was never published. Kepler learned about it from Jöstel's colleague Ambrosius Rhodius, with whom Kepler carried on a correspondence from 1601–1603. See Franz Hammer's note in *JKGW* II p. 456.

The technique to which Kepler refers was, in modern terms, the trigonometric identity

$$\sin a \cdot \sin b = \frac{1}{2}\cos(a-b) - \frac{1}{2}\cos(a+b).$$

Since the cosine had not been named as a separate function, there was no such simple way to state this relation then. It appears to have been an invention of Paul Wittich of Wroclaw; cf. Frisch's note in *JKOO* II pp. 438–9, and O. Gingerich and Robert Westman, *The Wittich Connection*, esp. p. 12 and the articles referred to on that page.

have happened that it had begun then, because of the clouds (this is what my note says). For less than two quarters later, the moon, not yet half obscured, stood fully erect, and a little after the fifth hour, half the diameter was eclipsed, the moon now being inclined. A little after the sixth hour, it withdrew behind the clouds with an extremely weak light, so much so that it seemed to diminish further, and about to lose all its light. The sun was at 2° 26′ Pisces. Therefore, this altitude of Spica would add 24 minutes to the town time. The beginning would fall a little before 5, the last phase at about 20 minutes past the sixth hour. And this was about the middle. For it was observed at Wandesburg not to be deeply eclipsed, of which you shall judge from Ch. 5 and 7. Further, the Tychonic observers refer the middle of this eclipse (with the help of computation of the observation of the beginning, for the end was beneath the earth) to $6^h\ 5^m$ at Hven. Therefore, the difference of meridians is about 15 minutes, or 4 degrees, probably.

Again, the difference of the meridians of Graz and Uraniburg.

On 25 May or 4 June 1602, at Prague in Bohemia, I observed the end of a lunar eclipse in this way. Tycho's clock was at hand, indicating the minutes and seconds. Once the beginning of the course was taken, by whatever means, I accommodated the clock to the setting sun's altitude of $4\frac{2}{3}$ degrees.

Tychonic clock	*Town clock*	*Altitude of the sun, going behind the mountain*	*Resulting true time*	
6.43		4° 40′	7.24	
6.52		Halo still seen.	7.33	
7.00		Trace of halo.	7.41	
		Clouds showed the color of the setting sun.		
7.01	24.00		7.42	
7.17	0.15		7.58	Sunset, by computation, as well as from the proportional extrapolation of the altitude previously noted. The moon rose in dense clouds.
8.02			8.43	Trace of the moon's light first seen.
8.13			8.54	About $\frac{1}{6}$ of the moon's round circumference was missing; the shadow was directed below Jupiter.
8.19			9.00	All had not yet come to an end.
8.21			9.02	The circumference was seen, but was pallid.

369

It came to a complete end one minute later, that is, $9^h\ 3^m$. The computation of Tycho showed the end as $9^h\ 19^m$, and that by the hourly motion, the equation of time, the radii of the moon and the shadow, and the difference of meridians of Prague and Hven established by Tycho, which can change, retaining the hypothesis the represents the actual moments of oppositions.

Problem 21

To compare the moon's visible place in longitude and latitude to the fixed stars, in the principal phases.

370 The instruction is chiefly about those things that are usefully observed in eclipses of the moon, so that they not be accidentally omitted: it has much in common with 19 above. Now this can be done in various ways. It is best if the eclipse is at the beginning of Cancer or Capricorn, and its middle coincides with the actual approach of the moon to the meridian; or in any sign, when at the middle the moon stands at the nonagesimal degree, and is not far removed in longitude from some fixed star. But if it was outside of those moments, provided that the middle of the eclipse is attended to, the remaining advantages will be able to be provided by consideration of the lunar parallaxes.

Direct a plane, perpendicular to the horizon, into the meridian, or the nonagesimal, or right at the moon's region at the middle of the eclipse. Then carefully note on the clock's indicator of minutes and seconds for the moments at which the moon's western edge, eastern edge, and the nearby fixed star approaches the meridian. For from the interval of times the difference of the intermediate spaces of the heavens is obtained. If it be the case that right at the middle of the eclipse the moon stood at the nonagesimal, the business will not require any further meandering,[50] and the right ascension of the centers of both the moon and the shadow is obtained, numbered from the nearby fixed star, consequently the distance of the fixed star from the sun's center as well. Nevertheless, other supports, lacking these conveniences, should not be spurned.

Tycho took the distances of the unbroken edge from nearby fixed stars frequently over the whole duration, before and after the middle, with sextants and armillaries.

It will also be helpful to take note of the track of lines, as Maestlin used to do in observations of the other celestial bodies.

It is plausible that Hipparchus used one of these methods. For, as Ptolemy relates in Book 3 Ch. 1, when he had ascertained that the spring equinox had occurred, on the morning of the 27th day of the month of Mechir in the year 32 of the third Calippic period, afterwards, in the observation of a certain lunar eclipse that coincided with the equinox, he related that Spica Virginis was at 24° 45′ Virgo.[51] (In this eclipse, the sun's position from the equator, and therefore on the zodiac, was given from the time, and from the sun's position, the opposite position of the shadow; from the shadow's position at the beginning and end of the eclipse, the moon's position).

The same observer first ascertained that the spring equinox had occurred on
371 the 29th of the month of Mechir, after midnight (on the 30th, following), in the

---

[50] When the moon is at the nonagesimal (the point on the ecliptic 90° from the eastern horizon) its parallax is at a minimum, and affects the latitude only.

[51] The date of this observation and the following one correspond to 24 March −145 and 23/24 March −134, respectively. See Ptolemy, *Almagest*, trs. Toomer pp. 138–9.

year 43 of the third Calippic period. And again, a coinciding eclipse of the moon, when the moon's position was compared to Spica, seemed to him to bring Spica back to 24° 45′ Virgo.

Regiomontanus and Peurbach recorded an eclipse of the moon at Vienna in these words. On 27 December 1460, "at the beginning of the eclipse, the star that they call Alramech[52] had a pre-meridan altitude of 7 degrees. At the beginning of the time interval, 17°; at the end of the interval, 28°. At the beginning of the eclipse, the moon was visually on the great circle passing through the head of the preceding of the Gemini and the brighter star of Canis Minor; at the end, on the other hand, it was above one circle passing through the head of the following of the Gemini and Canis Minor."[53]

From this, they conclude that the beginning was at $11^h\ 42^m$, the beginning of the time interval was $12^h\ 47^m$, and the end of the time interval $13^h\ 55^m$. To this record Tycho Brahe appended the note that if these things were exact, the positions of the stars could be verified from them.

On 29 November or 9 December 1601, we observed an eclipse of the moon here at Prague in this way. We used the Tychonic clock; the beginning of the course was arbitrary. We also used large quadrants for noting the arrival of stars at the meridian. The sun's position was 17° 48′ Aquarius.

Our clock	*True time*		
5.25$\frac{1}{2}$	5.00		
5.33$\frac{1}{4}$	5.21	Marcab in Pegasus was observed by the faithful Matthias Seiffard [a] at a certain azimuth, which he thought was the meridian. But once the error was noticed, it was immediately decided that the distance at that time from the meridian would be found out the next day. And it was found to be 17 minutes.	
5.35	5.23	From the following, the very beginning should be set back to here.	
5.37	5.25	The eclipse was now obvious. The shadow was about 15 degrees from below to the left, taken without an instrument	
5.50$\frac{1}{2}$	5.38	I observed the same Marcab and Scheat in Pegasus on the true meridian. [b] A silly confusion arising from making two stars out of one.	**372**
6.20	6.08	The line through the horns was parallel to the horizon. It was more than half eclipsed.	
6.29		6 by the town clock.	
6.56	6.46	Transit of the outermost star in the wing of Pegasus. [c] The circle through the centers was pointed towards the Horn of Taurus [d] and Capella.	
7.07	6.53$\frac{1}{2}$	The circle here fell upon the midpoint between the shoulders of Erichthonius, [e] towards the northern one. Therefore, it is about the middle. The line through the Horns of Taurus [f] cut the remainder of the moon such that $\frac{2}{3}$ of it would be left above to the south, $\frac{1}{3}$ below to the north, on the left.	
7.14		7 by the town clock.	

52 Arcturus: cf. Allen, *Star Names*, pp. 100–101.

53 Regiomontanus, *De torqueto* (Nürnberg 1544) 37^v–38^r. Although Kepler does not indicate it, this is a direct quotation, only slightly altered (except that he corrects the accusative *antecedentem* replacing it with the genitive *antecedentis*.).

8.08	7.54½	The crescent moon stood upright.
8.48	8.34½	End. The moon's shadow was about 70 degrees from the zenith, to the right.
8.51½	8.38	Transit of the bright star of Aries. [g]

a. Matthias Seiffard had been an assistant to Tycho Brahe before assisting Kepler. He also helped observe the lunar eclipse of 14/24 May 1603, described next. Cf. *JKGW* II p. 316
b. The right ascensions of the two stars differed by only six minutes of arc, so for timekeeping purposes they culminate together.
c. Gamma Pegasi, in Bayer's catalog.
d. Elnath or Beta Tauri.
e. Erichthonius is an alternative name for Auriga. The two shoulders are Alpha Aurigae (Capella) and Beta Aurigae.
f. Beta Tauri (Elnath) and Zeta Tauri. The former is to the north of the ecliptic, while the latter is to the south.
g. Alpha Arietis (Hamal).

The duration was $3^h\ 12^m$. In Tycho's lunar writings, there in fact exists a computed duration of $3^h\ 36^m$.[54] But an error in the computation slips in. For the minutes of the incidence, prescribed by rule and computed from the tables, come to more than $1^h\ 34^m\ 39''$.[55] Therefore, the whole duration was $3^h\ 9^m\ 18''$. A minimum difference from the observations. The middle falls at $6^h\ 59^m$. Tycho publishes $7^h\ 1^m$, again not impressively close. The eclipse nevertheless appeared somewhat smaller than that which is depicted in the lunar writings.

On 14/24 May 1603, the moon was seen with faint light from the south side, not far from bright fixed stars and Saturn. I append the observation of this, but I shall first give warning that we were occupied with the altitudes of Saturn and Jupiter, and did not note their arrival at the meridian with the degree of precision that would serve for the fixed stars. I used a clock with a pointer for the seconds,
373 again Tychonic.

Tychonic clock	*German town clock*	*Bohemian clock*	*True time*	
9.12½	10.00		9.53½	Arcturus on the meridian.
9.59				The moon, not yet eclipsed, stood in precisely the same vertical with the Heart of Scorpius, [a] from which it receded again to the west because of the first [i.e., diurnal] motion and the eastward parallax and the whole duration of time.
10.04		3.00		

[54] Tycho included the computation as an example in the *Progymnasmata*; see *TBOO* II pp. 141–143. The computed duration is on p. 143.

[55] In a letter to David Fabricius of 18 December 1604 (*JKGW* XV no. 308, p. 79), Kepler corrects this to $1^h\ 33^m\ 22''$, the consequent duration being $3^h\ 6^m\ 44''$, but adds, "by a correct clock, the minutes of time were $1^h\ 28^m$, the duration $2^h\ 56^m$." Similar remarks are found in a letter to Herwart von Hohenburg of 10 December 1604 (*JKGW* XV no. 302, pp. 68–9). The "rule" to which Kepler refers is the use of computed diurnal motions of the two bodies.

10.10			10.49	Jupiter on the meridian with altitude 24° 32′. And now the emission of the moon into my eyes, with glasses removed, was broken.
10.14	11.00			
10.18			10.59	To me it seemed the beginning. The lunar instrument gave an inclination of the shadow of 224$\frac{1}{2}$° below and to the east, with 270° standing at the bottom.
10.23			11.04	To others, too, something seemed to be eclipsed, the inclination being 228. The moon left the vertical of the Heart of Scorpius (but not with its entire body), as if it were retrogressing.
10.29$\frac{1}{2}$			11.10	Inclination 230$\frac{1}{2}$; the line from Saturn passed through the centers.
10.42			11.24	Head of the Serpent [b] on the meridian. Inclination 236$\frac{1}{2}$. 5 Digits were eclipsed. Now, the circle through the centers passed above Saturn.
10.48		3.45		
10.51			11.32	Inclination 242.
10.57				Half.
11.02$\frac{1}{2}$		4.00		Saturn on the meridian, altitude 21° 47$\frac{1}{2}$′.
11.09			11.49	Half the circumference was gone.
11.13	12.00		11.55	Inclination 244$\frac{1}{2}$. **374**
11.19			12.01	The moon's center on the meridian, the middle of its remaining bright part was 19° 7′ in altitude.
11.24			12.06	Inclination 255.
11.31			12.13	The horns were pointed at the Heart of Scorpius.
11.43			12.25	The moon was standing supine, the two horns directed equally downward.
12.02			12.44	The horns were in a straight line with Jupiter, which was now raised up somewhat more towards the west.
12.03		5.00		Putative middle.
12.23			1.05	Inclination 288$\frac{1}{2}$
12.47			1.29	Inclination 300. About $\frac{1}{3}$ of the diameter was gone.
1.03		6.00	1.45	Inclination 302$\frac{1}{2}$.
1.10			1.52	Still something missing.
1.14			1.56	Inclination 302 or 300. Some people thought it was over. It was considered by me to be not yet round.
1.26			2.08	[Broken] emission for me, as at the beginning at $10^{h}\ 10^{m}$.

a. Antares.

b. There is no *Caput Serpentis* in Brahe's star catalog; however, in his short list of the hundred brightest stars, listed by right ascension and declination, the one star from Serpens is *Lucida Colli Serpentis* (the "bright star of the neck"), Alpha Serpentis in Bayer's catalog. This is probably the star meant here. See *TBOO* III p. 376.

In the meantime, Saturn was also observed from the fixed stars, so that there might once be certainty about its place, because inclinations were described using it. Between Saturn and the Northern Pan of Libra[56] were 17° 22′. Between this and the Knee of Ophiuchus,[57] 13° 20′. From this and the meridian altitude, its place comes out to be 0° $2\frac{1}{2}'$ Aquarius, with latitude 2° $10\frac{1}{2}'$ N.[58]

The duration, from Tycho's computation, comes out to be $3^h\ 2^m$; it was seen by others to be $2^h\ 52^m$, by me, suffering from night blindness, $2^h\ 59^m$. And with the wavering that preceded and followed the bare moments, $3^h\ 19^m$.

The agreement of the three clocks is evidence of their constancy in numbering time intervals.

The middle was at $12^h\ 30^m$. Tycho's computation shows it to be at $12^h\ 27^m$ on the meridian of Hven, $12^h\ 32^m$ at Prague.

Since, therefore, it was half an hour past the hour of 12 at the middle, the
375 moon's center preceding the middle of the shadow by 15 minutes, which amounts to 1 minute in transit, it came out beautifully that the center of the moon was noted on the meridian at $12^h\ 1^m$.

In Tycho Brahe's observations, the most fitting of all for this procedure, both in the opportunity of the meridian, and the care of observation, is the eclipse of the moon on 3 March 1588.[59] For in the whole time of duration, especially at the nonagesimal, the distance of the edges from Cor Leonis and Spica Virginis were taken very frequently. However, this is more appropriately referred to the other part, which contains demonstrations of the moon's returns from eclipses.[60]

Problem 22

To note the times expeditiously with the ecliptic instrument

Nothing new is demonstrated, except that I explain the use of the instrument, and how one observer may carry out all the things, whatever they are, that have so far come up to be observed.

So, let the observer stand at a table, moving the crosspiece around to the sun's position with his arm, while either lifting the rule up or pushing it down, and where he wishes to note some moment with all its circumstances, and the pointer on the volvelle is correctly oriented to the ray, he makes a mark, both on the azimuthal quadrant, and on the crosspiece, and on the rule, where they

---

[56] β Librae.

[57] ζ Ophiuchi.

[58] Franz Hammer corrects this from 0° $46\frac{1}{2}'$, following Kepler's corrections in letters to Fabricius and Herwart (see the note above). The incorrect number had apparently come from Magini's ephemerides. For more on the errors in this passage, see Frisch's note in *JKOO* II p. 439, note 96.

[59] *TBOO* III pp. 257–264.

[60] Evidently, Kepler had planned a second part of the *Optics*, which never appeared. The material which he assembled for this purpose formed the basis for another projected (and uncompleted) work, which he called *Hipparchus*. The drafts for this work have been published in *JKGW* XX.i, especially sections VII and IX.

mutually intersect, and to these three marks he places the same sign. Afterwards, with the instrument unmoved, he writes down the number or sign of the marks, and adds what the pointer showed, what the transversal, what the shadow on the divided diameter. After that is done, he again applies himself to looking for another moment. Thus everything is done in the same place by the same hands and without wobbling, which should be considered a gain where experienced assistants are wanting. For inexperienced ones disrupt instead of helping.

Next, on the azimuthal square, the tangents of the azimuths will be found, and on the rule, the secants of the distance of the luminary from the zenith, and on the crosspiece, the tangents of the same arc, by the application of a proper measurement of units; and, for a test, these three are used to elicit the same moment of time. An example will follow at the end of the chapter.[61]

Protheorem for the Following Problem

That the path of that ellipse by which the sun, let through a slit, illuminates the floor, is a conic section

For the sun's path in the sky is a circle, a great circle at the equinoxes, and **376**
nearly a circle (for the joining, one from another, breaks up the full circularity to some extent), least in size at the solstices of the tropics, in between at intermediate places. But now both our opening and the whole globe of the earth is, within the limits of the senses, on an axis perpendicular to all those circles of the natural days. Therefore, the lines connecting the opening with all the points of the circle which the sun is passing along on any day, that is, the rays of the sun coming in through the same hole over an entire day, all joined together from all locations of the sun, form a right cone, by the *Conics* I 4 and Definition 8 of Book I. But by the same first definition, the same rays of the sun also form another cone inside the chamber or window, similar to the exterior one, since the angles, being vertically opposite,[62] are equal. But the floor is placed parallel to the plane of the horizon. And the sun's rays at whatever place, coming in through the opening, are extended to the floor. Let there be imagined a multitude of continuous bodies of the sun over the whole diurnal circle: therefore, the rays from all of them will go in simultaneously, that is, a conic surface will go in through the opening. And that is cut by the plane of the floor. Therefore, the common section, that is, the path which the solar ellipse describes over the whole day, is a conic section. When, therefore, the sun does not set, but grazes the horizon of the opening, the plane of the horizon is parallel to the lowest line of that cone, with the result that the section here is a parabola, by the *Conics* I 11. But where the sun does not set at all, and all the rays all around fall on the floor for 24 hours, the section is an ellipse, by 13 of the same book. Except right at the pole, where this section is a circle, by 4 of the same book.

Where the sun sets, however, the plane of the horizon does not receive all the rays of the whole circle (that is, not those that are below the horizon), and it is parallel to some plane that, drawn through the vertex of the cone, cuts its base,

---

[61] See p. 424, below.

[62] κατὰ κορυφὴν. Cf. Euclid, *Elements*, I 15.

namely, the circle of the natural day. Therefore, by the *Conics* I 12,[63] this section is a hyperbola, and to avoid confusion, I repeat to you that the continuation of the ellipses on the floor for the whole day would create the figure of a hyperbola.

377 The same is true of the ends or nodes of sundial gnomons, which serve as the opening. And as in our chamber, the two light-cones came together at the vertices, so in the art of sundials, the light-cone is joined to the shadow-cone at the common mark of the gnomon. I do not know whether this is remarked by the writers on sundials.

And since at our latitude all lines of this kind are hyperbolas (except when the sun makes the equinox, for then the path of the ellipses is a straight line, the most obtuse of all hyperbolas), the relation of the ellipses of the rays to their hyperbola should be noted. It is known from the preceding that if a perpendicular be dropped from the center of the opening, the axes (or, in common language, the longer diameters) of all the ellipses direct themselves to the point on the floor at which the perpendicular from the small opening falls. For the perpendicular and the ray and the axis of the ellipse are in the same plane of the vertical circle. But to the same point there also tends the axis of the hyperbola, which the ellipses of that day create. Therefore, when the shape and the axis of a hyperbola are given, the point on it is given, from which all the axes of the ellipses are extended, cutting the hyperbola.

Problem 23

To elicit the time of the phases from the extension of the ellipses

With a piece of paper laid motionless on the floor, capture or note the hyperbolic path of either or both vertices of the ellipse, and at the same time the drawn axis of a number of ellipses. For at the meridian hour, the axis of the ellipse falls directly on the hyperbolic path, in morning and evening most obliquely. And so, a tangent to the hyperbola being drawn, measure the angle it makes with the elliptical axis. Next, either with a compass or by computation, establish the cone for that day, and its hyperbolic section, and the point lying perpendicularly beneath the vertex of the cone, which will be on the axis of the hyperbola; likewise, the point of the hyperbola at which the observed angle is set up; and finally, the point on the cone's base at which falls the straight line that is drawn from the vertex to the point found on the hyperbola. For the arc between this point and the highest,
378 or the one through which the vertical drawn through the center of the base passes, measures the interval of time between the meridian and the phase.

Moreover, the duration of time, and the minutes, even without a clock, you will measure in this way. If you enclose the right side edge of the ellipse just marked with a stylus, and hold it for as long as the luminous ellipse is traversing the stylus, and becomes tangent to the stylus on its left side; at that moment, you will remember to note the axis of the ellipse, and with the stylus switch over to the right side, and so on. From the number of contiguous ellipses you will have the time. For the body of the sun at apogee takes up exactly half a degree,

---

63 Kepler has I 13 here; however, the proposition constructing and defining the hyperbola is I 12.

which measures nearly two minutes. But the figure of the ellipse on the plane is a little wider (because of the width of the opening) than that figure which would be created by an opening with a magnitude of a point.

For example, on 25 February or 7 March 1598, when I lacked suitable instruments, and the cloudy sky made me almost lose hope of observing, I was nevertheless beneath a roof, and I had opened up a crack of uncertain shape and size, directing my efforts to whatever occasions might arise. When it was now noticeably eclipsed, the shadow was standing exactly at the right; it was one or two digits; the ray on the paper was of a silver groschen in magnitude; not such as I would have been able to perceive accurately; for the sun was again hidden in the blink of an eye. On the town clock, it struck one quarter after ten. After an hour, another sneak view. Horn face down, on the ray turning a little to the west, and very faint; in the sky, therefore, facing up, and the eclipse on the north. The end was clear. At a little before 12 by the town's time, there were 6 digits in the shadow of the ray. The shadow on the lower left, if you were looking at the ray, on the northern parts when your face is turned to the sun. At a quarter past twelve in the town, the digits were about 4, the inclination about 23°. When the digits were about $3\frac{3}{4}$, the inclination appeared to be about 20°. When the digits were $3\frac{2}{5}$, with yet another clock striking one quarter, the inclination was about 21°. A little after, $3\frac{1}{4}$ digits; inclination about 22°. At half before one o'clock on the town clock, $1\frac{2}{3}$ digits; inclination about 25°. A little after, about $1\frac{2}{5}$ digits; inclination 21°, which seemed very accurate. The cause of the discrepancy was this, that, not well remembering the precepts of Maestlin, I neglected to bring to the observation the **379**
circles described on paper, their diameters divided into twelfths; therefore, when I mark the two circumferences with points, the sun meanwhile was going off, and I was not judging whether the ray was striking the paper perpendicularly. And so also, the moon's diameter appeared most of the time less, but sometimes greater or equal to the solar diameter. In fact, there was also this deficiency, that the indicator of the vertical was nothing more than the length of the paper. And so I acknowledged these obstacles, and therefore, turning to impromptu plans, I began to note the end of the eclipse with ellipses, having no doubt that by the computation that I am now presenting, both inclination and time could be obtained from them. Therefore, from this moment up to the end of the eclipse I captured nine ellipses, six of these at the very least touching each other. And thus there were more than twelve minutes left until the end. In fact, when it ended, a little later it struck the third quarter past twelve. I had a wooden quadrant, not greater than two thirds of a foot. So with this, I took the sun's distance from the zenith, after it had struck the third quarter: 54° minus.

On noon of that day the sun's position from Tycho was 16° 49′ Pisces,
declination 5° 14′ south. The altitude of the equator at Graz in Styria, from **380**
concurring observations, is 42° 58′. Therefore, the sun's meridian altitude was 37° 44′. The distance from the zenith 52° 16′, and with parallax, 52° 18′. Let A be the opening, the plane surface BC, the perpendicular AB, the cone on that day DAC with angle DAC 169° 32′, the cone's axis AE pointed towards the pole of the world, the hyperbolic section FCG, C the vertex, BC the axis, when in fact the sun is on the meridian, so that ABC is the plane of the triangle DAC through the cone's axis AE. Let side DA be extended until it meets BC at H.

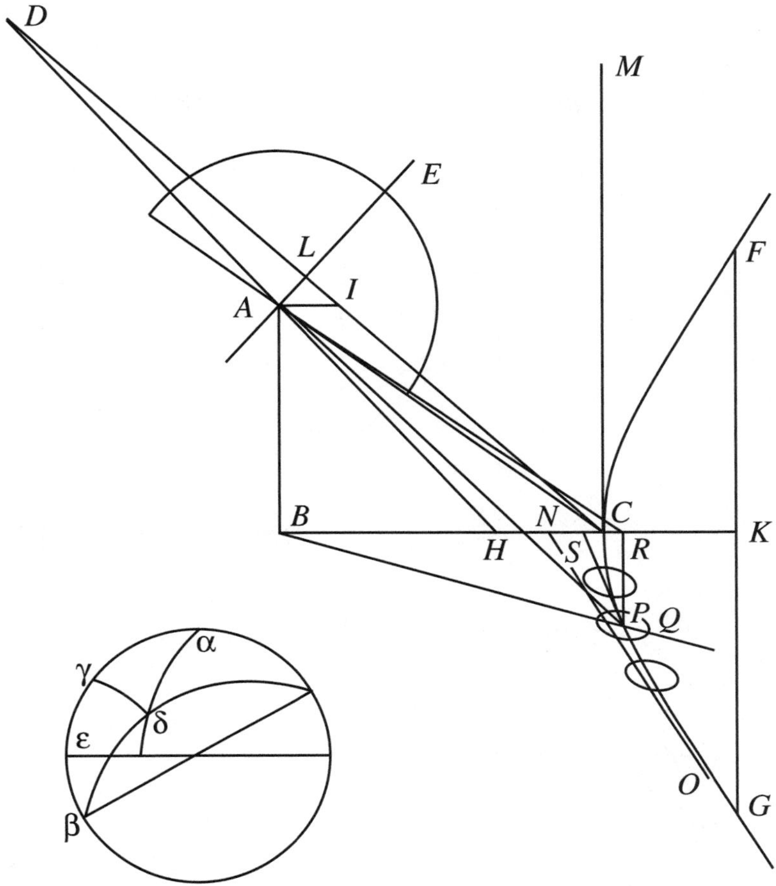

Therefore, HC is the transverse side of the figure, by *Conics* I 12. In numbers, where AB is 100,000, BC will be the tangent of the sun's distance from the zenith, 129,385; and, because the declination is 5° 14′, HAC will be 10° 28′. Therefore BAH is 41° 50′, whose tangent is 89,515. Therefore the remainder HC is 39,870. For the upright side [*latus rectum*] of the figure, AI must be drawn parallel to BC, and CM perpendicular to BC from C; and it must be done in such a way that as the square on AI is to the rectangle DI, IC, so is HC to CM.[64] In numbers: where AB is 100,000, AC is the secant of the angle 52° 18′, that is, 163,525. But because L is the center, the base DC, and the cone is right, therefore ALC is right, and ACL is 5° 14′. But AIL is the altitude of the equator, 42° 58′. Therefore, where AL is 100,000, AC is the secant of the complement of the declination, that is, of 84° 46′, 1,096,348, and CL the tangent of the same, 1,091,778. But AI is the secant of the altitude of the pole, 146,719, and IL is the tangent of the same, 107,362, with the result that DI is 1,199,140, and IC is 984,416. For that reason, the square on AI is to the

64 This is the way the length of the upright side CM is established in Apollonius I 12.

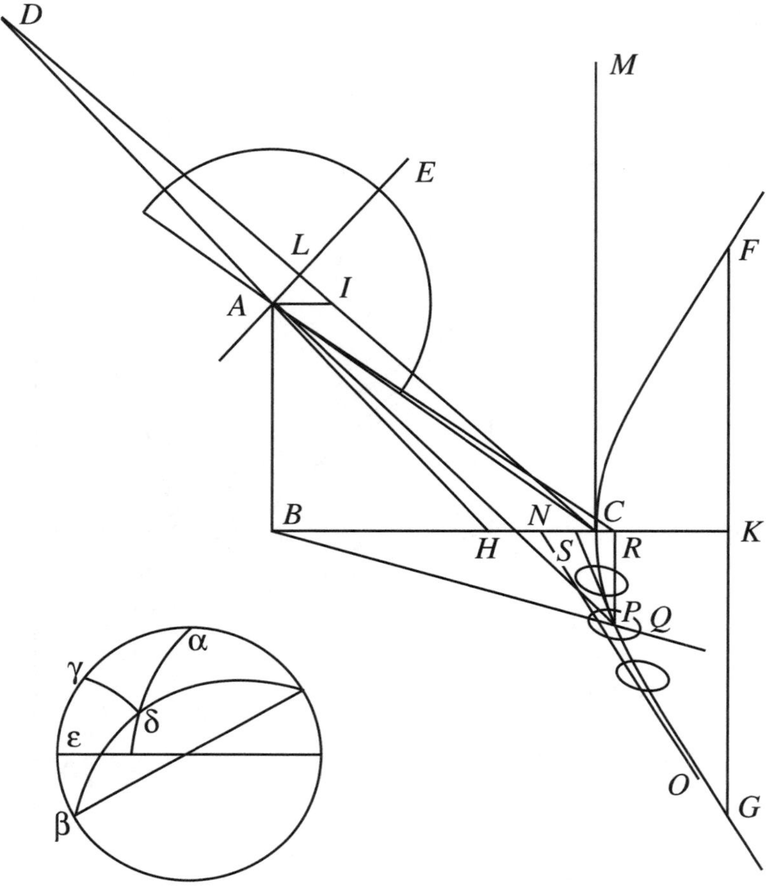

rectangle DI, IC approximately as 215 to 11,804. And HC should also be to CM in this ratio. Therefore, in the units of which HC was 39,870, CM will be 2,189,000, approximately. Therefore, the hyperbola is given since the sides of the figure are given. Therefore, by *Conics* II 1, the angle of the asymptotes is given. For when HC is divided in two at N, N will be the center, and NO the asymptote. Therefore, ONC is 82° 19′.[65] Thus the hyperbola of that day is obtained.

Now let PQ be the ellipse of illumination. I wish to know how much of an angle its axis PQ, that is, the line BQ, makes with the section or line that touches the section at P, at any hour of the day. If the sun were traversing a great circle, so that the path of the ellipses would be a straight line, such as CM, it is obvious that this very line would take the place of the tangent of the angle CAM as well as CBM, provided that in the former case CA, in the

[65] In the plane of the paper, draw CO from C perpendicular to HC, intersecting NO at O. By Apollonius II 1, $\frac{1}{4}HC, CM = CO^2$. But tan $ONC = \frac{CO}{CN}$, or, substituting, tan $ONC = \sqrt{CM/HC}$.

381 latter CB, would be the whole sine. But because the path itself is a curved line and is hyperbolic, while the sun's path is a smaller circle, therefore PR, applied ordinatewise, is the tangent of the angle PBR, with BR the radius, and of the angle PAR with AR the radius or the whole sine. But the arc of the great circle between the sun and the meridian, perpendicular to the meridian, is the measure of this angle PAR. Let it be proposed for us to continue the computation at these moments: $0^h\ 15^m$, $0^h\ 30^m$, $0^h\ 45^m$, $1^h\ 0^m$, after noon. Let the meridian βγ be set out, on it β the south pole. Now let the sun be at δ, a part of the great circle γδ. And because the time measures the angles at the pole, the angles γβδ will be 3° 45′, 7° 30′, 11° 15′, 15° 0′. But βγδ is right, and βδ is the complement of the sun's southern declination, that is, 84° 46′. From these knowns there is also sought the side βγ from the base and adjacent angle, which at the first moment becomes 84° $45\frac{1}{3}'$; at the second, 84° $43\frac{1}{3}'$; at the third, 84° 40′; at the fourth, 84° 35′. Let βε be the depth of the south pole, 47° 2′; the remainders εγ or BRA will be 37° $43\frac{1}{3}'$, 37° $41\frac{1}{3}'$, 37° 38′, 37° 33′. The complements of these, BAR, increased by parallax, are 52° $18\frac{2}{3}'$, 52° $20\frac{2}{3}'$, 52° 24′, 52° 29′. And so BR is 129,437; 129,592; 129,853; 130,244. But previously, BC was 129,385. Therefore, CR is 52; 207; 468; 859. So, since HC is 39,870, and its half, NC 19,935, NR will be 19,987; 20,142;
382 20,403; 20,974. Therefore, by Apollonius I 37, let the square of NC be divided by NR: the result is NS, the line between the center and the line touching the section at point P, at which the line extended out at right angles from R falls, namely, 19,883; 19,719; 19,478; $19{,}111\frac{1}{2}$. Therefore, if you subtract NS from NR, the remainder is SR, 104; 434; 925; 1682. But by Apollonius I 37, as HC is to CM, so will rectangle NR, RS be to the square of RP. Consequently, RP will be 10,676; 21,900; 32,170; 43,804. Previously, however, BR was 129,437; 129,592; 129,853; 130,244. Hence, the angles RPB are 4° 43′; 9° 36′; 13° 55′; 18° 35′. And their complements BPR, 85° 17′; 80° 24′; 76° 5′; 71° 25′. But in the triangles SRP, sides RP remain, and sides SR were given previously, hence the angles SPR are 0° $33\frac{1}{2}'$; 1° 8′; 1° 39′; 2° 12′; which, subtracted from BPR, leave the angles BPS sought, 84° 43′; 79° 16′; 74° 26′; 69° 13′. You can also, for a test, use other means to seek out the lines RP, by which this whole arrangement is governed, by seeking the arcs γδ of a great circle. For as the sine of the angle γ or the whole sine is to the sine of βδ, so are the sines of the angles β to the sines of the arcs γδ, which come out to be 3° 44′ 2″; 7° 28′ 6″; 11° 12′ 7″; 14° 56′ 7″. And these are in fact the angles PAR. And the secants AR of the angles BAR are given, namely, 163,565; 163,689; 163,895; 164,206. But these, multiplied by the tangents of the arcs γδ or the angles PAR give the lines PR, namely, 10,673; 21,459; 32,457; 43,800; not indeed much different from the preceding.[66]

In our eclipse, therefore, I would like these instructions to be used more familiarly. For the rest, when I took the ellipses, I had not yet thought about the ellipses' hyperbolic path. From this it happened that I did not keep the paper laid

[66] In the computations in this paragraph, Franz Hammer's corrections, which are many, have been adopted without comment.

on the floor carefully enough in the same place. Nevertheless, at the end of the eclipse I noted the two final ellipses more steadily. For I was watching for the end.

The drawing itself, examined with a compass, gives the angle BPS near the end of the eclipse as 70 degrees, with the diameters AB, CD of the two ellipses being connected on the two sides by the line BD, and, on consideration that the line BD cuts the hyperbola at B, D, the angle of the hyperbola itself with AB would have been a little greater, with CD a little less, but equal with some intermediate diameter. Therefore, at three minutes before 1 o'clock, the angle could have been 70°. Therefore, the end of the eclipse was slightly after $0^h\ 57^m$. And so, aside from the example of the procedure, you see that this can be
treated with sufficient accuracy. Two minutes after the end, 383
the town automaton struck quarter to one. The sun's altitude (but, as at noon, and with a small quadrant, unreliably), observed about three minutes after the end, indicates the $58\frac{1}{4}$th minute, which agrees not badly with my hyperbolic angle. For in the adjacent diagram, let αβ be 137° 2′, βδ 84° 46′, δα 53° 50′. Angle αβδ comes out to be 14° 34′. This is most certain, that the time was after noon. For the ellipses were now moving away from the opening. If we shall trust the constancy of the town clock over such a short space of time, which I think can be done safely, the duration will now be obvious. For at the time when I saw a defect of a digit or somewhat less, it struck precisely quarter past ten; therefore, since two minutes after the end it struck three quarters past twelve, the duration was therefore greater than $2\frac{1}{2}$ hours. Now in the thirteen final minutes of time the eclipse decreased by $1\frac{2}{3}$ digits, therefore, in eight minutes it increases or decreases a digit. And so if at the first view it was one digit eclipsed, the duration is $2^h\ 36^m$. If, however, the former was less, the latter will also be less.

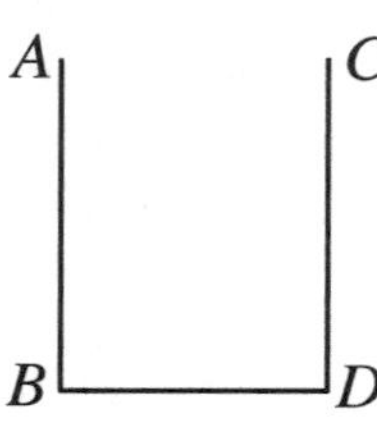

And since, by two indications, the end of the eclipse falls at hour $12^h\ 55^m$, when $2^h\ 36^m$ are subtracted, the beginning will fall at $10^h\ 19^m$ or more. Will we have confirmation from observation here too? Let us see. When on my quadrant the sun's distance from the zenith was shown as 59 degrees, there was not yet any trace of a defect. Computation shows it was eight minutes after ten. Then the sun went under the clouds, and stayed hidden for rather a long time, about $\frac{1}{2}$ an hour, until, on coming out, it appeared to be one digit eclipsed.

At Hven in Denmark, by a student of Tycho's who was on the island at that time for the purpose of this observation, the beginning was given as $10^h\ 3^m$, the end $12^h\ 32^m$.[67] The duration was $2^h\ 29^m$. The digits $9\frac{2}{3}$. And in fact the time ought to have been longer here than in Graz. For as the moon was to the north, it is in accord with this that at that elevation of the pole, which surpassed mine by 9 degrees, it appeared 5 minutes more immersed in the sun's body, and made its course through the sun's body that much longer. See, therefore, the fallacies of

---

67 In the *Tabulae Rudolphinae* (*JKGW* X p. 227), Kepler again gives the 1598 eclipse as an example, but notes that, since Tycho had left the island for good by that time, the observation was not made there. From the time of the eclipse's end, he deduces that these observations must have been made at Frankfurt an der Oder, farther to the east, by David Origanus.

vision above, from Chapter Six. Tycho himself, as he wrote to Maestlin, observed the beginning at Wandesburg in Hamburg with armillaries, at $9^{h}\ 52^{m}$, which at Hven would have been $10^{h}\ 4\frac{1}{2}^{m}$, since, by Hondius, the difference of meridians
384 is 3° 8′. Therefore, the middle at Hven was $11^{h}\ 17\frac{1}{2}^{m}$; at Graz, $11^{h}\ 37^{m}$ or more. Consequently, the difference of meridians is about 20 minutes, or 5 degrees, provided that everything that was used had been correctly established. Hondius's table makes it not much different, i.e., the difference of meridians is more than 4°.

Again, the difference of meridians between Uraniburg and Graz, but without consideration of the different parallax.

I shall now reveal Problem 17 with this example, and will show the particular inclinations that come out from my ellipses. First, I will find out the ratios of the diameters to each other, at the sun's altitudes of 37° and 36°, such as they were a little before and after the end of the eclipse at the time when I marked the ellipses. With an arc of 52° 45′, 53° 0′, 53° 15′, the differences of the tangents come out to be 1197, 1212. Therefore, the longer diameter is 2409, while the shorter, drawn through the axis of the cone (multiplying the secant of 53° by the tangent of 15′), is 1450. Again, with arcs 53° 45′, 54° 0′, 54° 15′, the differences of the tangents come out to be 1255, 1271. Therefore, the longer diameter is 2526, the shorter through the axis of the cone is 1486, a ratio of 17 to 10; the previous is a little less. The ratio in my sketches, which were only rough and very quick, comes out less, and therefore flawed, for the reasons stated in Problem 17; and this is the reason why I would suspect it must be corrected in this way, from the instructions in Prob. 17. So the longer diameter in the second of my ellipses was exactly this, expressed by the letters AB, one horn at the bottom apex D, the other on a line GL perpendicular to AB. Therefore, with BA divided at the midpoint at E, from E let the circle BG be described, and let LG cut this at G. And let the arc DG be bisected between the horns at H: DH will be the inclination, which now comes out to be $22\frac{1}{2}°$.

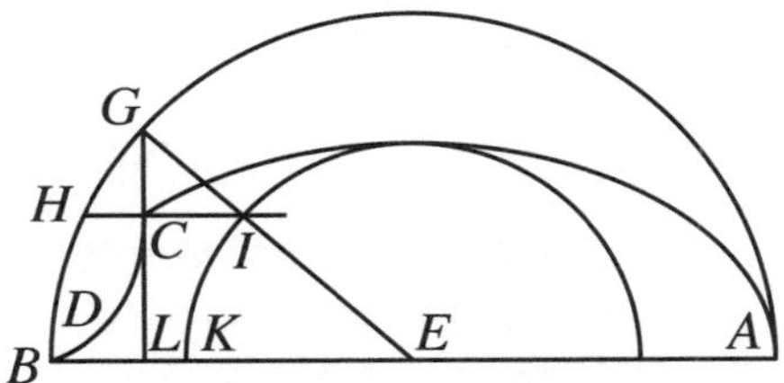

Suppose that you would like to know what the place of the other horn was. About center E with radius EI, the size of the shorter diameter in relation to the
385 longer one, draw a small circle IK, and connect EG, which will cut the circle at I; consequently, through I, parallel to BA, you should draw the line CI, which will cut line GL at C. Therefore, C was one horn, while D was the other. In my third ellipse, the horn was still about at the bottom, at the apex or vertex of the ellipse, the inclination by mechanical measurement being 20°; in the fourth, it was $19\frac{1}{2}°$; in the fifth, still less, while I judge that the other horn is always at the bottom vertex, although from the first it was gradually moving away from it. But in the sixth, where the lowness of the bottom vertex was now capable of being distinguished from the lowness of the lower horn, the inclination again comes out to be $22\frac{1}{2}°$. In the seventh, 26°, without a doubt erroneously. In the eighth, 20°. In the ninth and last, $22\frac{1}{2}°$, measured very carefully, right at the end, which by comparison with all of them is also the most true. For these inclinations vary slowly at the beginning and end, very quickly in the middle.

Please be indulgent, reader, for my setting before you observations that are not altogether consistent with themselves. For the eclipse was an important one, which was awaited with longing by certain astronomers for many years, but was only seen in a few places, the sky being cloudy. For that reason, I think it preferable that these reports of it, of whatever quality, be on record, rather than it be held to be completely unobserved. And the uncertainty is nonetheless not so great that they ought to be repudiated. For it is the nature of my candor that I lay out all the doubts. And the warning was not necessary everywhere; for at the end of the eclipse, two things that are unreliable in the observing, and about which, individually and separately, I can doubt whether I was sufficiently attentive, conspire almost for a single moment, perfectly sincerely and with no prejudice whatsoever, from the observation we set forth, before they were examined: the sun's altitude, and the angle between the ellipse and the hyperbola. And this consensus itself is not fortuitous, but they are consistent with observations in Saxony and Denmark, and with the actual revised computation of Tycho. And I do not know what can be accepted against four agreeing witnesses. Besides, I shall ask this same thing about the inclination in the following problems, whether they are in agreement with these times.

Problem 24

Given the time, the quantity of defect, the diameters of the luminaries, and the inclination to the vertical of the circle through the centers, to dig up the visible latitude of the moon from the sun, as well as the longitude

The problem is Maestlin's, but is made easier by our parallactic table. Let **386**
the beginning of the observation be given as $10^{\text{h}}\ 27^{\text{m}}$, when the defect was about one digit; the sun's diameter, from the above, is $30'\ 35''$. Let the moon's diameter at this distance from perigee of $49°\ 24'$ and eccentricity of 4336 be assumed to be $32'\ 44''$. Nor is there much uncertainty about this, as was argued above in Prob. 5 and 13. But the inclination of the circles is equal to a right angle, as was evident from observation. Let the meridian circle VP be set out, pole P, zenith V; let the sun be at S, the vertical be VS, the circle of declination PS. And through the center of the visible[68] sun S let the arc of a great circle go across to the center of the moon, and let this be SL. Its quantity is obtained thus. The sum of the radii is $31'\ 40''$; a digit is $2'\ 33''$, which, subtracted, leaves the distance of the centers. Therefore, SL is $29'\ 7''$. And let VSL be right. It is requisite to use this to examine in detail the moon's visible latitude from the visible ecliptic SE, that is, the arc LE, and the visible longitude ES. Therefore, by the theory of the *primum mobile*, let the angle ESV, and before this, the altitude of the nonagesimal degree from rising, which is also useful elsewhere, be sought. So, since the sun is at $16°\ 43'$ Pisces, its right ascension is $347°\ 47'$. When $23°\ 15'$,

---

[68] Here, as well as in the statement of the problem, Kepler uses the word *visibilis*, which has about the same meaning as its English cognate. It would perhaps be more idiomatic to substitute "apparent" for "visible," but this would obscure the connection Kepler evidently wished to bring out between vision and the position of the heavenly body. If he had meant the reader to understand this as "apparent," he could have used *apparens*, which also has a meaning close to its English cognate.

the times of the sun's distance from the meridian, are subtracted from this, the remainders, 324° 32″, are the right ascension of the meridian, with which it divides the heaven at 22° 10′ Aquarius, and its declination 14° 11′, which is MA. But AV is equal to the altitude of the pole, 47° 2′, therefore MV is 61° 13′. Now at this moment, 22° 31′ Gemini is rising. Therefore, 22° 31′ Pisces is at the nonagesimal degree, that is, at N; therefore, the arc MN is 30° 21′, and MNV is always right. Therefore, in the right triangle MNV, the base MV and the side MN are given. If you therefore will divide the secant of the former by the secant of the latter, there comes out the secant of NV, the arc sought, which is here 56° 4′, the distance of the nonagesimal from the zenith. The complement of this, 33° 56′, is the altitude of the nonagesimal, measuring the angle O between the ecliptic and the horizon, by which arcs the latitudinal parallaxes will later be taken, in accord with Chapter 9. For Copernicus's table is very concise,
387 and does not show these arcs adequately.
And so, once NV is found, in the right triangle SNV, with the right angle at N, the sides are given. For S is 16° 43′ Pisces, N is 22° 31′ Pisces, therefore SN is 5° 48′. Consequently, divide the tangent of NV, increased by the zeroes in the radius, by the sine of NS, and out will come the tangent of the angle NSV sought, 86° 7′, and VSM, 93° 53′, towards the moon, which, at the beginning, is normally on the western side, now closer to the meridian than is the sun. But from the observation, LSV is 90°. Therefore, the remainder LSE is 3° 53′.

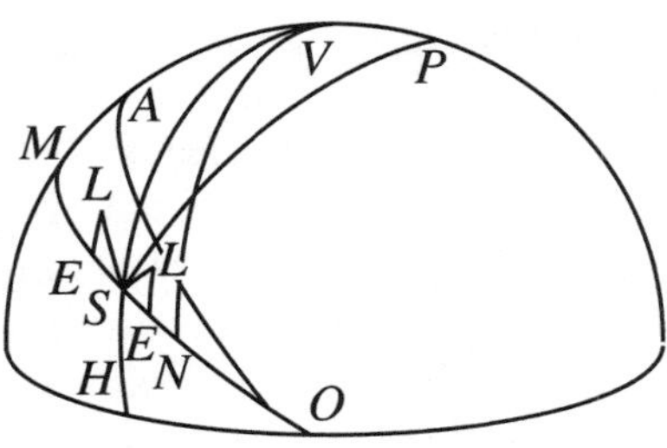

When a perpendicular is dropped from L to ME (let this be LE), a third triangle is given us, nearly plane, which is LES, in which the base LS and the angles are given, as a result of which the sides LE, the visible latitude of 1′ 57″ N, and ES, the visible longitudinal distance of the moon ahead of the sun, 29′ 3″, will not be unknown. But, so as to escape the tedium of multiplying the sines by the distance of the centers, *find the distance of the centers at the head of the parallactic table, and the angle between the ecliptic and the circle through the centers at the side of the table: the entry in the table will give the visible latitude. If you find the complement of this angle at the side, the entry will show the visible longitude.*[69]

A short cut for deriving the distance of the centers in longitude and latitude, using the parallactic table.

Problem 25

Given the visible latitude at a certain moment, to find quickly the visible latitude at another moment at a certain distance from the former one. It is further required that the distance of the moon from the center of the earth be known approximately, and the hourly motion of the moon, and angle and motion of latitude

[69] The parallactic table (in the pocket at the back of this book) gives the angle whose sine is the product of the sines of two other angles. It can thus be used to apply the sine rule for right spherical triangles, as Kepler does here.

In the example (after $10^h\ 27^m$, when the latitude is 2′ North), let the hour $12^h\ 55^m$, when the eclipse ended, be proposed for us. And let the hourly motion of the moon from the sun be 33′ 30″, from Tycho. Therefore, to $2^h\ 28^m$ is credited 1° 22′ 38″ of true motion of the moon from the sun. The moon also departed by about the same amount from the node. And since it is standing about at degree 10 from the node, its true latitude over an arc of that size increased 7′ 4″ to the north. But parallax too increases its visible latitude, which is clear thus. At $12^h\ 55^m$, the right ascension of the meridian is 2° 8′. Therefore, 24° 48′ Cancer is rising. The point on the meridian is 2° 17′ Aries, with declination 0° 51′ north. Consequently, the arc between the culminating point and the zenith (previously it was MV) is now 46° 11′. But MN is 22° 31′, the amount between the culminating point and the nonagesimal. Hence, VN comes out to be 41° 27′, which previously
was 56° 4′. With these two arcs, by the instructions in Ch. 9, I take from the *388*
parallactic table, under the heading of 55 radii (the amount now adopted for the moon's distance from earth), the parallaxes of latitude, first 51′ 28″, then 41′ 1″.[70] From this it is clear that between the moments adopted the parallaxes of latitude decreased 10′ 27″, or from the sun 10′ 22″; under the heading of 56 radii, the decrease would be only 10 seconds less, while there is the same amount of increase for the visible latitude. Therefore, 1′ 57″, 7′ 4″, and 10′ 22″, together make the sum of 19′ 23″. This, then, is the visible latitude deduced in this particular way at the end of the eclipse.

Now, by the converse of Problem 24, with the latitude, such as ought to appear, adopted among the givens, we shall establish the inclination, such as ought to be observed, so that we may compare it with our observed inclination. Let S now be beyond the meridian, with EL now closer to the eastern horizon. The end of the eclipse is when the circles of the luminaries touch each other.

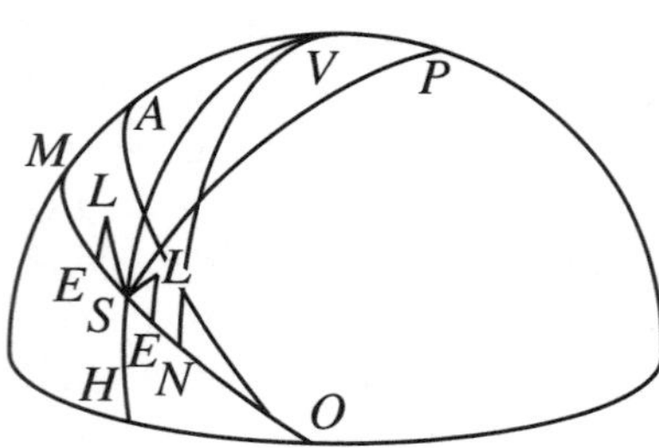

Therefore, SL is the sum of the radii, 31′ 40″, while EL is 19′ 23″. Hence, LSE comes out to be 37′ 39″. So VSN is sought in the following manner. First, VN has been found at this last moment to be 41° 27″. And SN between the sun and the nonagesimal is 37° 59′. Therefore, the angle between the ecliptic and the vertical comes out to be 55° 8′. I subtract LSN; there remains 17° 29′ between the circle passing through the centers and the vertical. It was observed to be $22\frac{1}{2}°$. The difference is slight. For conversely, if you will assume $22\frac{1}{2}$, the visible latitude will come out to be 17′ 6″, only $2\frac{1}{3}$ minutes less, and these minutes may be missed for various reasons. For example, because the shadow of the first moment did not stand exactly at the right, for this image was seen for barely a moment. Or because we are uncertain about the quantity of that defect, for who would observe these things precisely in such a small ray without an instrument? Or because, at the end, of the inexperience of a new art, and the inclinations being not everywhere perfect. Finally, because

---

[70] This should have been 41′ 8″.

I did not take the precautions of the second chapter here, on account of the smallness and the unknown quantity of the opening. For, as is evident from the last diagram, when the elliptical ray is diminished at the edges, the inclination will also be diminished. However that may be, I have taught by the example how much the observed inclinations contribute to finding the observed latitude. For the
389 visible longitude at the end of the eclipse, from the angle VSN, 55° 8′, let VSL, observed to be $22\frac{1}{2}°$, be subtracted, and the remainder, LSE, is 32° 38′. But it was computed to be 37° 39′; consequently, SLE in the former case is 57° 22′, in the latter, 52° 21′. The sines of these, multiplied by the sum of the radii, make, in the former case, 26′ 40″, in the latter, 25′ 4″, the moon's visible longitude beyond the sun; previously, it was 29′ 3″ before the sun. Therefore, in $2^h\ 28^m$, the moon's visible motion from the sun (the two longitudes taken together), is either 55′ 43″ or 54′ 7″, the former from the observed inclination, the latter from the computed one.

Problem 26

From a given visible longitude and latitude at certain moments, and with distances of the moon and sun from the center of the earth adopted from elsewhere, to establish the true longitude and latitude; likewise, the hourly motion as well, and the instant of true conjunction; or, on the contrary, assuming the hourly motion, to find the moon's distance from earth, approximately

Parallax, mixed with the true motion, establishes the visible motion; when separated, it leaves the true motion. Parallax, in turn, is brought about by the nearness of a celesial body to the earth's center. And so the problem is obvious enough from the conversion of the instructions for computing eclipses; only here, for the use of the instructions introduced in their place in Chapter 9 above, I repeat what should also be stated at the same time, in order that the use of the observed inclinations in establishing the moon's true position may be evident, and in order that the sun's eclipses may be freed from aspersion, by a demonstration that those eclipses contribute more to the investigation of the moon's motions than the lunar ones do. Let the distance of 1150 earth radii between the sun and the earth be adopted, as Tycho wished it, on which more elsewhere. This total, found at the head of our parallactic table, presents a horizontal refraction of about 3 minutes. The moon's distance from the earth, in turn, is taken as 55 radii, as above. Its horizontal parallax, taken proportionally from the head of our table, will be 62′ 30″. The solar parallax of 3′ subtracted from this leaves a horizontal parallax of the moon from the sun of 59′ 30″.[71]

Let it now be the first moment, when the distance between the nonagesimal and the zenith was 56° 4′, with which, under the heading of 59′, is taken exactly

[71] The solar parallax is, of course, much smaller than Kepler's figure, but the lunar distance corresponding to an apparent diameter of 32′ 44″ is $57\frac{1}{4}$. The decrease in the moon's parallax would thus be about equal to that of the sun's parallax, leaving the difference about the same as Kepler's.

48 minutes 57‴, while under the heading of 30″ is taken 24″ 53‴. The sum, 49′ 22″, is the latitudinal parallax of the moon from the sun.

But the apparent latitude of the moon from the sun was 1′ 57″ north. Therefore, when added to the parallax of latitude, this gives the true latitude, 51′ 19″. **390**

For the longitudinal parallax, with the altitude of the nonagesimal, 33° 56′, I extract under the same columns of 59′ and 30″ the maximum longitudinal parallax of 33′ 13″. Therefore, under the columns 33′ and 13″, through the moon's visible distance from the nonagesimal of 6° 17′ (the moon being at 16° 14′ Pisces to the sight, the nonagesimal at 22° 31′ Pisces), I extract, with a double entry, the correct longitudinal parallax of the moon from the sun, 3′ 37″ in this place; and this is westward, because the moon is east of the nonagesimal. Therefore, when 3′ 37″ of parallax are subtracted from 29′ 3″, the moon's visible distance from the sun, the remainder is 25′ 26″, the moon's true distance from the sun, to the west. At the other moment, when the distance between the zenith and the nonagesimal is 41° 27′, with this entered under the same columns of 59′ and 30″, as before, I extract the latitudinal parallax of the moon from the sun, 39′ 23″. Suppose that the visible latitude of 17′ 6″ north was found correctly in Prob. 25. Therefore, when this is added to the moon's parallax from the sun, the true latitude will be 56′ 29″. For the longitude at this moment, as before, I do as follows. Because NV is 41° 27′, the altitude of the nonagesimal is consequently 48° 33′. With this, entered under the headings of 59′ and 30″, I extract the new headings of 44′ 34″. Under these new columns, through the visible distance from the nonagesimal of 37° 31′ (for the moon is at 17° 17′ Pisces, the nonagesimal at 24° 48′ Aries) I extract the moon's longitudinal parallax from the sun, in this place 27′ 9″ to the west. Add this to the visible amount of superseding deduced from the observation, which was 26′ 40″, and the result is the moon's true longitude from the sun, 53′ 49″. Before, it was 25′ 26″ before the sun. Therefore, in its true motion, the moon over 2^{h} 28^{m} will have moved 1° 19′ 15″ from the sun; even less, if we use the computed inclination of the end. Above, from Tycho, we took this hourly motion as 1° 22′ 30″. If this be true, an error of estimation in the beginning of the eclipse would be indicated, and the moon would have very slightly (but not by a digit) encroached upon the sun; perhaps indeed a slightly larger diameter of the moon would be required, or the town clock might have been adjusted in mid interval. Finally, and in turn, all the observations being held correct, either Tycho's hourly motion would be too great, or the parallax too small. Meanwhile, when the arc of the moon's true motion is diminished, the increment of latitude computed **391**
from Tycho is also diminished, and thus the computed visible latitude of 19′ 22″ will approach closer to the observed value of 17′ 6″, nor would the duration be augmented excessively over that which was noted in Denmark.

From this variety, the diligent, talented, and circumspect astronomer easily sees which things fit together with which, and what things out of all of them may be sought most securely by the observations and may be adopted for the establishment of canons, and which, on the other hand, being rough and not fully known without a large error, may nonetheless be adduced. Finally, he sees how important it is for matters astronomical and geographical, to observe and note the inclinations of the phases exactly, and indeed optically, by an opening.

For the instant of true conjunction, it is requisite that you be certain of the moon's hourly motion. We shall therefore grant, with many conjectures in agreement, that in the beginning not a whole digit was observed, but nonetheless we shall subtract something from Tycho's hourly motion. For besides, I have a number of causes of this matter, which I am going to set forth in the second part, God willing.[72] Let the truth, then, be in the middle, and let an hourly motion of 32′ 50″ be adopted. For even if it had not been eclipsed at all in the first instant of time, there would nonetheless not result as great a true motion for 2 hours 28 ^{m} as results from Tycho's hourly motion, with the parallaxes, for which the evidence is more reliable, staying the same. So, at the end, at hour 12^{h} 55^{m}, the moon had passed 53′ 49″ beyond the sun's true position. If 32′ 50″ make one hour, what do 53′ 49″ make? The rule will produce 1^{h} 38^{m} 20″. Therefore, the instant of true conjunction at Graz will have been 11^{h} 17^{m}. Provided that he has the moment of the end correctly, we are plainly certain of the middle of the true conjunction within one minute.

If you should wish to work backwards from the adopted hourly motion to seek out the parallaxes, the road is not so flat, but is crabwise[73] or with a sidelong glance. For you must call upon Fortune, that you adopt that very parallax immediately at the beginning which you choose; that is, which would lead you back to the adopted hourly motion at the chosen place, by the method just explained. Should you err from the true parallax, the work has to be done again, and by comparison of the errors with the differences of the parallaxes, the
392 truth has to be palpated by a blind procedure, as if by *regula falsi*.

Problem 27

At a given elevation of the pole, the beginning and end, or moments, being visible, and the sun's place known, with hourly motion and diameters of the luminaries also chosen, and the qualitative motion of latitude (taken roughly), and finally the distances of the luminaries from the center of the earth, to investigate the instant of true conjunction and the true latitude; and from it the difference of meridians as well.

Let's play. For there is no reward for this work more certain than this pleasure. And, if it is allowed, we make it our practice even to work with song. I want to know the instant at which the true conjunction occurred in Denmark, so that a more reliable difference of meridians may be had. At the same time, I also desire to know whether an observer hindered by deceptions of vision will note the beginning earlier, the end later. For this will easily become evident if we obtain a greater latitude by our vision. We shall also not be forgetful that in our latitude at the end of the eclipse there is an uncertainty of two minutes. The altitude of the pole at Hven is 55° 54′ 45″. Let the moon's hourly motion be taken as 32′ 50″, as just now. The sum of the radii is 31′ 40″. We know that the moon is really to the

---

[72] See the footnote on p. 381, above.

[73] *Cancrina*. I have not found this word in the dictionaries, but derive it conjecturally from "cancer," crab, by way of musical theory. Cf. *Grove's Dictionary of Music and Musicians*, s.v. Cancrizans.

north, and is running away from the ascending node. Let the moon's horizontal parallax from the sun be 59′ 30″, as before. The beginning is $10^{h}\ 3^{m}$. The end, $12^{h}\ 32^{m}$. The sun's position at the beginning is 16° 43′ 27″ Pisces; at the end, 16° 49′ 42″.

Before everything, let the parallaxes be established. So, at the beginning, the sun's right ascension is 347° 48′. Subtracting from this 29° 15′, the time of the sun's distance from the meridian, following the line of the observation at Hven reviewed in Prob. 23, leaves the right ascension of the meridian, 318° 33′, with which it divides the sky at 16° 5′ Aquarius. And the declination of this point is 16° 5′ south. Hence, in the diagram of Prob. 24, since AM is 16° 5′, AV 55° 54′ 45″, therefore MV is 72°. Further, at this moment 18° 24′ Gemini is rising, and N is 18° 24′ Pisces.[74] Therefore, MN is 32° 20′, and NV is 68° 33′, its complement 21° 27′. By these, I take the moon's parallaxes from the sun, 56′ 22″ of latitude, a titular[75] 21′ 45″ of longitude. And since the sun is at 16° 43′ Pisces, and the nonagesimal is at 18° 24′, therefore, SN is 1° 41′: I increase it by about 30 minutes, by which the moon precedes the sun to the sight, so that it becomes 2° 11′, with which, from the columns of 21′ and 45″ I take the true parallax of the longitude as 0′ 48″ to the west. Let these be kept ready.

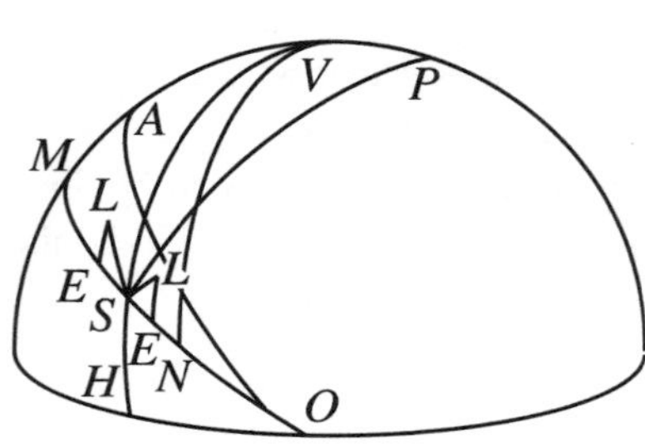

At the end of the eclipse, the sun's right ascension (at 16° 50′ Pisces) is 347° 54′. Add 8°, the time of the sun's distance from the meridian, the right ascension of the meridian will become 355° 54′, with which 25° 32′ Pisces is on the meridian. And the declination of this is 1° 47′ south, which is MA. Therefore, MV is 57° 43′. Further, 27° 43′ Cancer is rising at this moment, and N is at 27° 43′ Aries. Therefore, MN is 32° 11′, and NV 50° 44′. And its complement is 39° 16′, by which I take the parallaxes of the moon from the **393**

---

74 This rising degree is erroneous, and the error created difficulties in subsequent computations. I have calculated the rising degree, assuming Tycho's obliquity of the ecliptic (23° 31′ 30″), the elevation of the pole at Hven (55° 54′ 45″), and an arc on the equator of 48° 33′ between the vernal equinox point and the eastern horizon (which is consistent with Kepler's right ascension on the meridian, 318° 33′). The rising degree that results from these, as I have also confirmed using an independent method, is 28° $41\frac{1}{2}$′ Gemini. However, Hammer (*JKGW* II p. 458) gives the rising degree as 25° 13′ Gemini, professedly from the same spherical triangle (equator-ecliptic-horizon), though he did not give the numbers he used. Whichever of these is correct, it is clear that Kepler's number is wrong. The question is, in any case, moot, because the observation was made at Frankfurt an der Oder, not Hven (see footnote 67, above).

75 *titularem*; that is, relating to the head of the parallactic table. This is what Kepler elsewhere calls the "maximum parallax of longitude." He calls it *titularis* here because one must find these numbers (minutes and seconds separately) at the head of the table, and find the true parallax of longitude in the row of the moon's distance from the nonagesimal, as Kepler does below.

sun, 46′ 3″ of latitude, a titular 37′ 39″ of longitude. And since the sun is at 16° 50′ Pisces, and the nonagesimal at 27° 43′ Aries, *SN* is therefore 40° 53′. I decrease it by about 25 minutes, by which the moon follows the sun to the sight, so that it becomes 40° 28′, with which, under the heads 37′ and 39″ just found, I take the true parallax of longitude of the moon from the sun to be 24′ 26″ to the west. Let these again be kept ready.

Now, because the intervening time is 2^{h} 29^{m}, and the moon's hourly motion from the sun is 32′ 50″, therefore, the moon's motion is 1° 21′ 32″. By an arc of this quantity about the nodes, the latitude varies by 6′ 57″ at an approximate distance from the node of ten degrees (it being posited that the angle of latitude is 4° $58\frac{1}{2}'$). And because the moon is ascending to the north, therefore, at the end, the true northern latitude will by greater by 6′ 57″. But the visible latitude is also increased because of parallax. For at the beginning, the moon's parallax of latitude from the sun is 56′ 22″, at the end, 46′ 3″; the difference is 10′ 19″, which adds to the difference of the visible latitudes. I therefore add 6′ 57″: the result is 17′ 16″, the excess of visible latitude at the end over the initial amount. In the same way, since both the parallaxes of longitude retard the moon, I subtract the smaller, 0′ 48″, from the greater, 24′ 26″: there remains the retardation, 23′ 38″; this, in turn, subtracted from the moon's true motion, which was 1° 21′ 32″, leaves 57′ 54″, the resulting visible motion of the moon from the sun within the time of duration. Let a certain straight line *BA* be set out, representing the visible motion of the moon of 57′ 54″, and at right angles to it let *BC* be drawn, with length 17′ 16″, the difference of the visible latitudes; therefore, *CA*, being joined, will be the moon's visible path.
394 Upon *AC* let the isosceles triangle *AFC* be constructed, so that each leg may have 31′ 40″, the sum of the semidiameters, and from *A*, *C*, let circles of the moon's diameter be described, from *F* a circle of the sun's diameter, touching the diameters of the moon, and through *F* let straight line *DE* be drawn, parallel to *BA*, and let *CB* be extended to *D*, and *AE* be made equal to it. *DE* will be the visible ecliptic, and *AE*, *CD* the visible latitudes, from which afterwards the true ones are easily obtained by the mediation of the parallaxes.[76]

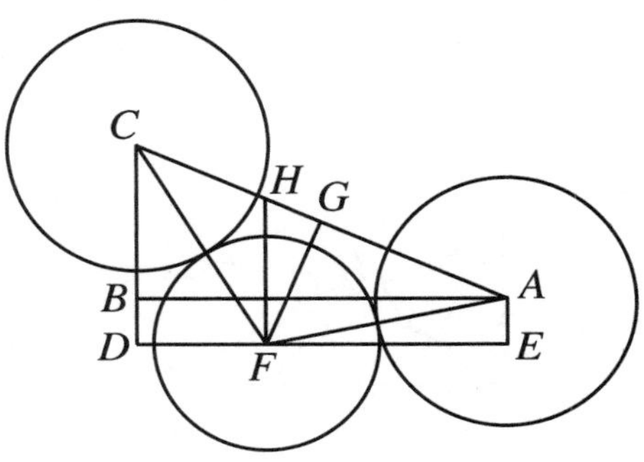

To *CB* there needs to be applied some line *BD* so contrived that when the squares on *CD* and *BD* or *AE* subtracted from those on *FC* or *FA*, there remain two squares whose roots together are equal to *ED*. If one prefers to extract what is sought by algebra, he will come to that equation where the cube and a constant

[76] Although the results of these computations are flawed because of errors in the data, these errors are irrelevant to Kepler's purpose here, which is to show the method for carrying out such computations. Hammer (*JKGW* II pp. 458–9) gives corrected values (as did Kepler in the *Tabulae Rudolphinae* pp. 110–11, *JKGW* X pp. 225–7).

number are equated to the squares and the positions [i.e., the first power of the variable].[77] For us, the geometric way is open. For as AB is to BC, so is the whole sine to the tangent of angle BAC, which becomes $16^\circ\ 36\frac{1}{2}'$. In turn, as the whole sine is to AB, so is the secant of angle BAC to AC, which becomes $1^\circ\ 0'\ 25''$, whose half, AG, is $30'\ 12\frac{1}{2}''$. Therefore, as AF is to AG, so is the whole sine to the sine of AFG, $72^\circ\ 33'$, whose complement FAG is $17^\circ\ 27'$. When BAG, $16^\circ\ 36\frac{1}{2}'$, is subtracted from this, the remainder, FAB or AFE, is $0^\circ\ 50\frac{1}{2}'$. Finally, as the whole sine is to the sine of angle AFE, so is AF to AE or BD, which is what is sought, which becomes $0'\ 28''$. Therefore, the visible latitude at the beginning of the eclipse is $0'\ 28''$ north, at the end $17'\ 44''$. When the latitudinal parallaxes are added, the former comes out to be $56'\ 50''$ north, the latter $1^\circ\ 3'\ 47''$ north. See how much greater it is than the one observed in Graz, $51'\ 50''$ at the beginning, $56'\ 30''$ at the end, or the sum $58'\ 20''$. And since this computation of ours is more rightly consistent with the computation of Tycho, who shows a true latitude at the middle of the eclipse of $54'\ 11''$, we evidently conclude that the usual error crept up on the observer, so that he did not take note of the beginning and end, but moments close to these, because of weakening vision in the sun's bright light. It should not trouble you, either, that I attribute to him an error of $5\frac{1}{2}$ minutes in the true latitude. For the principles by which I obtained this are set up thus, with the moon, especially, passing almost through the middle of the sun's body, and touching it from the slanting sides, so that if you will add even the slightest amount to the time, this sum just stated would be removed from the latitude, with the two small moons of our diagram settling deeper into the sun's body, as if their chains had been loosened.

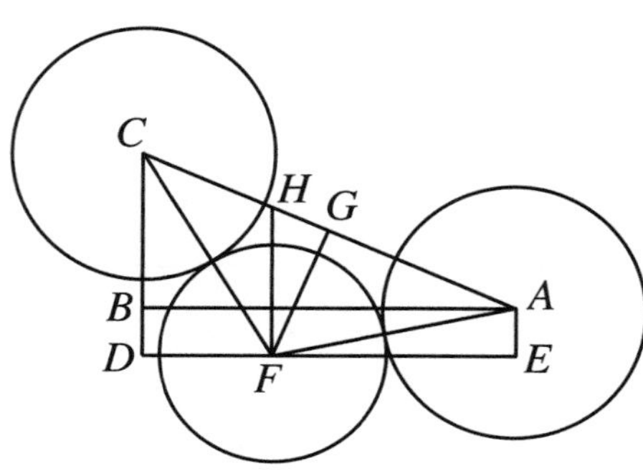

This also is evidence that the latitude was really smaller, that the observer 395
came up with $9\frac{2}{3}$ digits. And in fact it is certain, from Ch. 5, that whether it was contemplated through an opening or with eyes directed at the sun, the eclipse in either case was greater than $9\frac{2}{3}$ digits. If you compute the length FG from this diagram (for at G the eclipse is greatest) it will come out $9'\ 30''$, the distance of the centers under the maximum obscuration, and when the excess of the moon's diameter over the diameter of the sun is subtracted from this, the remainder is

---

[77] This is an equation of the form $x^3 + a = bx^2 + cx$. Kepler is using Cardano's terminology; cf. Cardano, *The Great Art*, trs. Witmer. However, the problem sketched out by Kepler reduces to the equation $2AB\sqrt{AF^2 - x^2} = AB^2 + CB^2 - 2CBx$ (where $BD = AE = x$), which is not a cubic equation. Hammer (*JKGW* II p. 459) explains that it is a cubic equation because the fourth powers of x cancel. This last statement is true, but the explanation ignores the fact that there are no third powers of x. Frisch (*JKOO* 2 p. 443) works out the solution without addressing the discrepancy between Kepler's statement about the cubic equation and his own analysis, which has no cubes. In any case, as the next sentence shows, Kepler chose another way to the solution.

$8' 25''$, the units in the free part of the sun, which make $3\frac{3}{10}$ digits. And so it would have been possible for only $8\frac{7}{10}$ digits to be in the shadow, when nonetheless it is certain that it was beyond 10 digits. On this point, see also Prob. 32 below. But suppose that the deception of the eyes is equal at the beginning and the end: the true conjunction at Hven is sought. And because AFE is $50\frac{1}{2}'$, EAF will be $89^\circ\ 9\frac{1}{2}'$. Consequently, as the whole is to AF, so is the sine of EAF to EF, the visible preceding of the moon, which remains $31'\ 40''$. When this is subtracted from $57'\ 54''$, which is ED, the remainder, FD, is $26'\ 14''$. When the parallax of longitude is subtracted from the former and added to the latter, the result in both cases is the moon's true preceding, $30'\ 52''$. Here the true excess is $50'\ 40''$. If a motion of the moon of $32'\ 50''$ gives an hour, what does 30′ 52 give? It comes out to be 56^{m} of an hour, to be added at the beginning. Therefore, the hour of $10^{\text{h}}\ 59^{\text{m}}$ is the instant of true conjunction; at Graz, the hour was $11^{\text{h}}\ 17^{\text{m}}$; the difference of meridians, 18^{m} or $4\frac{1}{2}$ degrees, about the amount that Hondius also makes it. And that is what was to be looked into in this case.

A more exact investigation of the difference of meridians between Uraniborg and Graz.

And thus if some eclipse of the sun were observed correctly in all of its circumstances in just a single location, but in other locations only the certain moments of the beginning and end, by this method a pronouncement can be made about the difference of meridians with perfect confidence.

Problem 28

If the ratio of the lunar diameter to the solar is observed through an opening, the proportion of the opening being unknown, but if the true ratio of the diameters is known from elsewhere, to reckon how much the observation of digits shall have erred from the truth, and the rest.

On 21 July 1590, my teacher Maestlin observed an eclipse of the sun at Tübingen beneath a broad and dark roof, the sun's ray being admitted through the roof tiles. I shall communicate the description of the observation as the au-
396 thor supplied it plentifully to me, so that I may use this example to display the application of a number of the problems above. We are missing the beginning. Therefore, when the sun was eclipsed by a semidiameter, its altitude above the eastern horizon was 26°; the inclination, observed as in Prob. 14, was 88°. The moon was higher than the sun on the ray, lower in the heavens.

About the middle of the eclipse, he measured the ratio of diameters in the way that is in Prob. 7, and found that where the sun was 24, the moon occupied 23, while the ratio of the distance of the centers to the sun's radius was as 59 to 88.

After the maximum obscuration, when it was again seen to be eclipsed by a semidiameter, the sun's altitude was found to be 33°. The inclination of the circle through the centers was $2\frac{1}{2}^\circ$ to the vertical, by which arc the moon was west of the sun, and south, as before.

When it was one fourth eclipsed, the sun's altitude was $37\frac{1}{4}^\circ$. The inclination of the circle through the centers was 19° to the vertical. Further, the moon was then made to be east of the vertical.

At the end of the entire eclipse, the sun's altitude was $41\frac{1}{4}$; the circle through the centers comprehended an angle of 30° with the vertical. The moon was to the east.

Let the ray $CBDI$ eclipsed by half its diameter be set out with centers B, A; and, because AB is the moon's radius, BI the sun's, the ratio of BI to BA, from observation, will be as 24 to 23; therefore, the whole AI will be 47. Further, since it was demonstrated above in Ch. 2 that the edge of a ray admitted through an opening is enlarged, accordingly, about the same centers B and A let the interior arcs be described, with radius BH for the former, AF for the latter, so that FB and IH are equal. Therefore, AF will be the radius of the moon, and BH the radius of the sun, which are obtained thus. Since the sun is one sign from apogee, as the whole sine is to 30 seconds, by which it is increased to the sight from apogee to its mean distance, so is the versed sine of its departure from apogee to the increase of this place, approximately. So the diameter is 30′ 4″, the half 15′ 2″.[78] Or, by our parallactic table: find the difference of the mean and least diameters (here 30″) at the head, the degree 397
of the complement of the distance from apogee (in this example) at the side. What the entry shows, subtract from the head; add the remainder to the minimum diameter. If the distance from apogee E exceeds a quadrant, the excess must be found at the side, and what the entry shows must be added to the head, and the sum added to the diameter at apogee. We will proceed in the same way with the moon. The moon was 40 degrees 50′ from apogee; by the complement of this, and the eccentricity of 4336, as in the above example, the diameter of the moon is obtained as 30′ 50″, the half 15′ 25″. Therefore, where AF is 15′ 25″, BH is 15′ 2″. And because BF and HI are equal, therefore, BH, FI are also equal. Therefore, the sum of AF and BH is equal to the sum of AF, FI, that is, AB, BI. Therefore, AI, 47 of Maestlin's units, is equal to the true measure of 30′ 27″. Therefore, if 47 gives 30′ 27″, what will 23 make in this measure? The rule produces 14′ 54″, which is AB; and AF is 15′ 25″; the obscured part is 15′ 32″, when it was considered by Maestlin to be eclipsed by half the diameter. And because BI is 15′ 32″, and AI is 30′ 27″, therefore the remainder AB, the correct visible distance of the centers (by Prob. 12), is 14′ 55″.

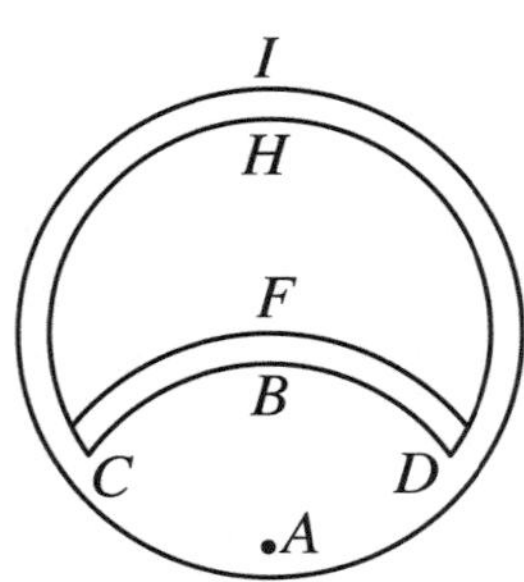

An easy way of varying the diameters of the luminaries using our parallactic table.

Problem 29

From the altitude of the sun or of a star with known declination, and a known altitude of the pole, to find easily and distinctly the time or elongation of the sun from the meridian

---

[78] The "whole sine" is, as usual, the number of units in the radius against which the trigonometric functions are compared. The "versed sine" here is $1 - \cos \frac{A}{2}$, where A is the sun's true anomaly. By this proportion, Kepler makes the change in the sun's apparent diameter proportional to the versed sines. This is a form of proportionality of which Kepler was fond; see, for example, his planetary speed rule in Ch. 20 of the *Mysterium cosmographicum* (trs. Duncan p. 201 and the notes on p. 249), and his refraction rule (Ch. 4 above).

We enter the fray in one form of triangle, with Wittich's procedure of adding and subtracting.[79] For although we do not diminish the labor, we nonetheless go up to the summit and have a view of our path before our eyes. The valleys, however, conceal Wittich. Because his demonstration is general, carried over from the circle to the sphere, this one of mine is especially adapted to the sphere.

Imagine that a plane parallel to the horizon passes through the center of that circle which at that moment the sun occupies in carrying out its diurnal motion. Then as the whole sine is to the sine of the altitude of the pole, so is the sine of the declination of the sun or of the star to the sine of the altitude or depth of this plane above [or below] the horizon, according to whether the sun was
398 to the north or south.[80] Next, consider that, as the ecliptic is inclined above the equator, the equator above the horizon, so is the parallel [upon which the sun travels] inclined above this plane, and the sines of the degrees beginning from the plane are proportional to the sines of the altitudes above the plane; those, that is, the greatest of which is the difference of the sines of the altitude of the plane and of the meridian altitude of the star; or, contrariwise [i.e., if the declination is south], the aggregate of the depth of the plane and of the meridian altitude of the star.[81] And so, subtract the sine of the altitude of the plane, to the sine of the

[79] For Paul Wittich (d. 1586), see Gingerich and Westman, "The Wittich Connection." Wittich never published any of his work; Kepler knew of it from Joost Bürgi, who had known Wittich when both were working at the court of William of Hesse.

[80] In the adjacent diagram, T is the earth, P the pole, NT the plane of the horizon projected on the meridian plane (the plane of the paper), TA the projection of the equator, and Z the zenith. The sun over the course of a day moves on a path that is very close to a circle parallel to the equator: let the projection of that circle be SH, H being its center. Kepler's imaginary plane therefore passes through H; let its projection be BC, and let K be the intersection of BC and ZT. Then the sine of arc NP, the polar elevation, is KT/HT, and the sine of arc AS, the sun's declination, is HT/ST, and the sine of arc BN, the altitude of Kepler's plane, is KT/ST. The relation stated follows directly.

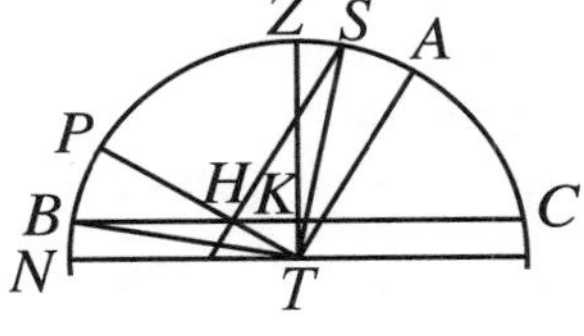

[81] The "parallel" is the circle about H, whose projection is SH: it is parallel to the equator AT. The "sines of the altitudes above the plane" are the lines SE and DL, which are in the same ratio as SH and DH, D being the projection of sun's position. Now imagine the sun's circle about H, perpendicular to the plane of the paper, passing through S, through the sun's position (call this D'), and intersecting Kepler's "plane" BC parallel to the horizon at a point behind H (call it H'). The line drawn from D' intersecting HH' at right angles is the sine of $H'HD'$ (with HH' or SH as radius), and is equal to DH, and this angle is the complement of the sun's distance from the meridian. But DH is equal to the perpendicular from D'; that is, the sine of the complement of arc $D'S$ is DH/HS, or DL/SE. And $SE = \sin STM - \sin CTM$, while DL is the observed altitude of the sun diminished by the sine of CTM.

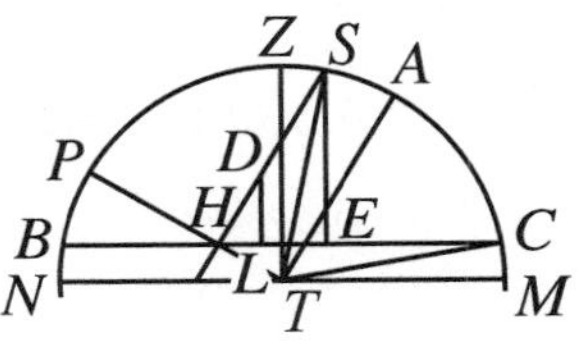

altitude add the sine of the depth, and of the meridian, and of the temporal or observed arc. Then, as the first of the two remainders or sums is to the second, so is the whole sine to the sine of the complement (if the star is higher than the plane) or of the excess over a quadrant (if it is lower) of the stars distance from the meridian. An example is in the eclipse under consideration, and at the first moment. Let the altitude of the pole at Tübingen be 48° 24′. For all tables make the latitude of Tübingen and Augsburg the same. But at the position of Augsburg, or Geggingen, a little farther south at half a mile or 2′, the altitude of the pole was found very accurately to be 48° 22′, as you see in vol. I p. 361 and following of Tycho's *Progymnasmata*.[82] In fact Maestlin too records this altitude at Tübingen in 1588 at the end of his book of spherics.[83] Let the sun's position be roughly 7° 30′ Leo, because the time is not known precisely; the declination of this is 18° 28′. Hence, the sine of the altitude of the plane is 23,687. And because the altitude of the equator is 41° 36′, I add the declination of 18° 28′: hence, the sun's meridian altitude is 60° 4′. The sun's observed altitude is 26°. The sines of these are 86,661 and 43,837; diminished by the altitude of the plane, the remainders are 62,974 and 20,150. And the latter divided by the former (increased by the zeroes in the radius), gives 31,997, the sine of the arc 18° 40′, whose complement, 71° 20′, measures the time of the distance from the meridian, $4^h\ 45\frac{1}{3}^m$. Therefore, 15′ 31″ were eclipsed at $7^h\ 14\frac{2}{3}^m$ am.

The altitude of the pole at Tübingen.

I shall also follow the same method at the other moment, when the sun's altitude was 33°, and in the two remaining ones, where the sun's altitude was $37\frac{1}{4}°$ and $41\frac{1}{4}°$. Except that here I shall mix in an addition/subtraction[84] in the first part, to avoid multiplication. The sun's position at $7^h\ 15^m$ is shown by Tycho's computation to be $7°\ 25\frac{1}{2}'$ Leo. Therefore, at the end of the eclipse, it is really at 7° 30′ Leo. The declination from $7°\ 25\frac{1}{2}'$ Leo to 7° 30′ Leo decreases 2 minutes. And thus we can safely use this declination unchanged. 399

Declination	18°	28′			
Alt. of equat.	41	36			
Sum	60	4	sine	86661	
Difference	23	8	sine	39287	
		Remainder		47374	
			Half	23687	altitude of the plane sought.
		Remainder		62974	Divisor.

Altitudes of the sun	33° 0′	37° 15′	41° 15′	
Sines	54464	60529	65935	
Altitude of the plane	23687	23687	23687	Sub.

82 *TBOO* II p. 348.

83 The *libellum sphaericum* would most likely be Book III of Maestlin's *Epitome*; cf. fol.)()(5v of the 1610 edition: "This [third] book is entirely and exclusively Spherical." However, Book III contains no computation of the polar altitude. At the end of Book IV (ed. 1610 p. 541), he gives the latitude of Tübingen as 48° 30′, but does not show the means of determining it; it is only included as part of an example.

84 *Prosthaphaeresis*. This is the method, described above, of using a trigonometric identity to substitute addition or subtraction for multiplication; see note 49, above.

Remainders	30777	36842	42248	Dividends
Results	48873	58503	67088	
Corresponding arcs	29° 15½′	35° 48′	42° 8′	
Complements	60° 44½′	54° 12′	47° 52′	
Hours	4.03	3.37	3.11½	
Time (am)	7.57	8.23	8.48½	

These are sufficient to set out this problem clearly. For the rest, we shall continue the example by Problems 24 and 26, first taking the visible latitude, longitude, and so on, then the true ones, at these four moments.

Sun's positions	7° 25½′	Leo	7° 27′ 10″	Leo	7° 28′ 12″	Leo	7° 29′ 14″	Leo
Right ascension	129.51		129.53		129.54		129.55	
Distance from meridian	71.20		60.45		54.12		47.52	
Rt. Asc. of Mer.	58.31		69. 8		75.42		82. 3	
Culminating	0.41	Gem.	10.44	Gem.	16.51	Gem.	22.43	Gem.
Declin. of these	20.22		22. 8		22.52½		23.19½	Sub.
Alt. of pole	48.24		48.24		48.24		48.24	
Sides MV in diag. of Prob. 24	28. 2		26.16		25.31½		25. 4½	
Obl. Asc. of Horosc.[a]	148.31		159. 8		165.42		172. 3	
Resulting rising deg.	7. 5	Vir.	14. 3	Vir.	19.34	Vir.	24.12	Vir.
On the Nonagesimal	7. 5	Gem.	14. 3	Gem.	19.34	Gem.	24.12	Gem.
400 Therefore, sides NM	6.24		3.19		2.43		1.29	
Computation shows NV	27.20		26. 4		25.23		25. 2	
Complements	62.40		63.56		64.37		64.58	
In triangle NSV NS is	60.20		53.24		47.54		43.17	Sun's dist. from nonag.
Therefore, by SV or by SN, NSV is found	30.43		31.36		32.35		34.15	

a. The "horoscope" is the same as the Ascendant, that is, the degree of the ecliptic that is on the eastern horizon at the given moment. Its "oblique ascension" is the degree of the celestial equator that is on the eastern horizon at the same moment; it is therefore 90° east of the Right Ascension of the Meridian (as comparison of the numbers in the table show). It is a useful number because it is equivalent to the arc between the equinox point and the horizon, and this, together with the altitude of the pole and the obliquity of the ecliptic, allow one to calculate the ascendant (as Kepler does in the next line of the table).

These things so far are from the motion of the *primum mobile*. Now, in the diagram of Problem 24, S in this particular instance is not between N, M, but N is closest to M—I note this to avoid confusion. And thus, if, facing the sun, you number the angle NSV from the upper part of the vertical to the ecliptic towards the right and west, 30° 43′ at the first moment, and the angle from the lower part of the vertical to the circle through the centers of the luminaries, 88° from observation, the remainder between the circle through the centers and the ecliptic will be 61° 17′. The moon is ahead of the sun to the south. At the second moment, the upper angle was 31° 36′, the lower was observed to be 2° 30′. Therefore, the remainder in the semicircle is 145° 54′ between the circle through the centers and the western part of the ecliptic; therefore, between this same circle and the eastern part (the moon being to the south, as before) is 34° 6′. Here, therefore, the moon is nearer to the east, and beyond the sun.

In the other two moments, because the inclinations are rather large, it is a good idea to find out whether the precaution of Problem 14 is necessary. The sun's altitude is 37° 15′. The inclination, 19°, is in fact in the plane of the instrument, which was set perpendicularly, not to the sun, but to the azimuthal point of the horizon.[85] Now when two planes are tangent to a sphere on the same great circle, or are equidistant from the tangent planes, their mutual intersection is a straight line, perpendicular to the plane of that great circle; and a line from the center of the sphere extended out through the plane of the great circle so as to meet that mutual intersection, will fall upon it at right angles. For by Euclid XI 4, a line perpendicular to a plane is perpendicular to all the lines of the plane. In the present undertaking, we have three such planes. For the great circle here is the vertical circle drawn through the center of the sun. Imagine that it is touched at the zenith by one plane: the plane of the horizon is therefore equidistant from it. Imagine that it is touched by another plane at its intersection with the horizon: we have said that the plane of the instrument was equidistant from that plane. Imagine thirdly that it is touched **401**
at the center of the sun by a third plane. Therefore, the intersection of this plane and the plane of the horizon falls outside the sphere. But the intersection of the plane of the instrument with the plane of the horizon falls within the sphere.

Let AXY be the plane of the horizon, STV that of the instrument, SXY the third plane,[86] A the center of the sphere, S the sun, TV, XY the intersections. In the plane TSV, let another line SV be set up at TS making the angle TSV 19°, and let AV be extended to XY and let SY be joined. And because ATV, ATS, AXY are right, TS will be the sine of the sun's altitude, 60,529. But TSV was found to be 19°. Therefore, TV is 20,842. Further, AT is the sine of the complement of the altitude, 79,600. But the angle TSX is equal to the sun's altitude, 37° 15′. Therefore, TX is 46,027. Therefore, the whole AX is 125,627. But as AT is to TV, so is AX to XY, which accordingly becomes 32,893. And because STX is right, and TSX is 37° 15′, therefore, as the whole is to ST, so is the secant TSX to SX, which becomes 76,051. Therefore, because SCY is right (for SX is in the plane of the vertical, XY in the plane of the horizon), as XS is to XY, so is the whole to the tangent XSY sought, which becomes 23° 23′ greater [than before].

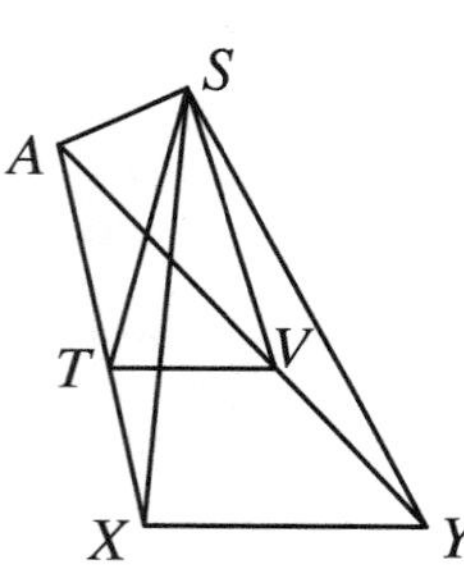

---

85 That is, the tablet upon which the inclination was measured was not set perpendicularly to the ray coming in from the sun, but was in a vertical plane set perpendicularly to a line coming in from the sun's azimuth. This would have the effect of distorting the inclination (just as it distorts the sun's image), making the measured inclination smaller than the true inclination.

86 This is the plane perpendicular to the ray coming from the sun; it is required to find the angle XSY which is the true inclination.

Thus at the end of the eclipse, when the sun's altitude was 41° 15′, the inclination on the instrument being 30°, the angle XSY was found right on the circumference of the sphere, by a similar process, to be 37° 31′. And this being so, this inclination is correct in both cases.

Since, therefore, at the third moment, the angle on the upper right between the ecliptic and the vertical is 32° 35′, as was proved above from the time, and the lower left angle is equal to it, while the angle of the vertical and the circle through the centers, likewise on the lower left, is 23° 23′, therefore, when the latter is subtracted from the former, there remains between the circle through the centers and the ecliptic 9° 12′, the moon being to the south.

At the fourth moment, the prior angle above was 34° 15′; the latter of the same conditions was 37° 31′ and greater. Therefore, when the former is subtracted from the latter, there remains between the ecliptic and the one through the centers 3° 15′, the moon now being to the north.

402 For the rest, in Problem 14 I advised that this method is risky, because of the wobbling of the hands. How much it will be this time will not be concealed in the outcome.

Now, with these angles found, we shall establish the visible longitude and latitude at all 4 moments. For in the small triangles LSE of Prob. 24, the bases LS and the angles are given, and consequently the sides LE of latitude, SE of longitude, as well. Moreover, the bases at the first and second moment are equal, because in both cases it appeared to be eclipsed by a semidiameter. Therefore the distance, or LS, was found above to be 14′ 55″. At the third moment, because one quarter of the enlarged ray appeared to be eclipsed, therefore, let this be subtracted from the sum of the radii: as, because the sun's radius is 24, the moon's 23, the sum 47, a quarter of the sun's diameter 12, therefore the remainder is 35 in the enlarged ray. But the centers remain in their places, by Problem 12. And the lunar radius is diminished by the same amount that the solar is increased; and so the sum of the radii also remains the same, as also in Problem 28. Therefore, the ratio of the sum to the distance of the centers is the same. But because the sum of the radii is really 30′ 27″, therefore, as 47 is to 35, so is the sum to the distance of centers, 22′ 40″. But at the end, the distance of the centers is equal to the sum of the radii, 30′ 27″. I nevertheless am employing this degree of precision only for the sake of the example. For as far as concerns the issue itself, I recall that in the midst of observing the openings were changed, and so the quantity will very likely have been different in the later times; and not everything corresponds to this exactly. But I carry on with the example.

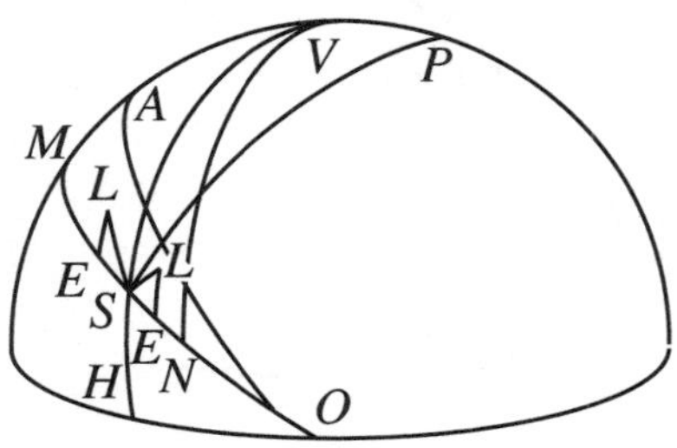

Now, with the bases known, since they are hardly different from straight lines, we shall multiply them by the sines of the angles, and, the last five digits being dropped, the latitudes will come out, with the sines of the complements being the visible longitudes of the moon from the sun.

	First moment before the sun		*Second after the sun*		*Third*			*Fourth*		
	7′.	10″.	12′.	21″.	22′.	23″.		30′.	24″.	
Latitude	13.	3.	8.	22.	3.	37.	South	1.	44.	North

The same thing will result if you will follow the short cut appended to Problem 24.

Up to this point, we have treated the example by Problem 24. Now, to give full and sufficient attention to Problem 26 also, let us also consider parallaxes. **403** And because Tycho Brahe gave the moon a distance of $56\frac{1}{2}$ earth semidiameters at its mean distance, therefore, using an eccentricity of 4336, which I consider close to the truth, at the anomaly of this eclipse the moon will $58\frac{1}{2}$ semidiameters distant from earth. Consequently, the horizontal parallax is 58′ 54″, while the sun's is 2′ 58″. Therefore, the moon's parallax in excess of the sun's is about 56 at the horizon.

So, by the arcs *NV* found above, under the heading of 56′ I take the parallaxes of latitude 25′ 42″; 24′ 36″; 24′ 1″; 23′ 42″. The southern visible latitudes subtracted from these, and the northern ones added, make the true latitudes 12′ 39″, 16′ 14″, 20′ 24″, 25′ 26″.

Then, using the complements of *NV* I take new headings from the parallactic table, 49′ 50″; 50′ 20″; 50′ 36″; 50′ 45″. And beneath these heads, using the visible distances of the moon from the nonagesimal (or the approximate distances), that is, 60′ 13″; 53′ 36″; 48′ 15″; 43′ 47″; I take the parallaxes of longitude of the moon from the sun, 43′ 15″; 41′ 31″; 37′ 45″; 37′ 7″. I increase these by the moon's observed precessions [ahead of the sun], and diminish them by the separations, because the eclipse is to the east: the results are the true longitudes of the moon before the sun, 50′ 25″; 29′ 10″; 15′ 22″; 4′ 43″. And thus the true conjunction followed the end of the eclipse. And because the hourly motion of the moon from the sun resulting from this is 29′ 16″ (Tycho gives 27′ 56″), therefore, dividing the remaining 4′ 43″ by this shows about 10 minutes by which the true conjuction falls after the end of the eclipse, that is, at $8^{\mathrm{h}}\ 58\frac{1}{2}^{\mathrm{m}}$ a.m.

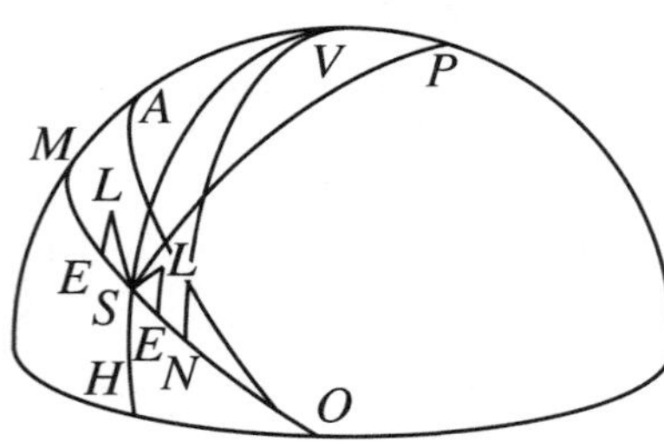

But about this true longitude one must have a little doubt, and about the latitude not a little. For within half an hour, that true latitude can vary by barely 4 minutes; here, there are almost 13. And even if we use the last uncorrected inclinations of 19° and 30°, there will still be 9 minutes in the variation of latitudes. And so I attribute this to the shaking of the hands and to the error-prone method of observing. Tycho's computation at the meridian of Hven, puts the moment of true conjunction back at an apparent time of $9^{\mathrm{h}}\ 2^{\mathrm{m}}$. Therefore, the difference of meridians would be 1 degree, excessively small. Let us therefore consult the

Difference of meridians of Tübingen and Uraniburg

observation at Uraniburg, so that we may also at the same time have something more certain about the moon's latitude.[87] At 6^h 53^m, the eclipse was just noticed; at 9^h 0′, the whole sun was shining. The observers judged that the eclipse had begun at 6^h 50^m, and ended at 8^h 58^m, so that the duration would have been 2^h 8^m. Pictures were also added, the first of which represents the moon a little higher
404 than the sun. And at 7^h 15^m, the luminaries were in equal balance. At 8^h 21′ it was added, that the sun was one fifth eclipsed, and that the horns, head foremost, were turned equally downward. There is no more trustworthy notation. We shall consider the last first. One fifth of the diameter is 6 minutes. Therefore, the distance of the centers is 25′ 27″, and that is right on the vertical. But the time of 8^h 21^m (so that we might take this short cut too) shows the altitude of the sun, and the altitude shows the combined parallax, or the parallax of longitude-latitude,[88] which, added to the moon's altitude, immediately establishes the true altitude. From this, afterwards, by the mediation of the angle of the vertical with the ecliptic, the true latitude and longitude of the moon is given immediately.[89]

Declination of the sun	18°.	28′.		
Alt. of equat.	34.	5.		
	52.	33.	79388	
	15.	37.	26920	
			52468	
			26234	altitude of the plane.
			53154	elevation of the meridian sun above the plane.

The distance from noon of 8^h 21^m is 3^h 39^m; the degrees, 54° 45′.

The complement is 35° 15′.

Therefore, as the whole sine is to 53,154, so is the sine of the complement, 57,715, to the sun's elevation above the plane at that time.

53154	32.	7				
	54.	45.				
	86.	52.	99851			
	22.	38.	38483			
			61368			
Dropped from the sun to the plane			30684			
			26234			
On the horizon			56918	34°.	42′.	Sun's altitude.
				55.	18.	Complement

To this distance of the sun from the zenith, the visible distance of the centers is added (because the moon is lower), making 55° 43′ 27″. This, under the heading of 56′ in the parallactic table shows a parallax of 46′ 16″. The visible distance subtracted from this leaves 20′ 49″, the true distance of centers.

405 But as the sine of the complement of the sun's altitude is to the sine of the angle 54° 45′, so is the sine of the altitude of the equator to the sine of the

87 The observations are in *TBOO* XII pp. 8–10.

88 μηκοπλατῆ. See Ch. 9 p. 325.

89 The computations below follow the same procedure as the one used earlier in this same problem. Kepler therefore thought it unnecessary to label each step. As before, multiplications are avoided by using the addition/subtraction method.

angle between the ecliptic and the meridian, 33° 49′, the complement of which is 56° 11′.[90] These, under the headings of 20′ and 49″ (which is the distance of centers) shows the moon's true latitude, 11′ 33″ north, the longitude 17′ 14″ before the moon.

I was expecting that the Tychonic observations would help Maestlin on the latitude. But I go away more uncertain than before. For Tycho's computation, founded upon lunar eclipses, shows a latitude of 16′ 47″ at the middle of the eclipse. At this moment, therefore, it ought to have become greater, because it is after the middle. Give me a greater eclipse at this moment, and I shall construct a greater latitude. And so you see how much the sense of vision deceives. Perhaps, too, for $\frac{1}{3}$ they wrote $\frac{1}{5}$. For how could it happen that for Maestlin the eclipse, in restoring a quarter of the diameter, took no more than 25 minutes of time; for the Tychonic observers, in restoring a fifth, fully 35^{m}. And so a third is 10^{m}; previously it was 6^{m}; from this we have constructed a latitude of about 15′ 30″. I, for my part, drive these things home, though with some uncertainty, for no empty reason. For I wish to lay it out before astronomers, how frequent are the opportunities for being deluded, and, in turn, how great and how greatly desired is the usefulness, if diligence be applied in observations of this kind.

But because at Tübingen the time of the end has no evident error, let us also examine the observed end at Uraniburg. Let the moon's true latitude at the end be 17′ north, and the end be exactly at 9^{h} 0^{m}. The degrees of distance from the meridian, 45°. The right ascension of the meridian, 84° 55′. The culminating degree 25° 20′ Gemini, with declination 23° 26$\frac{1}{2}$′. This, subtracted from the altitude of the pole, 55° 54$\frac{1}{2}$′, leaves 32° 28′, the side MV, in the diagram of Prob. 24. And because 174° 55′ is rising, the co-rising point on the ecliptic is therefore 26° 37′ Virgo, and the nonagesimal is 26° 37′ Gemini. Therefore, the side NM is 1° 17′, and NV by computation is 32° 26′, showing a latitudinal parallax (under the heading of 56′) of 30′ 3″; its complement, 57° 34′, shows a heading of 47′ 17″. Let the moon be exactly at 8° Leo, to the sight. It is therefore 41° 23′ from the nonagesimal, which, from the headings of 47′ and 17″ shows a longitudinal parallax of 31′ 15″. The latitude, in turn, is placed at 17′ north, the latitudinal parallax 30′ 3″. Therefore, the visible latitude is 13′ 3″ south. But the visible distance of centers is 30′ 27″. Therefore, from the base and the side, **406**
the remaining side of visible longitude is 27′ 31″. And the parallax of longitude is 31′ 15″. When the former is subtracted from the latter, the remainder is 3′ 44″, the interval between the true positions of the sun and the moon at Hven in Denmark. If we use the same latitude at Tübingen, and subtract 23′ 42″ from the parallax of latitude, the visible latitude will be 6′ 42″ south. And the base of the small

90 This computation uses the sine rule for spherical triangles to find the angle at the sun in the sun-zenith-pole triangle; i.e., the angle between the vertical circle and the circle of the sun's declination. This angle is not the same as the angle between the ecliptic and the meridian. The next step in the computation shows that what Kepler needed here, and thought he had found, was the angle between the ecliptic and the vertical circle through the sun. Because the position is not far from the solstice, the difference was not great enough to make the results obviously wrong.

triangle is 30′ 27″. Therefore, the side of longitude is 29′ 42″. But the parallax of longitude is 35′ 7″. And so the difference of the moon's positions is 1′ 31″, which make 3 minutes of time. Therefore, at Tübingen, at $8^{h}\ 51\frac{1}{2}^{m}$ (with 3 minutes added at the end), the moon was at the place where it was at Uraniburg at 9^{h} 0′. The difference of meridians is $8\frac{1}{2}$ minutes, in degrees 2° 8′. Less, too, if at Hven the eclipse had ended at $8^{h}\ 58^{m}$. And we do not differ much using this way: at Tübingen, the defect was seen equally [i.e., equal in size] at $7^{h}\ 14^{m}$ and $7^{h}\ 57^{m}$. Therefore, the middle of the eclipse is close to the middle time, that is, at $7^{h}\ 36^{m}$. But at Hven it began at $6^{h}\ 50^{m}$, ended at $9^{h}\ 0^{m}$. Half the duration is $1^{h}\ 5^{m}$. Therefore, the middle is at $7^{h}\ 55^{m}$. But the parallax of longitude at Tübingen exceeds the Danish by 4′. Therefore, the eclipse at Tübingen really happens earlier by 8 minutes of time. And thus, when the difference of parallaxes is removed, the middle at Tübingen would be $7^{h}\ 44^{m}$; the difference of meridians, 11^{m} of time or 2° 45′ of the equator. And because at Hven the moon has 3′ 44″ left to reach the sun, and thus almost 8^{m} of time, the true conjunction is at an apparent time of $9^{h}\ 8^{m}$. Tycho places it at $9^{h}\ 2^{m}$ apparent, a difference of 6^{m} of time, 3′ minutes of the moon's motion. From this, the certainty of the Tychonic computation at this place is evident. Below, in Prob. 32, there is more on this eclipse.

A more exact difference of the meridians of Tübingen and Uraniburg. Differently.

Problem 30

From the time and inclinations of the principal phases, correctly observed, to give the angle of visual latitude, or the angle which the moon's visual path makes with the ecliptic. Where this angle is portentous; and a warning about Ptolemy, Reinhold, and their disciples

This is the very thing we have just set out in both of the preceding examples, and we have set it out several times. In the diagram of Prob. 27, given the moments of time and the inclinations, and the distance of centers FA, FC, that is, once the phase is designated,
407 the moon's visual longitude from the sun, FE, FD, will not be unknown; likewise, the visual latitude EA, DC, by Problem 24.

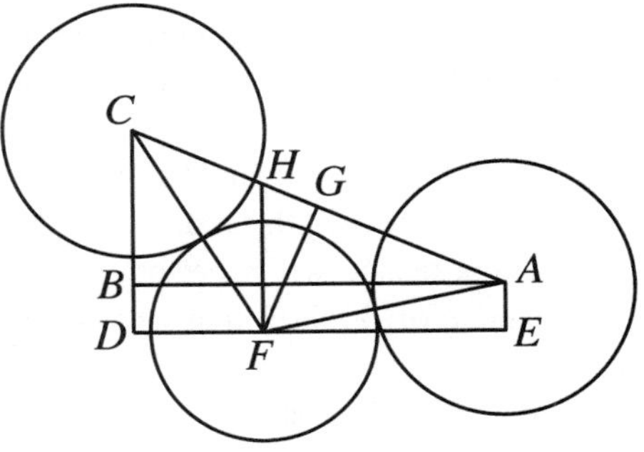

Therefore, when the smaller latitude EA is subtracted from the greater DC, if they were in the same direction; added, if they were in different directions; the side BC will be obtained. But BA and DE are equal, and DE is composed of the moon's amount of preceding [praecessio] EF and its excess [superatio] FD. Thus, with the sides about the right angle being given, the angle BAC will not be unknown. So, in the eclipse of 1598, at the beginning the moon's latitude was 1′ 57″ south. The moon was 29′ 3″ in front of the sun. At the end, the latitude was either 17′ 6″ or 19′ 21″, and the moon was beyond the sun by either 26′ 40″ or 25′ 4″. Consequently, *AB* is 54′ and *BC* is approximately 20′, and *BAC* is about 21′.[91]

In the other example, we shall proceed gradually, following, not the mistrusted visual latitude of the end, such as is taken from the example, but the one that followed from the presupposed right latitude. Therefore, when the moon was 7′ 10″ before the sun, the visual latitude was 13′ 3″ south. At the end, we have calculated that it ought to have been at a distance of 29′ 42″ beyond the sun, with a visual latitude of 6′ 42″. Subtracting the latter from the former gives 6′ 21″, which will now be side *BC*. When in turn the longitudes are added, they make *BA* 36′ 52″. Therefore, angle *BAC* is about 10°. See another, mechanical way below, in Prob. 31.

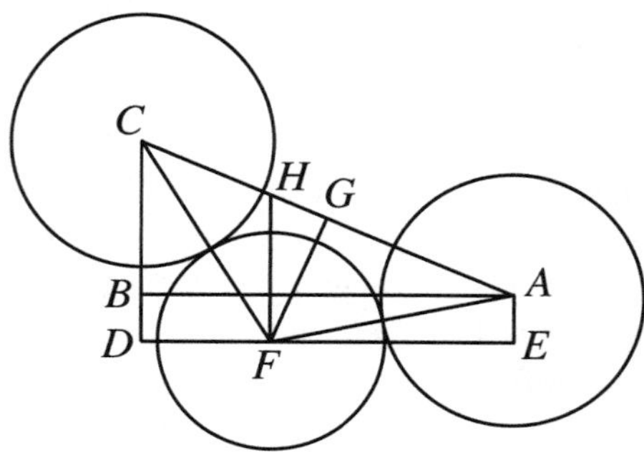

Further, since the angle of true latitude is not greater than 5°, it cannot be said in how great a confusion I was stuck for how long, while Ptolemy and Reinhold presented me an absurd undertaking. For although Maestlin warned me in letters about the parallaxes which bring this about, still turning before my eyes were the method of Ptolemy, Precepts 63, 64, and 65 of the *Prutenics*, and the examples of Magini, all of whom treat parallaxes first, then at length proceed to this angle, or to the initial and final visual latitude; and this they seldom vary by more than 6 minutes from the beginning to the end of the greatest eclipses.Arguments also entered in: if the authorities were to neglect parallax, how could the beginnings and ends of eclipses come out right? Indeed, how could Ptolemy use these correctly to construct the inclinations, to which he attributes a great deal in the foretelling of events, so much so that he asserts that the regions indicated are those towards which the inclinations are directed?

And so I was of the erroneous opinion for a very long time, that some miraculous **408** inequality lay hidden in the moon's motion, not noticed by the authorities, and that this shows itself chiefly around eclipses, when the moon is passing across the ecliptic.

And thus I have thought it worth while to warn others as well, if any perchance will be sailing this sea, to beware of this ledge.

Such a person will therefore note first, that Ptolemy openly asserts, in the last chapters of Book 6, that he was not going to follow the highest precision in this undertaking; and consequently judged that these inclinations in the directions of the winds should not be very precisely noted.[92] But Reinhold, whom

---

91 Previously (at the end of Prob. 25, on p. 391 above), Kepler had given the latitude, correctly, as north. In Problem 32 below, he recomputes the angle *BAC* using the correct latitude and finds it to be 15° 12′ 40″. Again, evidence of haste.

92 Cf. Ptolemy, *Almagest* VI 11, trs. Toomer p. 314.

Magini followed, does not give visual latitudes of the moon at the beginning and end despite his having called them "visual": note this well. For the examples of eclipses, and the truer method of computing proposed in Prob. 25, and the discussion of Ptolemy himself on the variation of parallaxes, cry out in objection. Reinhold enters the table of the moon's latitude through the moon's motion, corresponding in minutes and seconds to the incidence and emergence, and from this asserts that he is referring to the visual latitudes. But that table of latitude was constructed from a constant angle of 5 degrees.[93] It would therefore be required that the parallax of latitude not vary from beginning to end, if the visual latitude could truly be taken by Reinhold's method. But Ptolemy himself asserts, and the tables bear witness, that the parallaxes of latitude change substantially at every moment. It is therefore not exactly the visual latitude, but quasi-visual, that is taken by Precept 64. For the parallax of latitude is indeed applied, but it is not the proper one, but one carried over from the mean to the ends of the eclipses; and it is not the true diagram of the eclipse that is constructed following Precept 65 of the *Prutenics*, but one nearly true. For most of the time, the visible path of the moon is twice, thrice, or four times more oblique to the ecliptic; and consequently, the maximum darkening differs considerably from the instant of visual conjunction in longitude, and the times of incidence and emergence vary.

I would also add here an example from Walther's observations, which finally led me back to the path, after straying for a very long time. In this, I learned that the same thing that happened, to my wonder, at conjunctions and at the nodes, is sometimes seen at the time between the months, when the moon is halved, thirty-seven degrees removed from the node.

409 Thus, Bernard Walther relates that on 12 January 1482, in the morning, two and a half hours before sunrise, the moon, now nearly bisected, was seen by him to move around towards Saturn, which it would also cover later, when it was around the meridian. The description is as follows. *When the moon was at its last quadrature, or thereabouts, it was in any case darkened from the direction of the west. And when it was first noticed, which was about $2\frac{1}{2}$ hours before sunrise, Saturn was to the east, and, as it seemed, to the south of the moon, distant by two moons. Afterwards clouds had come in the way, so that I was unable to see the beginning [of the occultation]. I judged it for certain, however, that the moon was going to overtake Saturn with its southern horn. However, after I saw Saturn again, it was distant on the diameter by two digits or thereabouts from the northern horn; and then, in turn, it began to reappear. They had now passed across the meridian. But at the time that I judged to be the middle of this eclipse, I took the moon's altitude, when it was almost on the meridian line, and found 32 degrees.*

What I set down above, that the moon was at first going to overtake Saturn with the southern horn, DOES NOT APPEAR POSSIBLE, CONSIDERING THE

---

93 This is the inclination of the moon's true path to the ecliptic. However, parallax intervenes to make the apparent angle different.

MOON'S PATH.[94] *But this was in fact very obviously apparent, that Saturn was at a distance of two digits or thereabouts on the moon's diameter from the northern horn.* He is worthy, and a well-known author, and the observation careful, and especially there is that, on account of which we have introduced this: that we should apply no less diligence in weighing it.

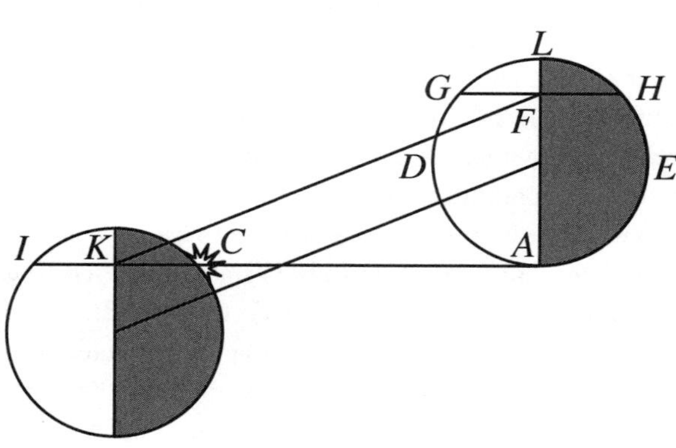

Therefore, let the moon's diameter or the section dividing the shining part from the dark, be AL, and let E be the darkened part at the western side of the ecliptic (or the great circle between the sun and the moon), D at the eastern side towards the sun. Let a line AC be drawn at right angles to LA tangent to the moon at A, and let it be AC. Since, then, the moon is a little more than a quadrant away from the sun, a great circle through AL extended will pass through the poles both of the ecliptic and of the great circle drawn through the centers of the luminaries, and will cut both at right angles. And the arcs of those circles about this place of a quadrant[95] will be nearly parallel, but AL also cuts line AC at right angles. Therefore, AC is in a way parallel to both the ecliptic and to the circle through the luminaries. Let AC be made twice AL, in accord with Saturn's having been seen at the beginning at two moons' distance from the moon, and let the star Saturn be placed at C, which, in this place, will really be seen as south of the moon at AL (because AC is parallel to the ecliptic) and in any event in such **410**
a position that if the moon's center were to approach the sun directly towards D, it would appear that it would catch Saturn at C exactly with its horn A. Now, when Saturn was uncovered, because it was then seen to be two digits from the northern horn in the moon's diameter, let the diameter AL be divided into 6 parts, and let LF be a sixth, or two digits, and through F let a straight line be drawn at right angles to LA, cutting the circumference at G, H, and let a line equal to GH be further extended on AC, and let this be CI, and through points C, I let the diagram or circle of the moon be described. For since the moon, as before, is situated at quadrature, the diameters or boundaries of the luminous part in the

[94] Emphasis added by Kepler. He also made some minor changes, such as altering *vel citra* (or beyond) to *vel circa* (or thereabouts). The quotation is from Ioannes Regiomontanus, *Scripta de torqueto* (Nürnberg 1544) 49v–50r (*Observationes factae per doctissimum virum Bernardum Waltherum*).

[95] The "circles" are the ecliptic and a great circle through the moon and the sun, which is within a few degrees of the ecliptic. These two circles intersect at the sun, and since the moon is about a quadrant from the sun, the parts of these circles in that region are at their greatest distance from each other, and are nearly parallel.

two diagrams will remain approximately parallel, and will be at right angles to the same CA. And thus will this observation of Walther have been delineated geometrically.

Now let CI be bisected at K, and let K, F be joined. And let the diameter LA in some way be given a measure, such as 200,000. LF will be the sixth part, that is, 33,333. Consequently, FA is 166,667, and FH or KC, 74,528, while CA is 400,000. And the whole KA is therefore 474,528. But as KA is to the whole sine, so is AF to the tangent of angle FKA. Therefore, angle FKA is $19^\circ\ 21'$; namely, the angle that the visual path of point F, which is FK, makes with the ecliptic, to which KA is nearly parallel here. Here, then, although the moon is not carried by such a precipitous motion in latitude as when right at the nodes, still, it is simultaneously tending towards the descending node while Sagittarius and Capricorn and Aquarius are rising (through which signs the angle between the ecliptic and the horizon is decreasing), and in turn, the latitudinal parallax is increasing. As a result of these things, by both its true motion and its visual semblance it is simultaneously carried to the south. Therefore, something happened even to the experienced astronomer Walther, of such a sort that not only seemed to him a marvel, but even led him to doubt the trustworthiness of his own
411 eyes.

Problem 31

Whether it is possible for the beginning of some solar eclipse to decline towards the east, and for the end of another to decline to the west; but for the beginning of an eclipsed moon to be in the west, and at another time for the end to be in the east

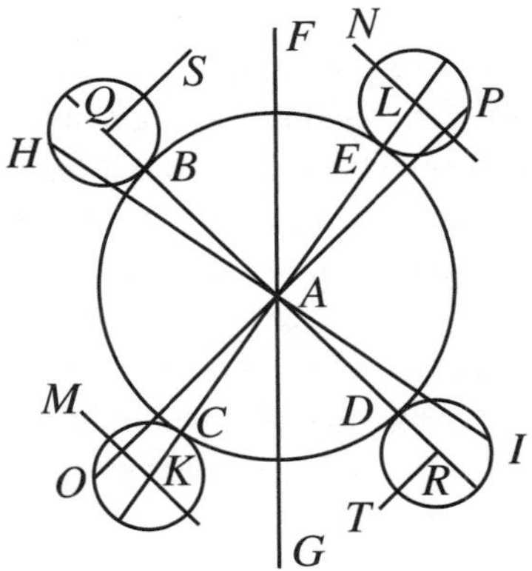

This is among the paradoxes proposed by Pliny.[96] For it befits the sun to be eclipsed by the ordinary way from the west in the beginning, and to be made full from the east; and it befits the moon, on the contrary, to begin from the east, and to end from the west. Let us begin from the moon. About center A let the circle of the earth's shadow $BCDE$ be described; throught A let the straight line FG be drawn representing the vertical; through the same let the arc of the ecliptic HI closest to the west also be drawn, so that FAH may be the angle between the ecliptic and the vertical. And let KL be drawn at right angles to HI. Should the center of the moon be at K or L, it will be the moment of true conjunction. Then, accordingly, let the sum of the radii of the moon and the shadow equal the moon's true latitude, decreasing southern at K, so that the moon's path may be KM towards the ecliptic; increasing northern at

96 *Natural History* II 10 57, trs. Rackham pp. 206–7.

L, so that the moon's path may be *LN* away from the ecliptic. In both cases, then, some small part of the moon will be eclipsed (for in this particular matter as well, the authorities require this slight correction); for the moon, a little past *K* and a little before *L*, becomes closer to the center *A* because of the obliquity of the moon's path with respect to the ecliptic; and in that point of its path upon which a perpendicular from *A* falls, the eclipse is greatest. And, for the sake of a measure: as the secant of 5 degrees (the maximum latitude), 100,382, is to the radius, so is the sum of the radii, which we shall take as 60 minutes, to the distance of the centers at a maximum eclipse, 59′ 46″. Thus 14″ fall into eclipse, a small part, certainly; not as much as a hundredth. Nevertheless, we are following the highest points here. Nothing prevents the occurrence of the same thing we are demonstrating here, even when the moon at *K* has taken a little of the shadow.

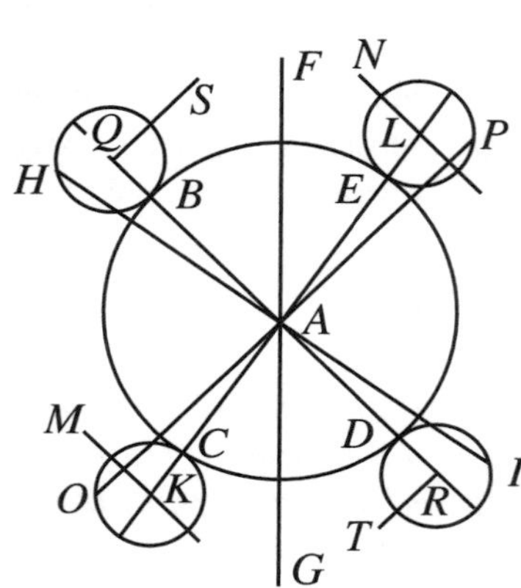

Now the beginning of the eclipse will be at *K*, the end at *L*, and the point of contact
of the circles *C*, or the index of the eclipse, 412
will verge to the west at the right, as well as the whole duration; while the end *E* will verge towards the east. Also, both can take place on an eastern arc. For let *OP* be an arc close to rising, and *FAP* be the angle, while *QR* is the perpendicular, and the moon at *Q* is northern descending, its path being *SQ*, while at *R* it is southern, likewise descending, or moving away from the node, by the path *RT*. Then, in the former case, the eclipse *B* at the beginning verges to the west at the right, but the final defect *D* will verge to the east.

Such an eclipse of the moon was seen most recently on 8/18 November 1603. Since this began about 65 degrees from the zenith, counting to the left, it ended with the shadow not quite reaching the zenith, but still verging to the left, towards the east. For the eclipse was in the southeastern quadrant, with the moon descending on the meridian.

It began ten minutes after the right shoulder of Aquarius culminated,[97] while already half an hour before, the light was observed to be becoming pallid at that very part, or somewhat to the lower left. And when the sun was at 25° 55′ Virgo. The time is thus demonstrated to be $6^h\ 21^m$.

It ended three minutes after the head of Andromeda culminated.[98] Therefore, the time was $8^h\ 17^m$; the duration, $1^h\ 56^m$. The computation of Tycho gives $2^h\ 10^m$. I in fact considered that about the middle less than a quarter was missing, but others argued that more than a quarter was hidden. Of the circumference, certainly less than $\frac{1}{3}$, more than $\frac{1}{4}$, was considered to be gone. These things are nevertheless in agreement: that the defect was smaller, and the duration shorter,

---

97 α Aquarii.

98 Alpheratz, α Andromedae.

than in the computation, and the shadow did not reach to the zenith. The middle is 7^{h} 19^{m}, which very minute the computation of Tycho shows at the meridian of Hven for the true conjunction in longitude, which differs somewhat from the middle of the eclipse.

The darkened circumference was seen, while the very bright Pleiades, distant by a few degrees, were hardly seen: so brightly was the moon illuminated, even in the shadow—which you should ascribe to the above.

As for the computation, it is fully consistent with this phenomenon. For at 8^{h} 17^{m}, the angle between the vertical and the ecliptic is 62° 48′, whose sine, multiplied by the sum of the radii, 60′ 16″ (for Tycho makes the radius of the shadow 44′ 6″, of which in its place), shows the latitude at this moment,
413 53′ 40″, the amount it would have been had the shadow ended exactly at the zenith. Now, since it declined slightly to the left, the angle between the ecliptic and the line passing through the centers is also slightly larger. Let it be larger by 5 degrees, which is the amount of the declining to the left. Therefore, the latitude at this moment is 55′ 48″. At the middle, then, an hour earlier, it is about 53′ 20″, and subtracting 44′ 6″ from this leaves 9′ 14″ of the moon's body in the shadow, a little more than 3 digits. Thus, therefore, the computation also requires this phenomenon. Tycho Brahe also saw something closely similar.[99] For on 19 Oct. 1594, at 5^{h} 56^{m} in the morning, the moon began to be darkened at its top edge, or, as the picture was captured in the other observatory, a little to the right, although this eclipse appeared quite large to those farther to the west. The moon therefore began to be darkened from the west, and was again made full from the same side (although beneath the earth, at Hven).

But the account and the amount are more obvious in solar eclipses, because of the parallaxes. For if you will adopt an angle of 20 degrees for the moon's visual path, as it had more than once come out to be, the secant, 106,418, divided into the sum of the radii—let this be taken as only 30—makes it $28\frac{1}{5}$; the remaining $1\frac{4}{5}$ units is hardly different from a digit in quantity. And so, when at the very instant of conjunction, the visual sum of the radii is equal to the visual latitude of the moon,[100] it can still be eclipsed by a digit.

It is, moreover, sufficient for the furthering of the demonstration to recount again the extremes of the occasions upon which this happens to occur. So again, as in the lunar eclipses, and as much as for the moon's true motion, one needs the moon's latitude to be either northern decreasing, and the eclipse in the east, or southern decreasing, and the eclipse in the west, if the eclipse of the sun is to begin from the east. Again, if it is to end from the west, the moon's latitude should be either northern increasing, the eclipse being in the west, or southern increasing, the eclipse being in the east.

---

99 See *TBOO* XII pp. 317–8.

100 Here Kepler refers to the following endnote.

To p. 413. Therefore, note that the true conjunction in longitude is one thing, the apparent conjunction in longitude is another, and the apparent conjunction according to nearness is yet another. For the moon, at the same visual longitude as the sun, is nevertheless not at the greatest visual nearness, unless the eclipse is exactly central.

But as far as concerns parallaxes, at whose charge is the governance of the case, those things demonstrated in Chapter Nine must be considered, and Copernicus's table of the the angles of the horizon must be inspected.[101] Thus when Aries is rising, the angles begin to increase, up to Libra, and when Cancer is rising they increase maximally. At that time, as a consequence, the parallax **414** of latitude is decreasing maximally, and the moon, in whatever sign it is above the horizon, is carried visibly to the north more than it is by its true motion to the south, and much more if it is ascending in its true motion. Then, as a consequence, when the sun stands to the west in a small southern eclipse, what was proposed can possibly happen, that the sun begins to be eclipsed from the east. On the contrary, a small northern eclipse standing to the east will be able to be filled from the west.

On the contrary, when Capricorn is rising, from Libra to Aries, the angle of the horizon is decreased, the parallax is increased, [and] the moon in its visual motion is carried to the south. Therefore, northern eclipses that stand in the eastern quadrant will begin to be eclipsed from the east; southern eclipses in the eastern quadrant will end from the west. Even so, this cause is so obvious that it is valid almost at the meridian itself, with the assistance of the others. For at the meridian the moon is driven swiftly backwards by a visual illusion. And so it appears to ascend to the north by almost as much, at least in suitable signs.

On 12 or 22 July 1599, in the morning, right at sunrise, a slight eclipse was seen by Tycho at Prague, almost at the vertex of the solar body.[102]

Now the moon, though with its true motion it descended towards the downward-leading node, was on the contrary carried by a visual illusion mostly to the north, with the parallaxes of latitude decreasing (for Leo was rising). Therefore, the inclination was a minimum, and the end looked towards the east. So, in southern latitudes, where the eclipse is small, the eclipse must have stood from the west from beginning to end. On 20 May 1593, at Zerbst, the sun was observed to be eclipsed by two digits from below; at the beginning, the eclipse verged slightly to the left, at the end more so. It therefore began from the east, when the moon in its motion of latitude was struggling up towards the north. At Hven it was not seen to be eclipsed at all, as the computation of time shows at $2^h\ 40'$.[103]

On 16 February 1588 at Hven, at $1^h\ 32'$ after noon, the sun began to be eclipsed; it ended at $2^h\ 51'$. At the beginning, the eclipse is depicted to decline from the vertical at about 36° to the right, which, while it was ascending, **415** still did not come to the vertex of the solar body.[104] For at the end, the inclination to the right and west is depicted as still 12 or 15 degrees. Leo was rising. The moon's true northern latitude was decreasing, but to the sight it was increased more, with the parallaxes of latitude diminished. On the other hand,

---

101 *On the Revolutions* II 10, trs. Rosen p. 76.

102 *TBOO* XIII p. 160.

103 *TBOO* XII pp. 267–8. This includes observations at Rostock and Zerbst, as well as Tycho's computation.

104 *TBOO* XI pp. 245–6.

through the great increments of longitudinal parallax (for both the heading in the table and the fractional part of the heading were increasing, the former as the angle of the horizon was increasing, and the latter because the moon was close to the nonagesimal), the moon was greatly retarded in its motion from west to east.

But so that no doubt may remain, I shall here compute the inclinations, such as ought to have been seen at the beginning and the end. The sun was at 7° 17′ Pisces | 7° 20′ Pisces. The right ascension was 339° 0′ | 339° 3′. The distance in time from the meridian was 23° 0′ | 42° 45′. Therefore, the right ascension of the meridian was 2° 0′ | 21° 48′. The degrees on the meridian are 2° 11′ Aries | 23° 34′ Aries, the northern declination of which is 0° 52′ | 9° 12′. The altitude of the pole is 55° 55′. Therefore, the altitude of the culminating points is 34° 57′ | 43° 17′. But 1° 56′ Leo | 15° 6′ Leo are rising. And the quadrangle[105] of these is on the nonagesimal. Therefore, between the nonagesimal and the culminating point is 29° 45′ | 21° 32′. Hence, the distance of the nonagesimal from the zenith is 48° 43′ | 42° 31′. Under the headings of 57′ 20″, the moon's parallax from the sun, they select the parallaxes of latitude, 43′ 5″ | 38′ 44″. But their complements, 41° 17′ | 47° 29′, elicit the headings of 37′ 49″ | 42′ 16″. And because the sun or moon is distant from the nonagesimal by about 54° 39′ | 67° 46′, by these, under the headings found, the parallaxes of longitude are elicited, 31′ 1″ | 39′ 6″. But through the same distances of the sun from the nonagesimal, the angles between the ecliptic and the vertical come out to be 54° 24′ | 44° 44′. Now, let the sun's radius be taken as 15′ 20″; the moon's, 15′ 58″; the sum, 31′ 18″, the base of the small triangle. From Tycho's computation, let the latitude also be taken as 1° 8′ | 1° 5′. When the parallaxes of latitude are subtracted from these, the remainders are 24′ 55″ | 26′ 16″, one side of the small triangle about the right angle. From the base and the side are given the angles opposite the latitude, 52° 45′ | 57° 3′. And, so as to make the agreement evident, from the same procedure come the sides of the longitude, 19′ 0″ | 17′ 21″, the former before, the latter after the sun. The former, subtracted from the parallax of longitude, the latter added,
416 each to its own, give the true longitudes beyond the sun, 12′ 1″ | 56′ 27″. Therefore, in 1^{h} 19^{m} the moon's true motion is 44′ 26″, so the hourly motion is nearly 34′, slightly greater than is correct, because, just as the beginning is always observed later than is correct, the end is observed earlier than the true. But to the angles: let AB be the vertical, CB the ecliptic. Angle CBA is 54° 24′ | 44° 44′; let BE be the arc through the centers; and EBD is 52° 45′ | 57° 3′. Therefore, ABD at the beginning of the eclipse is 125° 36′, as a result of which ABE is 72° 51′. But at the end, because ABC is 44° 44′ and the greater EBC is

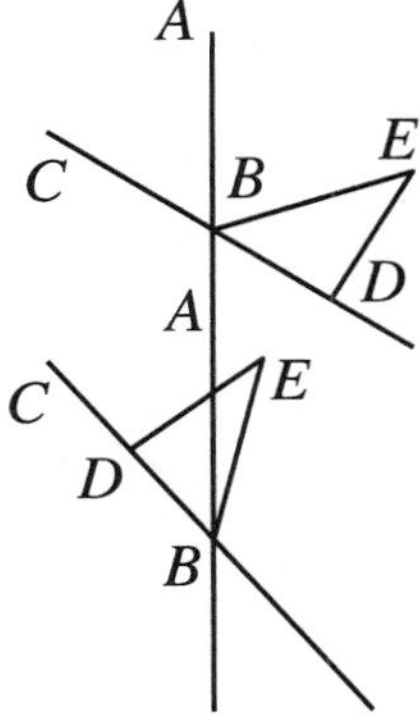

[105] This is astrological terminology, meaning that the nonagesimal is 90° removed from the rising degree.

57° 3′, therefore EBA is the excess, 12° 19′, E still standing to the right of the vertical, or towards the west, which was what was to be sought out through the computation. But that the first angle comes out twice as great as mine, which was taken from Tycho's diagrams, I think happened because of an erroneous picture, or because in measuring I had by mistake said the simple angle instead of the double.

A completely similar eclipse was seen on 23 Sept. or 3 Oct. 1595.[106] For Maestlin, in the disputation on eclipses published in 1596, in thesis 53, describes it thus. "In the beginning, a little past noon, the edge of the eclipsed sun declined from the vertical, not to the west, but 9 degrees to the east; it was, moreover, eclipsed by 2 digits and a half, no more, by the testimony of careful observation." This was at Tübingen.

At Strassburg, as I have found in Tycho's observations, it was noted by a certain observer to have begun before eleven o'clock, finished at one o'clock precisely. All the diagrams show that the shadow always declined to the left from the vertex.

I also observed the same eclipse at Graz; it began for me exactly at the vertex, when the sun's distance from the zenith was $51\frac{1}{6}°$, by a small wooden quadrant. When there appeared to me to be three digits missing on the diameter, the sun's distance was 55° from the zenith. A little later, the eclipse appeared to decline by 15 degrees. The moon's diameter appeared to be less than the solar on the ray. At Uraniburg, they observed its end at $2^h\ 5^m$. Four digits were eclipsed.

In this eclipse, the true and the visual motions of the moon were in the same 417
direction. For the moon was revolving towards the downward-bearing node. And with Sagittarius and Capricorn rising, the angle of the horizon was decreasing very rapidly, and the parallax of latitude was increasing.

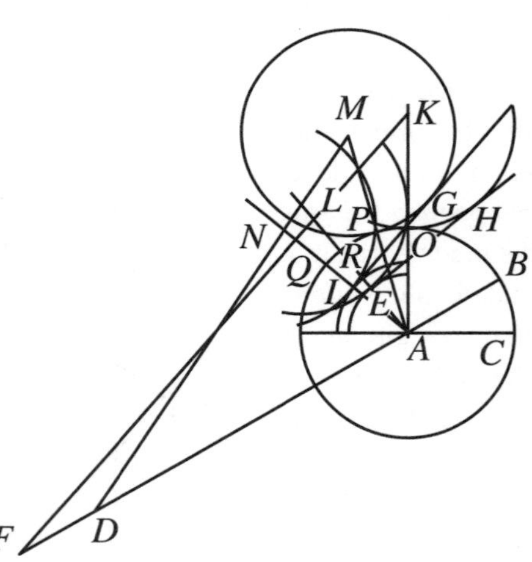

And because it was nowhere observed under adequate circumstances, it will be in vain to call it back to computation. Only, from these givens, we shall seize the occasion of also solving Problem 30 differently. For the angle is great. About center A let the circle OC of the sun's body be described, on which let OA represent the vertical, AC the horizon, AB the ecliptic, and let OAB be 67 degrees. For if OA were the meridian, it would cut AB at A, the 10th degree of Libra (which is the sun's position) at an angle of 66° 48′. But now OA is beyond the meridian towards the

[106] Cf. *TBOO* XII pp. 368–72.

nonagesimal, and is therefore a little straighter. Because for Maestlin the beginning inclined 9 degrees to the east, therefore, let OP be 9°, and let a straight line be sent out from A through P, while from P let the moon's diameter PM be extended, and about center M let the circle of the moon's diameter PG be drawn. And because the maximum obscuration was $2\frac{1}{2}$ digits, let the radius AQ be divided so that QI is $2\frac{1}{2}$, IA $3\frac{1}{2}$, and about center A with radius AI let a small circle be delineated, to which a straight line is tangent that also is tangent to the circle PG, and let the point of tangency be I. Then let AI be extended, and from I let the moon's radius IN be extended, let MN be joined and extended to the common intersection with the ecliptic BA: let this be D. MDA will be the visual angle of the moon's path, about 27 degrees, as mechanics shows. For because in the beginning the moon's center is certainly at M, it will therefore go either on a higher path MN, and will therefore not conceal $2\frac{1}{2}$ digits, or a lower one, and will therefore conceal more. Therefore, it will seem to go only on path MND.

From my observation, the angle comes out a little smaller, for as it is only the visible angle, it is thus different in different places. And for me, the altitude of the pole was smaller, and also the east was closer. For in Prob. 27 above, there
418 was 4° 30′ of longitude between Hven and Graz. Further, in Prob. 29, between Tübingen and Hven was 2° 45′. Therefore, between Tübingen and Graz is about 7° 15′, approximately.

Since for me, then, it began at the vertex, let AO be extended, and let OK be the moon's radius, OH its circle, RE three digits, EH tangent to the moon's body at E, EL the moon's radius, and let the points KL be joined, and let these and the ecliptic be extended until they meet at F. They will make an angle of about 20°. However, I am not sure whether the defect might not have come out greater. For clouds covered half of the time, and they concealed both the end, and perhaps also the maximum defect.

Problem 32

Whether the visual path of the moon's center is a straight line; that is, whether from observation of the beginning and end the quantity of the maximum defect may be obtained in the usual way without error. Here is treated the correction to be appended to Problem 30, and the usual precepts about forming a plat of an eclipse.

First, this is sufficiently clear: that a space so small as the little amount that the moon makes up in three hours is equivalent to a straight line, even though a curve is drawn around the center of the earth. In fact, that is not in question now, but rather, whether the moon's center is seen on the same great circle for the whole time of the duration of the eclipse? I say that it is not necessary that this always happen. For the path can become curved. For since the true latitude and the parallax in longitude and latitude combine for the moon's path—though as regards the true latitude, about the nodes it is nearly proportional to the spaces traversed, and so this does not affect the straight line at all. For in equal times it is very nearly uniformly borne off to the sides, and also uniformly carried forward to the east. But because, by a legitimate cause, the parallaxes of longitude vary nonuniformly, and more rapidly in proportion to their nearness

to the nonagesimal, it can as a result happen that the true latitude, increasing uniformly, is applied to the nonuniformly increasing visual longitudes, and a bending of the path results from this. This happens much more so as a result of the parallaxes of latitude. For the signs of the zodiac rise nonuniformly, and over three hours it can happen that the angle of the horizon increases slowly at the beginning, quickly at the end; and therefore (following what was shown in ch. 9) the parallaxes of latitude will vary slowly at the beginning, quickly at the end ***419***
(and likewise the observed latitude), whence again it happens that the moon's composite path becomes curviform. However that may be, the slight difference that exists as a result does not appear perceptible, except around the meridian and the nonagesimal. Consequently, one must proceed using examples.

In the eclipse of 1598,[107] I was convinced that about the middle of the darkening, as if at a single moment of time, when the sun was penetrating the clouds, I saw a rather large defect, i.e., the remaining horn being exceedingly thinned. It is in fact impossible that I saw a defect of this quantity, if we should state that the moon's visual path is straight, and should make a diagram based on this; much less so, if those things be taken into account which Tycho had annotated for us above in this observed eclipse. For it ought to have been greater at Hamburg and in Denmark than in Styria, because it was to the north. Further, the horns were in front on the ray, upturned in the sky; but for a short time inclined by an estimated 10 degrees. And Jessenius had almost persuaded me, in stating that he had seen it very close to central at Torgau. When I wrote to Maestlin in 1598 on this subject, he, while not asserting anything, ascribed everything that I had truly observed to a doubtful observation of the beginning and to the parallaxes, giving support rather vividly to this very problem.[108] However that may be, this too is pertinent, that, since this image was seen fleetingly, I might have set the paper askew on the ray, whence the elliptical section of the cone produced the image of a longer, and thus thinner, horn. We shall nevertheless consult the computation. In Problem 25, the moon's visual path was $55' \, 43''$; the latitude at the beginning, $1' \, 57''$ north; at the end, $17' \, 6''$; the difference is $15' \, 9''$. In the diagram of Prob. 27, duplicated here, BC is $15' \, 9''$; BA is $55' \, 43''$; therefore, BAC is $15° \, 12' \, 40''$, and AC $57' \, 44''$ approximately, while AG is $28' \, 52''$ and AF $31' \, 40''$. Hence, FAG is $24° \, 16' \, 35''$, and the distance of the centers FG at maximum darkening is $13' \, 1''$. But the radius of the moon is $16' \, 22''$. Therefore, it goes past the sun's center by $3' \, 21''$,

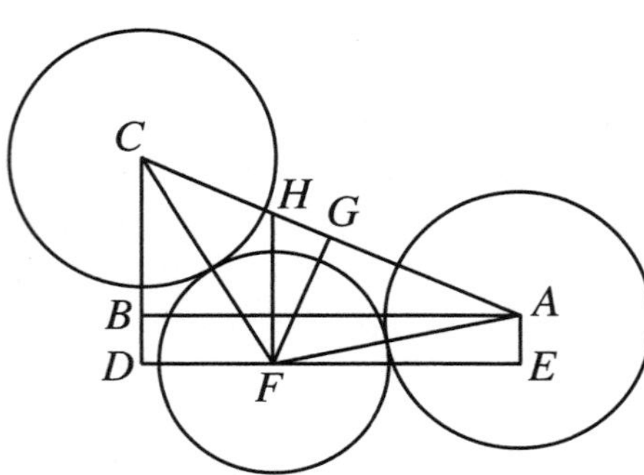

[107] This is the eclipse whose observation Kepler describes in Problem 23.

[108] Kepler to Maestlin, 15 March 1598, no. 89 in *JKGW* XIII p. 179; Maestlin to Kepler, 2/12 May 1598, no. 97 in *JKGW* XIII p. 208.

420 which, added to the sun's radius of 15′ 18″, makes 18′ 39″, of which 30′ 35″ are 12 digits. Therefore no more than $7\frac{2}{5}$ digits would have been eclipsed. Even less by the computed latitude of the end, and this in the case where we assume that some imperceptible amount was eclipsed at the beginning. But if we should say that about one digit was eclipsed, the centers will be brought a little closer at the middle, by about 12″.

Let me then make a test through parallaxes at maximum darkening, the point of which is in a somewhat different place than at the point of visual conjunction. A line perpendicular to *DE*, set up from *F*, cuts *AC* at *H*. And *H* is the visual place of conjunction, while *G* is that of maximum darkening, because *FH* is longer than *FG*, since the angle *H* is acute, *G* right. Nor should it trouble you that the authorities measure the quantity of greatest darkening at *H*, for they do it inaccurately because it matters little. *G*, however, is the midpoint between *C*, *A*, inasmuch as *FC*, *FA* are taken as equal. On the supposition that the visual motion is also proportional to the time, the time of greatest darkening would have been 11^{h} 41^{m}, or a little before. In 1 hour 14^{m}, the moon's motion is 40′ 29″, to which corresponds a change in latitude of 3′ 24″. But at the beginning, the true latitude was 51′ 19″, as in Prob. 26. Therefore, it is now 54′ 43″. The sun's right ascension is 347° 51′. Subtract 4° 45′, which make 19 minutes' distance from the meridian, and there remains 343° 6′, the right ascension of the meridian, with which it divides the sky at 11° 40′ Pisces. Its declination is 7° 12′ 45″; I add this to the elevation of the pole, 47° 2′. *MV* will be 54° 14′ 45″. The rising sign is 9° 33′ Cancer.[109] Hence, *MN* is 27° 53′. And *VN* is 48° 38′, showing, under the [parallactic table] headings of 59′ 30″, a latitudinal parallax of 44′ 40″, which, subtracted from the true latitude, 54′ 43″, leaves the visual latitude, 10′ 3″. The complement of *VN*, 41° 22′, shows a heading of 39′ 20″. And because the moon is visually closest to the sun at 16° 46′ Pisces, *NS* will be 22° 47′, which from the headings found shows a longitudinal parallax of 15′ 14″.

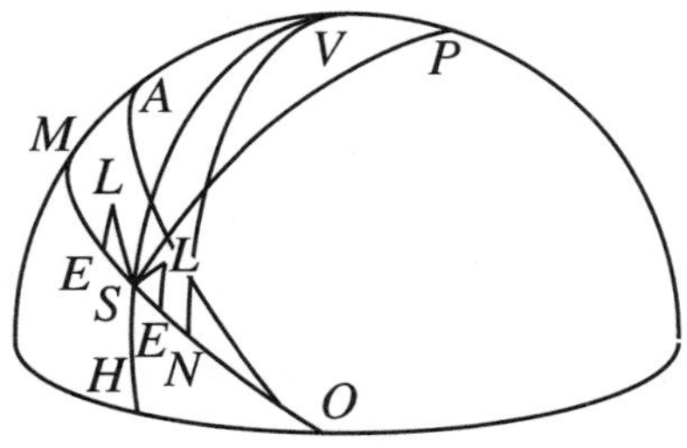

But the moon's true motion in 2^{h} 28^{m} was found to be 1° 19′ 15″; at the half, therefore, it was 39′ 37″. At the beginning, though, in its true motion it was 25′ 26″ before the sun. Therefore, it is now 14′ 11″ after the sun, truly, and by parallax visibly 1′ 14″ before the sun. And in the diagram repeated here, *FGH* and *CBA* are similar triangles, with the result that *HFG* is also 15° $12\frac{2}{3}$′. And

109 Although Franz Hammer (*JKGW* II p. 461) notes an error in the computation of the rising point, stating that it should have been 18° 39′ Cancer, my recomputation, using Kepler's right ascension and polar elevation, shows a rising degree of 9° $36\frac{1}{2}$′ Cancer, not much different from Kepler's. Hammer cites an article on this subject by Lalande, titled "Erreur de Kepler. Sur la courbure de l'orbite apparente de la Lune," in *Connoissance des Temps*, VIe Année (1798) pp. 238–243. Lalande claims that the supposed "curvature" of the path is the result of Kepler's errors.

thus, where FG is 10′ 3″ (if you suppose that the moon is now at point G of maximum darkening, with this visible latitude), the moon's longitudinal distance **421** GH from the visible point of conjunction would be 2 minutes 45″. But because the moon is only 1′ 14″ before H, it has therefore now gone past the point G of maximum darkening by 3 minutes of time. And since the visible latitude is increasing, 3 minutes of time earlier it was less than 10′ 3″. And so the difference between the visual latitude obtained from the computation of parallaxes, and that obtained from the drawing of the beginning and end and the middle proportional [to them], is more than 3 minutes, or $1\frac{1}{5}$ digits. The eclipse was therefore more than $8\frac{3}{5}$ digits. And thus it has been proven that this is sufficiently evident in eclipses occurring at midday.

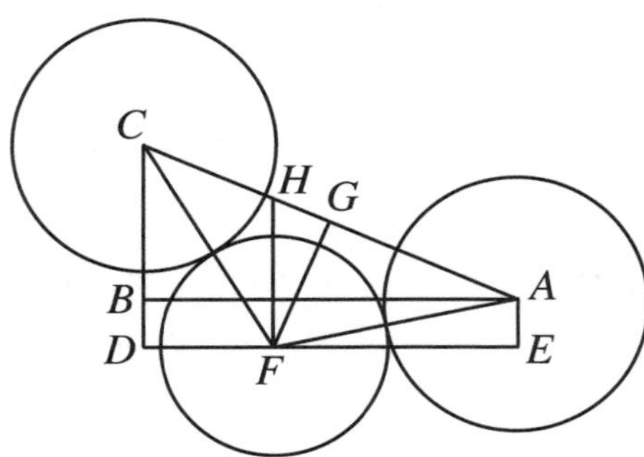

As a result, in Problem 27 above, if you leave this to the Tychonic observers, that the error of the eyes is equal, not only at the beginning and end, but also right at the middle, there will nonetheless be a cause for $9\frac{2}{3}$ digits having appeared, more, that is, than is derived from the proportion of the beginning and the end.

In the eclipse of 1590,[110] when it was near maximum darkening, the distance of the centers was to the radius of the sun (including the "fringe") as 59 to 88, and where the sun is 24 the moon is 23. Therefore, where the distance of centers is 59 the moon's diameter is $84\frac{1}{3}$. And so in these units, the sum of the radii is $172\frac{1}{3}$. Therefore (following the example of Prob. 28 and 29), as $172\frac{1}{3}$ is to 59 so is 30′ 27″ to 10′ 25″, the required distance of the centers in the usual units. And because the moon's radius is 15′ 25″, the excess of 5′ reaches beyond the half of the sun. There were accordingly eight digits eclipsed. We shall see what follows if we proceed proportionally, as if the moon's path were a straight line.

Since, then, a semidiameter was seen to be eclipsed at two moments, one before, the other after the middle, and since the distance of the centers was 14′ 55″, and the longitude accordingly was 7′ 10″ before the sun, and 12′ 21″ after the sun, therefore, the sum or visual path was 19′ 31″. The latitude for the former was 13′ 3″, for the latter 8′ 22″, south, so that the difference was 4′ 41″. Consequently, AC is 18′ 7″, but AG 9′ 4″, and the distance GF of the centers is 11′ 52″. The difference from the observations is 1′ 27″. As a result, the digits would have been only $7\frac{2}{5}$. And so, if those things hold true that have been established by the observations, in this eclipse as well the amount eclipsed was perceptibly greater than that which was derived from the diagram and the moments of equal eclipse at the middle.

Further, so that this may be established in doubtful instances, and so that **422**

110 *TBOO* XII pp. 8–10, and other observations mentioned in Problems 28–9 above.

the reader may at once see almost all the things I have hitherto presented in a crippled way for various reasons—these things are not so because of the difficulty of observation, or because they relate to Plato's *Republic*, but, once care is taken, they can be brought to complete perfection—I now add two very clear instances of the two most recent solar eclipses, by which I shall display an example of almost the entire art presented in these 32 Problems.

First Example

In 1600, on June 30 or 10 July, at Graz in Styria, I was outdoors with the wooden instrument whose complete description appears in Prob. 1, with a surrounding curtain. And since the instrument had not yet been divided into its scales, I carved in 15 marks with a knife, and identified them in sequence with numbers; and whatever was seen at the individual marks on the panel, I recorded separately on paper. But I shall show you most candidly the series of observations, described from the paper slip. For I abhor the "additio-subtractions," lacking artfulness and untrustworthy, that I wish had not beset Ptolemy's principal observations.[111] I, in contrast, write these things as if I considered myself persuaded that they were going to fall into the hands of posterity. Consequently, so that a free and open judgment may be left to it either way, nothing will either be contrived or concealed.

Notes of Azimuths and Altitudes.	*Angles of the vertical with the circle through the centers, numbered from below towards the west up to the index (which stood opposite the shadow on the ray; therefore, on the same side as the moon in the sky).*	*Digits on the Shadow.*	
1.	$62\frac{1}{2}$ or $72\frac{1}{2}$	0	Beginning. The rule covered the notations of the tenths.
2.	58	$2\frac{1}{3}$	
3.	$52\frac{1}{2}$	3	
4.	46		
5.	34	more than 4	
6.	13 approximately		
423	Towards the east.		
7.	4	6	approximately.
8.	22 approximately	6	I thought more.
	Very pallid and blurry.		
9.	32		
10.	47	5	approximately.
11.	59	less than 5	
12.	64	4	approximately.

[111] Kepler here uses the Greek phrase τὰς προσθαφαιρέσεις ἀτέχνους to describe Ptolemy's "corrections." This refers to Ptolemy's practice of suppressing the actual observations and presenting only adjusted positions based on them.

Notes	*Angles of the vertical with the circle through the centers*	*On the Shadow*	
13.	64. 15′ precisely	3. 55′	
14.	80	more than 2	by estimate, not by measurement.

Since now (on the ray) I could no longer perceive whether any part was eclipsed or not (though it was still eclipsed by about $\frac{1}{2}$, and the pallor alone was responsible), the vertical angle was not yet 90°.

15. A little before the end, it was not yet 90 on the vertex.

Note on these notes, indicators of the azimuths and altitudes: at the end, the crosspiece or horizontal rule was hanging down, and was giving too great an altitude.

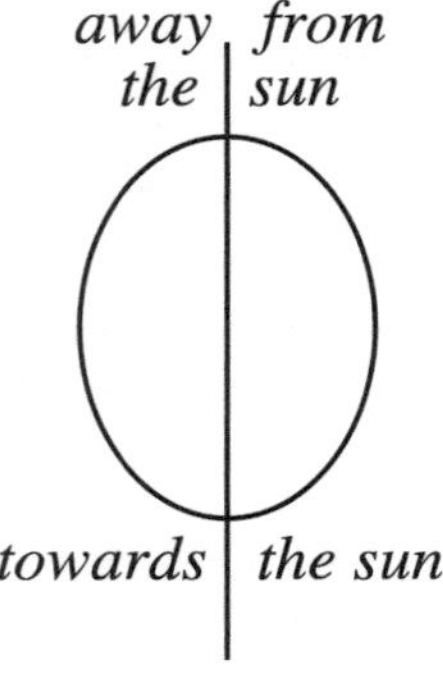

Then all the configurations through the cracks in the roof had this form on the plane of the horizon.[112]

On Azimuths and Altitudes at these 15 Moments

When the observation was finished, my azimuthal quadrangle[113] was divided, on one side, set up on the meridian, into 2,000 units, and on the other, which looked east and west, into 1,200 of the same units in each direction. In the preceding days I had somehow by estimation pointed this rectangle at the meridian, and once it was secured thus, I sought in the usual way, by altitudes of the sun before and after noon, equal on both sides, the true meridian of the azimuthal plane, and found that the meridian of the instrument inclined to the west, by 1 degree 4 minutes. And when that arc was subtracted from the arcs found by the **424**
numbering of divisions, the remainders were the true azimuths, or the directions

[112] In the 1604 edition, the diagram accompanying these words had, at the top, the letter "a" followed by the sun symbol, and at the bottom, the word "*ad*" followed by the sun symbol. However, since the woodcut was not very clear, both later editions misunderstood Kepler's meaning and, reading the "a" in "*ad*" as spurious, put "A [sun sign]" at the top and "D [sun sign]" at the bottom, as if the letters were merely labels for the ends of the ellipse.

[113] This is the base of the instrument, upon which the crosspiece pivots, carrying the rest of the instrument.

on the horizon of the verticals passing through the sun. And so that I might have no doubt about the projection of the meridian found, I sought the altitude of the pole from the distance of the verticals that showed the same altitude of the sun, and from both the sun's declination and the altitude being known. In the triangle between the pole P, the zenith V, the sun S, SV was given from the sun's altitude, PS from the sun's declination, and SVP from half the sum of the two azimuths. As a result, the complement VP of the pole's altitude was also given, which came out to $47^\circ\ 10'$, which of course is for the setting up of this operation and of the wooden instrument, very closely approaching the true altitude of $47^\circ\ 2'$. And so the meridian was certain. The following is the sequence of numerical measures.

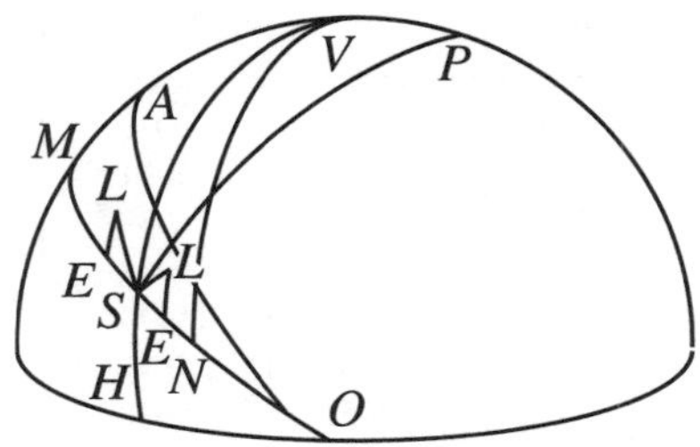

		On the Line towards the east.	*Resulting reduced arcs*		
	noon.	$37\frac{1}{3}$.	1. 4.	0. 0.	Towards the west.
Moment	1.	751.	19. 31.	20. 35.	
of time	2.	1171.	29. 17.	30. 21.	
		On the Line towards the south			
	3.	1830.	32. 11.	33. 15.	
	4.	1613.	35. 35.	36. 39.	
	5.	1374.	40. 4.	41. 8.	
	6.	1075.	47. 5.	48. 9.	
	7.	1003.	49. 3.	50. 7.	
	8.	874.	52. 52.	53. 56.	
	9.	828.	54. 20.	55. 24.	
	10.	745.	57. 6.	58. 10.	
	11.	678.	59. 28.	60. 32.	
	12.	637.	60. 58.	62. 2.	
	13.	524.	65. 21.	66. 25.	
	14.	474.	67. 23.	68. 27.	
	15.	431.	69. 11.	70. 15.	

425 However, domestic affairs hindered me from dividing the crosspiece, the post, and the rule at the same time, and counting the divisions intercepted at the moments of observation. For only a few days before, I had returned from Bohemia, and I had meanwhile gotten the instrument ready, and now I was immediately preparing myself for a new journey to Prague with my family. And so when I returned to Styria in 1601, I carefully and at leisure examined the instrument, which I had left behind, and which was found undamaged. And so, where the post or perpendicular had 5,040 units, other parts had the following:

	Hypotenuse or Rule	*Base or Crosspiece*	*This Yields a Perpendicular*	*Corresponding Distance of Sun from the Zenith on the Hypot.*	*On the Base*	*These Altitudes Give an Azimuth*	*The above Azimuths Therefore err by*
1	5604	2450	5040	25. 56.	25. 56.	20. 18.	17. +
2	5679	2615	5041	27. 27.	27. 25.	30. 21.	0.
3	5706	2677	5039	27. 58.	27. 58.	33. 8.	7. +
4	5740	2750	5038	28. 36.	28. 37.	36. 2.	37. +
5	5813	2896	5040	29. 53.	29. 53.	41. 8.	0.
6	5957	3173	5041	32. 12.	32. 12.	48. 32.	23. –
7	6012	3265	5048	33. 2.	32. 56.	50. 40.	33. –
8	6116	3468	5038	34. 30.	34. 32.	54. 30.	34. –
9	6160	3537	5043	35. 6.	35. 3.	55. 48.	24. –
10	6266	3723	5040	36. 27.	36. 27.	58. 46.	36. –
11	6369	3893	5041	37. 42.	37. 41.	60. 15.	17. +
12	6444	4014	5042	38. 33.	38. 32.	62. 52.	50. –
13	6692	4393	5047	41. 8.	41. 5.	67. 19.	54. –
14	6850	4635	5044	42. 38.	42. 36.	68. 24.	3. +
15	6964	4797	5048	43. 38.	43. 35.	71. 15.	60. –
			should have been 5040 everywhere				

It is accordingly apparent that when the crosspiece was moved along, with the pivot of the post adhering to it quite firmly, the instrument followed, pulled from its mounting by force, the bindings slacking, and that this happened after moment 5. The error in time, however, was very slight. And on the other hand, in the last moments, the pole from which I had hung the curtain hindered the rule somewhat, which had encountered it in moving across, preventing the rule from following the sun's descent in elevation. So if at the end of the time you establish the time from the altitude, and again from the azimuth, you will find a difference of 4 minutes. For the altitude gives a time of $2^h\ 59^m\ 36^s$, the azimuth $2^h\ 55^m\ 23^s$. And since it is certain that both the altitude and the azimuth are slightly in error, the mean of $2^h\ 57\frac{1}{2}^m$, taken as true, will be no more than one minute from the truth itself. In the others, I stand by the altitudes, with the azimuths giving their testimony from a distance. This I did with all the more diligence, both because the foundations of lunar demonstrations will be able to be laid upon this eclipse as upon a cornerstone, and because I had no little disagreement with the late Tycho Brahe about the time of the beginning, who, having grasped some slight opportunity, called the whole record of times into doubt. But I have no doubt that if his life had continued to a time when I could have showed him this consensus, he would have submitted. The times, then, along with the necessary data from the first motion,[114] are as according to the previous problems. The sun's declination was 22° 17′, and 16′ because of parallax. The pole's altitude was 47° 2′. **426**

114 The daily rotation of the earth.

	Moment	*Times*	*Culminating points*	*On the Nonagesimal*	*Its Distance from the Zenith*	*Sun's from the Nonagesimal*	*Angle between Vertical & Ecliptic*	*Between Ecliptic and Circle through Centers*
	1	12. 37. 44.	27. 2. Canc.	21. 31. Canc.	25. 40.	3. 22.	83. 2.	20. 32.[a]
	2	12. 58. 16.	1. 59. Leo	24. 24. Canc.	26. 16.	7. 14.	75. 41.	17. 41.
	3	1. 4. 20.	3. 27. Leo	26. 32. Canc.	26. 46.	8. 22.	73. 53.	21. 23.
	4	1. 10. 52.	5. 3. Leo	27. 45. Canc.	27. 5.	9. 35.	71. 58.	25. 58.
	5	1. 22. 56.	8. 0. Leo	29. 57. Canc.	27. 39.	11. 46.	68. 44.	34. 44.
	6	1. 42. 40.	12. 54. Leo	3. 35. Leo	28. 37.	15. 24.	64. 4.	51. 4.
427	7	1. 48. 16.	14. 18. Leo	4. 38. Leo	29. 0.	16. 26.	62. 58.	66. 58.
	8	1. 59. 36.	17. 10. Leo	6. 42. Leo	29. 39.	18. 30.	60. 51.	82. 51.
								To the E.
	9	2. 3. 36.	18. 10. Leo	7. 27. Leo	29. 54.	19. 15.	60. 10.	87. 50.
	10	2. 13. 8.	20. 36. Leo	9. 13. Leo	30. 29.	21. 0.	58. 40.	74. 20.
	11	2. 21. 32.	22. 47. Leo	10. 45. Leo	30. 58.	22. 32.	57. 26.	63. 34.
	12	2. 26.	23. 56. Leo	11. 35. Leo	31. 19.	23. 21.	56. 55.	58. 50.
	13	2. 42.	28. 4. Leo	14. 32. Leo	32. 26.	26. 18.	55. 7.	44. 53.
	14	2. 53. 20.	1. 3. Virg.	16. 38. Leo	33. 10.	28. 23.	53. 58.	between 46. 2.
								and 36. 2.
	15	2. 57. 30.	2. 18. Virg.	17. 31. Leo	33. 36.	29. 16.	53. 40.	between 46. 20.
								and 36. 20.

a. Kepler rightly doubts (in the paragraph following the next table) the accuracy of this angle. The error arises from the incorrect azimuth of 20° 18′ given for the first moment in the next table, which vitiates this entire row of the table.

Therefore, from the beginning and end, the middle at Graz was estimated by me to be at $1^h\ 47\frac{1}{2}^m$, the duration $2^h\ 20^m$. For the Tychonic observers, in the castle of Benatek, which is 5 German miles to the northeast of Prague, the middle was $1^h\ 46\frac{1}{2}^m$, the duration $2^h\ 1^m$. The digits were judged by them to be about 5, while to me they appeared $6\frac{1}{3}$ while still on the ray, and thus after removal of the shell, $7\frac{1}{2}$.

And Tycho Brahe ought not to have feared, lest his observation be undermined by mine, nor ought he to have confuted mine by his, or to have considered it absurd that the beginning was seen by me, farther to the east, at $12^h\ 37\frac{1}{2}^m$, and by him, farther west, at $12^h\ 46^m$. For I was also 3 degrees farther south, and saw the sun more greatly eclipsed by about the same number of units, and thus saw the beginning earlier and the end later; and the darkness beneath the curtain helped me in distinguishing the beginning and the end: optical principles explained above deceive the eyesight itself somewhat both as the duration and as to the quantity. Nor were they able, as I was, to use a compass, subjecting themselves to error in mere guesswork.

Maestlin, at Tübingen, very greatly hindered by clouds, said, "I gathered
428 this one thing, that the sun was noticeably more than half eclipsed." Indeed, some noticed the same thing at Wittenberg, which is 2 degrees to the north. Of course, this is in accord with the difference of observers.

On the estimation of the moon's diameter in this eclipse, see Prob. 12 above.

Now, as you see both from this table and from the digits, the maximum darkening was between $2^h\ 0^m$ and $2^h\ 3\frac{1}{2}^m$; closer to the latter. Therefore, the visual conjunction with respect to longitude was about $2^h\ 3^m$, nearer by half an hour to the end than to the beginning. For the increments of longitudinal parallax decrease after noon. And because the sun is at 18° Cancer, its radius will be 15′ 1″. But the moon's simple anomaly, 8 signs 14° 30′, through an eccentricity of 4336, indicates a radius of 16′ 10″, where it is 15′ 15″ at apogee. Therefore, the sum of the radii is 31′ 11″. This is the distance of centers at the beginning and end of the eclipse.

But in order to obtain this same thing at the other moments, note that in these problems it has been more than once repeated from Ch. 2 that the digits eclipsed on the enlarged ray number the true approach of the centers. But the digits eclipsed on the ray are twelfths of the enlarged ray, and the enlarged ray of the sun, as pointed out in Prob. 2 above, had $105\frac{1}{2}$, more correctly 106.[115] Therefore, at the second moment, and the others in which the digits were noted, these units are given. When, therefore, they are subtracted from the half-ray, 53, along with the radius of the opening, $8\frac{1}{4}$, the remainder is the units remaining to the center of the sun, and the number of minutes to which this amounts will be obtained from the unshelled ray, which is $89\frac{1}{2}$. For if $89\frac{1}{2}$ are 30′ 1″, how much, then, will the remainder to the center be? Whatever comes out here, added to the moon's radius, establishes the true distance of centers. And if the sum of the units from the radius of the opening and the digits together exceeds the units of the radius of the ray, the excess, reduced to minutes of arc, will have to be subtracted from the moon's radius, so that the true distance of centers may again be obtained.

With these things set up thus, and with the angle adopted which we had found most recently, the visual longitude and latitude at all moments is established, which I present in the following table. **429**

Moment	*Units Eclipsed*	*Remainders to the Center*	*Value*	*Distance of the Centers*	*Apparent South Latitude*	*Apparent Longitude before Sun*
1	2. approx.	$42\frac{3}{4}$ approx.	14.′16.″	30.′26.″ approx.	10.′40.″	28.′31.″
					or 5. 34.	or 29. 56.
2	$20\frac{1}{2}$	$24\frac{1}{4}$	8. 8.	24. 18.	7. 22.	23. 10.
3	$26\frac{1}{2}$	$18\frac{1}{4}$	6. 7.	22. 17.	8. 7.	20. 45.
5	35 beyond	$9\frac{1}{4}$ minus	3. 12.	19. 22. minus	11. 1.	15. 55.
7	53 approx.	$8\frac{1}{4}$ approx.	2. 42.	13. 28. approx.	12. 24.	5. 16.
8	53 I saw more	$8\frac{1}{4}$ plus	2. 42.	13. 28 minus	13. 23.	1. 40.
						Past Sun
10	44 approx.	$\frac{3}{4}$ approx.	0. 11.	16. 21. approx.	15. 44.	4. 26.
11	44 below	$\frac{3}{4}$ plus	0. 11.	16. 21. plus	14. 43. +	7. 17.

[115] These are the seventy-seconds of an inch, which Kepler also calls "digits". See footnote 7, above.

12	35	$9\frac{3}{4}$	3. 12.	19. 22.	16. 43.	9. 58.
	$34\frac{1}{2}$	$10\frac{1}{4}$	3. 22.	19. 32.		
13	18 beyond	$26\frac{3}{4}$	8. 54.	25. 4.	17. 40.	17. 42.
14	$4\frac{1}{2}$ approx.	$40\frac{1}{4}$	13. 25.	29. 35.	between 21. 18.	20. 34.
					and 17. 24.	23. 56.
15	0	$44\frac{3}{4}$	15. 0.	31. 11.	between 22. 34.	21. 34.
					and 18. 28.	25. 8.

First, you see what use it is to have many moments other than the beginning and the end. For unless another moment, and the rest after it, had immediately supported the first with their consensus, I would have been left within 5 minutes of initial latitude in that doubt in which I had fallen from the gap in the numbering of the inclination, in passing over an entire decade of degrees.

Second, you also see, in the eleventh moment, how easily error is produced. For it is in the notes that there were less than 5 digits in the defect. And so here, from the supposition of 5 digits precisely, the visual latitude would become decidedly retrograde and decreasing. As a result, it is good to seek this latitude more from the measurement of the observation.

430 Third, it is clearly evident that the maximum darkening was at moment 7 and 8, when the moon was not yet visibly in conjunction with the sun with respect to longitude. The digits on the ray were $6\frac{1}{3}$; in the sky, after removal of the shell, $7\frac{1}{2}$.

Fourth, the angle of visual latitude is notable, as is evident from the first and thirteenth moments, which are reliable. For the difference in latitudes is $12' 6''$, while the moon visibly traverses $47' 38''$ of longitude. The angle of visual latitude is therefore $14° 16'$. For the parallax of latitude was increasing, and at the same time the moon was descending southwards in its true motion. The reason why this angle was not larger is the altitude of the sign, in which the parallaxes are small, and the increments themselves are slight.

Fifth, the moon's visual path was curved, as is evident from moments 1, 8, and 13. For if 1 and 8 are compared, over $28' 16''$ of longitude the moon was carried up $7' 49''$ in altitude. Therefore, the angle of latitude, $15° 27'$, is greater than before, from moments 1 and 13.

The direction in which this eclipse is useful to us in testing the moon's motions, for the sake of which we have put in so much work in examining it, will follow in the second part [of the *Optics*], to which place the parallaxes of this eclipse are also deferred.

Another example

On 14/24 December 1601, at Prague in Bohemia, we saw the sun eclipsed on the north in the afternoon, in the following manner.

I had brought my panels with me from Styria along with a foot measure, with which I prepared a rule of the same length as the one in Styria. The room was very dark. The rule with the panel rested on a crosspiece, and was set upright upon it in the plane of the vertical, as was described in Prob. 15, since the thickness of the wall did not bear the whole apparatus of the instrument, described in the first problem. Further, at the end the opening also had to be changed, since the sun struck the south wall too obliquely.

The times were noted, not from the sun's altitude, but more conveniently from the Tychonic clock, with a hand for minutes and seconds. We had made a test of this clock on the night before, as to whether it was of the right speed, by observing transits of the fixed stars, and on the preceding noon, when the hand was made fast at the point of 12^h, we held up its motion until the sun's center, by the Tychonic quadrants, had come to the meridian. Further, by a small brass quadrant of a foot and a half, we took note rather crudely of the azimuths of the sun between the initial observations. **431**

There follows a true and reliable transcript of the paper on which the observations were recorded

Time	*Azimuth on the Small Quadrant*		*Vertical Inclination. Scale from beneath to the West on the Instrument: thus, with the Heavens*
12.	1. Towards the east, approximate		
12. 59. 0.	12. 30. to the west		
1. 11. 40.	15. 20.		
1. 17. 30.	. . . .	Beginning.	
1. 18. 45.	. . . .	. . . .	70. more: read 72, as in the next.
1. 23. 15.	. . . .	. . . .	72.
1. 24. 30.	. . . .	1 digit . . .	72. carry down from preceding.
1. 25. 15.	18. 30.	. . . .	75. uncertain.
1. 30. 0.	. . . .	. . . .	72.
1. 35. 30.	21. 40.	2 digits . .	72.
1. 42. 0.	. . . .	. . . .	72. 15. more certain.
1. 44. 30.	23. 0.	3 digits . .	76.
2. 20. 30.	. . . .	$6\frac{1}{3}$ digits . .	86.
2. 23. 30.	. . . .	. . . .	90.
			Upwards and back.
2. 30. 0.	. . . .	$7\frac{1}{3}$ digits . .	84.
2. 43. 30.	. . . .	$7\frac{2}{3}$ digits . .	79.
2. 53. 0.	. . . .	8 digits . .	19.
3. 0. 0.	. . . .	8 digits minus Let it be $7\frac{2}{3}$	Not yet zero.
3. 6. 0.	. . . .	. . . .	7.
3. 9. 30.	. . . .	$6\frac{2}{3}$ digits .	14.
3. 15. 0.	. . . .	6 digits approx.	
3. 21. 0.	The sun skimmed the visual horizon, with a few degrees of altitude.	$5\frac{1}{2}$ digits . .	25. approximately. For the sun's upper limb was behind a cloud, the lower limb behind a mountain.

My small moon subtended 30′ where the sun was 31′, but the shadow perceptibly surpassed it. See Prob. 13.

The five azimuths, increased by one degree, by which the instrument de-
432 clined from the meridian to the east, show the times.

	Less than the truth
12. 56. 50.	2′. 10″.
1. 8. 44.	2′. 56″.
1. 22. 36.	2′. 39″.
1. 35. 0.	0′. 30″.
1. 42. 23.	2′. 2″.

This gives evidence that the declination of the instrument was a little greater, about 1$\frac{1}{2}$° to the east. And it was impossible to distinguish half a degree's difference on such a small instrument and in this procedure, so that the sun may equally illuminate the instrument on both sides.

There follow the requisites from the first motion

Times	*Culminating Point*	*On the Nonagesimal*	*Distance of the Nonagesimal from the Zenith*	*Distance of the Sun from the Nonagesimal*	*Angle between the Zenith and the Ecliptic*	*Angle between the Ecliptic and the Circle through the Centers*
						South
1. 17$\frac{1}{2}$.	20. 45. Capr.	15. 44. Aquar.	70. 4.	42. 51.	76. 9.	6. 9. less, read 4. 9.
1. 18$\frac{3}{4}$.	21. 17. Capr.	16. 41. Aquar.	69. 50.	43. 48.	75. 44.	5. 44. less, read 3. 44.
1. 23$\frac{1}{4}$.	22. 6. Capr.	18. 8. Aquar.	69. 39.	45. 15.	75. 15.	3. 15.
1. 24$\frac{1}{2}$.	22. 26. Capr.	19. 40. Aquar.	69. 21.	46. 46.	74. 52.	2. 52.
1. 35$\frac{1}{2}$.	25. 2. Capr.	23. 0. Aquar.	68. 37.	50. 6.	73. 17.	1. 17.
						North
1. 44$\frac{1}{2}$.	27. 10. Capr.	26. 22. Aquar.	67. 58.	53. 27.	71. 59.	4. 1.
2. 20$\frac{1}{2}$.	5. 51. Aquar.	8. 24. Pisc.	64. 46.	65. 29.	66. 47.	19. 13.
2. 30.	8. 11. Aquar.	11. 17. Pisc.	63. 53.	68. 21.	65. 30.	30. 30.
2. 43$\frac{1}{2}$.	11. 32. Aquar.	15. 8. Pisc.	62. 36.	72. 22.	63. 43.	37. 17.
						Now to the East
2. 53.	13. 54. Aquar.	17. 44. Pisc.	61. 40.	74. 47.	62. 31.	81. 31.
3. 0.	15. 39. Aquar.	19. 35. Pisc.	60. 59.	76. 38.	61. 39.	between 80. 39. and 61. 39.
3. 9$\frac{1}{2}$.	18. 5. Aquar.	22. 3. Pisc.	60. 3.	79. 5.	60. 30.	46. 30.
3. 15.	19. 38. Aquar.	23. 25. Pisc.	59. 31.	80. 27.	59. 22.	between 45. 52. and 34. 52.
3. 21.	21. 1. Aquar.	24. 26. Pisc.	58. 54.	81. 57.	59. 8.	34. 8.

433 The diameter of the sun with its border reached 110 units. That of the moon or shadow reached 75, on which see Prob. 13. Half the sum is 92$\frac{1}{2}$. So, whatever the amount of the eclipse will have been, will be subtracted from this, and the true distance of the centers in our units will remain. However, this should be compared, not to the quantity 110 of the diameter with the border, but to the unshelled diameter, 93$\frac{1}{2}$, following this line of reasoning: If 93$\frac{1}{2}$ make 31 minutes, what is our remainder? Now these distances of the centers, according to its [sic] angles with the ecliptic, and their complements, will be deduced by our parallactic table, in longitude and latitude, in the following table.

Times	*Distance of Centers*	*Apparent Southern Latitude*	*Apparent Longitude*
		read	read
1. 17. 30.	30. 40.	3′. 17″. minus. 2′. 14″.	30′. 30″. plus. 30′. 35″.
1. 24. 30.	27. 38.	. . . . 1. 26.	. . . . 27. 36.
1. 35. 30.	24. 33.	. . . . 0. 32.	. . . . 24. 32.
		North	
1. 44. 30.	21. 33.	. . . . 1. 30.	. . . . 21. 30.
2. 20. 30.	11. 25.	. . . . 3. 46.	. . . . 10. 47.
2. 30. 0.	8. 25.	. . . . 4. 16.	. . . . 7. 15.
2. 43. 30.	7. 24.	. . . . 4. 30.	. . . . 5. 53.
			Beyond the Sun.
2. 53. 0.	6. 22.	. . . . 6. 19.	. . . . 0. 56.
3. 0. 0.	6. 22. +	Between 6. 18.+ and 5. 36. +	Between 1′. 2″.+ and 3′. 1″. +
	Suppose 7. 24.	Suppose between 7. 19. and 6. 31.	Between 1. 12. and 3. 38.
3. 9. 30.	10. 28.	. . . . 7. 35.	. . . . 7. 26.
3. 15. 0.	12. 26. appr.	Between 8. 56. and 7. 7.	Between 8. 37. and 10. 12.
3. 21. 0.	13. 58.	. . . . 7. 50.	. . . . 11. 33.

Here, because the eclipse is very large, and at the middle the distance of centers is small, the moments of greatest darkening and of visual conjunction therefore hardly differ in longitude. The latter was at $2^{h}\ 51^{m}$, the former a little before. The eclipse, accordingly, was slightly greater than 8 digits on the ray. Subtract the distance of centers, 6′ 22″, and less than that, from the moon's radius, 15′ 15″, and there remains 8′ 53″ and more, and the sun's radius, added 434
to this, makes 24′ 23″ and more eclipsed, which, in the sky, are about $9\frac{1}{2}$ digits or more than that. And the image of the eclipse was precisely this on the ray, where the inner lines show the truest image of the eclipse after the unshelling of the ray.

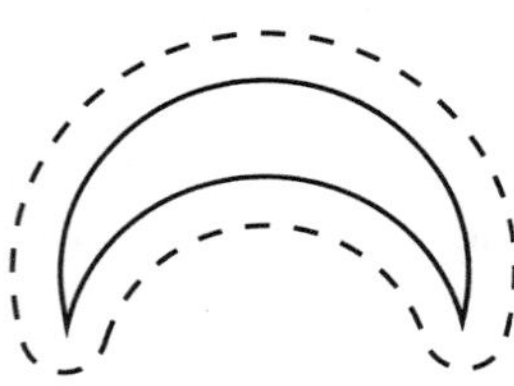

The moon's path is again curved. For if you add 30′ 35″ and 0′ 56″, the former from $1^{h}\ 17\frac{1}{2}^{m}$ and the latter from $2^{h}\ 53^{m}$, the longitudinal path is 31′ 31″. When the latitudes are added, the latitudinal path is 8′ 33″. The angle is therefore 15° 11′. If you do the same thing with $1^{h}\ 17\frac{1}{2}^{m}$and $3^{h}\ 21^{m}$, the angle will become 13° 26′. So, up to this point, the eclipse is always greater at the middle than is gathered from the beginning and end.

[In the 1604 edition, Kepler's endnotes followed here. They have been included as footnotes (or appendices, when they are lengthy) at the appropriate places in the text.]

Conclusion

And so up to this point we have dealt with the deceptions of vision, and, treating the subject through the procedures and precautions of observing, we have brought it down pretty much within the limits of books 4, 5, and 6 of Ptolemy's Almagest. Time, then, to sound the retreat, so as not to attempt all in everything.

Besides, if God should prolong my life and vigor, I shall set forth the use of these observations in another book, which you might consider either as the second part of this work, or as an appendix. I shall everywhere interlace the three books of Ptolemy just mentioned with new problems, highly ingenious and delightful, on the second movables,[116] *and I shall more briefly and more completely show how to use observations, fewer in number and easily comparable, to investigate the same things that Ptolemy investigated. This seems all the more necessary in that Tycho Brahe's lunar theory was published without the provision of demonstrations, in the form of the book of* Progymnasmata,[117] *as was noted in its appendix, which form allowed the entire subject only a few pages.*[118] *And because the chief aim of this book will be to investigate the sizes and distances of the three bodies of the Sun, Moon, and earth, while indeed, as is clear from Theon, Hipparchus worked through the same material in a separate book, to which he gave this very title; so therefore, that it may be happy and fortunate, the name of the book shall be* Hipparchus.[119] *Farewell, reader, and to follow through my efforts with your good wishes, remember that*

He who has begun well has half of the achievement.[120]

---

116 *Secunda mobilia.* These included the sun, moon, and planets, as distinguished from the *primum mobile*, the outermost body that performed the daily rotation of the heavens. Kepler is of course using conventional terminology here, and does not mean to endorse the cosmology from which these terms came.

117 That is, preliminary exercises.

118 The appendix, which appears on pp. 320–23 of volume III of *TBOO* (not volume II, as stated in the note in *JKGW* II p. 462), was written by Kepler. In it (p. 321) Kepler explains that, although Brahe had not originally planned to include the lunar theory, he expanded the work in the course of printing so as to include a brief account of the moon. The "few pages" were actually more than fifty.

119 Hipparchus's book *On the Sizes of the Sun and Moon* is lost. Its title is known from Adrastus's commentary on the *Timaeus*, from which Theon of Smyrna quoted extensively in his own commentary on the *Timaeus*. Kepler's *Hipparchus* was never completed, though some of the ideas and themes of that work appeared in other books that Kepler did publish. For an account of the *Hipparchus*, and a transcript of the surviving manuscripts, see *JKGW* XX.1 pp. 181–268 and 501–16.

120 Horace, *Epistolae*, I.2, 40.

[Kepler's] Index of Memorable Things and Questions, and of Passages Adduced from the Authorities

[Page numbers are those of the 1604 edition (in the inner margins). Kepler originally included line numbers; these have been omitted. Entries have been sorted by English alphabetical order.]

[1] Entries in square brackets are cross references provided by the translator.

D

E

[2] The eclipses of 1600 and 1601 are listed twice in Kepler's index, with the page numbers shown here.

T

U

V

X

Y

Z

Bibliography

This bibliography does not cover the substantial secondary literature on Kepler's *Optics*. Its purpose is to provide bibliographical information about the works and editions cited in the text and footnotes. Additional bibliographical information is provided in the notes, especially where the work is mentioned only once in the course of the book. Included herein are also the abbreviations by which frequently cited works are referenced. —Translator.

Albategnius [Al-Battānī], *Opus astronomicum*, ed. Alfonso Nallino, 3 Vols. (Milan, 1899–1907).

Alhazen (i.e., ibn al-Haitham), *Alhazeni Arabis libri septem*, in *Thesaurus opticae.*

Allen, Richard Hinkley, *Star Names: Their Lore and Meaning* (New York: Dover, 1963 and other eds.).

Apollonius of Perga, *Conics I–IV*, R. Catesby Taliaferro and Michael N. Fried, trs. (Santa Fe: Green Lion Press, 2013).

Apollonius of Perga, *Apollonii Pergaei conicorum libri quatuor,* F. Commandino, ed. and trans. (Bologna, 1566).

Archimedes, *Archimedis opera omnia cum commentariis Eutocii,* J. L. Heiberg, ed., 2 vols. (Leipzig: Teubner, 1880–1).

Archimedes, *The Works of Archimedes,* T. L. Heath, ed. and trans. (Cambridge: Cambridge University Press, 1897 etc.).

Aristotle, *On the Soul,* trs. Joe Sachs (Green Lion Press, 2001).

Aristotle, *De Anima,* see *On the Soul.*

Aristotle, *Mechanica*, in *Minor Works*, trans. W. S. Hett (Loeb Classical Library, 1980).

Aristotle, *Meteorologica*, trs. H. D. P. Lee (Loeb Classical Library, 1952).

Aristotle [pseudo-], *Problems*, 2 Vols., trs. W. S. Hett (Loeb Classical Library, 1961 and 1957).

Aristotle, *On the Heavens,* trs. W. K. C. Guthrie (Loeb Classical Library, 1960).

Aristotle, *On Sense and Sensibles,* in *On the Soul,* trs. W. S. Hett (Loeb Classical Library, 1957).

Aristotle, *Physics*, in Joe Sachs, *Aristotle's Physics: A Guided Study* (New Brunswick, N. J.: Rutgers University Press, 1995).

Brahe, Tycho, *Tychonis Brahei Dani Opera Omnia,* 15 vols. (Copenhagen: Libraria Gyldendaliana, 1913–29).

Bruhns, Carl, *Die astronomische Strahlenbrechung in ihrer historischen Entwickelung* (Leipzig, 1861). A still-useful account of Kepler's treatment of atmospheric refraction.

Cardano, Girolamo, *The Great Art; or, the Rules of Algebra*, trs. T. Richard Witmer (Cambridge, MA: MIT Press, 1968).

Caspar, Max, *Kepler*, trs. C. Doris Hellman (New York: Dover, 1993).

Cleomedes, *De mundo: Kleomedous Kyklikes theorias meteoron biblia duo = Cleomedis de motu circulari corporum caelestium libri duo,* H. Ziegler, ed. (Leipzig: Teubner, 1981).

Cleomedes, *De mundo, sive circularis inspectionis meteorum libri duo,* Basel, 1547. Latin translation of the preceding.

Cleomedes, *Théorie elementaire = De motu circulari corporum caelestium*, trs. and ed. by Richard Goulet (Paris: J. Vrin, 1980). French translation of the preceding.

Copernicus, *On the Revolutions,* trs. Edward Rosen (Baltimore: Johns Hopkins University Press, 1992).

Delambre, J. B. J., *Histoire de l'astronomie moderne*, 1821. Facsimile reprint, ed. I. Bernard Cohen (New York: Johnson Reprint Corp. 1969).

Diogenes Laertius, *Lives of Eminent Philosophers,* R. D. Hicks, trans., 2 vols. (Loeb Classical Library, 1938).

Du Cange *Glossarium mediae et infimae latinitatis,* facsimile reprint (Graz: Akademische Druck- und Verlagsanstalt, 1954).

Euclid, *Elements*, trs. T. L. Heath (Green Lion Press, 2002).

CGOO: *Claudii Galeni opera omnia,* 20 vols. in 21, ed. C. G. Kühn (Leipzig 1821–33, reprinted Hildesheim: Olms 1964–5).

Gilbert, William *De magnete magneticisque corporibus et de magno magnete telluris physiologia nova,* London 1600, trs. P. Fleury Mottelay (New York: Wiley, 1893; Dover 1968).

Gingerich, Owen and Robert S. Westman, "The Wittich Connection: Conflict and Priority in Late Sixteenth-Century Cosmology," *Transactions of the American Philosophical Society* **78** Part 7 (1988).

Grant, Edward, *Planets stars & orbs: The Medieval Cosmos, 1200–1687* (Cambridge: Cambridge University Press, 1994).

HAMA: Neugebauer, Otto, *A History of Ancient Mathematical Astronomy* (Berlin and New York: Springer Verlag, 1975).

Hellman, C. Doris, *The Comet of 1577: Its Place in the History of Astronomy* (New York: Columbia, 1944; reprinted 1971 by AMS Press, New York).

JKGW: Johannes Kepler *Gesammelte Werke*, see Kepler.

JKNA: Johannes Kepler, *New Astronomy*, see Kepler.

JKOO: *Joannis Kepleri Astronomi Opera Omnia*, see Kepler.

Julius Obsequens, *Prodigiorum liber*, trs. A. C. Schlesinger, in Livy Vol. XIV pp. 238–319 (Loeb Classical Library, 1967).

Kepler, Johannes, *Astronomia nova* ([Heidelberg] 1609).

Kepler, Johannes, *De fundamentis astrologiae certioribus,* Prague [1601]; English translation in J. Bruce Brackenridge, ed., and Mary Ann Rossi, trans., "Johannes Kepler's *On the More Certain Foundations of Astrology*, Prague, 1601," Proceedings of the American Philosophical Society **123** (April 1979).

Kepler, Johannes, *Les Fondements de l'Optique Moderne: Paralipomènes a Vitellion* (1604), trs. by Catherine Chevalley (Paris: J. Vrin, 1980) French translation of the front matter and chapters 1–5 of the *Optics*.

Kepler, Johannes, Johannes Kepler *Gesammelte Werke,* 24 Vols. (Munich: C. H. Beck, 1937–).

"Kepler: De Modo Visionis," trs. A. C. Crombie, in *Mélanges Alexandre Koyré,* Vol. I, *L'aventure de la science*, pp. 135–72 (Paris: Herrmann, 1964). English translation of selections from Chapter 5 of the *Optics*.

Kepler, Johannes, *Mysterium cosmographicum* (Tübingen 1596), in JKGW I.

Kepler, Johannes, *New Astronomy,* trs. W. H. Donahue (Cambridge: Cambridge University Press, 1992). New revised edition: *Astronomia Nova* (Green Lion Press, 2015).

Kepler, Johannes, *Joannis Kepleri Astronomi Opera Omni*a, 8 vols., ed. Christian Frisch (Frankfurt and Erlangen: Heyder und Zimmer, 1858–71).

Kepler, Johannes, *Kepler's Somnium: the Dream, or Posthumous Work on Lunar Astronomy,* trs. E. Rosen (Madison: University of Wisconsin Press, 1967).

L&S: *A Greek-English Lexicon,* H. G. Liddell and R. Scott, eds., revised by H. S. Jones (Oxford: Oxford University Press, 1940 etc.).

Landau, Barbara R., *Essential Human Anatomy and Physiology* (Glenview, Illinois: Scott, Foresman & Co., 1976).

Lattis, James, *Between Copernicus and Galileo: Christopher Clavius and the Collapse of Ptolemaic Cosmology* (Chicago: University of Chicago Press, 1994).

Lindberg, David C., *Theories of Vision from Al-Kindi to Kepler* (Chicago: University of Chicago Press, 1976).

Livy, *History*, trs. Evan T. Sage, Vols. X and XI (Loeb Classical Library, 1965).

Livy, Pseudo-, *see* Julius Obsequens.

Macrobius, *The Saturnalia,* translated with an Introduction and Notes by Percival Vaughan Davies (New York: Columbia University Press, 1969).

Maestlin, Michael, *Disputatio de eclipsibus solis et lunae.* (Def. Marcus ab Hohenfeld). Tübingen 1596.

Maestlin, Michael, *Epitome astronomiae,*Tübingen 1610.

Martianus Capella, *The Marriage of Philology and Mercury,* Vol. 2 of *Martianus Capella and the Seven Liberal Arts*, translated by W. H. Stahl and R. Johnson with E. L. Burge (New York: Columbia University Press, 1977).

James, G., and James, R. C., eds., *Mathematics Dictionary* (New York: Van Nostrand 1949).

Neugebauer, Otto, see HAMA.

OLD: *Oxford Latin Dictionary,* P. G. W. Glare, ed. (Oxford: Oxford University Press, 1982).

Opticae thesaurus. Alhazeni Arabis libri septem, nunc primum editi. Eiusdem liber de crepusculis et nubium ascensionibus. Item Vitellonis Thuringopoloni libri X. F. Risner, ed. (Basel 1572). Reprinted 1972 by Johnson Reprint Co. (New York and London), David C. Lindberg, ed.

Pappus of Alexandria, *Commentary on Books 5 and 6 of the Almagest,* A. Rome, ed. (Studi e Testi 54, Vol. I) (Rome, 1931).

Partington, J. R., *A History of Greek Fire and Gunpowder* (Cambridge: W. Heffer, 1960; reprinted Baltimore: Johns Hopkins University Press, 1999).

Pauly-Wissowa: Pauly, August, *Real-Encyclopädie der klassischen Altertumswissenschaft,* ed. G. Wissowa (Stuttgart 1894–1972).

Pedersen, Olaf, *A Survey of the Almagest* (*Acta Historica Scientiarum Naturalium et Medicinalium*, no. 30) (Odense: Odense University Press, 1974).

Placita Philosophorum, in *Doxographi Graeci* ed. H. Diehls (Berlin 1879) pp. 273 ff.

Plutarch, *The Face in the Moon,* see *Moralia* Vol. XII.

Plutarch, *Moralia*, Harold Cherniss and William C. Helmbold, trs., 15 Vols. (Loeb Classical Library, 1949–).

Porta, Giovanni Baptista, *Magia naturalis* (Naples 1589).

Proclus Diadochus, *Hypotyposis astronomicarum positionum,* Latin trs. by Georgio Valla, Basel 1541 and 1551, in Ptolemy, *Opera*; Greek text in TLG.

Ptolemy, Claudius, *Almagest*, trs. by G. J. Toomer (New York: Springer Verlag 1984).

Ptolemy, Claudius *Claudii Ptolemaei Pelusiensis Alexandrini Omnia quae extant opera... ,* Latin trs. by George of Trebizond, Basel 1541 and 1551. Also contains Proclus, *Hypotyposis*, on pp. 377–428 in the 1541 edition, pp. 333–378 in the 1551 edition.

Ptolemy, Claudius, *Tetrabiblos*, (Loeb Classical Library, 1940 and 1980).

Regiomontanus, Joannes, *Scripta clarissimi mathematici M. Ioannis Regiomontani de torqueto...* (Nuremberg 1544, reprinted Frankfurt: Minerva, 1976).

Rosen, Edward, *Kepler's Somnium,* see Kepler, *Somnium.*

Shapiro, Alan, "Archimedes's measurement of the Sun's apparent diameter," *Journal for the History of Astronomy* vi (1975), 75–83. Contains a translation of the passage from Archimedes cited on p. 227 above.

Straker, Stephen M., *Kepler's Optics: A Study in the Development of Seventeenth-Century Natural Philosophy,* (Doctoral Dissertation, Indiana University, 1970).

Straker, Stephen, "Kepler, Tycho, and the *Optical Part of Astronomy*: The genesis of Kepler's theory of pinhole images," in *Archive for History of Exact Sciences* xxiv (1981), 267–293.

Swerdlow, N. M., "Shadow Measurement: the *Sciametria* from Kepler's *Hipparchus*—a translation with commentary," in P. M. Harman and Alan Shapiro, eds., *The Investigation of Difficult Things: Essays on Newton and the history of the exact sciences in honour of D. T. Whiteside* (Cambridge: Cambridge University Press, 1992, 19–70. Translation from manuscript sources of part of the book Kepler promised to publish (p. 433, above).

TBOO: *Tychonis Brahei Dani Opera Omnia,* see Brahe, Tycho.

Temkin, Owsei, *Galenism: Rise and Decline of a Medical Philosophy* (Ithaca, N.Y.: Cornell University Press, 1973).

Thesaurus, see *Opticae thesaurus.*

TLG: *Thesaurus Linguae Graecae: Canon of Greek Authors and Works,* Luci Berkowitz and Maria C. Pantelia, eds. (Irvine: University of California, 1999–). CD-ROM. and online at https://stephanus.tlg.uci.edu/.

Translator's Index